Lecture Notes in Computer Science   16550

Founding Editors

Gerhard Goos
Juris Hartmanis

Editorial Board Members

Elisa Bertino, *Purdue University, West Lafayette, IN, USA*
Wen Gao, *Peking University, Beijing, China*
Bernhard Steffen, *TU Dortmund University, Dortmund, Germany*
Moti Yung, *Columbia University, New York, NY, USA*

The series Lecture Notes in Computer Science (LNCS), including its subseries Lecture Notes in Artificial Intelligence (LNAI) and Lecture Notes in Bioinformatics (LNBI), has established itself as a medium for the publication of new developments in computer science and information technology research, teaching, and education.

LNCS enjoys close cooperation with the computer science R & D community, the series counts many renowned academics among its volume editors and paper authors, and collaborates with prestigious societies. Its mission is to serve this international community by providing an invaluable service, mainly focused on the publication of conference and workshop proceedings and postproceedings. LNCS commenced publication in 1973.

Maribel Acosta · Marieke van Erp ·
Sebastian Rudolph · Olaf Hartig · Blerina Spahiu ·
Anisa Rula · Daniel Garijo · Francesco Osborne
Editors

The Semantic Web

23rd European Semantic Web Conference, ESWC 2026
Dubrovnik, Croatia, May 10–14, 2026
Proceedings, Part II

 Springer

Editors
Maribel Acosta
Technical University of Munich
Heilbronn, Germany

Sebastian Rudolph
TU Dresden
Dresden, Germany

Blerina Spahiu
University of Milano-Bicocca
Milan, Italy

Daniel Garijo
Universidad Politécnica de Madrid
Madrid, Spain

Marieke van Erp
KNAW Humanities Cluster
Amsterdam, The Netherlands

Olaf Hartig
Linköping University
Linköping, Sweden

Anisa Rula
University of Brescia
Brescia, Italy

Francesco Osborne
KMi, The Open University
Milton Keynes, UK

ISSN 0302-9743　　　　　　ISSN 1611-3349　(electronic)
Lecture Notes in Computer Science
ISBN 978-3-032-25158-9　　　ISBN 978-3-032-25159-6　(eBook)
https://doi.org/10.1007/978-3-032-25159-6

Preface

These two volumes contain the main proceedings of the 23rd edition of the European Semantic Web Conference (ESWC 2026). ESWC is a major venue for discussing the latest scientific results and innovations related to the semantic web, knowledge graphs, and web data.

This year ESWC's Research track addressed the theoretical, analytical, and empirical aspects of the Semantic Web, semantic technologies, knowledge graphs, and semantics on the Web in general. The Resources track welcomed resource contributions that were, on the one hand, innovative or novel, and on the other hand sharable and reusable (e.g. datasets, knowledge graphs, ontologies, workflows, benchmarks, frameworks), and provide the necessary scaffolding to support the generation of scientific work and advance the state of the art. The In-Use track focused on contributions that reuse and apply state-of-the-art semantic technologies or resources in real-world settings.

The main scientific program of ESWC 2026 contained 55 papers selected out of 216 submissions (124 Research, 63 Resources, 29 In-Use). After desk rejects, 206 submissions were reviewed (116 Research, 63 Resources, 27 In-Use), resulting in 28 papers accepted in the Research track, 17 in the Resources track, and 10 in the In-Use track. The overall acceptance rate was 25.5% before desk rejects (22.6% Research, 27.0% Resources, 34.5% In-Use) and 26.7% after desk rejects (24.1% Research, 27.0% Resources, 37.0% In-Use). As in previous years, the review process included a rebuttal phase, allowing authors to respond to reviewers' comments and clarify technical contributions. This interaction helped ensure a fair decision process. The program chairs are grateful to the 48 senior PC members, the 248 PC members, and the 47 external reviewers for providing their feedback on the scientific program, and to all other community members who contributed to reviewing. Each paper received an average of 4.1 reviews, with the Research track being double-anonymous, and the Resources and In-Use tracks being single-anonymous. We adopted ACM's terminology in all calls to improve ESWC's Diversity, Equity, and Inclusion principles in naming review guidelines.

We welcomed invited keynotes from three world-renowned speakers, spanning industry and academia: Ian Horrocks (University of Oxford), Isabelle Augenstein (University of Copenhagen), and Atanas Kiryakov (Graphwise).

ESWC 2026 featured 13 workshops and 7 tutorials. The workshops covered a wide variety of topics, ranging from well-established research areas such as knowledge graph construction and neuro-symbolic artificial intelligence to emerging topics such as LLM integration and agentic AI. Some workshops focused on specific application domains, including biomedicine, digital humanities, and sustainability, while others addressed cross-cutting themes such as data quality, semantics in dataspaces, and policy-based access control. The tutorials covered data interoperability, decentralised personalisation, agent-based learning systems, and FAIR Ontology Engineering.

The conference also offered other opportunities to discuss the latest research and innovation work, including a poster and demo session, workshops and tutorials, a PhD symposium, an industry track, and a project networking session. We thank Claudia d'Amato and Sebastian Neumaier for organising the Workshop and Tutorials track. We are also thankful to Pasquale Lisena and Marta Sabou for successfully running the Posters and Demos track. We are grateful to Maria-Esther Vidal and Victor de Boer for coordinating the PhD Symposium, which welcomed 18 PhD students who had the opportunity to present their work and receive feedback in a constructive environment. Thanks go to Lars Heling and Stefan Schmid for their management of the Industry Track, which welcomed submissions from several large industry players. We also thank Valentina Presutti and Beatriz Esteves for increasing the networking potential of ESWC by running the Project Networking Session.

We would like to express our special thanks to Tabea Tietz, Pierre Monnin, and Ignacio García for their outstanding work as Web and Publicity Chairs. We are also grateful to Eleni Ilkou and Maria Angela Pellegrino for their contributions as Metadata Chairs, and to Francesco Osborne for preparing this proceedings volume with Springer. We thank Pieter Colpaert for his contribution as Student Mentoring Chair, and María Poveda-Villalón for her role as Programme Coordinator. We would like to express our gratitude to the Pitea Group led by Marija Komatar, and the Jožef Stefan Institute for their activities in organising the conference locally and their invaluable support. Our sincere thanks go to our sponsors for their generous contributions to ESWC 2026, and to our Sponsor Chairs, Albert Meroño Peñuela and John Domingue, for successfully securing their support. Finally, we are grateful to the Semantic Web Science Association (SWSA), with special mention to John Domingue, Elena Simperl, and Edward Curry, as well as the ESWC 2025 Organising Committee, for their invaluable support and advice.

This year's conference continues to celebrate excellence through two distinguished honours. The Best Paper Awards recognise the strongest contributions across the main tracks. We are also proud to continue the ESWC Dieter Fensel Visionary Award, established in 2025 to spotlight bold and forward-thinking ideas presented at the conference.

As we reflect on both the past and the future, and our role as researchers and technologists, our thoughts go out to all those impacted by war around the world.

March 2026

Maribel Acosta
Marieke van Erp
Sebastian Rudolph
Olaf Hartig
Blerina Spahiu
Anisa Rula
Daniel Garijo
Francesco Osborne

Organization

General Chair

Maribel Acosta — Technical University of Munich, Germany

Research Track Program Chairs

Marieke van Erp — KNAW Humanities Cluster, The Netherlands
Sebastian Rudolph — TU Dresden, ScaDS.AI, Germany

Resources Track Program Chairs

Olaf Hartig — Linköping University, Sweden
Blerina Spahiu — University of Milano-Bicocca, Italy

In-Use Track Program Chairs

Anisa Rula — University of Brescia, Italy
Daniel Garijo — Universidad Politécnica de Madrid, Spain

PhD Symposium Chairs

Maria-Esther Vidal — Leibniz University of Hannover, TIB, Germany
Victor de Boer — Vrije Universiteit Amsterdam, The Netherlands

Poster and Demo Chairs

Pasquale Lisena — EURECOM, France
Marta Sabou — WU Wien, Austria

Workshops and Tutorials Chairs

Claudia d'Amato	University of Bari, Italy
Sebastian Neumaier	St. Pölten University of Applied Sciences, Austria

Industry Track Program Chairs

Lars Heling	SAP, Germany
Stefan Schmid	Bosch Research, Germany

Project Networking Chairs

Valentina Presutti	University of Bologna, Italy
Beatriz Esteves	Ghent University - imec, Belgium

Student Mentoring Chair

Pieter Colpaert	Ghent University - imec, Belgium

Sponsor Chairs

Albert Meroño Peñuela	King's College London, UK
John Domingue	Open University, UK

Program Coordinator

María Poveda-Villalón	Universidad Politécnica de Madrid, Spain

Metadata Chairs

Eleni Ilkou	TIB, Germany
Maria Angela Pellegrino	University of Salerno, Italy

Web and Publicity Chairs

Tabea Tietz	FIZ, Karlsruhe Institute of Technology, Germany
Pierre Monnin	Université Côte d'Azur, Inria, CNRS, I3S, France
Ignacio García	Technical University of Munich, Germany

Proceedings Chair

Francesco Osborne	Open University, UK

Program Committee

Nathalie Abadie	LASTIG, Université Gustave Eiffel, IGN-ENSG, France
Soheil Abadifard	Kansas State University, USA
Alessandro Adamou	Bibliotheca Hertziana - Max Planck Institute for Art History, Italy
Shqiponja Ahmetaj	TU Wien, Austria
Mehwish Alam	Télécom Paris, Institut Polytechnique de Paris, France
Alsayed Algergawy	FIZ Karlsruhe, Leibniz Institute for Information Infrastructure, Germany
Manzoor Ali	Paderborn University, Germany
Bradley Allen	University of Amsterdam, The Netherlands
Elvira Amador-Domínguez	Universidad Politécnica de Madrid, Spain
Rafael Angarita	Paris Nanterre University - LIP6, Sorbonne University, France
Amin Anjomshoaa	Vienna University of Economics and Business, Austria
Gianluca Apriceno	Fondazione Bruno Kessler, Italy
Julián Arenas-Guerrero	Universidad Politécnica de Madrid, Spain
Dörthe Arndt	TU Dresden, Germany
Natanael Arndt	German National Library, Germany
Ghislain Atemezing	European Union Agency for Railways, France
Nathalie Aussenac-Gilles	IRIT CNRS, Université de Toulouse, France
Carlos Badenes-Olmedo	Universidad Politécnica de Madrid, Spain
Tania Bailoni	Fondazione Bruno Kessler, Italy
Booma Sowkarthiga Balasubramani	Microsoft, USA
Pierpaolo Basile	University of Bari Aldo Moro, Italy

César Bernabé   University of Illinois Urbana-Champaign, USA
Abraham Bernstein   University of Zurich, Switzerland
Christian Bizer   University of Mannheim, Germany
Peter Bloem   Vrije Universiteit Amsterdam, The Netherlands
Carlos Bobed   University of Zaragoza, Spain
Fernando Bobillo   University of Zaragoza, Spain
Pieter Bonte   KU Leuven, Belgium
Loris Bozzato   Università degli Studi dell'Insubria, Italy
Janez Brank   Jožef Stefan Institute, Slovenia
Christoph Braun   Karlsruhe Institute of Technology, Germany
Carlos Buil Aranda   Philip Morris International, Switzerland
Grégoire Burel   Open University, UK
Davide Buscaldi   LIPN, UUniversité Sorbonne Paris Nord, France
Jean-Paul Calbimonte   University of Applied Sciences and Arts Western Switzerland HES-SO, Switzerland
Pablo Calleja   Universidad Politécnica de Madrid, Spain
Cinzia Cappiello   Politecnico di Milano, Italy
Antonella Carbonaro   University of Bologna, Italy
Irene Celino   Cefriel, Italy
Davide Ceolin   CWI, The Netherlands
Javad Chamanara   TIB, Leibniz University of Hannover, Germany
Pierre-Antoine Champin   W3C/Inria, France
William Charles   IRIT, France
Victor Charpenay   Mines Saint-Étienne, France
David Chaves-Fraga   Universidade de Santiago de Compostela, Spain
Jiaoyan Chen   University of Manchester, UK
Gong Cheng   Nanjing University, China
Yashrajsinh Chudasama   Leibniz University Hannover, Germany
Philipp Cimiano   Bielefeld University, Germany
Michael Cochez   ELLIS Institute Finland & Åbo Akademi University, Finland
Pieter Colpaert   Ghent University – imec, Belgium
Diego Conde-Herreros   Universidad Politécnica de Madrid, Spain
Oscar Corcho   Universidad Politécnica de Madrid, Spain
Julien Corman   Free University of Bozen-Bolzano, Italy
Marco Cremaschi   University of Milano-Bicocca, Italy
Claudia d'Amato   University of Bari, Italy
Mathieu d'Aquin   LORIA, CNRS, University of Lorraine, France
Jennifer D'Souza   TIB Leibniz Information Centre for Science and Technology, Germany
Enrico Daga   Open University, UK
Jérôme David   Université Grenoble Alpes - Inria, France

Jacopo de Berardinis	University of Liverpool, UK
Victor de Boer	Vrije Universiteit Amsterdam, The Netherlands
Stefano De Giorgis	Vrije Universiteit Amsterdam, The Netherlands
Ben De Meester	Ghent University – imec, Belgium
Christophe Debruyne	Université de Liège, Belgium
Emanuele Della Valle	Politecnico di Milano, Italy
Elena Demidova	University of Bonn, Germany
Gayo Diallo	University of Bordeaux, France
Anastasia Dimou	KU Leuven, Belgium
Daniel Dobriy	Vienna University of Economics and Business and Dobriy AI GmbH, Austria
Milan Dojchinovski	Czech Technical University in Prague, Czechia and InfAI/DBpedia Association, Germany
John Domingue	Open University, UK
Mauro Dragoni	Fondazione Bruno Kessler, Italy
Xuemin Duan	KU Leuven, Belgium
Michel Dumontier	Maastricht University, The Netherlands
Fajar J. Ekaputra	Vienna University of Economics and Business, Austria
Andreas Ekelhart	SBA Research, Austria
Basil Ell	Universität Bielefeld, Germany
Vadim Ermolayev	Ukrainian Catholic University, Ukraine
Pavlos Fafalios	Technical University of Crete and FORTH-ICS, Greece
Nicola Fanizzi	Università degli Studi di Bari "Aldo Moro", Italy
Alessandro Faraotti	IBM, USA
Daniel Faria	INESC-ID, Instituto Superior Técnico, Universidade de Lisboa, Portugal
Catherine Faron	Université Côte d'Azur, France
Anna Fensel	Wageningen University & Research, The Netherlands
Javier D. Fernández	F. Hoffmann-La Roche AG, Switzerland
Jesualdo Tomás / Fernández-Breis	Universidad de Murcia and IMIB Pascual Parrilla, Spain
Sebastián Ferrada	Universidad de Chile
Erwin Filtz	Siemens AG Österreich, Austria
Manuel Fiorelli	Tor Vergata University of Rome, Italy
Giorgos Flouris	ICS-FORTH, Greece
Flavius Frasincar	Erasmus University Rotterdam, The Netherlands
Ken Fukuda	National Institute of Advanced Industrial Science and Technology, Japan
Naoki Fukuta	Shizuoka University, Japan
Mohamed H. Gad-Elrab	Bosch Center for Artificial Intelligence, Germany

Fabien Gandon	Inria, France
Aldo Gangemi	Università di Bologna & CNR-ISTC, Italy
Raúl García-Castro	Universidad Politécnica de Madrid, Spain
Sandra Geisler	RWTH Aachen University, Germany
Anna Lisa Gentile	IBM Research, USA
Pouya Ghiasnezhad Omran	Australian National University, Australia
Martin Giese	University of Oslo, Norway
Birte Glimm	University of Ulm, Germany
Jorge Gracia	University of Zaragoza, Spain
Paul Groth	University of Amsterdam, The Netherlands
José Manuel Gómez Pérez	expert.ai, Italy
Peter Haase	metaphacts, Germany
Torsten Hahmann	University of Maine, USA
Andreas Harth	Friedrich-Alexander-Universität Erlangen-Nürnberg and Fraunhofer IIS, Germany
Mounira Harzallah	LS2N, University of Nantes, France
Oktie Hassanzadeh	IBM, USA
Manfred Hauswirth	TU Berlin, Fraunhofer FOKUS, Weizenbaum Institute, Germany
Ivan Heibi	University of Bologna, Italy
Nathalie Hernandez	IRIT, Université Toulouse Jean Jaurès, France
Daniel Hernández	University of Stuttgart, Germany
Sven Hertling	University of Mannheim, Germany
Daniel M. Herzig	Digital Science, UK
Ryohei Hisano	University of Tokyo, Japan
Rinke Hoekstra	Elsevier, The Netherlands
Aidan Hogan	Universidad de Chile
Laura Hollink	Centrum Wiskunde & Informatica, The Netherlands
Andreas Hotho	University of Würzburg, Germany
Wei Hu	Nanjing University, China
Andreea Iana	University of Mannheim, Germany
Luis-Daniel Ibáñez	University of Southampton, UK
Ryutaro Ichise	Institute of Science Tokyo, Japan
Enrique Iglesias	Forschungszentrum L3S, Germany
Ana Iglesias-Molina	Johnson & Johnson, USA
Prateek Jain	The Tradedesk Inc., USA
Ernesto Jiménez-Ruiz	City St George's, University of London, UK
Jan-Christoph Kalo	University of Amsterdam, The Netherlands
Eduard Kamburjan	IT University of Copenhagen, Denmark
Maulik R. Kamdar	Optum Health, USA

Timotheus Kampik	SAP, Umeå University, Sweden
Tomi Kauppinen	Aalto University, Finland
Takahiro Kawamura	University of Tokyo, Japan
C. Maria Keet	Meaningfy, Belgium
Mayank Kejriwal	Information Sciences Institute, USA
Evgeny Kharlamov	Bosch AI, Germany and University of Oslo, Norway
Sabrina Kirrane	Vienna University of Economics and Business, Austria
Tomáš Kliegr	Prague University of Economics and Business, Czech Republic
Matthias Klusch	Deutsches Forschungszentrum für Künstliche Intelligenz, Germany
Haridimos Kondylakis	FORTH & University of Crete, Greece
Stasinos Konstantopoulos	NCSR Demokritos, Greece
Roman Kontchakov	Birkbeck, University of London, UK
Manolis Koubarakis	National and Kapodistrian University of Athens, Greece
Tobias Käfer	Karlsruhe Institute of Technology, Germany
Jose Emilio Labra Gayo	Universidad de Oviedo, Spain
Frédérique Laforest	Laboratoire d'Informatique en Image et Système d'Information UMR CNRS 5205, France
Christoph Lange	Fraunhofer Institute for Applied Information Technology FIT and RWTH Aachen University, Germany
Davide Lanti	Free University of Bozen-Bolzano, Italy
Anh Le-Tuan	TU Berlin, Germany
Maxime Lefrançois	Mines Saint-Étienne, France
Huanyu Li	Linköping University, Sweden
Ke Liang	National University of Defense Technology, China
Pasquale Lisena	EURECOM, France
Vanessa Lopez	IBM, USA
Claudenir M. Fonseca	University of Twente, The Netherlands
Maria Maleshkova	Helmut-Schmidt-Universität/Universität der Bundeswehr Hamburg, Germany
Maria Vanina Martinez	Artificial Intelligence Research Institute (IIIA - CSIC), Spain
Jose L. Martinez-Rodriguez	Autonomous University of Tamaulipas, Mexico
Margherita Martorana	Vrije Universiteit Amsterdam, The Netherlands
Patricia Martín-Chozas	Universidad Politécnica de Madrid, Spain
Miguel A. Martínez-Prieto	University of Valladolid, Spain
Simon Mayer	University of St. Gallen, Switzerland

Barbara McGillivray	King's College London, UK
Lionel Medini	LIRIS, Université Claude Bernard Lyon 1, France
Nandana Mihindukulasooriya	IBM Research AI, USA
Daniel Miranker	University of Texas at Austin, USA
Pascal Molli	University of Nantes - LS2N, France
Pierre Monnin	Université Côte d'Azur, Inria, CNRS, I3S, France
Gabriela Montoya	Nantes Université, France
Alba Catalina Morales Tirado	Open University, UK
Boris Motik	University of Oxford, UK
Summaya Mumtaz	University of Oslo, Norway
Raghava Mutharaju	IIIT-Delhi, India
Tuan-Phong Nguyen	VNU University of Engineering and Technology, Vietnam
Werner Nutt	Free University of Bozen-Bolzano, Italy
Andrea Giovanni Nuzzolese	CNR - Institute of Cognitive Sciences and Technologies, Italy
Sarah Rebecca Ondraszek	FIZ Karlsruhe, Leibniz Institute for Information Infrastructure, Germany
Femke Ongenae	Ghent University - imec, Belgium
Francesco Osborne	Open University, UK
Julian Padget	University of Bath, UK
George Papadakis	National and Kapodistrian University of Athens, Greece
Pierre-Henri Paris	Paris-Saclay University, France
Peter Patel-Schneider	Independent Researcher, USA
Evan Patton	Massachusetts Institute of Technology, USA
Terry Payne	University of Liverpool, UK
Maria Angela Pellegrino	Università degli Studi di Salerno, Italy
Bernardo Pereira Nunes	Australian National University, Australia
Silvio Peroni	University of Bologna, Italy
Catia Pesquita	Universidade de Lisboa, Portugal
Alina Petrova	University of Oxford, UK
Rafael Peñaloza	University of Milano-Bicocca, Italy
Lydia Pintscher	Wikimedia Deutschland, Germany
Dimitris Plexousakis	Institute of Computer Science, FORTH and University of Crete, Greece
Axel Polleres	Vienna University of Economics and Business, Austria
Riccardo Pozzi	Università degli Studi di Milano-Bicocca, Italy
Nicoleta Preda	Versailles Saint-Quentin-en-Yvelines University, France
Tiago Prince Sales	University of Twente, The Netherlands

Disha Purohit	TIB Leibniz Information Centre for Science and Technology, Germany
Joe Raad	Paris-Saclay University, France
David Ratcliffe	Microsoft, USA
Simon Razniewski	ScaDS.AI, TU Dresden, Germany
Georg Rehm	Deutsches Forschungszentrum für Künstliche Intelligenz, Germany
Achim Rettinger	Trier University, Germany
Juan L. Reutter	Pontificia Universidad Católica, Chile
Artem Revenko	Graphwise, Austria
Mariano Rico	Universidad Politécnica de Madrid, Spain
Giuseppe Rizzo	LINKS Foundation, Italy
Sergio José Rodríguez Méndez	Australian National University, Australia
Philipp D. Rohde	TIB Leibniz Information Centre for Science and Technology, Germany
Edelweis Rohrer	niversidad de la República, Uruguay
Maria Del Mar Roldan-Garcia	Universidad de Málaga, Spain
Oscar Romero	Universitat Politècnica de Catalunya, Spain
Marco Rospocher	Università degli Studi di Verona, Italy
Catherine Roussey	INRAE, France
Jože M. Rožanec	Jožef Stefan Institute, Slovenia
Anisa Rula	University of Brescia, Italy
Marta Sabou	Vienna University of Economics and Business, Austria
Harald Sack	FIZ Karlsruhe, Leibniz Institute for Information Infrastructure & KIT Karlsruhe, Germany
Tomer Sagi	Aalborg Universitet, Denmark
Angelo Salatino	Open University, UK
Muhammad Saleem	University of Leipzig, Germany
Md Kamruzzaman Sarker	Bowie State University, USA
Fatiha Saïs	Paris-Saclay University, France
Ralf Schenkel	Trier University, Germany
Andrea Schimmenti	University of Bologna, Italy
Daniel Schwabe	Pontifical Catholic University of Rio de Janeiro, Brazil
Mario Scrocca	Cefriel, Italy
Oshani Seneviratne	Rensselaer Polytechnic Institute, USA
Patricia Serrano Alvarado	LS2N - University of Nantes, France
Barış Sertkaya	Frankfurt University of Applied Sciences, Germany
Sarah Binta Alam Shoilee	Vrije Universiteit Amsterdam, The Netherlands
Pavel Shvaiko	Trentino Digitale SpA, Italy
Leslie Sikos	Edith Cowan University, Australia

Ana Claudia Sima   SIB Swiss Institute of Bioinformatics,
                              Switzerland
Gerardo Simari   Universidad Nacional del Sur and CONICET,
                              Argentina
Kamal Singh   Télécom Saint-Étienne, France
Kuldeep Singh   Eka Labs, Germany
Hala Skaf-Molli   University of Nantes - LS2N, France
Thibaut Soulard   LISN, France
Marc Spaniol   Université de Caen Normandie, France
Steffen Staab   University of Stuttgart, Germany & University of
                              Southampton, UK
Bram Steenwinckel   Ghent University, Belgium
Nadine Steinmetz   University of Applied Sciences Erfurt, Germany
Armando Stellato   University of Rome Tor Vergata, Italy
Simon Steyskal   Siemens AG Austria, Austria
Umberto Straccia   ISTI-CNR, Italy
Chang Sun   Maastricht University, The Netherlands
Zequn Sun   Nanjing University, China
Vojtěch Svátek   Prague University of Economics and Business,
                              Czech Republic
Danai Symeonidou   INRAE, France
Ruben Taelman   Ghent University – imec, Belgium
Lionel Tailhardat   Orange Research, France
Annette Ten Teije   Vrije Universiteit Amsterdam, The Netherlands
Steffen Thoma   FZI Research Center for Information Technology,
                              Germany
Ilaria Tiddi   Vrije Universiteit Amsterdam, The Netherlands
Tabea Tietz   FIZ Karlsruhe, Leibniz Institute for Information
                              Infrastructure, Germany
Konstantin Todorov   LIRMM/University of Montpellier/CNRS, France
Dominik Tomaszuk   University of Bialystok, Poland
Riccardo Tommasini   INSA Lyon - LIRIS, France
Sebastian Tramp   eccenca GmbH/AKSW e.V., Germany
Cassia Trojahn   Université Grenoble Alpes, France
Antonis Troumpoukis   NCSR Demokritos, Greece
Solenn Tual   LASTIG, Université Gustave Eiffel, IGN-Géodata
                              Paris, France
Gabriele Tuozzo   University of Salerno, Italy
Yannis Tzitzikas   University of Crete and FORTH-ICS, Greece
Takanori Ugai   Fujitsu Limited, Japan
Chukwudi Uwasomba   Edge Hill University, UK
Ludger van Elst   Deutsches Forschungszentrum für Künstliche
                              Intelligenz, Germany

Frank van Harmelen	Vrije Universiteit Amsterdam, The Netherlands
Jacco van Ossenbruggen	VU Amsterdam, The Netherlands
Maria-Esther Vidal	Leibniz University Hannover, Germany
Serena Villata	CNRS - Laboratoire d'Informatique, Signaux et Systèmes de Sophia-Antipolis, France
Fabio Vitali	University of Bologna, Italy
Domagoj Vrgoc	Pontificia Universidad Católica de Chile
Abdul Wahid	Data Science Institute, University of Galway, Ireland
Laura Waltersdorfer	TU Wien, Austria
Ruijie Wang	University of Zurich & SIB Swiss Institute of Bioinformatics, Switzerland
Shenghui Wang	University of Twente, The Netherlands
Xiaxia Wang	University of Oxford, UK
Thilini Wijesiriwardene	University of South Carolina, USA
Xander Wilcke	Vrije Universiteit Amsterdam, The Netherlands
Josiane Xavier Parreira	Siemens AG Österreich, Austria
Bo Xiong	Stanford University, USA
Nadia Yacoubi Ayadi	Université Claude Bernard (Lyon 1), France
Jicheng Yuan	Technische Universität Berlin, Germany
Fouad Zablith	American University of Beirut, Lebanon
Ondřej Zamazal	Prague University of Economics and Business, Czech Republic
Veruska Zamborlini	KNAW Humanities Cluster, The Netherlands & Federal University of Espirito Santo, Brazil
Pengyu Zhang	University of Amsterdam, The Netherlands
Xiaowang Zhang	Tianjin University, China
Richard Zijdeman	International Institute of Social History, The Netherlands
Antoine Zimmermann	Mines Saint-Étienne, France
Remzi Çelebi	Maastricht University, The Netherlands

Additional Reviewers

Baloni, Raghvi	Giesteira Cotovio, Pedro
Brambilla, Marco	Giordano, Luca
Di Pierro, Davide	Hau, Christoph
Diliso, Ivan	Henselmann, Daniel
Freund, Michael	Hoang Thi Pham, Thi
Gautheron, Thibault	Hu, Wenxin
Georgakopoulos, Antonios	Hu, Yujia
Ghizzota, Eleonora	Illich, Moritz

Ionov, Maxim
Korini, Keti
Li, Xue
López-Otal, Miguel
Mamie, Noah
Michel, Franck
Mumtaz, Ummara
Mutlu, Osman
Navas-Loro, María
Nédelec, Brice
Omeliyanenko, Janna
Palmirani, Monica
Pardal, Nina
Patkos, Theodore
Pfister, Jan
Plas, Konstantinos

Qu, Yuanwei
Rincon Yanez, Diego
Schmid, Sebastian
Severin Andermatt, Pascal
Sideri, Sophia
Taghzouti, Yousouf
van den Bergh, Laurin
Vassiliou, Giannis
Wang, Xinyi
Wehr, Thomas
Welt, Michael
Wunderle, Julia
Xu, Ziwei
Yang, Duo
Zamprogno, Giacomo

Sponsors

Gold Sponsors

A part of Digital Science, **metaphacts** is a semantics and AI company delivering innovative solutions that help global enterprises transform data into consumable, contextual, and actionable knowledge—supporting customers across industries in building semantic layers, creating digital twins, and developing trustworthy AI apps for knowledge discovery.

Graphwise enables organizations to unlock ROI for enterprise AI by delivering the most comprehensive and trusted industry solution in the field of knowledge graphs and semantic AI technologies. As enterprises pour millions into AI investment, Graphwise delivers the critical knowledge graph infrastructure to ensure enterprises are ready to realize the technology's full potential, is trusted, and can be implemented at scale. Graphwise, which is the result of the merger between tech visionaries Ontotext and Semantic Web Company, has over 200 employees worldwide, with offices located across North America, Europe and APAC.

Student Support

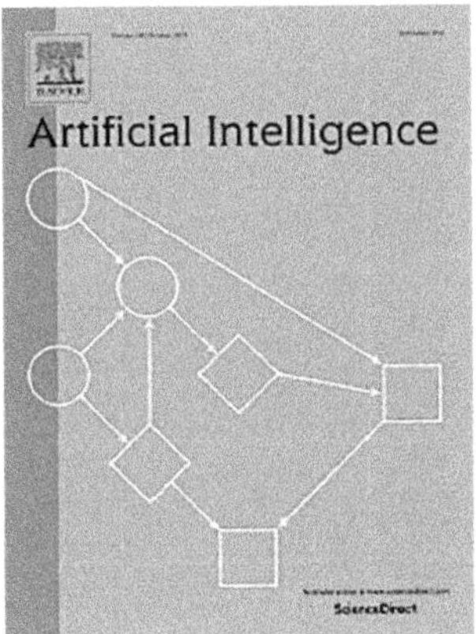

The **Journal of Artificial Intelligence** (AIJ) is a broadly scoped venue covering all areas of artificial intelligence, including automated reasoning, commonsense reasoning, computer vision, constraint processing, ethical AI, knowledge representation, machine learning, multi-agent systems, natural language processing, planning and action, and reasoning under uncertainty.

Best Paper Award

Springer is part of Springer Nature, a leading global research, educational and professional publisher, home to an array of respected and trusted brands providing quality content through a range of innovative products and services. Springer Nature is the world's largest academic book publisher, publisher of the world's most influential journals and a pioneer in the field of open research. The company numbers almost 13,000 staff in over 50 countries and has a turnover of approximately €1.5 billion. Springer Nature was formed in 2015 through the merger of Nature Publishing Group, Palgrave Macmillan, Macmillan Education and Springer Science+Business Media.

Contents

In-Use

Resources

A Linked Open Data Service and Semantic Portal to Study the Assembly Minutes and Prosopography of the League of Nations (1920–1946)

Petri Leskinen[1,2(✉)] [iD], Eero Hyvönen[1,2] [iD], Alexandre Lionnet[1,3],
Blandine Blukacz-Louisfert[4], Pierre-Étienne Bourneuf[1] [iD], Davide Rodogno[1] [iD],
Grégoire Mallard[1] [iD], and Florian Cafiero[1,3] [iD]

[1] Center for Digital Humanities and Multilateralism, Geneva Graduate Institute,
Geneva, Switzerland
{eero.hyvonen,alexandre.lionnet,pierre-etienne.bourneuf,davide.rodogno,
petri.leskinen,gregoire.mallard,florian.cafiero}@graduateinstitute.ch
[2] Department of Computer Science, Semantic Computing Research Group (SeCo),
Aalto University, Espoo, Finland
{petri.leskinen,eero.hyvonen}@aalto.fi
[3] École nationale des chartes, Paris, France
[4] UN Library and Archives Geneva, Geneva, Switzerland

Abstract. This paper presents a new Linked Open Data (LOD) service and a semantic portal on top of it available on the Semantic Web: LEAGUE OF NATIONS SAMPO. This paper shows, how this system can be used for Digital Humanities (DH) research, application development, and can form a basis for a larger LOD intrastructure. In our case, the system is targeted on studying the prosopography and activities of the League of Nations (LoN) (1920–1946), the fore-runner of the United Nations, and is the first concrete step of a larger initiative "Minutes of Multilateralism" for publishing and using a cloud of Knowledge Graphs (KG) about mutually interlinked international organizations in Geneva and beyond. LEAGUE OF NATIONS SAMPO is based on 27 000 pages of minutes of LoN assembly meetings, a prosopographical knowledge graph about some 3100 people mentioned in the minutes, and contextualizing data about the real world. For the first time, this wealth of historical documentation is now openly available as FAIR LOD for DH research and practical application development, as demonstrated by the new LEAGUE OF NATIONS SAMPO portal.

Keywords: knowledge graphs · digital humanities · information retrieval · data analysis

1   Minutes of Multilateralism: Studying Cross-National Organizations

International organizations have produced vast corpora of assembly minutes, plenary debates, committee reports, and voting records. These "minutes of multilateralism" are a core documentary infrastructure for accountability and global governance [3]. Yet scholarship on transparency and record keeping shows that access is uneven and highly shaped by how documents are recorded, registered, and exposed. [9,19] In practice, many minutes remain locked in scanned volumes or PDFs with heterogeneous metadata and without interoperable, machine-readable representations for searching, browsing, analyzing, and visualizing the contents for Digital Humanities research [5].

Our broader vision is to turn such dispersed materials into a sustainable Linked Open Data (LOD) ecosystem that supports transparency, critical scholarship, and public engagement. In line with the FAIR principles [30] and ongoing work on datafying diplomatic and parliamentary activities [2,13], this ecosystem should allow users to trace who spoke, on what issues, in which capacities, and how agendas evolved over time, while remaining explicit about uncertainties, biases, and curatorial choices.

The ultimate goal in this work is to create and align with each other minutes data from several related international organizations, with a focus on Geneva-based ones, such as the League of Nations, United Nations, Inter-Parliamentary Union (IPU), and Red Cross. Integrating data from several cross-national organizations, enriched by contextual historical information from related external data sources, such as the Lonsea database[1], Dodis[2], Metagrid[3], and Wikidata, would finally allow for cross-institutional studies based on, e.g., shared prosopographical data about international diplomats and ontologies about politics. For example, the same people have typically been involved in the activities of several organizations[4] and historical events.

In this paper, we present the first concrete step toward this "Minutes of Multilateralism" agenda through the case of the League of Nations (LoN). Its Assembly minutes (1920–1946) offer a rich, bounded laboratory of early multilateralism that makes a valuable source of knowledge for research also by itself. Thanks to the Total Digital Access to the League of Nations Archives (LONTAD) project, [28,29] these materials are comprehensively digitized and accessible online for humans to read, but not available as Findable, Accessible, Interoperable and Re-usable FAIR data for research and application development: the minutes are available mostly as PDFs with limited structured metadata, which constrains large-scale querying, linking, and computational analysis.

---

[1] Lonsea: https://universe.unibas.ch/projects-collaborations/47560.

[2] Dodis: https://www.dodis.ch/.

[3] Metagrid: https://www.metagrid.ch/.

[4] According to the Lonsea project data the Greek representative Nicolas Politis (1872–1942) has had nine different roles in eleven organizations.

Building on LONTAD, we transformed 27 000 pages of Assembly minutes related to over 3 000 mentioned representatives and other people into a Linked Open Data service and semantic portal LEAGUE OF NATIONS SAMPO. The data was the enriched from external data sources using linked data principles of the Web of Data [8]. This work re-uses and adapts the "Sampo" model of LOD-based semantic portals[5] [11] successfully, used in cultural heritage and parliamentary contexts, notably the system *ParliamentSampo – Parliament of Finland on the Semantic Web* to publish parliamentary speeches and political prosopography. [13,18,26]. The ParliamentSampo experience shows how such infrastructures can enable prosopographical research, longitudinal analyses of political discourse, and civic transparency, and suggests that similar approaches can render multilateral institutions more legible to both experts and the wider public.

The new LoN LOD created, that includes prosopographical and LoN assembly minutes-based data, has been published as linked open data (by CC BY 4.0) with documentation, using W3C best practices[6] (e.g., content negotiation), and as a SPARQL endpoint on the LDF.fi platform[7] [14] Also a data dump on Zenodo with a DOI for reference is available[8]. The portal on top of the data service has been opened for feedback from the research community[9] and the software is available under the MIT License in Github[10] for further development.

In the following, Sect. 2 explicates the research problems and hypotheses of our work, after which the data transformation pipeline for the LOD service is presented (Sect. 3). The portal to test and demonstrate how the LOD service can be used in practice is presented in Sect. 5. In conclusion, contributions of the paper are summarized and directions for further research outlined.

2   Research Problems and Hypotheses

The assembly minutes of international organizations have been openly available as human readable images with minimal metadata but not as FAIR data for searching, browsing, and data-analysis in Digital Humanities research. This paper therefore seeks answers to the following research questions:

1. How to represent assembly minutes documents in a semantic form suitable for Digital Humanities research?
2. How to transform the PDF documents into that format?
3. How to enrich the data with biographical data about the actors involved to facilitate prosopographical research about the international representatives and politicians involved?

---

[5] Sampo portal series: https://seco.cs.aalto.fi/applications/sampo/.

[6] Best practices for publishing linked data by W3C: https://www.w3.org/TR/ld-bp/.

[7] League of Nations KG LOD service: https://ldf.fi/dataset/lon/.

[8] LoN LOD as data dump: https://doi.org/10.5281/zenodo.18867718.

[9] LEAGUE OF NATIONS SAMPO portal online: https://minutes.ldf.fi.

[10] LEAGUE OF NATIONS SAMPO portal software: https://github.com/Semantic Computing/lon-minutes-web-app.

4. How to search, browse, and analyze the data as needed by researchers in humanities.

Our research hypothesis is to create an ontological data model of minutes, of the underlying actors, and contextualize these data by knowledge about the real world. The Linked Data paradigm [8] could be used for representing, interlinking, enriching, and publishing the data. As for the web service model and software tools, the Sampo model and Sampo-UI framework [11] could be adapted and re-used as exemplified by the ParliamentSampo system for publishing minutes of parliamentary discussions [13].

3   Data Transformation from Primary Sources to a LOD Service

This section explains the data transformation and LOD publication pipeline created for LEAGUE OF NATIONS SAMPO.

3.1   Data Transformation Pipeline

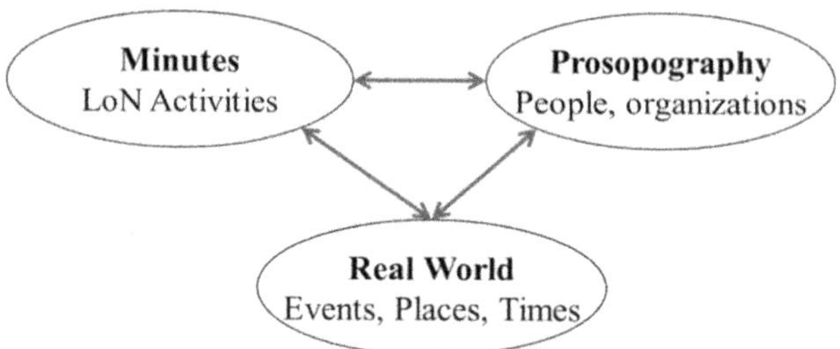

Fig. 1. Interlinked components of the LEAGUE OF NATIONS SAMPO knowledge graph.

The data underlying LEAGUE OF NATIONS SAMPO falls in three parts (cf. Fig. 1). Firstly there is the primary data about the LoN minutes provided by LONTAD as PDF documents with very little metadata. To FAIRify these documents, this material was OCR'd, its named entities were tagged (Subsect. 3.1), the entities were disambiguated and linked, and the metadata schema used was populated (Subsect. 3.3). Secondly, proposographical data was extracted from seven data sources (listed in Table 1) to facilitate semantic search, browsing, and data analysis of the Minutes. Here dataset specific Python scripting were used for the Resource Description Framework (RDF) transformations. Thirdly, in the same vein, real world knowledge about the places, keywords, and events pertaining the underlying real world was assembled to give historical context for the Minutes. An event-centric ontological model was adopted for data harmonization, where actors in different roles participate in events that occur in different

Table 1. Presence of the 3000 person entities in the interlinked data sources of LEAGUE OF NATIONS SAMPO used for enriching the primary minutes dataset LONTAD.

#	Data Service	People	Domain
1	Wikidata	1845	Knowledge graph and data service underlying Wikipedias
2	Lonsea[a]	1686	Database about the history of the League of Nations
3	VIAF.org	1500	Virtual International Authority File system combining other authority services of national libraries etc.
4	Wikipedia	1289	English or French Wikipedia
5	GND[b]	1187	Integrated authority file system of the German National Library
6	DODIS[c]	337	Diplomatische Dokumente der Schweiz
7	Metagrid.ch[d]	314	Linking service designed to connect metadata and resources of research projects throughout the humanities and social sciences

[a] Lonsea: http://www.lonsea.de/
[b] GND: https://www.dnb.de/
[c] Dodis: https://www.dodis.ch/
[d] Metagrid API service: https://metagrid.ch/docs/enhancer/

locations at different times. Finally, the data was published in a SPARQL endpoint that is used directly for DH research and application development, such as the LEAGUE OF NATIONS SAMPO portal.

3.2   Data Transformation: OCR and NER

Due to the complex layouts and sometimes multilingual documents we had to handle, our approach has been twofold (Fig. 2).

First, a subset of pages was labeled for layout regions (columns, main text, marginalia, page headers/footers). These annotations serve to fine-tune a YOLO-based detector [24] that identifies text blocks and, crucially, distinguishes one- and two-column areas. On each page, the detected regions are ordered top-to-bottom and left-to-right within each column, allowing us to reconstruct the correct reading sequence across complex layouts.

In parallel, another subset was transcribed to create an OCR ground truth used to benchmark candidate engines; PERO-OCR [21] was selected and applied to the full corpus. On the complete dataset, YOLO layout regions and PERO-OCR line outputs are overlapped to associate each line with its column and structural context. The result is serialized into XML that preserves page structure, segment boundaries, and links to images.

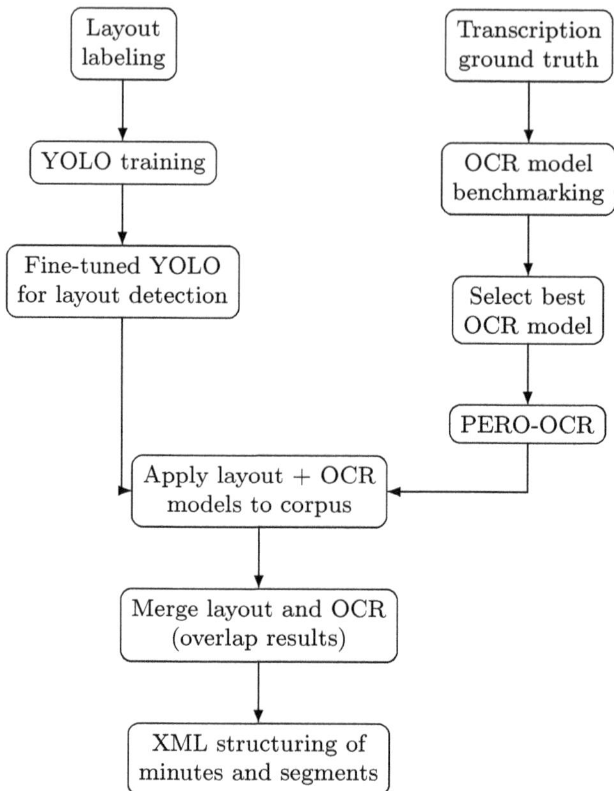

Fig. 2. End-to-end workflow for layout detection, OCR, and XML structuring of the LON Assembly minutes.

Named Entity Recognition (NER) is then applied using transformer-based multilingual models in the XLM-R family, [4] implemented via Flair [1] and matched against curated authority lists.

3.3   Creating the Knowledge Graph and SPARQL Endpoint

The model schema for the RDF dataset is depicted in Fig. 3. The schema is based on CIDOC–CRM standard[11] as well as on the Bio CRM [27] data schema. A similar schema has previously been used, for example, for parliamentary data [18]. The model facilitates to modeling and enriching actor data with related events, such as holding a specific position or belonging to a specific organization. The resources of the minutes (`:Minute`) and the references (`:Reference`) produced in the NER process form the core of the data. Each (`:Minute`) contains the fields of the speech text both as a plain text (`:content`) and in HTML format containing the tagging for showing it in an online browser

---

[11] CIDOC-CRM model: https://cidoc-crm.org.

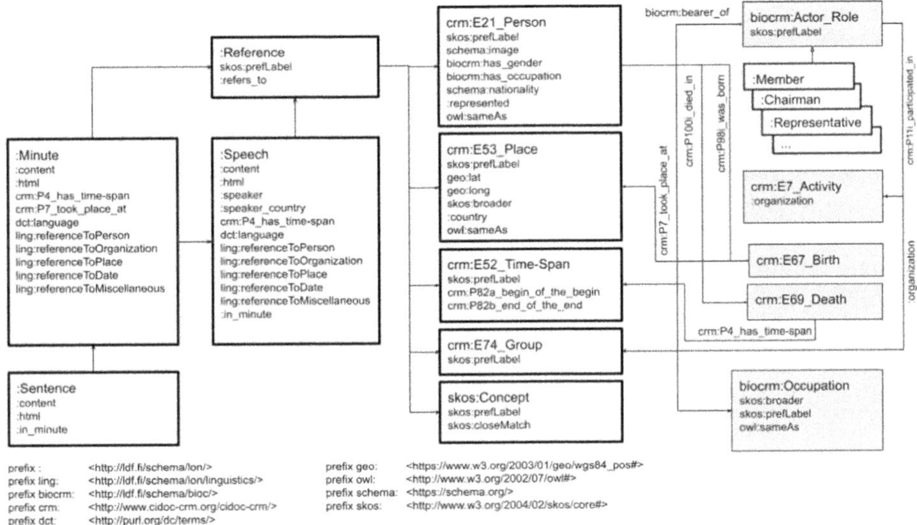

Fig. 3. RDF data schema.

(`:html`). In the RDF different types of named entities are modeled using classes: 1) (`crm:E21_Person`) for individual people, 2) (`crm:E74_Group`) for organizations, 3) (`crm:E53_Place`) for locations, 4) (`crm:E52_Time-Span`) for temporal expressions, and 5) (`skos:Concept`) for miscellaneous references. Furthermore, the schema of Bio CRM is applied to model temporal group memberships (`crm:E7_Activity`) with different roles (`biocrm:Actor_Role`).

Besides the minute speeches, the recognized individual people are the second main focus in the data publication. In the NER results for people there were basically three types of references:

1. Named individuals: "M. Andreas Oldenburg" or "général Nemours
2. Individuals mentioned by title: "Danish representative" or "Secrétaire général"
3. Anonymous groups or individuals: "technical officers", "aggressor" or "women".

In Named Entity Linking (NEL) the main focus has been in disambiguating and linking the entities of the first type to external databases. In average an individual was referred in the minutes texts using 2.59 different literal forms. As an example the Salvadoran representative José Gustavo Guerrero (1876–1958) was referred in 24 different ways (Mr. Guerrero, M. Gustavo Guerrero, Dr. J. Guerrero, J. G. Guerrero, etc.) including the ones containing typos.

The references containing person names or geographical locations are disambiguated against data extractions from external databases, e.g., League of Nations Search Engine (Lonsea)[12] and Wikidata. Furthermore, Wikidata iden-

---

[12] League of Nations Search Engine: http://www.lonsea.de/ NB. the site might be occasionally unavailable.

tifiers and Metagrid API service[13] are used to add linkage to databases such as GND, VIAF, and DODIS. Finally, the prosopographical person data is enriched with name variations, places and times of birth and death, nationalities, occupations, and images. In addition, the LONSEA database provided information about the memberships in the sub-organizations of the League of Nations. The ontology of places represents the geographical locations mentioned in the minutes, the places related to the biographical details of the people, and the nationalities. The hierarchy in the ontology is wide-spread, reaching from the level of a single building to towns, countries, and finally continents.

In the conversion process the named entity linking was performed using Python modules SPLink[14], Nameparser[15], Deep-translator[16], and PolyFuzz[17] taking into consideration the time of speech as well as the other named entities mentioned in the same minutes or in their sentences, e.g. locations and nationalities. The most of the false or missing links were caused by

1. not having the biographical data available in our chosen external sources,
2. name variations or differing practices in spelling (specially with East Asian person names), or
3. multiple candidates, e.g., in cases when only the family name of a person is mentioned.

3.4   Enriching the Data from External Sources

After transforming the LONTAD minutes data into RDF form it was enriched from related data sources by data linking, This was done by aligning person, organization, and place entities with corresponding data entries in seven external data services listed in Table 1. The table quantifies, as an example, that most of the 3200 key people in LONTAD are also present in one or more related web services. This means that lots of additional data about the people could be extracted into the LEAGUE OF NATIONS SAMPO KG, making the data richer than in LONTAD.

As the data comes from several sources, provenience information about the aggregated triples are included in the KG. This metadata is useful for making the data transparent to the end user. In our case, 1) provenience data can be for verifying results in data analyses from the primary sources, 2) for showing data sources to the end user in application user interfaces, and 3) also to find automatically possible inconsistent pieces of shared data across the contributing data services, as suggested and shown in [17].

---

[13] Metagrid widget api reference: https://metagrid.ch/docs/widget/reference/.
[14] SPLink: https://moj-analytical-services.github.io/splink/index.html.
[15] Nameparser: https://pypi.org/project/nameparser/.
[16] Deep-translator: https://pypi.org/project/deep-translator/.
[17] PolyFuzz: https://maartengr.github.io/PolyFuzz/.

3.5   LOD Service

The enriched KG was published on the Linked Data Finland platform LDF.fi[18] [15] using the best publishing practices of the W3C, including a SPARQL endpoint and related services, such as content negotiation, resolving URIs, RDF browsing etc. [15]. LDF.fi also allows publication of schemas alongside the actual data as well as automatic data documentation[19]. In LDF.fi each dataset is described using VoID[20], where a rating of 1–8 stars can be given, extending Tim Berners-Lee's 5-start model [14] (the addtional stars are for publishing the schemas, validating the data against the schemas, and for truthfulness of the data). Based on the VoID metadata, a homepage for the dataset with a SPARQL endpoint, associated LOD services, and instructions for re-using the data are automatically created. LDF.fi is part of a national Finnish LOD infrastructure [12].

4   Using the LOD Service for Digital Humanities Research

The LOD service is useful for several purposes. 1) It can be used directly by scripting for DH research. 2) It provides a basis on which applications such as portals can be built. 3) It can be used to enrich other related KGs as envisioned by the Minutes of Multilateralism agenda. In this section, demonstrational illustrations of the use case (1) are given, (2) is discussed in Sect. 5, and (3) in conclusions as a topic for further research.

A common way to use an LOD endpoint is by a Google Colab or Jupyter notebook or in R-Studio. One can first query the available endpoint for the needed information, and then visualize the results using datasheets, time series, charts, or networks. An example in our case study is given in this notebook[21]. This notebook was created using Python modules, such as SPARQLWrapper for database query, and Pandas, NumPy, and Matplotlib for data analysis and visualization.

An output created the example notebook is depicted in Fig. 4. The continuous lines show the average age of the General Assembly representatives during the years 1920–1946 for males and females with blue and red, respectively. Since the total number of female representatives is very low, 50 out of the total 1174, even a minor change in the number of members affects the average. The dots in the background illustrate the ages of individual representatives.

Fig. 4. Ages of the General Assembly Representatives.

5   League of Nations on the Semantic Web: The Portal

This section presents first the generic portal UI model supported by the Sampo-UI framework and then illustrates by examples how the portal is used. The examples demonstrate some of the benefits of the Sampo-UI model and data linking in contrast to traditional legacy systems, typically based on a single data-silo and without FAIR data available for searching, browsing, and analyzing the contents.

5.1   Portal Model for Minutes Data

To test and demonstrate the approach above, a web application LEAGUE OF NATIONS SAMPO was built by using the Sampo-UI framework [16,23][22] and the domain specific ParliamentSampo framework [13] on top of it. The idea of the "Sampo framework" is to take an existing Sampo in a domain of interest (here speeches and documents related to assembly meetings), with its ready to use user interface (UI) model and knowledge graph as a starting point. The UI specification is a set of JSON specifications that are then modified *declaratively* for the new UI. In addition, the underlying SPARQL queries are adapted to access the more or less different KG of the new Sampo. This approach allows for extremely rapid prototyping and software development, if the framework used (in our case ParliamentSampo) is mostly fit for the new purpose. Re-using an existing framework requires only modest programming skills, but for a more experienced developer, also the framework can be modified and extended with new components and functionalities as needed.

---

[22] Sampo-UI home: https://seco.cs.aalto.fi/tools/sampo-ui/; Github: https://github.com/SemanticComputing/sampo-ui.

The Sampo-UI consists of two main components: (1) a client-side interface built using the well-established React[23] and Redux[24] libraries and (2) a Node.js[25] back-end developed with the Express[26] framework. This framework allows developers to reuse components, such as faceted search, data tables, data analyses, and visualisations through a specification-based configuration.

An ambitious idea behind Sampo-UI is to provide a generic approach for accessing KGs of different kinds in *any* SPARQL endpoint on the Web, based on the core concepts of the RDF[27], the foundation of the Semantic Web: Classes, Properties, and Individuals (Instances). (Cf., e.g., work on the Nobel Price Sampo [6] at the University of Latvia, based on an external SPARQL endpoint provided by the Nobel Foundation in Sweden.) Each class, i.e., entity type of interest, can have an *application perspective*, which is a faceted semantic search view for filtering and exploring individuals, i.e., instances, of the perspective class. Each perspective involves two main types of UI components:

1. One for faceted search [7], earlier called view-based search [22], to filter instances of the perspective class using semantic properties and associations of the class, including dynamically updated hit counts to direct search and prevent dead-end 'no hits' situations in browsing. Sampo-UI integrates this search paradigm into the world of the Semantic Web ontologies, as suggested in [10].
2. Visual components on separate tabs for the filtered search results through tools such as maps, timelines, networks, and tables for semantic exploration.

5.2   LEAGUE OF NATIONS SAMPO Portal

The LoN data publication can be used for two main purposes.

1. For straight-forward DH research and analyses on top of the SPARQL endpoint, using tools such as the YASGUI editor with visualizations [25] and Google Colab for scripting.
2. For developing applications, such as portals, that can be used without programming skills.

As a use case of the latter option, this section presents the portal LEAGUE OF NATIONS SAMPO in order to demonstrate and evaluate the feasibility of the vision and agenda of "Minutes of Multilateralism" presented in Sect. 1.

The landing page of the portal (cf. Fig. 5) presents three application perspectives for searching Speeches, People, Places, Minutes. Below them shortcut links to example searches and visualizations are presented: one to the minutes where the ownership question of the Åland Islands between Finland and Sweden was

---

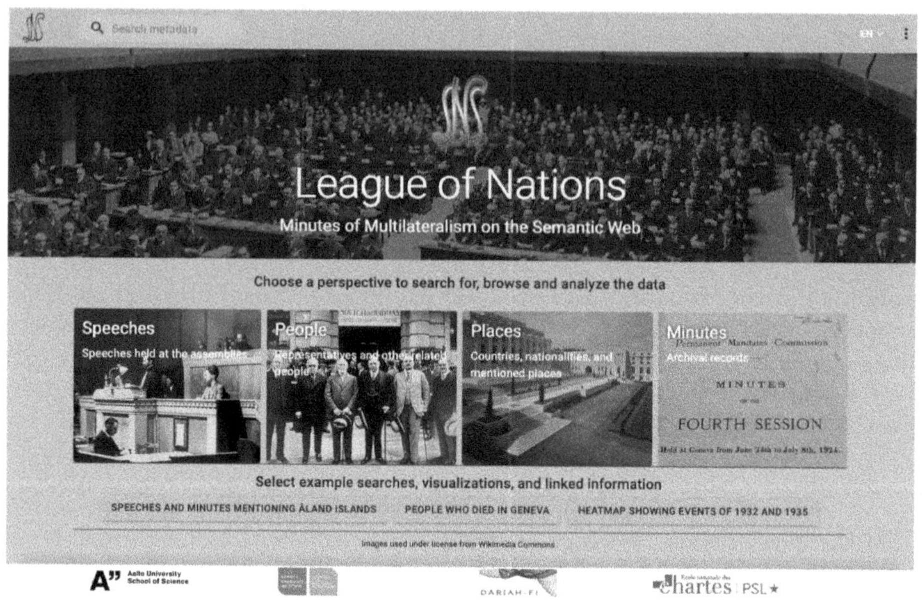

Fig. 5. Landing page of League of Nations Sampo

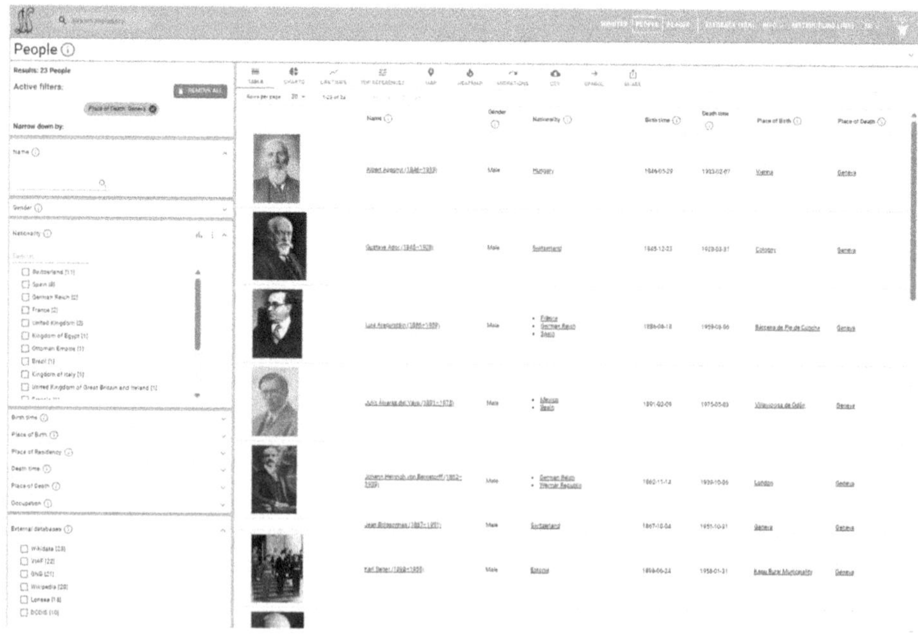

Fig. 6. Application perspective for people mentioned in the LoN minutes

discussed (in the Minutes perspective) and one to a visualization about people mentioned in the minutes who died in Geneva (in People perspective), depicted in Fig. 6 and one opening the heatmap in Places perspective showing most regularly mentioned places with highlights, e.g., at Manchuria and Ethiopia. On the upper bar, links to perspectives, instructions, project information, and to a feed-back channel are always readily available.

In Fig. 6, 11 facets are shown on the left with the facets Nationality and External databases opened. The result set includes 23 people listed on the right with images from Wikidata and links to metadata values for more information, especially to the *instance pages* (homepages) of each individual. In Sampo-UI each individual instance can be associated with a "homepage" where information about the individual is gathered by data linking and reasoning. In this case, the individuals represent a wide cross-national spectrum of nationalities and the data has been harvested from a variety of data sources. These sources are shown and available for search on the open External databases facet on the left low corner, and include Wikidata, Virtual International Authority File service VIAF.org provided by OCLC in the United States, Gemeinsame Normdatei GND register of the German National Library, Wikipedia, Lonsea, and Dodis. For example, the extensive Lonsea database knows 18 out of the 23 people, which demonstrates the added value earned by the linked data approach for enriching data.

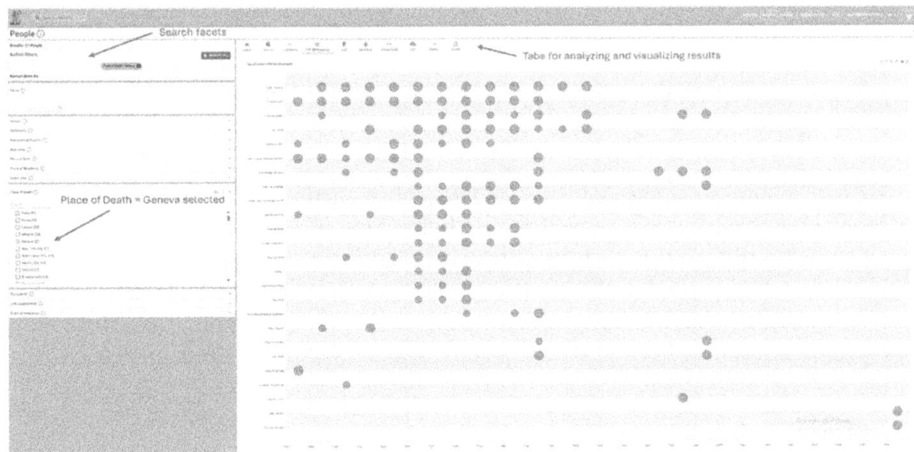

Fig. 7. Visualization tab for illustrating the amount of mentions of people in the minutes on a timeline. Tab TOP REFERENCES is selected.

Above the search result, a list of data-analytic visualization tabs are available for Digital Humanities research in addition to the default TABLE view: CHARTS, LIFETIMES, MAP, HEATMAP, and MIGRATIONS, In addition the CVS tab can be used to import the results into a CSV table for external tools (e.g., R) for further analysis. The SPARQL tab opens a YASGUI interface for querying the SPARQL endpoint and SHARE tab to share the URLs of the visualization. For example, Fig. 7 depicts on the timeline of the TOP REFERENCES tab how often the people who died in Geneva are mentioned in the minutes.

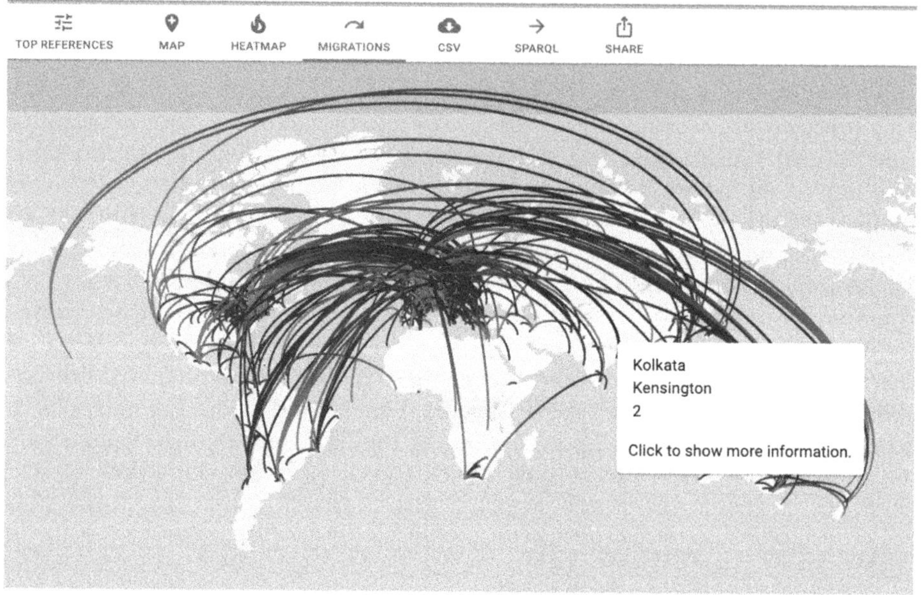

Fig. 8. MIGRATIONS tab for illustrating movements of LoN people on a map.

In Fig. 8 shows on the MIGRATIONS tab a visualization of how the international LoN people have moved from their place of birth (blue end of an arc) to place of death (red end of the life line). Clicking on an arc open an popup window showing the related locations and the involved people. In the example case two people were born in Kolkata and died in Kensington, UK. Because it was still colonial era, very few of the representatives originate from Africa.

6   Conclusions

This paper presented how an existing Sampo framework with its portal UI design on parliamentary speeches and prosopography could be re-used for a related system on assembly minutes of an international organization, the League of Nations. Based on the OCR'd minutes, NER data, and related external data sources for enriching the data, the first version of the LEAGUE OF NATIONS SAMPO data service and portal could be created in a couple of months, because the ParliamentSampo framework and the Sampo infrastructure could be adapted and re-used in this new analogous application case.

The new resource is available openly on the Web as a LOD service, as data dumps, and as a semantic portal for DH research, application development, and as a basis for an infrastructure about international organizations, diplomats, and their psosopography. The system can be used for enriching further related datasets, such as the General Assembly Minutes of the United Nations (UN), Inter-Parliamentary Union IPU, and Red Cross. By moving on from publishing PDF documents, as customary in current legacy systems, to publishing also the underlying data as FAIR KGs in RDF form adds value substantially to these historically important documents.

As a proof of concept, this paper demonstrated by examples how the data service can be used directly for DH research and for application development. Work is currently underway for extending the infrastructure next to the General Assembly Minutes of the United Nations, the follow-up organization of LoN after the Second World War.

A key practical challenge in systems such as LEAGUE OF NATIONS SAMPO is sustainability: how to maintain the data service and the portal in the future. For example, how are possible changes propagated to LEAGUE OF NATIONS SAMPO if data in the primary sources are updated or what happens if the services are terminated? In order to minimize dependency risks pertaining to the primary data sources, LEAGUE OF NATIONS SAMPO copies essential data into its own KG. If the primary sources provide a functioning API, such as the SPARQL endpoint in the case of Wikidata, it can be used for dynamic updates and for providing back links for further information in the original data owner's website, such as a Wikipedia and Dodis.ch. In our case, many of the data sources are not expected to change much, and the case is in this respect easier than, e.g., ParliamentSampo where new speeches may come in on a daily basis and the prosopography changes substantially after every new election of the Parliament. We anticipate that LEAGUE OF NATIONS SAMPO has the potential of becoming a new primary source of LoN data in the larger Minutes of Multilateralism infrastructure to be maintained directly in native RDF form. For this purpose, some earlier Sampos in use, such as BookSampo and OperaSampo, have used the SAHA editor on top of the SPARQL endpoint [20].

To addresss the sustainability challenge to start with, the data service and portal software of LEAGUE OF NATIONS SAMPO are included as a demonstrator in the FIN-CLARIAH/DARIAH-FI infrastructure programme (2022–2029). Furthermore, the data service and portal are available as Docker containers[28], which makes it very easy to re-install the services if needed on different server platforms or locally on a personal computer, including compatible versions of the underlying software packages.

Acknowledgments. We thank the LONTAD archives project, UN Library & Archives Geneva, Lonsea project, Dodis, and Metagrid for fruitful collaborations. Support of the Finnish Digital Humanities research infrastructure initiative FIN-CLARIAH/DARIAH-FI funded by the Research Council of Finland and NextCon-

---

[28] Docker containers: https://www.docker.com/.

nectionEU is acknowledged. Computational resources of CSC – IT Center for Science have been used in the research.

References

1. Akbik, A., Bergmann, T., Blythe, D., Rasul, K., Schweter, S., Vollgraf, R.: FLAIR: an easy-to-use framework for state-of-the-art NLP. In: Proceedings of the 2019 Conference of the North American Chapter of the Association for Computational Linguistics (Demonstrations), pp. 54–59 (2019)
2. Cafiero, F.: Datafying diplomacy: how to enable the computational analysis and support of international negotiations. J. Comput. Sci. **71**, 102056 (2023). https://doi.org/10.1016/j.jocs.2023.102056
3. Cafiero, F., Cointet, J.P., Mallard, G.: Digital accountability can re-legitimate multilateralism. HAL preprint (2025). https://hal.science/hal-05396546v1, preprint, HAL Id: hal-05396546v1. Accessed 3 Dec 2025
4. Conneau, A., et al.: Unsupervised cross-lingual representation learning at scale. In: Proceedings of the 58th Annual Meeting of the Association for Computational Linguistics, pp. 8440–8451 (2020)
5. Drucker, J.: The Digital Humanities Coursebook. An Introduction to Digital Methods for Research and Scholarship. Routledge (2021)
6. Grislis, N.K., Čerāns, K., Grasmanis, M., Rantala, H., Hyvönen, E.: How to add a user interface on top of an external sparql endpoint: case nobel prize sampo. In: SEMANTiCS-PDWT 2025, Posters, Demos, Workshops, and Tutorials at SEMANTiCS 2025, vol. 4064. CEUR Workshop Proceeding (2025). https://ceur-ws.org/Vol-4064/PD-paper17.pdf
7. Hearst, M., Elliott, A., English, J., Sinha, R., Swearingen, K., Lee, K.P.: Finding the flow in web site search. CACM **45**(9), 42–49 (2002)
8. Heath, T., Bizer, C.: Linked Data: Evolving the Web into a Global Data Space. Morgan & Claypool, Palo Alto, California (2011). http://linkeddatabook.com/editions/1.0/
9. Heikkonen, S.: Transparency materialised: how registers can regulate access to documents. Eur. Law Open **3**(1), 1–24 (2024). https://doi.org/10.1017/elo.2024.7
10. Hyvönen, E., Saarela, S., Viljanen, K.: Application of ontology techniques to view-based semantic search and browsing. In: Bussler, C.J., Davies, J., Fensel, D., Studer, R. (eds.) ESWS 2004. LNCS, vol. 3053, pp. 92–106. Springer, Heidelberg (2004). https://doi.org/10.1007/978-3-540-25956-5_7
11. Hyvönen, E.: Digital humanities on the semantic web: sampo model and portal series. Semant. Web **14**(4), 729–744 (2023). https://doi.org/10.3233/SW-223034
12. Hyvönen, E.: How to create a national cross-domain ontology and linked data infrastructure and use it on the semantic web. Semant. Web (2024). https://doi.org/10.3233/SW-243468
13. Hyvönen, E., et al.: Publishing and using parliamentary linked data on the semantic web: parliamentsampo system for parliament of finland. Semant. Web **16**(1) (2025). https://doi.org/10.3233/SW-243683
14. Hyvönen, E., Tuominen, J.: 8-star linked open data model: extending the 5-star model for better reuse, quality, and trust of data. In: Posters, Demos, Workshops, and Tutorials of the 20th International Conference on Semantic Systems (SEMANTiCS 2024), vol. 3759. CEUR Workshop Proceedings (2024). https://ceur-ws.org/Vol-3759/paper4.pdf

15. Hyvönen, E., Tuominen, J., Alonen, M., Mäkelä, E.: Linked data Finland: a 7-star model and platform for publishing and re-using linked datasets. In: Presutti, V., Blomqvist, E., Troncy, R., Sack, H., Papadakis, I., Tordai, A. (eds.) ESWC 2014. LNCS, vol. 8798, pp. 226–230. Springer, Cham (2014). https://doi.org/10.1007/978-3-319-11955-7_24

16. Ikkala, E., Hyvönen, E., Rantala, H., Koho, M.: Sampo-UI: a full stack JavaScript framework for developing semantic portal user interfaces. Semant. Web **13**(1), 69–84 (2022). https://doi.org/10.3233/SW-210428

17. Leskinen, P., Ahola, A., Rantala, H., Tuominen, J., Hyvönen, E.: Consistency checking in a cloud of interlinked cultural heritage knowledge graphs – first results of using the samposampo data service and portal (2025, submitted for review)

18. Leskinen, P., Hyvönen, E., Tuominen, J.: Members of parliament in Finland knowledge graph and its Linked Open Data service. In: Further with Knowledge Graphs. Proceedings of the 17th International Conference on Semantic Systems (SEMANTiCS 2021), pp. 255–269. IOS Press (2021). https://doi.org/10.3233/SSW210049

19. Casadesús de Mingo, A., Cerrillo-i Martínez, A.: Improving records management to promote transparency and prevent corruption. Gov. Inf. Q. **35**(4), 624–632 (2018). https://doi.org/10.1016/j.giq.2018.09.008

20. Mäkelä, E., Hyvönen, E.: SPARQL SAHA, a configurable linked data editor and browser as a service. In: Presutti, V., Blomqvist, E., Troncy, R., Sack, H., Papadakis, I., Tordai, A. (eds.) ESWC 2014. LNCS, vol. 8798, pp. 434–438. Springer, Cham (2014). https://doi.org/10.1007/978-3-319-11955-7_62

21. Novotný, V., Horák, A., et al.: PERO-OCR: an open-source optical character recognition system (2022). https://pero-ocr.fit.vutbr.cz. Accessed 10 Nov 2025

22. Pollitt, A.S.: The key role of classification and indexing in view-based searching. Technical report, University of Huddersfield, UK (1998). http://www.ifla.org/IV/ifla63/63polst.pdf

23. Rantala, H., Ahola, A., Ikkala, E., Hyvönen, E.: How to create easily a data analytic semantic portal on top of a SPARQL endpoint: introducing the configurable Sampo-UI framework. In: VOILA! 2023 Visualization and Interaction for Ontologies, Linked Data and Knowledge Graphs 2023, vol. 3508. CEUR Workshop Proceedinbgs (2023). https://ceur-ws.org/Vol-3508/paper3.pdf

24. Redmon, J., Divvala, S., Girshick, R., Farhadi, A.: You only look once: unified, real-time object detection. In: Proceedings of the IEEE Conference on Computer Vision and Pattern Recognition (CVPR), pp. 779–788 (2016)

25. Rietveld, L., Hoekstra, R.: The YASGUI family of SPARQL clients. Semant. Web **8**(3), 373–383 (2017). https://doi.org/10.3233/SW-150197

26. Sinikallio, L., et al.: Plenary debates of the parliament of Finland as Linked Open Data and in Parla-CLARIN markup. In: Proceedings of the 3rd Conference on Language, Data and Knowledge (LDK 2021). OASIcs, vol. 93, pp. 1–17. Schloss Dagstuhl–Leibniz-Zentrum für Informatik (2021). https://drops.dagstuhl.de/opus/volltexte/2021/14544/

27. Tuominen, J., Hyvönen, E., Leskinen, P.: Bio CRM: a data model for representing biographical data for prosopographical research. In: BD2017 Biographical Data in a Digital World 2017, Proceedings, pp. 59–66. CEUR WS Proceedings (2018). http://ceur-ws.org/Vol-2119/paper10.pdf

28. UN Library & Archives Geneva: LONTAD: Total digital access to the league of nations archives (2022). https://libraryresources.unog.ch/lontad, digitization project (2017–2022) providing comprehensive online access to the League of Nations archives

29. Wells, C.M.: Total digital access to the league of nations archives: digitization, digitalization, and analog concerns. In: Archiving Conference, vol. 16, pp. 12–16. Society for Imaging Science and Technology (2019)
30. Wilkinson, M.D., Dumontier, M., Aalbersberg, I.J., et al.: The FAIR guiding principles for scientific data management and stewardship. Sci. Data **3**, 160018 (2016). https://doi.org/10.1038/sdata.2016.18

OntoPFAS: An Ontology for the Forever Chemicals

Davide Di Pierro[1]([✉])[iD], Lylia Abrouk[2,3][iD], Alexis Guyot[4][iD],
Danai Symeonidou[5][iD], Pierre Labadie[6][iD], and Benjamin Lysaniuk[7][iD]

[1] Université de Montpellier, Montpellier, France
davide.di-pierro@umontpellier.fr
[2] LIRMM, Université de Montpellier, Montpellier, France
lylia.abrouk@lirmm.fr,lylia.abrouk@u-bourgogne.fr
[3] LIB, Université Bourgogne Europe, Dijon, France
[4] LIS, Aix-Marseille Université, Marseille, France
alexis.guyot@lis-lab.fr
[5] MISTEA, INRAE & Institut Agro, Montpellier, France
danai.symeonidou@inrae.fr
[6] CNRS UMR EPOC, Talence, France
[7] CNRS UMR PRODIG, Aubervilliers, France
benjamin.lysaniuk@cnrs.fr

Abstract. *Per- and polyfluoroalkyl substances (PFAS)* are synthetic chemical compounds used widely for fundamental everyday products such as textiles, household products, cosmetics, or food processing, thanks to their resistance to heat, water, and oil. However, it has recently been acknowledged that they represent a threat to human health and the environment, given their ability to persist and spread across multiple living organisms (including mammals, plants, and other fauna). They have been linked to different types of health issues, such as tumors and cancers. Given some recent availability of data, this manuscript presents *OntoPFAS*, the first formal ontology related to PFAS and their relations with the environment and human exposure. The result of this work extensively reuses existing knowledge from different domains, is capable of answering relevant questions, and is compliant with the first PFAS measurements information available. We describe the reasons for this work, its design methodology, the quality of the proposed conceptualization, and the impact in terms of applications for several communities. The resource is available under a free access license: https://github.com/davide1797/PFAS.git.

Keywords: ontology · pfas · knowledge engineering

1 Introduction

Per- and polyfluoroalkyl substances (PFAS) are of concern because of their high persistence and their impact on human and environmental health [13]. There are plenty of industrial uses that, by themselves, may deserve their own classification. Among the most well-known uses are fire-fighting foams and applications in the textile, clothing, automotive, construction, and electronics industries. Although the use of PFAS in many cases cannot be avoided, given their connection to critical sectors, it is becoming evident that alternatives should be sought. In addition, studies demonstrate their tendency to persist and present toxic properties in the environment [1], while the examination and removal processes remain costly.

The first unambiguous definition of PFAS has been provided by Buck et al. [4], who defined them as substances containing the moiety $-C_nF_{2n+1}$ ($n > 0$) that basically summarises the *per- and polyfluoroalkyl* aspect. More recent definitions [7,23] have added the need for the chain to be attached with a *functional group*, responsible for its toxic properties. PFAS can move through water or air during the production or discharge processes. Through sources of flows (lakes, rivers), water becomes one of the major sources of exposure. Becoming part of the food chain, PFAS are therefore found in fauna (including humans) and vegetation. In the vicinity of industrial sites, conversely, PFAS tend to accumulate in soil. Among the more suspected effects on humans, we cite increased cholesterol levels, cancer, effects on fertility, and foetal development. They are also suspected of interfering with the endocrine (thyroid) and immune systems [2].

In support of this research field, the *CNRS Humanities & Social Sciences* has recently published a large dataset about PFAS contamination: the *PFAS Data Hub (PDH)*[1]. It consists of about 1M measurements (mainly across Europe), with information about production facilities, presumptive contamination sites, and users. These data come from 143 datasets, including 15 scientific articles and 91 datasets from authorities, and are, to the best of our knowledge, the unique sort of measurements dataset. It is the result of independent assessments that do not have (or share) semantics, do not correlate data, and have no kind of inference.

While several established ontologies provide valuable domain-specific vocabularies, none of these is designed to represent PFAS as an integrated and multidimensional phenomenon. Some resources capture only one facet of the problem: the chemical structure, the environmental context, or the exposure outcomes. This gap highlights the need for a dedicated ontology bridging these domains and enabling coherent, semantically rich reasoning about PFAS across datasets.

The result is - as an extension of the preliminary work [9] - *OntoPFAS*, an ontology to model the domain of PFAS, their chemical elements, the interaction with the environment, and the human exposure. The ontology is compliant with the PFAS Data Hub, but will not be limited to that. Since the available data cannot capture some of the mentioned aspects (like the effects of the exposure), we also provide concepts to anticipate possible future data. The resource is fully accessible and interoperable, and its degree of formality allows reusability, traceability, and the potential use of any artificial intelligence (AI) tool.

The remainder of this paper is organised as follows. Section 2 discusses existing relevant ontologies in the field and the main construction methodologies.

---

[1] https://pdh.cnrs.fr/en/.

Section 3 shows the requirements and the development of the methodology for the ontology construction. Section 4 presents and discusses the main building blocks of OntoPFAS, the logical rules proposed, and a usage example. Section 5 shows how to map the PFAS Data Hub onto the ontology. Section 6 provides different criteria to assess and validate our work. Finally, Sect. 7 provides some final considerations, summarises its impact, and concludes the manuscript.

2   Related Work

Domain Ontologies. We start from the informal chemical classification of Buck et al. [4] in which the authors provided clarifying definitions of the different PFAS subgroups, helpful in creating the correct associations between the fundamental chemical properties and the labels associated with them. The concepts of **perfluoroalkyl**, **polyfluoroalkyl**, and **polymer** are here defined, put into correlation, and many instances for each of them are given. This introductory classification has been updated by Buck et al. [5] later, with a focus on commercially relevant PFAS. Among the thousands of PFAS officially recognised, only less than 6% are commercially relevant, providing the opportunity to work with a much smaller subset. For a more general chemical overview, the *ChEBI* ontology [8] is a reference in the literature. Specific PFAS subclasses are covered here. To represent an **organismal entity** that accumulates in PFAS, we reuse the same concept from *PCO* [31]. Generic properties of the **environment** and its subtypes are available in the *ENVO* [6]. Finally, *EXO* [24] represents many aspects of the exposure, from the **exposure stimulus** to the **exposure outcome** that applies to a **human individual**.

General-Purpose Ontologies. We are interested in the origin of the PFAS. This aspect is strictly linked to the **product**s represented in the general *Schema.org* [15] vocabulary. The main concept in the PFAS Data Hub is the **observation** (measurement), which we reuse from the *SOSA* [18] ontology. An observation is located at a point, which is a **spatial thing** in the *GEO* vocabulary [32]. A measurement comes from a **dataset**, reused from the *DCAT* [33].

Construction Approaches. Here, we focus on manual ontology construction techniques: a set of strategies, guidelines, and tools to support ontology construction (generally) supervised by experts. Some of the first conceived methodologies are still among the most adopted nowadays. In *Methontology* [11], the authors presented a milestone for ontology engineering, representing the basis for modern ontology construction, in the context of the waterfall software engineering approach. *Ontology Development 101* [25] introduced many general guidelines and details to avoid the so-called *pitfalls* (fallacies) in ontology construction. The spread of agile development influenced ontology construction. Borrowing some practices from the software engineering field, *eXtreme* programming has been embedded into the ontology construction process [17]. Subsequent refinements led to *SAMOD* [27], in which the separation of roles between domain experts and ontology experts is much more integrated and functional within the context of agile development. Latest developments brought the modern *Linked Open Terms (LOT)* [28] methodology, which also comprises many guidelines for the development in the context of Linked Open Data. Current research is exploring *Large Language Models (LLMs)* in collaboration with manual ontology construction techniques, in what is referred to as *collaborative ontology engineering*. The first results and considerations are promising [22], but also limitations and challenges are present [10,21], and an agreed methodology is not available yet.

3   Methodology

To build OntoPFAS, we followed the Linked Open Terms methodology. This choice is mostly justified by the need for agile development in a field where the requirements and data are expected to update continuously in the coming years. Its main agile alternative, SAMOD, fits best when new terms are to be defined, whereas LOT is considered more suitable when reusing much existing knowledge. LOT also guarantees the respect of the Linked Open Data (LOD) principles. Within the methodology, four main steps are depicted: (i) ontological requirements specification; (ii) ontology implementation; (iii) ontology publication, and (iv) ontology maintenance. The core of the workflow is based on the iteration of the sequence ⟨requirements elicitation, implementation, evaluation⟩. The strategy proposes several ways of defining requirements, especially by following the *competency questions (CQs)*, which are questions the ontology must answer [14]. By following LOT, the result mostly follows the *FAIR (Findable, Accessible, Interoperable, Reusable)* principles[2].

Ontology Requirements Specification. In this phase, the scope and objectives of the ontologies are clarified, agreed upon, and elicited. In agreement with experts (a chemist and a geographer), we identified several use cases in which they would be interesting. These use cases went beyond the data available in the PFAS Data Hub. We then decided to move from use cases and examples of interactions to competency questions that guided our development process. Not only are experts cooperative in finding relevant scenarios, but they also contributed in providing existing glossaries, ontologies, or definitions of terms in domains in which a generic ontology designer is likely to commit errors in the semantics of concepts. The result of this phase is discussed in Sect. 4.1.

Ontology Implementation. In this phase, five main activities are performed: (i) term filtering; (ii) subclass refinements; (iii) properties elicitation; (iv) rule definitions, and (v) formalization. Phase (i) consists of filtering the relevant terms in existing resources and understanding their definitions. In our case, an agreement between experts was needed about how concepts would be represented. For instance, the *water* can be represented at a chemical level as the composition of hydrogen and oxygen, but from a geographer's point of view, it is basically a

---

[2] https://www.go-fair.org/fair-principles/.

means of transportation for organismal entities, which best represents the scope of this work. Phase (ii) requires an understanding of the concepts. Especially for PFAS classification, although we have some references, many ontological insights came from the chemistry expert. Phase (iii) consists of declaring all the necessary properties among the concepts to satisfy the previously defined objectives. By properties, we mean both relationships among concepts (object properties) and qualifying data on concepts (data properties). Phase (iv) consists of defining factual knowledge or definitions in the form of rules. Rules allow specifying, for instance, *disjointness* between concepts, matching chemical properties with a concept, or uncovering implicit knowledge. Finally, phase (v) allows changing the representation from an informal, graphical, or text-oriented one to a machine-readable and shareable one. Many user-friendly tools allow formal modelling: we opted for *Protégé*, which offers tools for conceptualization and rule elicitation. The result of this phase is discussed in Sect. 4.2. The evaluation phase is part of the iterative process, and is discussed in Sect. 6.

Ontology Publication and Ontology Maintenance. Finally, we published the resource on GitHub under the CC-BY-4.0 license to allow users to propose changes or variations in a software versioning environment. It is comprehensive of (i) the requirements (in the form of CQs) with their corresponding answers (in the form of *validated*[3] SPARQL queries), (ii) graphical conceptualizations, (iii) ontology source, (iv) the results of the evaluation, (v) the documentation, and (vi) and some examples of some PDH instances (see Sect. 5). We provided URIs in compliance with the W3C Permanent Identifier Community Group[4]. We used pyLODE[5] to generate the documentation. For maintenance, the issue tracker allows us to collect suggestions or corrections as *issues*, which will then be evaluated by the experts to guarantee future enrichment and improvement. We will periodically populate the ontology by injecting new PDH instances. A copy is also available on a platform for the environment domain: *AgroPortal*[6].

4 OntoPFAS

4.1 Requirements

There are different types of requirements to consider for this ontology.

Functional requirements are collected from the needs and objectives of the experts, but we also asked them to share their vision on the possible future developments of this research. They expressed their goals through specific use cases and examples of expected interaction with a "virtual" ontology-based platform. This phase has been delivered regardless of the data available on the PDH. Finally, we discussed existing additional PFAS literature and agreed on

---

[3] https://sparql.org/query-validator.html.
[4] https://w3id.org/.
[5] https://github.com/RDFLib/pyLODE.
[6] https://agroportal.lirmm.fr/.

secondary use cases in which other aspects are addressed. Additional resources included scientific papers, websites, and official regulations. After this process, we translated scenarios and interactions into CQs and agreed on them. As a result, *30* questions have been depicted. The full list is available on our repository[7].

PFAS data hub requirements are conceived to make our ontology compliant with the first extensive dataset on PFAS measurements. Although the hub collects more than 100 datasets (in tabular form), they are divided into 4 different schemas of information (Known PFAS User, PFAS production facility, Presumptive, and Measurement). Operatively, the ontology must represent the four schemas, mapping them onto some concepts and properties.

LOD requirements includes a set of good practices in ontology construction. First of all, the resource must be available in a formal language. We adopt standard languages like OWL/RDF. It should also include as many existing concepts as possible. Not only does this save time, but it also helps in modelling the knowledge correctly and avoiding ambiguity, given that existing concepts are, in principle, validated or in use by the community.

FAIR requirements are finally expected for every software result. The resource is made publicly available on the Web to make it *findable*, the GitHub protocols guarantee it will be *accessible*, the imports of existing concepts and the sharing of its concepts make it *interoperable*, and the fact that it can be easily reused by AI systems, other ontology engineers, or non-technical practitioners make it *reusable*. To facilitate the correct understanding, the standard metadata (author, license, etc.) are provided. The reuse is subject to a non-commercial license.

4.2   Construction

After the requirement specification, LOD indicates the *ontology conceptualization* step. This phase can be done with different levels of formality and tools. For our purposes, we needed a tool that was both user-friendly - to discuss with the experts -, and comprehensive of all the ontology representation elements. We leveraged the online version of draw.io[8]. It is quite a standard tool for ontology conceptualization, and indeed, there is also a library to translate from the visual representation expressed through the Chowlk notation to OWL[9]. We did not exploit this functionality since, for importing from many ontologies, we found Protégé more suited. This step is, in fact, done in parallel with the *ontology reuse* one. The experts suggested comprehensive ontologies to fetch resources concerning the environment or chemistry, while the ontology group made them familiar with resources concerning generic aspects like provenance,

---

[7] https://github.com/davide1797/OntoPFAS/blob/main/ontology/competency_questions.csv.
[8] https://www.drawio.com/.
[9] https://chowlk.linkeddata.es/.

source, or everyday concepts. To capture more complex aspects of the PFAS spread, we also designed the necessary concepts to express their movements. In the end, we agreed to apply a *soft* reuse - limited to some terms - of the following ontologies: *ChEBI*[10] for detailed references of chemical entities; *ENVO*[11] to represent environmental entities and features; *EXO*[12] to represent exposure basic agents and outcomes; *PROV-O*[13] for generic properties modelling the use of products; *RO*[14] for generic causal relationships; *SOSA*[15] to represent features of sensor data or general observations; *Schema.org*[16] to represent generic entities or actions; *GEO*[17] to represent geospatial information; *Allotrope*[18], aligned with BFO, for basic semantics of chemical relationships; *DCMI Metadata Terms*[19] for general-purpose metadata vocabulary; *OCQA*[20] to represent legislative entities; *D-CAT*[21] to represent metadata about different kinds of resources; *SIO*[22] to represent scientific connections and processes; *BFO*[23] for basic ontology construction like the *partOf* property; *PCO*[24] to represent organismal entities.

The conceptualization is split into three sub-areas: *PFAS*, *Exposure*, and *Measurements* to simplify both the discussion and the visualizations. Some classes are evidently at the intersection. For readability, we report full URIs only in the pictures. Our ontology IRI is https://w3id.org/OntoPFAS#.

PFAS. The main class is `Pfas`, which follows the latest Buck definition provided above. We recall that there is no formal classification of the PFAS subclasses. Conversely, there are many informal classifications that list many specific combinations. There is no general agreement on how in-depth to categorise subclasses of PFAS. This is why we found a compromise between three factors: generality, impact, and chemical nature. For generality, we represented the subclasses that are the most comprehensive, and generally represented in every categorisation. Among these, we can mention `Polyfluoroalkyl`, `Perfluoroalkyl`, `PolymerPfas`, `NonPolymerPfas`, `LongChainPfas`, and `ShortChainPfas`. These classes are also strongly recognisable by their chemical properties. In Buck et al. [5], the PFAS mostly correlated with commercial products are also listed (`Pfaa`, `Pfsa`, `Pfca`, `Pfoa`, and `Pfos`). Representing these classes is practical to instantiate

---

[10] https://www.ebi.ac.uk/chebi/.
[11] https://sites.google.com/site/environmentontology/.
[12] https://bioportal.bioontology.org/ontologies/EXO.
[13] https://www.w3.org/TR/prov-o/.
[14] https://bioportal.bioontology.org/ontologies/OBOREL.
[15] https://www.w3.org/TR/vocab-ssn/.
[16] https://schema.org/.
[17] https://www.w3.org/2005/Incubator/geo/XGR-geo-ont-20071023/.
[18] https://bioportal.bioontology.org/ontologies/AFO.
[19] https://www.dublincore.org/specifications/dublin-core/dcmi-terms/.
[20] https://sebseis.github.io/OCQA/.
[21] https://www.w3.org/TR/vocab-dcat-3/.
[22] https://ontobee.org/ontology/SIO.
[23] https://basic-formal-ontology.org/.
[24] https://bioportal.bioontology.org/ontologies/PCO.

the PDH instances, which come from everyday products. For impact, we intend to distinguish classes that may be relevant to represent human exposure. This is why we are particularly interested in properties like `isHydrophilic` (moves through water) and `isEther` (moves through air), or even `UltraShortChainPfas` (moves easily, through plants or water). For chemical nature, we intend to represent crucial PFAS processes like a `Precursor` (quite often a `Fluorotelomer`) that `evolves` into a PFAA, and other fundamental aspects like the `numberOfCarbons` property, or the `FunctionalGroup` present in the PFAS. We recall that the carbon atoms are responsible for the weight, i.e., the mobility, while the functional group is responsible for the toxicity. Although some experts do not consider a precursor as a PFAS, for our scope, it helps in the modelling, and our experts approved it. All the concepts mentioned so far are `Chemical entities`. Finally, PFAS can also move thanks to `OrganismalEntities transporting` them. The modelling is shown in Fig. 1.

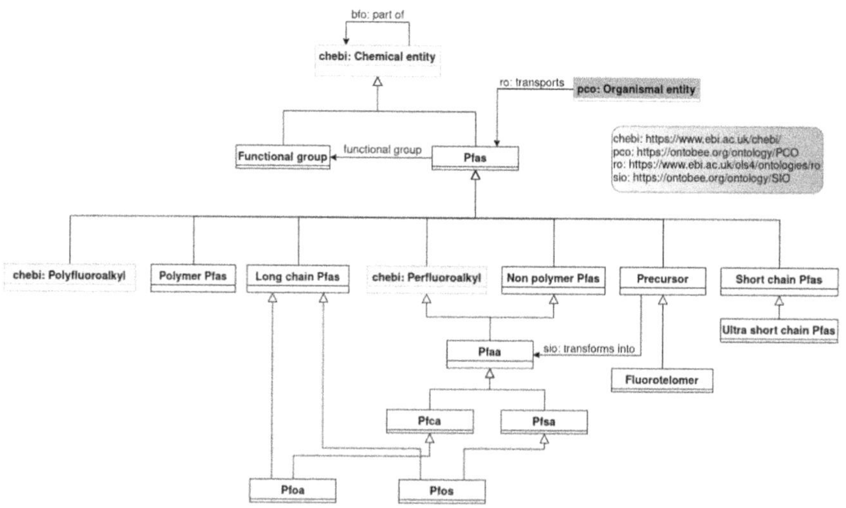

Fig. 1. Conceptualization of PFAS in OntoPFAS.

Exposure. This section is the most uncertain among the three concerning the instances. To the best of our knowledge, a collection of data about the effects of PFAS on humans is not available yet, unlike, for instance, on plants [16,19]. In this effort, the `exo` ontology is a valid and generic resource to guarantee a large *coverage* of terms when data on this aspect will be present. The two main concepts are `ExposureStimulus`, defined as an interaction in which an agent causally affects an organism and interacts during an exposure event, and `ExposureOutcome`, the result of a stimulus, under different forms: `Dose`, `BiologicalResponse` (`Disease`, `Phenotype`, `Symptom`, `Anatomy`), and `ExposureIntervention`. The stimulus is caused by the presence of an agent in a `Place`, `affecting` it. This makes the agent a possible `exposureStimulusOf` all the `Persons` located in that place, especially for the `VulnerablePersons`. A place can be specialised into `City` and `Continent` (see the Measurements section). The exposure is `stimulatedBy` at least one PFAS, which may be `contained` in products that spread in the environment. These products are generally `used` or `produced` by some `Organizations`. The position in which organizations are `locatedIn` has an impact on the understanding of the PFAS lifecycle. Organizations are typically recognised by a `source`, i.e., a `Dataset`. The modelling is shown in Fig. 2.

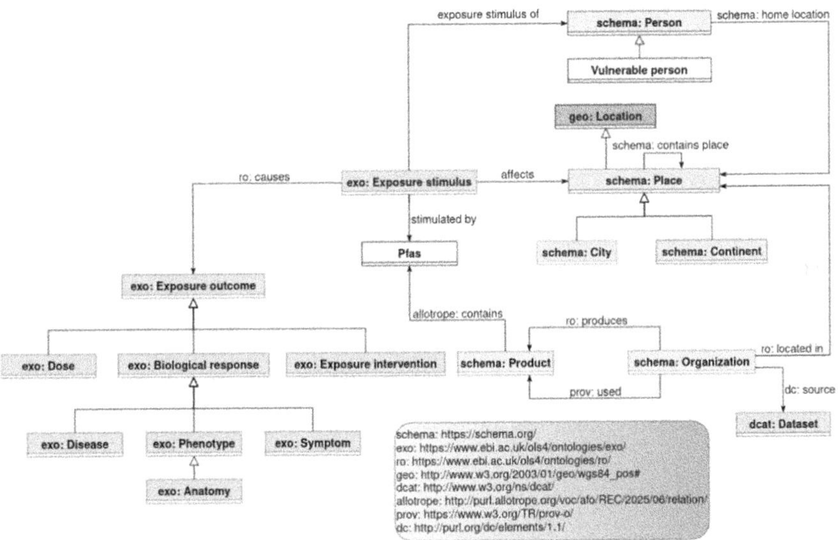

Fig. 2. Conceptualization of the exposure in OntoPFAS.

Measurements. This section mostly represents the PDH data. The main concept here is `Observation`, containing the necessary properties to represent quantities. The datatype property `hasSimpleResult` represents the collected data value, `resultTime` indicates the instant of time when the observation took place, `site` indicates the presence of a proximal industrial site, `valueRange` indicates data approximation by specifying a range ($<$, $\leq$, $\geq$, $>$), and `unit` the unit measure (in general ng/L). An observation that needs to be further evaluated is a `Presumptive` measurement. Observations measure the presence of PFAS (`measuredPfas`), `measuredIn` a certain `Location`. Locations may be `(directly)ConnectedTo` each other and constitute a flow for lightweight molecules. Locations are present in (`locationInEnvironment`) an `EnvironmentalZone`, specifying the morphological information of the territory (ground, air, water, biota). Every environment is `surroundedBy` others in the vicinity. Then, there is a subtle distinction between the environment of the location where the measurement took place and the environment of the PFAS observation (e.g., a solid material flowing into water). The latter aspect is represented by the `matrix` of the observation. We also represent the environments in which organizations are present (`activityInEnvironment`). Every PFAS has a certain general `PfasMobility` degree under certain environmental conditions (`mobilityInEnvironment`). Finally, the presence of PFAS *should* be restricted according to some *local* `Regulation`. Current regulations exist in three areas of influence: the USA, Europe, and China [3]. Each regulation `restricts` the limits of PFAS in some environments (`validInEnvironment`). The modelling is shown in Fig. 3, and the source is available at https://github.com/davide1797/OntoPFAS/blob/main/ontology/pfas.rdf and in AgroPortal[25].

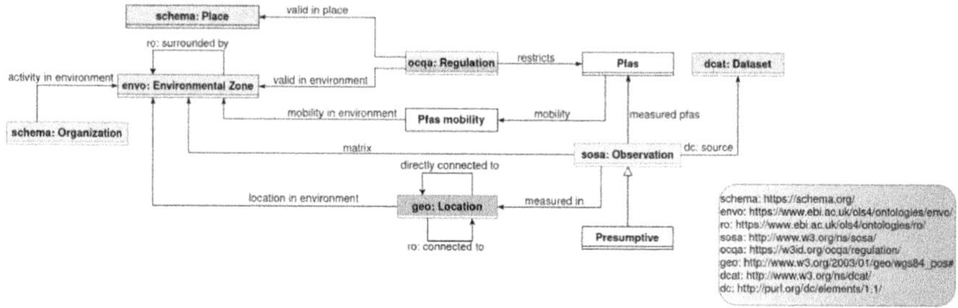

Fig. 3. Conceptualization of the measurements in OntoPFAS.

Rules. The ontology is supported by rules providing definitions, disjunctions, and equivalence axioms. We report references for (i-iii), (iv, v) are provided by the experts, while (vi) expresses that a stimulus affecting a place is an exposure source for all its inhabitants.

i UltraShortChainPfas $\equiv$ (Pfca or Pfsa) and (*numberOfCarbons* some [< 4]) (cf. [20]).

ii ShortChainPfas $\equiv$ (Pfca and (*numberOfCarbons* some [< 8])) or (Pfsa and (*numberOfCarbons* some [< 6])) (cf. [26]).

iii LongChainPfas $\equiv$ (Pfca and (*numberOfCarbons* some [> 7])) or (Pfsa and (*numberOfCarbons* some [> 5])) cf. [26]).

iv NonPolymerPfas $\sqcap$ PolymerPfas $\equiv$ $\perp$.

v Perfluoroalkyl $\sqcap$ Polyfluoroalkyl $\equiv$ $\perp$.

vi *exposureStimulusOf* $\supseteq$ *affects* $\circ$ *homeLocationOf*.

---

[25] https://agroportal.lirmm.fr/ontologies/ONTOPFAS.

Usage Example. Here, we exemplify the use of OntoPFAS in one of its most useful tasks: answering the CQ (xi): "Are there new PFAS that have replaced previous ones in a location?". In other terms, it is asked for the list of PFAS for every measurement in a given location, sorted by date of measurement. Then, a filter should recognise which PFAS are no longer present in that location. Let **sosa:observation_0**, **sosa:observation_1**, **sosa:observation_2** be instances of observations; they record which are the **measuredPfas** associated with the measurements. Every observation has a **sosa:resultTime**, which is the parameter of grouping and sorting. Measured PFAS are, hence, grouped according to the dates associated with the observation. An example of the output of the query is shown in Table 1, with the associated subgraph shown in Fig. 4.

Table 1. Example of output for CQ (xi).

Old PFAS of location_X	Current PFAS of location_X
{pfas_2, pfas_3}	{pfas_0, pfas_1}

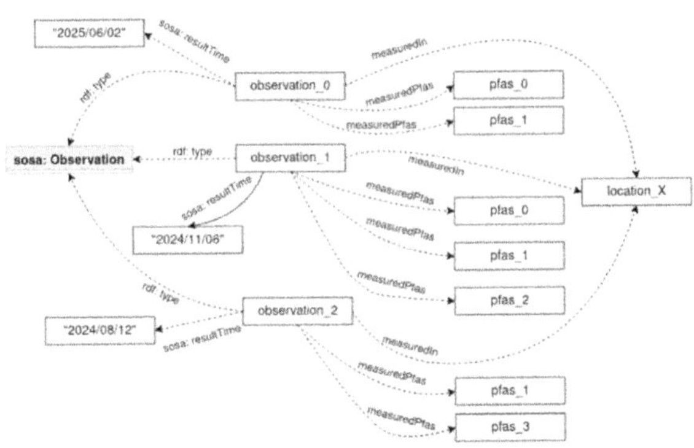

Fig. 4. Graph associated with the usage example.

5 PFAS Data Hub Mapping

There is a full literature about how to express the transformation of columns into classes and properties in knowledge graphs. Discussing the *languages* to express this mapping is out of the scope of this work, while we show *how* information is mapped. We recall that we have four schemas to consider: Known PFAS User, Presumptive, Measurement, and PFAS Production Facility. We use **ontopfas** as namespace for https://w3id.org/OntoPFAS.

Known PFAS User. This schema represents generic users of PFAS; i.e., organizations using products containing them. It consists of the following columns:

- *category* can be neglected since it is only a label to indicate that the data refer to organizations, instances of https://schema.org/Organization;
- *lat* expresses the latitude of a site. It is mapped with the property of the Location class https://www.w3.org/2003/01/geo/wgs84_pos#lat, and the organisations are connected to their location with the property of RO http://purl.obolibrary.org/obo/RO_0001025;
- *lon*, analogously, represents the longitude of a Location and is mapped onto https://www.w3.org/2003/01/geo/wgs84_pos#long;
- *name* is naturally mapped onto https://schema.org/name;
- *city* is mapped onto an instance of https://schema.org/City;
- *country* is mapped as an instance of https://schema.org/Country;
- *type*, *sector* and *source_type* are mapped, respectively, to the properties **ontopfas#type**, **ontopfas#sector**, and **ontopfas#sourceType**;
- *data_collection_method* is the dataset storing the information, mapped as https://www.w3.org/TR/vocab-dcat-3/#Property:resource_title of the class http://www.w3.org/ns/dcat#Dataset;
- *source_text*, *source_url*, and *details* are properties of Organization mapped, respectively, onto **ontopfas#sourceText**, **ontopfas#sourceUrl**, and **ontopfas#mapsLink**.
- *dataset_id* and *dataset_name* are internal identifiers, and we neglect them.
- *pfas_values*, *unit*, *pfas_sum*, *matrix*, *date*, and *year* are not used in this schema, but will be in others.

According to the creators of the dataset, this portion contains type B (AFFF) firefighting foam manufacturing facilities using lists and information from the ECHA restriction report on firefighting foams, KEMI 2015, and IPEN 2018. They finally identified 231 users.

PFAS Production Facility. This schema refers to the PFAS producers. Also referred to as generic organizations, they mostly present the same schema of Known PFAS User. We report only columns mapped differently:

- *pfas_values* contains the list of PFAS produced by the organisation. Every item on the list is mapped onto **ontopfas#Pfas** (or one of its subclasses when applicable) while the property expressing the production is http://purl.obolibrary.org/obo/RO_0003000.
- *unit* represents the unit measure of the quantity expressed in the *pfas_value* column. It is mapped onto the property **ontopfas#unit**.
- *pfas_sum* represents the overall quantity of PFAS produced by the organisation. This value is computable, so we decided not to represent it.
- *details* contains some comments on the status of the organisation. We decided to map it onto the property **ontopfas#status**.

- *matrix*, mapped as **ontopfas#activityInEnvironment**, refers to the environment in which the organisation is present. The most common are biota, air, soil, and water. These are all instances of the (Environmental zone) class **http://purl.obolibrary.org/obo/ENVO_01000408**. All the possible values are listed in the European Pollutant Release and Transfer Register (E-PRTR)[26].
- *date* refers to the latest date at which the update has been done on an organisation; we do not represent this aspect in the ontology.
- *year* is neglected for the same reason.

Every line is an instance of https://schema.org/Organization. The creators collected European facilities producing PFAS during the FPP, with the OSINT techniques, and the list was updated in May 2025. It counts 20 producers.

Presumptive and Measurement. are based on the same schema information. The sole semantic difference is the fact that presumptive measurements are the result of screening that *indicates* the presence of PFAS, but further analysis is necessary. We report only the new mappings:

- *name* indicates the name of a nearby site or known organisation. It is mapped onto the property **ontopfas#site**.
- *pfas_values* collects the PFAS detected in the measurement (or presumptive). For every item in the list, an **ontopfas#Pfas** instance is created, and the measurement is connected to the instances through **ontopfas#measuredPfas**. Quantities are stored in http://www.w3.org/ns/sosa/ hasSimpleResult.
- *date* is mapped as http://www.w3.org/ns/sosa/resultTime, to indicate the measurement instant of time.
- *year* is mapped as *date*, and used only in the absence of the previous column.

Every line of measurement (or presumptive) is associated with an instance of http://www.w3.org/ns/sosa/Observation (or **ontopfas#Presumptive**). A presumptive is considered a subcase of a measurement since it is an indirect PFAS measurement that also shares the same properties.

6   Metrics, Evaluation and Potential Impact

By answering the competency questions, we satisfied the *functional requirements*. By mapping the ontology concepts onto the PDH dataset, we respected the *PFAS Data Hub* requirements. *FAIR requirements* are generally satisfied by following our ontology methodology construction described in Sect. 3, but a more precise evaluation can be done. For the *LOD requirements*, we report the static quantitative and qualitative metrics/evaluations.

---

[26] https://www.eea.europa.eu/data-and-maps/data/member-states-reporting-art-7-under-the-european-pollutant-release-and-transfer-register-e-prtr-regulation-23/european-pollutant-release-and-transfer-register-e-prtr-data-base.

Metrics. Ontology metrics provide a glimpse of the balance of the modelling. Although highly dependent on imports, they can make some missing conceptualisation apparent. *OntoMetrics*[27] is an online tool and a comprehensive set of metrics. Table 2 shows our values, demonstrating a general balance between classes, relationships, and attributes. Our values have also been compared[28] with related ontologies, showing an overall balance with known references.

Table 2. Ontology metrics.

Metric	Value
Logical axioms count	207
Class count	42
Object property count	57
Data property count	31
DL Expressivity	SRI(D)
Attribute Richness	0.738095
Inheritance Richness	0.714286
Relationship Richness	0.677419
Class/relation ratio	0.451613
Annotation assertion axioms count	462

Evaluation. A qualitative evaluation is based on *pitfalls*, i.e., misconception errors. Poveda-Villalón et al. [29] introduced the *OntOlogy Pitfall Scanner (OOPS)*[29], an automatic evaluation, which lists a set of major and minor errors. In our case, only minor issues have been found. This is caused by the imported ontologies, while we did not introduce new ones.

The evaluation based on the work by Vrandečić et al. [30] consists of guaranteeing many positive features. We report here those applicable in this work that are not covered by FAIR requirements; *accuracy*: the ontology has been developed in collaboration with the experts who have provided their requirements and validated our results; *clarity*: all the proposed terms are compliant with the literature on the PFAS and are provided with metadata supporting the clarity (especially labels and definitions); *adaptability*: OntoPFAS can be easily extended to other perspectives (health above all), and the collaboration with the experts will guarantee a higher level of compliance with these domains; *completeness*: with respect to the predefined objectives, this ontology answers all the competency questions, and a use case example has been shown; *efficiency*: the

---

ontology is light both with respect to the logical axioms and the expressivity; *conciseness*: with the reuse of many ontologies, we introduced only the necessary concepts and properties; *consistency*: the ontology in both versions (w/ and w/o instances) are logically consistent, tested with the HermiT 1.4.3.456 reasoner[30]; *compliance to expertise*: the use of terms taken from the literature makes the resource accessible for other experts; all the major decisions are supported by quoted papers or our experts, belonging to different communities (chemistry for environmental science, and geography).

FAIR Score. Finally, to deepen the FAIR aspect, we report the evaluation of our ontology by leveraging the *FAIR Ontology Pitfall Scanner (FOOPS)*[31] [12]. Our ontology reports an overall score of *90/100*, being penalised mostly for the absence of some metadata that cannot be introduced now. We report at least 75% in every applicable evaluation, apart from the one demanding publisher info. The full evaluation results are available here[32].

Potential Impact. At this stage, the resource is being adopted by the experts with placeholders to construct examples of use cases and answers. A web demo is in development to provide the users with visualisations and query answering support. An area or cluster-based visualization presents a summary of the quantity of measurements (Fig. 5). Several filters allow the partial visualization of the various aspects (e.g., presumptive, specific PFAS, specific regions or environments). By now, the demo does not take advantage of the semantics of concepts, relying only on flat data, but it represents a visual connection for experts and helps us in understanding their needs and extensions.

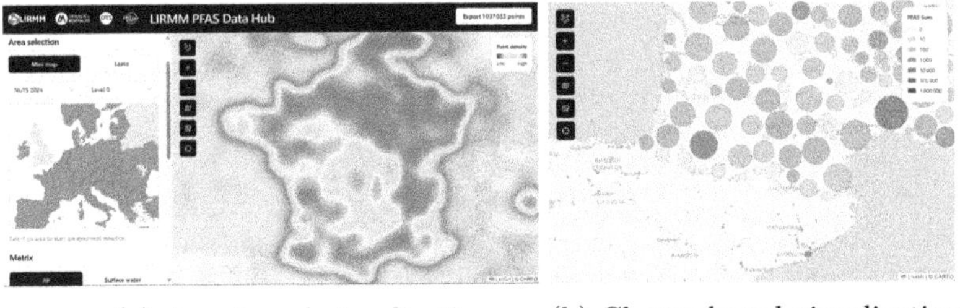

(a) Area-based visualization   (b) Cluster-based visualization

Fig. 5. Demo of the PFAS application.

---

[30] http://www.hermit-reasoner.com/.
[31] https://foops.linkeddata.es/FAIR_validator.html.
[32] https://github.com/davide1797/OntoPFAS/blob/main/evaluation/.

7   Conclusions

In this paper, we presented OntoPFAS, an ontology to represent PFAS according to their chemical properties, measurements, and the main interactions with the environment and people. Being compliant with the PDH, it is capable of representing an extensive dataset of measurements in Europe. Its main contribution is the possibility to help experts in understanding when and why certain areas are polluted, and to forecast the future spread. The time dimension is expressed by the measurement date and the degree of mobility of PFAS within certain environments. The causal dimension is provided by the presence of polluting companies or the connections between environments. While being the first to contribute to this field with a formal conceptualization, we will continue to publish extended versions, also with the introduction of the medical field, in parallel with the growing interest of both society and the scientific communities.

Acknowledgments. This work benefits from the conjoint support of the institute ExposUM and from the Mission pour les Initiatives Transverses et Interdisciplinaires (MITI) of the CNRS.

References

1. Abunada, Z., Alazaiza, M.Y., Bashir, M.J.: An overview of per-and polyfluoroalkyl substances (PFAS) in the environment: source, fate, risk and regulations. Water **12**(12), 3590 (2020)
2. ANSES: Pfass: very persistent chemicals. https://www.anses.fr/en/content/pfass-very-persistent-chemicals (2025). Accessed 03 Dec 2025
3. Brennan, N.M., Evans, A.T., Fritz, M.K., Peak, S.A., von Holst, H.E.: Trends in the regulation of per-and polyfluoroalkyl substances (PFAS): a scoping review. Int. J. Environ. Res. Public Health **18**(20), 10900 (2021)
4. Buck, R.C., et al.: Perfluoroalkyl and polyfluoroalkyl substances in the environment: terminology, classification, and origins. Integr. Environ. Assess. Manag. **7**(4), 513–541 (2011)
5. Buck, R.C., Korzeniowski, S.H., Laganis, E., Adamsky, F.: Identification and classification of commercially relevant per-and poly-fluoroalkyl substances (PFAs). Integr. Environ. Assess. Manag. **17**(5), 1045–1055 (2021)
6. Buttigieg, P.L., Morrison, N., Smith, B., Mungall, C.J., Lewis, S.E., Consortium, E.: The environment ontology: contextualising biological and biomedical entities. J. Biomed. Semant. **4**(1), 43 (2013)
7. Cui, J., Gao, P., Deng, Y.: Destruction of per-and polyfluoroalkyl substances (PFAs) with advanced reduction processes (ARPs): a critical review. Environ. Sci. Technol. **54**(7), 3752–3766 (2020)
8. Degtyarenko, K., et al.: ChEBI: a database and ontology for chemical entities of biological interest. Nucleic Acids Res. **36**(suppl_1), D344–D350 (2007)
9. Di Pierro, D., Abrouk, L., Guyot, A., Symeonidou, D., Labadie, P., Lysaniuk, B.: OntoPFAS: Ontologie des PFAS et de leur exposition. PFIA 2025 p. 80 (2025)
10. Doumanas, D., Bouchouras, G., Soularidis, A., Kotis, K., Vouros, G.: From human- to LLM-centered collaborative ontology engineering. Appl. Ontol. **19**(4), 334–367 (2024)

11. Fernández-López, M., Gómez-Pérez, A., Juristo Juzgado, N.: Methontology: from ontological art towards ontological engineering (1997)
12. Garijo, D., Corcho, O., Poveda-Villalón, M.: Foops!: an ontology pitfall scanner for the fair principles. In: ISWC (Posters/Demos/Industry), p. 0 (2021)
13. Glüge, J., et al.: An overview of the uses of per-and polyfluoroalkyl substances (PFAS). Environ. Sci. Processes Impacts **22**(12), 2345–2373 (2020)
14. Grüninger, M., Fox, M.S.: The role of competency questions in enterprise engineering. In: Rolstadås, A. (ed.) Benchmarking — Theory and Practice. IAICT, pp. 22–31. Springer, Boston, MA (1995). https://doi.org/10.1007/978-0-387-34847-6_3
15. Guha, R.V., Brickley, D., Macbeth, S.: Schema. org: evolution of structured data on the web. Commun. ACM **59**(2), 44–51 (2016)
16. He, Q., et al.: Phytoextraction of per-and polyfluoroalkyl substances (PFAS) by weeds: effect of PFAS physicochemical properties and plant physiological traits. J. Hazard. Mater. **454**, 131492 (2023)
17. Hristozova, M., Sterling, L.: An extreme method for developing lightweight ontologies. In: Workshop on Ontologies in Agent Systems, 1st International Joint Conference on Autonomous Agents and Multi-Agent Systems, (Bologna, Italy, 2002) (2002)
18. Janowicz, K., Haller, A., Cox, S.J., Le Phuoc, D., Lefrançois, M.: SOSA: a lightweight ontology for sensors, observations, samples, and actuators. J. Web Semant. **56**, 1–10 (2019)
19. Karamat, A., Tehrani, R., Foster, G.D., Van Aken, B.: Plant responses to per-and polyfluoroalkyl substances (PFAS): a molecular perspective. Int. J. Phytorem. **26**(2), 219–227 (2024)
20. Lenka, S.P., Kah, M., Padhye, L.P.: A review of the occurrence, transformation, and removal of poly-and perfluoroalkyl substances (PFAS) in wastewater treatment plants. Water Res. **199**, 117187 (2021)
21. Lichtnow, D., Fleischmann, A.M.P., do Nascimento, L.V., Machado, G.M., de Oliveira, J.P.M.: Pipeline for ontology construction using a large language model: a smart campus use case. In: Proceedings of the 27th International Conference on Enterprise Information Systems, ICEIS, pp. 4–6 (2025)
22. Lippolis, A.S., et al.: Ontology generation using large language models. In: Curry, E., et al. (eds.) ESWC 2025. LNCS, vol. 15718, pp. 321–341. Springer, Cham (2025). https://doi.org/10.1007/978-3-031-94575-5_18
23. Mahinroosta, R., Senevirathna, L.: A review of the emerging treatment technologies for PFAS contaminated soils. J. Environ. Manage. **255**, 109896 (2020)
24. Mattingly, C., McKone, T., Callahan, M., Blake, J., Cohen Hubal, E.: Exo: an ontology for exposure science. Nature Proceedings 1–1 (2011)
25. Noy, N.F., McGuinness, D.L., et al.: Ontology development 101: a guide to creating your first ontology (2001)
26. Olympia, W.: Per-and polyfluoroalkyl substances chemical action plan
27. Peroni, S.: Samod: an agile methodology for the development of ontologies. In: Proceedings of the 13th OWL: Experiences and Directions Workshop and 5th OWL Reasoner Evaluation Workshop (OWLED-ORE 2016), pp. 1–14 (2016)
28. Poveda-Villalón, M., Fernández-Izquierdo, A., Fernández-López, M., García-Castro, R.: Lot: An industrial oriented ontology engineering framework. Eng. Appl. Artif. Intell. **111**, 104755 (2022)
29. Poveda-Villalón, M., Suárez-Figueroa, M.C., Gómez-Pérez, A.: Validating ontologies with OOPS! In: ten Teije, A., et al. (eds.) EKAW 2012. LNCS (LNAI), vol. 7603, pp. 267–281. Springer, Heidelberg (2012). https://doi.org/10.1007/978-3-642-33876-2_24

30. Vrandečić, D.: Ontology evaluation. In: Staab, S., Studer, R. (eds.) Handbook on Ontologies. IHIS, pp. 293–313. Springer, Heidelberg (2009). https://doi.org/10.1007/978-3-540-92673-3_13
31. Walls, R.L., et al.: Meeting report: advancing practical applications of biodiversity ontologies. Stand Genomic Sci. **9**(1), 17 (2014)
32. World Wide Web Consortium (W3C): Wgs84 geo positioning: a basic geo (wgs84 lat/long) vocabulary (2009). https://www.w3.org/2003/01/geo/wgs84_pos
33. World Wide Web Consortium (W3C): Data catalog vocabulary (dcat) — version 3 (2024). https://www.w3.org/TR/vocab-dcat-3/

Neuro-Symbolic Stream Reasoning
with Kolibrie

Volodymyr Kadzhaia$^{(\boxtimes)}$ and Pieter Bonte$^{(\boxtimes)}$

Department of Computer Science, KU Leuven Campus Kulak, Kortrijk, Belgium
{volodymyr.kadzhaia,pieter.bonte}@kuleuven.be

Abstract. The rapid growth of the Internet of Things (IoT) has created vast networks of interconnected devices generating continuous data streams, offering opportunities for real-time analytics across domains like smart cities, healthcare, and transportation. On the one hand, traditional knowledge representation methods provide interpretability, precise semantic modeling, and structured reasoning but often struggle with scalability, making predictions, and handling uncertainty. On the other hand, machine learning models offer powerful predictive capabilities and efficient handling of large, uncertain datasets, but typically lack interpretability, adaptability, and require extensive labeled data. Neuro-symbolic AI tries to use the best of both symbolic reasoning and machine learning—combining interpretability and structured inference from symbolic approaches with the predictive power and generalization capabilities of machine learning. Existing neuro-symbolic approaches, such as DeepProbLog and NeurASP, although successful in general logical programming contexts, they do not natively support streaming data or querying over RDF. This paper introduces Kolibrie, a hybrid stream reasoning engine developed in Rust that integrates both at the language and system level (1) SPARQL for querying, (2) Datalog-style rules for symbolic inference, (3) RSP-QL semantics for RDF stream processing and (4) machine learning predictions.

Resource type: Software framework
License: Mozilla Public License Version 2.0
DOI: https://doi.org/10.5281/zenodo.17808656
URL: https://github.com/StreamIntelligenceLab/Kolibrie

Keywords: Stream Reasoning · Neuro-Symbolic AI · SPARQL

1 Introduction

Over the past decade, the Internet of Things (IoT) has expanded significantly, creating vast networks of interconnected sensors and devices that continuously generate massive data streams [4]. This streaming data has enormous potential for real-time analytics and actionable insights in a variety of fields, including smart cities, healthcare monitoring, industrial automation, and intelligent transportation systems [25].

M. Acosta et al. (Eds.): ESWC 2026, LNCS 16550, pp. 39–57, 2026.
https://doi.org/10.1007/978-3-032-25159-6_3

Knowledge Graphs (KGs) have become popular as powerful tools for representing heterogeneous, multi-relational data in an interpretable format. A KG typically structures information as entities connected by explicit semantic relationships, facilitating easy querying and integration of diverse data sources [19]. Standard knowledge representation frameworks such as RDFS, Web Ontology Language (OWL) [11] and Notation3 Logic (N3Logic) [12] are widely used to formalize complex domains and enable deductive reasoning. Although they offer clear interpretability and precise semantic modeling, these methods have significant drawbacks when dealing with IoT data: they often struggle to scale in rapidly changing data environments, enforce rigid schemas, and are unable to handle uncertainty or incomplete information efficiently [27].

Machine learning (ML), particularly deep neural networks, have significantly advanced capabilities for pattern recognition, predictive modeling, and decision-making from large datasets. However, despite their effectiveness, ML models often suffer from critical limitations such as a lack of interpretability, heavy reliance on labeled data, fail to incorporate domain knowledge and limited adaptability in changing environments [3, 19].

Neuro-symbolic AI is a paradigm to address these challenges by integrating symbolic reasoning methods with statistical learning approaches, using the strengths of both domains [19]. Following Hitzler et al. [23], we focus specifically on the neural-symbolic logic family, where symbolic rules and probabilistic logic programming frameworks are combined with neural models to support logical inference and learning. Symbolic reasoning contributes structured, interpretable inference mechanisms, while neural networks offer powerful pattern recognition and generalization capabilities. Neuro-symbolic approaches, such as DeepProbLog [29] and NeurASP [39], have demonstrated potential in tasks like knowledge graph completion, reasoning over uncertain information, and handling incomplete datasets. However, these methods are not designed to operate over data streams and therefore do not support continuous, real-time reasoning, which is needed in IoT settings.

To effectively use neuro-symbolic approaches within dynamic and real-time environments, stream reasoning, i.e. a technique dedicated to applying inference mechanisms to continuous data streams, emerges as a critical enabling technology [14]. Stream reasoning refers to the study of how logical inference, query processing, and knowledge-graph techniques can be applied to data that arrives continuously and evolves over time. Its goal is to maintain conclusions incrementally as new events appear, rather than recomputing results from scratch [13]. In this work, we focus on a particular branch of stream reasoning, i.e. RDF Stream Processing (RSP), which specifically targets reasoning over RDF data streams. Stream reasoning involves applying inference techniques to continuous data streams, including processing streams of RDF data, enabling dynamic reasoning capabilities [14, 18]. Systems like C-SPARQL and CQELS provide engine and query language capabilities, while efforts like RDF Stream Processing Query Language (RSP-QL) aim to unify various RSP languages such as C-SPARQL and CQELS under a common semantic framework [8, 17]. Although these developments hold great promise, many research challenges remain, such as improving the systems' scalability, environment adaptability, and providing good explanations of their decision-making processes [14]. Yet, traditional RSP engines primarily focus on streaming query evaluation and lack integrated symbolic reasoning and machine-learning capabilities.

Addressing these issues leads to the research question:

1. How can we realize a reusable framework that integrates symbolic rules and machine learning predictions for continuous reasoning over RDF streams?

The article is structured as follows. Section 2 reviews related work and Sect. 3 distills the requirements. Section 4 presents Kolibrie's architecture. Section 5 reports performance results. We provide a discussion in Sect. 7, and finally conclude with a roadmap for the future in Sect. 8.

2   Related Work

RDF Stream Processing (RSP) extends SPARQL with stream-procesing operators such as sliding windows, timestamps, and continuous queries, allowing SPARQL to operate over unbounded RDF streams [17]. Early RDF Stream Processing (RSP) systems, such as C-SPARQL [8], CQELS [26], were influenced by Data Stream Management System (DSMS) [5,7], which introduced continuous query evaluated incrementally over unbounded data streams using window operators. Following this vision, languages such as C-SPARQL [8] and CQELS-QL [26] extended SPARQL with constructs for defining and querying upon windows. Later, RSP-QL [17] unified these approaches, providing a common semantics for query processing and a formal notion of result correctness over dynamic data. RSP4J [35] was introduced as an API for the development under RSP-QL semantics. However, these systems focus primarily on query evaluation, they lack support for symbolic inference, and learning.

While traditional RSP systems excel at managing rapidly incoming data, they still face significant scalability challenges when handling the large volumes and high velocity typical of real-world IoT scenarios [18]. Additionally, although RDF inherently supports heterogeneous data integration, RSP systems struggle with highly diverse data formats beyond structured RDF, such as unstructured or semi-structured data, as well as uncertainty and incompleteness in real-world streaming environments [14]. In IoT scenarios, this lack of native support for multimodal data (e.g., video, audio, raw sensor time series) and machine-learning-based perception becomes a bottleneck, since many applications must combine such signals with RDF streams in a unified pipeline [31]. Current RSP stacks fall short for highly dynamic and uncertain environments: they lack native multimodal handling, learning-aware operators, and semantics to integrate predictive outputs with symbolic reasoning.

Beyond pure query evaluation, recent work by Le-Phuoc [31] demonstrates a neural-symbolic extension of CQELS, called Autonomous Stream Fusion (ASF), which integrates deep neural network inference directly into the RSP pipeline.

Their architecture couples CQELS with an agent-based stream fusion framework, where OpenCV and Deep Neural Networks components are embedded allowing to query both video and sensor streams declaratively. However, this approach does not offer a general-purpose neuro-symbolic RSP language, lacks a mechanism for expressing and composing pipelines using arbitrary bring your own ML models, models and rules do not integrate in a unified fashion, and it does not address model selection across heterogeneous IoT workloads. Furthermore, to the best of our knowledge, the solution is not open source, making it infeasible as a basis for a reusable framework.

From a static data querying perspective, the commercial triple store Stardog[1] enables some ML support through built-in classification and regression models that can be learned and executed on the data in the triple store. However, the used models are fixed, i.e. one cannot bring their own model, they do not seamlessly integrate with logical rules and have no support for streaming data.

From a neuro-symbolic AI perspective, both DeepProbLog [29] and NeurASP [39] provide expressive ways to combine symbolic rules with neural components. DeepProbLog interprets neural outputs as probabilistic facts [29], while NeurASP embeds pretrained neural networks into ASP as probabilistic atoms [39]. However, neither system provides native support for RDF, SPARQL, supports processing of continuous data streams or can easily switch between ML models.

3 Requirements

To realize a *reusable framework that integrates symbolic rules and machine learning predictions for continuous reasoning over RDF streams* must satisfy several critical requirements.

Such a system must provide both language-level abstractions for expressing neuro-symbolic reasoning patterns and system-level mechanisms for efficient execution over continuous data streams. At the language level, users need to declaratively express both symbolic rules and ML predictions in a unified manner, while the system must handle streaming semantics, exploit modern hardware, and maintain a modular architecture that supports extensibility and long-term maintenance. These requirements were identified through a literature-driven gap analysis of existing SPARQL, RSP, and neuro-symbolic systems (Sect. 2), complemented by an analysis of recurring design challenges in IoT-style continuous reasoning scenarios. This allows to define the following requirements:

R_1 A declarative language that allows to seamlessly integrate symbolic inference and ML predictions.

R_2 From a language perspective, support explicit stream-oriented constructs (e.g. windowing, sliding ranges) for reasoning over continuous RDF data, and from a system perspective provide such mechanisms.

---

[1] https://docs.stardog.com/machine-learning.

R$_3$ Exploit modern hardware (e.g., SIMD instructions, multi-threading, GPU acceleration) to scale reasoning and query evaluation.

R$_4$ Integrate symbolic reasoning, stream processing, and ML predictions in a single execution pipeline.

R$_5$ Enable dynamic model selection at runtime, choose between multiple ML models based on logical premises.
In Kolibrie, model choice is driven by ML Schema metadata and resolved by the ML handler at runtime, while keeping models cached across ticks/windows to avoid repeated loading.

R$_6$ A modular and extensible architecture that allows components to be reused, replaced, or extended independently.

R$_7$ An open-source implementation that is actively maintained, enabling reproducibility and long-term sustainability.

The literature reviewed above shows that different families of systems address only fragments of these requirements, as summarized in Table 1.

Table 1. Coverage of requirements R$_1$–R$_7$ by related approaches. For R$_7$ we checked if the repo (if available) was updated in the last year.

Tool/System	R$_1$	R$_2$	R$_3$	R$_4$	R$_5$	R$_6$	R$_7$
C-SPARQL [8]	–	✓	–	–	–	–	–
CQELS [26]	–	✓	Partial	–	–	–	–
RSP4J [35]	–	✓	–	–	–	✓	✓
SPARQL engines	–	–	Partial	–	–	–	✓
DeepProbLog	✓	–	Partial	–	–	–	–
NeurASP	✓	–	Partial	–	–	–	–
ASF [31]	Partial	✓	Partial	✓	–	–	–
Kolibrie	✓	✓	✓	✓	✓	✓	✓

4 Kolibrie: Language and System

This section details how Kolibrie realizes a reusable framework that integrates symbolic rules and ML predictions for continuous reasoning over RDF streams.

4.1 Kolibrie Language

Kolibrie introduces a hybrid query language that extends SPARQL with explicit constructs for stream reasoning, rule-based inference, and ML predictions. At the language level, Kolibrie provides a declarative language in which symbolic inference and ML predictions are expressed uniformly as rules whose results become ordinary RDF triples (R_1), and it adds explicit stream-oriented constructs stream operators and window clauses for continuous reasoning over RDF streams (R_2), while still supporting standard SPARQL queries over static graphs.

Inspired by RSP languages [16], which extend SPARQL with continuous queries and window semantics over RDF streams, Kolibrie offers a single declarative syntax in which SPARQL queries, Datalog-style rules, stream rules, and ML-augmented rules coexist and can be composed. In particular, Kolibrie treats the conclusions of any rule as RDF triples that can be consumed by other rules (of any type) as well as by ordinary SPARQL queries, thereby realizing a neuro-symbolic RSP language rather than a pure streaming query language.

At the rule level, Kolibrie follows the idea of SPARQLog [15] and 'SPARQL in N3' [6], allowing Datalog-style [28] rules to be written using familiar SPARQL CONSTRUCT/WHERE blocks. This keeps rule definitions close to standard SPARQL queries while enabling rule-based inference. For the fragment that aligns with Datalog, Kolibrie also supports an N3 Logic [12] rules syntax, allowing users to define rules in the Notation3 Logic style when desired. In Kolibrie, all rules share the same syntactic skeleton, but can play different roles:

- *Logical rules* over static RDF graphs (no stream operator, no `ML.PREDICT`);
- *Stream rules* that use a stream operator (`RSTREAM`, `ISTREAM`, `DSTREAM`) and one or more `FROM NAMED WINDOW` clauses;
- *ML rules* that invoke a pre-trained model via `ML.PREDICT`;
- *Hybrid rules* that combine streaming, logic or ML in a single rule.

Crucially, the results of any rule are materialized as RDF triples and can serve as input to other rules and SPARQL queries. This chaining of rule outputs is central for R_1 and R_2: symbolic inferences, stream-based inferences, and ML predictions all become first-class facts in the evolving knowledge graph.

A Kolibrie rule consists of:

1. a `RULE` naming the inferred predicate and its arguments;
2. an optional `STREAM` clause specifying how RDF streams are windowed;
3. a `CONSTRUCT` clause that defines which RDF triples are produced;
4. a `WHERE` clause that specifies the graph pattern used as rule conditions;
5. an optional `ML.PREDICT` clause for calling pre-trained ML models.

Rules without a stream clause are evaluated over static data; rules with a window clause are evaluated over windowed streams, following RSP-QL semantics.

In the remainder of this subsection we use a single running example in a traffic scenario to illustrate how these rule types interact.

Example 1. Logical rule over static data: We first define a purely symbolic rule over a static RDF graph that classifies roads with high vehicle counts as instances of the class `ex:HighTrafficArea`:

```
PREFIX ex: <http://example.org/>

RULE :HighTrafficArea :-
  CONSTRUCT {
    ?road a ex:HighTrafficArea .
  }
  WHERE {
    ?road ex:vehicleCount ?c .
    FILTER (?c > 100)
  }.
```

Listing 1.1. Simple Kolibrie rule over static data

This rule is a standard Datalog-style rule written in SPARQL CONSTRUCT/WHERE syntax: whenever the condition in the WHERE clause matches, the engine materializes the triple `?road a ex:HighTrafficArea`.

Example 2. Stream rule using derived classes: We next illustrate a continuous streaming query that consumes the results of the previous rule. Kolibrie follows the RSP-QL style for continuous queries via the REGISTER clause. The query below registers an RSTREAM that operates over a sliding window on the `:trafficS` stream:

```
REGISTER RSTREAM <http://example.org/RecentSpeedStream> AS
SELECT ?road (AVG(?speed) as ?avgSpeed)
FROM NAMED WINDOW :w ON :traffic [RANGE PT5M STEP PT30S]
WHERE {
  WINDOW :w {
    ?road a ex:HighTrafficArea.
    ?obs  ex:featureOfInterest ?road ;
          ex:measuredSpeed  ?speed .
  }GROUP BY ?road}
```

Listing 1.2. Kolibrie rule with sliding window

This continuous query is evaluated over a 5-min sliding window that advances every 30 s. Within each window, the query identifies average speed occurring on high traffic roads.

Example 3. ML-augmented rule and chaining. Kolibrie integrates pre-trained ML models via the ML.PREDICT clause. This rule predicts congestion levels based on historical data, using features derived from recent observations on the traffic stream. Note that ML.PREDICT can also be used on static rules.

```
PREFIX ex: <http://example.org/traffic#>

RULE :PredictCongestionLevel :-
  RSTREAM
  FROM NAMED WINDOW :w ON :traffic [RANGE PT5M STEP PT30S]
  CONSTRUCT {
    ?road ex:congestionLevel ?level .
  }
```

```
 9    WHERE {
10      ?road  a ex:HighTrafficArea .
11      ?obs   ex:featureOfInterest  ?road ;
12             ex:avgVehicleSpeed ?avgSpeed ;
13             ex:vehicleCount    ?vehicleCount .
14    }
15    ML.PREDICT(
16      MODEL "congestion_model",
17      INPUT {
18        SELECT ?road ?avgSpeed ?vehicleCount
19      },
20      OUTPUT ?level
21    ).
```

Listing 1.3. Hybrid Kolibrie rule with windowing and ML.PREDICT

Conceptually, the rule works as follows:

1. The WHERE clause matches roads in high-traffic areas and binds feature variables such as ?avgSpeed and ?vehicleCount from recent observations in the window.
2. The ML.PREDICT clause calls the external model "congestion_model" on those features. The INPUT { SELECT ...} block specifies which currently bound variables are passed to the model. In many cases, this block contains only a SELECT clause (as shown here); an additional WHERE inside the INPUT block is optional and only needed when a local subquery is required.
3. The model returns a predicted congestion level bound to ?level, which is then used in the CONSTRUCT clause as an ordinary variable.

This running example illustrates that:

- a *logical* rule (:HighTrafficArea) derives a class from static data;
- a *stream* rule (:RecentHighTrafficObservation) uses that class inside a window over a stream;
- a *hybrid neuro-symbolic* rule (:PredictCongestionLevel) combines windowed data, earlier symbolic inferences, and ML predictions, and materializes the results as RDF triples.

The output of ML-augmented rules can immediately serve as input to any other rule, be as a symbolic rule, another stream rule, or a further ML rule, and can also be queried directly using standard SPARQL. In this way, Kolibrie offers a unified, declarative language for continuous reasoning that combines RDF streams, rule-based inference, and ML predictions.

To make these extensions precise, the following EBNF summarizes the concrete syntax supported by our parser for RULE, window clauses, and ML.PREDICT (literals in quotes):

```
1  <rule> ::=
2  "RULE" <rule-head> ":-" [ <stream-type> ] { <from-named-window> } <construct-
       clause> <where-clause> [ <ml-predict> ] [ "." ] ;
3  <rule-head> ::= <predicate> [ "(" <var-list> ")" ] ;
4  <var-list> ::= <var> { "," <var> } ;
5  <predicate> ::= IRI | CURIE | ":" IDENT ;
6  <var> ::= "?" IDENT ;
7  <stream-type> ::= "RSTREAM" | "ISTREAM" | "DSTREAM" ;
```

```
 8  <from-named-window> ::= "FROM" "NAMED" "WINDOW" <win-id> "ON" <stream-id> <
        window-spec> ;
 9  <window-spec> ::= "[" <duration> [ "STEP" <duration> ] [ "REPORT" <report-
        strategy> ] [ "TICK" <tick-strategy> ] "]" ;
10  <construct-clause>::= "CONSTRUCT" "{"<triple-template>* "}";
11  <where-clause> ::= "WHERE" "{" <graph-patterns> "}" ;
12  <ml-predict> ::= "ML.PREDICT" "(" "MODEL" <quoted-string> "," "INPUT" "{" "
        SELECT" <var-list> [ <where-clause> ] "}" "," "OUTPUT" <var> ")" ;
```

Listing 1.4. Kolibrie EBNF for RULE, window clauses, and ML.PREDICT (parser-accurate)

4.2 System Design

CLI	Python Binding	Rust API	HTTP API
Query Builder API			
Parser & Validator			
Volcano Optimizer			
Query Executor			
Index Manger SPO POS OSP PSO OPS SOP	Reasoning Abox/TBox Rules Forward chain Backward chain	RSP Engine (Streams) Windows S2R/R2S Ticks	ML Handler Mode Cache MLSchema Predictions
Dictionary		Triple Store	

Fig. 1. High-level Kolibrie architecture.

Kolibrie is a high-performance stream reasoning engine built in Rust. Figure 1 presents the system architecture, which exposes multiple front-end interfaces, applies a Volcano-style optimizer [22], and executes queries in a batch-oriented, materializing fashion. The architecture integrates symbolic rules, RDF stream processing, and ML predictions in a unified pipeline, directly addressing the requirements R_3–R_6 summarized in Sect. 2.

A central design objective is reusability. Kolibrie is structured so that new languages, new ML backends, or alternative triple stores can be plugged in without changing the core execution pipeline. Figure 1 illustrates these abstractions: every layer hides its implementation behind a stable interface, allowing new components to be added or replaced independently (R_6). At the top layer, Kolibrie exposes four interfaces: a CLI, a Rust API, a Python binding, and an HTTP API.

Volcano-Style Logical and Physical Optimization. Kolibrie adopts the Volcano architecture for optimization. Given a parsed query or rule, the optimizer produces a logical plan consisting of standard relational operators (scans, filters, projections, joins) extended with RSP operators (windows, S2R/R2S) [17] and an `ML.PREDICT` operator for invoking pretrained machine-learning models.

A set of rewrite rules performs three tasks: (i) simplifying graph patterns, (ii) recognizing and optimizing common query patterns, and (iii) reordering joins based on selectivity heuristics. The optimizer then produces a physical plan by selecting concrete operator implementations (e.g., hash join, sort-merge join).

The physical operators chosen by the optimizer are configured for processing batches of bindings, multi-threading, and, when beneficial, optional GPU kernels for joins. GPU-enabled joins are only available if configured by the user; the optimizer does not yet make dynamic CPU/GPU choices, although a cost-based GPU dispatcher is part of future work.

Vectorized, Materializing Execution Model. Although Kolibrie's optimizer follows Volcano, the execution model does not use Volcano's iterator interface (`open-next-close` protocol) [21]. Instead, each physical operator implements a Rust method `execute_with_ids()`, which:

1. recursively executes its children;
2. receives batches of bindings;
3. applies its physical transformation (join, filter, projection, etc.);
4. returns batches of materialized results.

This design has several advantages:

- Data parallelism: operators process whole batches at a time, enabling SIMD acceleration and parallelization via Rayon [32];
- Dictionary IDs: all operators work on compact integer IDs until the final projection stage, minimizing string manipulation;
- Flexible execution: the architecture supports both pull-based query evaluation (where operators request data from their children) and push-based stream processing (where data is pushed through the pipeline as it arrives), enabling efficient handling of both static queries and continuous streams.

Dictionary and Multi-index Triple Store. Kolibrie's storage layer consists of:

- a dictionary mapping RDF terms to 32-bit identifiers[2];
- an in-memory triple store backed by six permutation indexes (SPO, POS, OSP, PSO, OPS, SOP), similar to modern RDF systems [30,38].

The Index Manager selects the most selective index based on the bound variables in each triple pattern. The compressed ID representation minimizes memory usage and supports fast scans - important for R_3 and R_4. This in-memory implementation is the default configuration, but thanks to Kolibrie's modular architecture (R_6), one can easily introduce a persisted data source by implementing alternative storage backends that conform to the same interface, enabling disk-based or hybrid storage strategies when needed.

---

[2] Although 32-bit IDs are used in the current implementation, Kolibrie's dictionary is fully configurable and can be compiled with 64-bit or 128-bit identifiers for large-scale deployments.

Reasoning Engine. Kolibrie represents the ABox as dictionary-encoded triples stored in a unified index, and maintains a set of dynamic rules in a dedicated rule index that maps premise patterns to rule identifiers. All rules are indexed by their premise patterns. Evaluation follows a family of bottom-up procedures including semi-naive evaluation. The key point is that the rule engine reuses exactly the same join, scan, and filter operators as the SPARQL executor, made possible by Kolibrie's modular design (R_6). This unifies symbolic reasoning and query processing in a single physical engine (R_4), and ensures that system optimizations automatically benefit both.

On top of this, the engine offers a family of bottom-up forward-chaining [1] procedures. A basic variant `infer_new_facts` repeatedly evaluates all rules over the current fact set using an optimized hash-join procedure, while semi-naive evaluation is available through `infer_new_facts_semi_naive`: in each iteration, only premises that contain at least one triple from the current delta are considered, and new inferences are added both to the global fact set and to the next delta. A parallel variant `infer_new_facts_semi_naive_parallel` exploits Rust's data-parallel primitives to distribute rule evaluation over cores, combining per-thread delta contributions into a global fixpoint. In all cases, rule bodies are matched against facts using unification over dictionary IDs, and conclusions are materialized as new triples and inserted back into the unified index.

RSP Engine and Window Management. Kolibrie's RSP engine maintains windows over streams, supports sliding and tumbling windows. Windowed graphs are exposed to the query executor as ordinary relations represented with dictionary IDs, allowing the physical plan to apply joins, filters, aggregates, window and ML operators uniformly. Kolibrie provides the same operator abstractions as RSP4J [35], supporting RSP-QL semantics.

Window contents are exposed to the query executor as ordinary relations over dictionary IDs, so the optimizer and executor can handle them using standard operators. The optimizer computes join ordering and index selection once; the same physical plan is reused for every window instance, enabling per-window efficient evaluation.

Neuro-Symbolic Integration via `ML.PREDICT`. The ML Handler manages a registry of pretrained models together with an ML schema [20] describing their expected inputs and outputs. During planning, each `ML.PREDICT` clause becomes a physical ML operator that:

1. executes its `INPUT { SELECT ... }` subquery to produce features;
2. feeds those features to a selected model, supporting both tabular features extracted from RDF and references to external multimodal data sources (e.g., video streams, images, audio) accessible via URLs or file paths;
3. receives prediction outputs;
4. converts predictions into RDF triples or bindings using dictionary IDs.

Under forward chaining, prediction outputs are simply new facts and become part of the materialized knowledge base. Under query evaluation, ML operators act like ordinary physical operators inside the plan.

Multimodal data support is enabled through ML schema, which can specify that models expect non-RDF inputs such as video frames, images, or sensor time series. For instance, a video-based object detection model can be invoked by passing a camera's access URL (stored as an RDF property) directly to the model. The `INPUT` clause can reference URLs pointing to video streams (e.g. `?device hasAccessURL ?url`), and the ML Handler resolves these references, fetches the appropriate data, and feeds it to the model. This allows Kolibrie to declaratively integrate perception models over multimodal streams such as detecting motion in camera feeds or analyzing audio from microphone sensors while maintaining the same uniform, rule-based interface used for tabular features.

Model selection is expressed declaratively in the rule `WHERE` clause: depending on which conditions hold (e.g., light level, road context, sensor reliability), different rules containing different `ML.PREDICT` operators may fire, effectively selecting different models based on the logical premises. This is consistent with the rule-based design and implements requirement R_5 as intended.

The ML schema is used to describe, in RDF, which features a model expects (types, shapes, and modalities) and which outputs it produces (labels, scores, detections). At planning time, this metadata allows us to type-check `ML.PREDICT` clauses and automatically construct the correct feature vectors or multimodal batches from RDF data. At runtime, it guides the translation of raw predictions back into RDF terms, so that different models can be swapped or combined declaratively while preserving a uniform, explainable interface to the knowledge base. The ML schema can be obtained either by training the model directly within Kolibrie, or registering a pretrained model with predefined ML schema.

5   Performance Evaluation

Kolibrie is not designed to be the fastest engine for any single task, but to provide competitive performance across all tasks it supports: (i) SPARQL querying, (ii) reasoning, (iii) RDF stream processing (RSP), and (iv) neuro-symbolic ML integration. Consequently, our evaluation is not meant to be exhaustive; it is intended to demonstrate that Kolibrie performs reasonably well on these various dimensions, and in particular on their combination, which to the best of our knowledge no existing open-source engine supports in a single system. Due to space restrictions, we focus our evaluation on only SPARQL querying and reasoning. The evaluation of RSP can be distilled from those, as we currently employ a C-SPARQL-style of RSP, which means that we extract the content of the window and then employ reasoning and querying upon the window content, which at that time can be seen as a static dataset. Given that the performance of the neuro-symbolic part depends largely on the externally used models and backend, we do not claim a fair cross-engine comparison of ML inference. Instead, we provide an end-to-end scenario together with a reproducible benchmark that reports both pipeline-level latency/throughput and an `ML.PREDICT` breakdown (model compute versus orchestration and Rust→Python overhead). This makes the combined behavior directly verifiable while keeping the evaluation focused on Kolibrie's integrated execution model (Sect. 6).

All experiments were run on Fedora Linux on a 64-bit x64 machine with an AMD Ryzen 7 6800H with Radeon Graphics (3.20 GHz) and 32 GB RAM.

5.1 Querying Evaluation

Workload (WatDiv) and Setup. We use the WatDiv benchmark [2] on the 10M dataset (uncompressed size 58,558,746 bytes) and execute the full query mix. This covers a range of query shapes (linear, star, snowflake, and complex joins), which gives a representative view of SPARQL query performance. We compare Kolibrie against Apache Jena [24], Blazegraph [33], Oxigraph [34][3], and QLever [9]. These engines are widely used, open-source SPARQL systems with published benchmarks, making them suitable and reproducible baselines. Each engine was executed with its default configuration. For every query and engine we run 20 iterations and report the average execution time.

Result. Figure 2 summarizes the average run times (log scale) for all WatDiv queries and engines over 20 runs. The figure reports the query evaluation times, without considering the loading of the data for any of the engines. Kolibrie's load performance is currently not yet on par, as we have not optimized this part yet. On the 10M dataset, loading took on average 58.6 s, compared to 18.1 s for Oxigraph and 43.1 s for Jena.

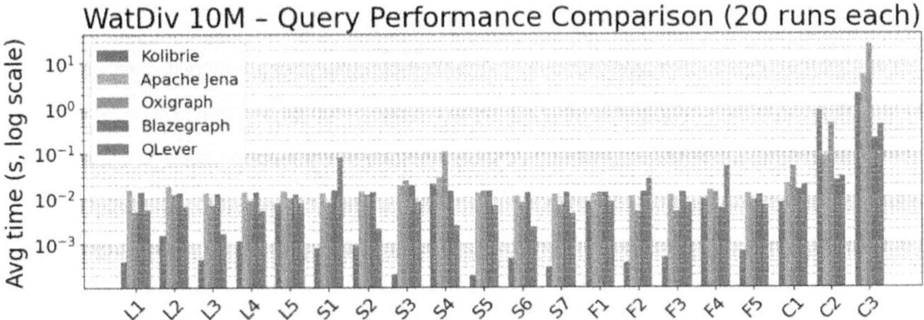

Fig. 2. WatDiv 10M: average query runtimes over 20 runs (log scale).

Kolibrie demonstrates competitive query performance across the WatDiv workload. For most queries, Kolibrie achieves runtimes comparable to established engines. On complex queries Kolibrie outperforms Oxigraph and QLever, though Jena and Blazegraph show better optimization for some patterns. These results show that Kolibrie provides robust SPARQL evaluation.

---

[3] Oxigraph was configured with RocksDB backend.

5.2 Reasoning Evaluation

Workload (Deep Taxonomy) and Setup. We use a modified version of the EYE reasoner's deep taxonomy benchmark[4], designed to stress-test transitive reasoning over deep class hierarchies. The original benchmark contained 10,000 levels; we extended it to include depths of 10, 100, 1,000, and 10,000 to evaluate scalability. The benchmark defines a regular hierarchy where each class at level i has three superclasses at level $i+1$. The reasoning task infers all transitive types of `:TestVariable` (initially typed as `:A1` and `:N1`) via the rule: *If X rdf:type C and C rdfs:subClassOf D then X rdf:type D.*

We compare Kolibrie against Apache Jena (custom `GenericRuleReasoner`) and EYE [36] (backward-chaining). Jena required `-Xmx4G -Xss1G` to avoid stack overflow at deep hierarchies. Each configuration ran 20 times; we report mean reasoning time (excluding loading).

Result. Figure 3 shows average reasoning times (log scale) across hierarchy depths. Kolibrie scales near-linearly from 1 ms (depth 10) to 5 s (depth 10,000) using semi-naive evaluation, which processes hierarchies level-by-level while maintaining a delta set of new facts.

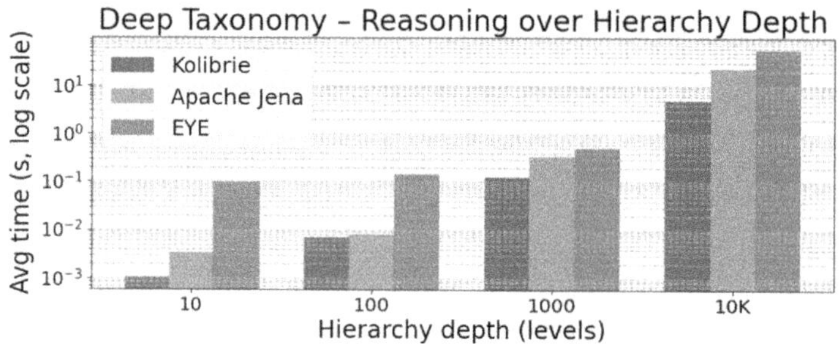

Fig. 3. Deep taxonomy: average reasoning time over 20 runs (log scale).

Apache Jena performs competitively at shallow depths but degrades significantly depth 10,000 (5.6x slower than Kolibrie). EYE's backward-chaining strategy shows the highest runtimes (62 s at depth 10,000, 2.3x slower than Kolibrie). All systems produced correct results. Kolibrie's semi-naive evaluation proves highly effective for deep hierarchical reasoning while maintaining the unified architecture supporting SPARQL, RSP, and ML integration (R_4, R_6).

6   Use Case

We showcase Kolibrie through a neuro-symbolic fraud detection scenario[5], composing a sliding window, symbolic rules, and an ML classifier into a single declar-

---

[4] https://github.com/eyereasoner/eye/tree/master/reasoning/deep-taxonomy.

[5] Full implementation at https://github.com/StreamIntelligenceLab/Kolibrie/blob/main/kolibrie/examples/real_scenario/fraud_detection_system.rs.

ative pipeline. Transactions arrive as RDF triples on `:transactionStream`, backed by a knowledge graph pre-populating account profiles and a merchant risk ontology (e.g., cryptocurrency exchanges, wire transfers).

A `REGISTER RSTREAM` query opens a 5-minute sliding window (step: 60 s), computing a per-account transaction count within the window, which is materialized as `ex:windowVelocity` in the knowledge graph. Six symbolic rules then derive binary fraud-signal flags from raw transaction features: five rules fire on individual attributes (transaction velocity, amount, merchant risk score, foreign origin, and their combinations), while a sixth rule fires specifically on the window-derived velocity to capture burst patterns invisible to single-transaction rules. These flags, combined with the raw transaction attributes and a per-account counter of recent fraud verdicts, form a 14-dimensional feature vector for a `GradientBoostingClassifier`; its output $P(\text{fraud}) \in [0, 1]$ is materialized as `ex:mlFraudScore` in the knowledge graph. Two further symbolic rules then consume this ML-derived fact: one escalates borderline ML scores when window velocity is simultaneously elevated; the other flags accounts whose accumulated verdict history indicates a repeated fraud pattern. A deterministic symbolic fusion rule produces the final verdict: FRAUD, SUSPICIOUS, REVIEW, or CLEAR.

Averaged over 60 `ML.PREDICT` calls, per-call latency was ~138 ms, decomposing as: ~101 ms for model inference, ~32–40 ms for postprocessing (converting predictions to RDF-compatible values), and <0.003 ms for the Rust→Python bridge, with model computation accounting for 73% of total wall-clock time.

7   Discussion

Kolibrie was designed with the requirements introduced in Sect. 2 in mind, and its architecture shows how neuro-symbolic stream reasoning can be realized in practice. The language directly meets the need for a declarative language that integrates symbolic inference with ML predictions (R_1), while providing explicit support for stream-oriented constructs such as sliding windows and time-varying RDF graphs (R_2).

From a systems perspective, the execution model, multi-index triple store, SIMD-enabled operators, and GPU acceleration ensure that Kolibrie exploits modern hardware to achieve scalable query evaluation (R_3). The unified execution pipeline, in which rules, windows, and ML predictions are all treated as first-class operators, fulfills the requirement of integrating symbolic reasoning, stream processing, and predictive components in a single framework (R_4).

The `ML.PREDICT` operator shows how pretrained models can be invoked during rule evaluation without breaking the declarative abstraction. Its output is materialized as dictionary-encoded RDF facts and becomes available to both further rules and SPARQL queries. Because different rules may invoke different models depending on which logical premises hold, Kolibrie also supports dynamic model selection at runtime (R_5), allowing predictions to adapt to changing environmental context without imperative control flow. The modular design and architecture achieves (R_6). Kolibrie is actively maintained, and its community is growing (targetting R_7).

Regarding scalability, Kolibrie's current limits arise from three sources: (i) memory pressure from materializing window contents and all inferred facts in-memory over long time horizons or large window sizes; (ii) coordination overhead in multi-stream and multi-window joins, where synchronization across concurrent streams adds latency; and (iii) ML inference cost when `ML.PREDICT` is in the reasoning loop, which depends heavily on the chosen model and backend. Throughput and latency therefore depend strongly on window size, slide interval, and result selectivity. Addressing these limits via incremental view maintenance, memory-tiered storage, and native ML optimization is part of the future roadmap (Sect. 8).

8   Conclusion and Roadmap

This work introduced Kolibrie, a neuro-symbolic stream reasoning engine that unifies: (i) SPARQL-style querying over RDF graphs, (ii) rule-based symbolic inference, (iii) RSP, and (iv) the invocation of pretrained machine-learning models through a declarative `ML.PREDICT` operator.

A key contribution of Kolibrie is its explicit treatment of machine-learning predictions as first-class facts within the reasoning process. Importantly, so far Kolibrie itself does not provide neuro-symbolic approaches to train models: it relies on pretrained models whose outputs are incorporated into the fixpoint computation. The system therefore preserves explainability by keeping symbolic rules explicit, while still benefiting from predictive signals when needed. Combined with RSP-style window semantics and a batching-oriented execution engine, Kolibrie demonstrates that neuro-symbolic reasoning can be realized efficiently in streaming environments. The most promising directions for future research involves extending Kolibrie's language and execution model:

1. SPARQL 1.1 completeness: Although Kolibrie implements core SPARQL features[6], it does not yet support the full SPARQL 1.1 specification. Completing these operators is essential for enabling full compatibility with existing SPARQL and RSP workloads.

2. Enhanced stream reasoning and window management: Kolibrie presently adopts the basic RSP-QL semantics for windowing. Extending this with out-of-order event handling and incremental view maintenance would enable the engine to support richer and more realistic continuous reasoning scenarios.

3. Richer symbolic reasoning: Future extensions include more expressive stream reasoning paradigms, e.g. datalogMTL [37] and LARS [10].

4. neuro-symbolic extensions: Kolibrie integrates pretrained machine-learning models via `ML.PREDICT`, but it does not yet support neural predicate learning or constraint learning.

---

[6] Currently, the following SPARQL 1.1 features are missing or incomplete: Currently, the following SPARQL 1.1 features are missing or incomplete: Property Paths, Update operations (DELETE, LOAD, CLEAR, CREATE, DROP, COPY, MOVE, ADD), MINUS, EXISTS/NOT EXISTS, SERVICE (federation), complete blank node support, and named graphs.

5. System-level optimization: At the execution level, future versions of Kolibrie will extend the hardware-aware optimizer to reason about CPU vs. GPU execution costs, and incorporate memory-tiered strategies suitable for edge devices. These improvements aim to strengthen Kolibrie's ability to run efficiently across heterogeneous computing environments.

Kolibrie demonstrates that neuro-symbolic stream reasoning can be implemented within a unified, efficient, and declarative framework. By unifying SPARQL querying, rule-based symbolic inference, RDF stream processing, and ML predictions in a single system, Kolibrie realizes a reusable framework that integrates symbolic rules and machine learning predictions for continuous reasoning over RDF streams, directly addressing the research question in Sect. 1. The system achieves this through a declarative language that seamlessly combines these paradigms, a modular architecture that supports independent extensibility of components, and execution pipeline that treats symbolic inferences and ML predictions uniformly as RDF facts. This unified approach enables adaptive, explainable reasoning in dynamic IoT environments while maintaining competitive performance across querying, reasoning, and stream processing tasks.

Acknowledgment. The authors acknowledge the use of ChatGPT for polishing the text.

References

1. Al-Ajlan, A.: The comparison between forward and backward chaining. Int. J. Mach. Learn. Comput. **5**(2), 106–113 (2015)
2. Aluç, G., Hartig, O., Özsu, M.T., Daudjee, K.: Diversified stress testing of rdf data management systems. In: International Semantic Web Conference, pp. 197–212. Springer, Cham (2014)
3. Alzubaidi, L., et al.: Review of deep learning: concepts, cnn architectures, challenges, applications, future directions. J. Big Data **8**(1), 53 (2021)
4. Aouedi, O., et al.: A survey on intelligent internet of things: Applications, security, privacy, and future directions. IEEE Commun. Surv. Tutorials (2024)
5. Arasu, A., Babu, S., Widom, J.: Cql: A language for continuous queries over streams and relations. In: Lausen, G., Suciu, D. (eds.) International Workshop on Database Programming Languages, pp. 1–19. Springer, Cham (2003). https:// doi.org/10.1007/978-3-540-24607-7_1
6. Arndt, D., Van Woensel, W., Tomaszuk, D.: Sparql in n3: Sparql construct as a rule language for the semantic web. In: Hogan, A., Satoh, K., Dağ, H., Turhan, AY., Roman, D., Soylu, A. (eds.) International Joint Conference on Rules and Reasoning, pp. 209–226. Springer, Cham (2025). https://doi.org/10.1007/978-3-032-08887-1_13
7. Babu, S., Widom, J.: Continuous queries over data streams. ACM SIGMOD Rec. **30**(3), 109–120 (2001)
8. Barbieri, D.F., Braga, D., Ceri, S., Valle, E.D., Grossniklaus, M.: C-sparql: a continuous query language for RDF data streams. Int. J. Semant. Comput. **4**(01), 3–25 (2010)

9. Bast, H., Buchhold, B.: Qlever: A query engine for efficient sparql+ text search. In: Proceedings of the 2017 ACM on Conference on Information and Knowledge Management, pp. 647–656 (2017)

10. Bazoobandi, H.R., Beck, H., Urbani, J.: Expressive stream reasoning with laser. In: d'Amato, C., et al. (eds.) International Semantic Web Conference, pp. 87–103. Springer, Cham (2017). https://doi.org/10.1007/978-3-319-68288-4_6

11. Bechhofer, S., et al.: OWL Web Ontology Language Overview. W3c recommendation, W3C (2004). https://www.w3.org/TR/owl-features/

12. Berners-Lee, T.: Notation 3 Logic. W3c design issues, W3C (2005). https://www.w3.org/DesignIssues/Notation3.html

13. Bonte, P., et al.: Grounding stream reasoning research. Trans. Graph Data Knowl. (2024)

14. Bonte, P., Callé, C., Curé, O., Kondylakis, H., Tommasini, R.: Languages and systems for RDF stream processing, a survey. VLDB J. **34**(4), 1–34 (2025)

15. Bry, F., Furche, T., Marnette, B., Ley, C., Linse, B., Poppe, O.: Sparqlog: Sparql with rules and quantification. In: de Virgilio, R., Giunchiglia, F., Tanca, L. (eds.) Semantic Web Information Management: A Model-Based Perspective, pp. 341–370. Springer, Cham (2009). https://doi.org/10.1007/978-3-642-04329-1_15

16. Cyganiak, R., Wood, D., Lanthaler, M.: RDF 1.1 Concepts and Abstract Syntax. W3c design issues, W3C (2014). https://www.w3.org/TR/rdf11-concepts/

17. Dell'Aglio, D., Della Valle, E., Calbimonte, J.P., Corcho, O.: RSP-QL semantics: a unifying query model to explain heterogeneity of RDF stream processing systems. Int. J. Seman. Web Inf. Syst. (IJSWIS) **10**(4), 17–44 (2014)

18. Dell'Aglio, D., Della Valle, E., van Harmelen, F., Bernstein, A.: Stream reasoning: a survey and outlook: a summary of ten years of research and a vision for the next decade. Data Science **1**(1–2), 59–83 (2017)

19. DeLong, L.N., Mir, R.F., Fleuriot, J.D.: Neurosymbolic ai for reasoning over knowledge graphs: A survey. IEEE Trans. Neural Netw. Learn. Syst. (2024)

20. Esteves, D., Ławrynowicz, A., Panov, P., Soldatova, L., Soru, T., Vanschoren, J.: ML Schema Core Specification. W3c design issues, W3C (2016). https://ml-schema.github.io/documentation/ML%20Schema.html

21. Graefe, G.: Volcano/spl minus/an extensible and parallel query evaluation system. IEEE Trans. Knowl. Data Eng. **6**(1), 120–135 (2002)

22. Graefe, G., McKenna, W.J.: The volcano optimizer generator: extensibility and efficient search. In: Proceedings of IEEE 9th international conference on data engineering, pp. 209–218. IEEE (1993)

23. Hitzler, P., Sarker, M.: Neuro-symbolic ai= neural+ logical+ probabilistic ai. Neuro-Symbolic Artif. Intell. State Art **342**, 173 (2022)

24. HP Labs: Apache jena. https://jena.apache.org/, a free and open source Java framework for building Semantic Web and Linked Data applications

25. Kumar, S., Tiwari, P., Zymbler, M.: Internet of things is a revolutionary approach for future technology enhancement: a review. J. Big data **6**(1), 1–21 (2019)

26. Le-Phuoc, D., Dao-Tran, M., Xavier Parreira, J., Hauswirth, M.: A native and adaptive approach for unified processing of linked streams and linked data. In: Aroyo, L., et al. (eds.) International Semantic Web Conference, pp. 370–388. Springer, Cham (2011). https://doi.org/10.1007/978-3-642-25073-6_24

27. Liu, L., Wang, Z., Tong, H.: Neural-symbolic reasoning over knowledge graphs: a survey from a query perspective. ACM SIGKDD Explor. Newsl **27**(1), 124–136 (2025)

28. Maier, D., Tekle, K.T., Kifer, M., Warren, D.S.: Datalog: concepts, history, and outlook. In: Declarative Logic Programming: Theory, Systems, and Applications, pp. 3–100. Association for Computing Machinery and Morgan & Claypool (2018)
29. Manhaeve, R., Dumancic, S., Kimmig, A., Demeester, T., De Raedt, L.: Deepproblog: Neural probabilistic logic programming. Adv. Neural Inf. Process. Syst. **31** (2018)
30. Neumann, T., Weikum, G.: Rdf-3x: a risc-style engine for RDF. Proceedings of the VLDB Endowment **1**(1), 647–659 (2008)
31. Nguyen-Duc, M., Le-Tuan, A., Hauswirth, M., Le-Phuoc, D.: Towards autonomous semantic stream fusion for distributed video streams. In: Proceedings of the 15th ACM International Conference on Distributed and Event-based Systems, pp. 172–175 (2021)
32. Rayon Contributors: Rayon. https://github.com/rayon-rs/rayon, rayon: A data parallelism library for Rust
33. Systap LLC: Blazegraph. https://blazegraph.com/, blazegraph DB is a ultra high-performance graph database supporting Blueprints and RDF/SPARQL APIs
34. Thomas Pellissier Tanon: Oxigraph. https://github.com/oxigraph/oxigraph, oxigraph is a graph database implementing the SPARQL standard
35. Tommasini, R., Bonte, P., Ongenae, F., Della Valle, E.: Rsp4j: an api for rdf stream processing. In: Verborgh, R., et al. (eds.) European Semantic Web Conference. pp. 565–581. Springer, Cham (2021). https://doi.org/10.1007/978-3-030-77385-4_34
36. Verborgh, R., De Roo, J.: Drawing conclusions from linked data on the web: the eye reasoner. IEEE Softw. **32**(3), 23–27 (2015)
37. Walega, P., Cuenca Grau, B., Kaminski, M., Kostylev, E.: Datalogmtl: Computational complexity and expressive power. In: Proceedings of the Twenty-Eighth International Joint Conference on Artificial Intelligence. International Joint Conferences on Artificial Intelligence (2019)
38. Weiss, C., Karras, P., Bernstein, A.: Hexastore: sextuple indexing for semantic web data management. Proc. VLDB Endowment **1**(1), 1008–1019 (2008)
39. Yang, Z., Ishay, A., Lee, J.: Neurasp: embracing neural networks into answer set programming. arXiv preprint arXiv:2307.07700 (2023)

ArtKB: A Multimodal Art Knowledge Base for Cultural Heritage

Giacomo Blanco[1]([✉])[iD], Tommaso Monopoli[1][iD], Federico D'Asaro[1,3][iD],
Ruben Peeters[2][iD], Xuemin Duan[2][iD], Anastasia Dimou[2][iD],
and Giuseppe Rizzo[1][iD]

[1] LINKS Foundation, Turin, Italy
`{giacomo.blanco,tommaso.monopoli,giuseppe.rizzo}@linksfoundation.com`
[2] KU Leuven – Flanders Make@KULeuven – Leuven.AI, Leuven, Belgium
`{ruben.peeters,xuemin.duan,anastasia.dimou}@kuleuven.be`
[3] Dipartimento di Automatica e Informatica (DAUIN), Politecnico di Torino,
Turin, Italy
`federico.dasaro@polito.it`

Abstract. The increasing availability of heterogeneous cultural heritage (CH) data calls for semantic, scalable, and operational infrastructures capable of integrating symbolic knowledge with multimodal representations. Existing CH knowledge graphs (KGs) are typically built from individual institutions' datasets and focus predominantly on semantic modelling. They seldom incorporate the results of AI methods, such as visual representations or neural retrieval methods, which limit their ability to support cross-collection interoperability and advanced multimodal retrieval and sustain continuous, automated data enrichment. In this paper, we introduce ArtKB: a modular end-to-end knowledge base (KB) that unifies semantic modelling, multimodal representation, and operational access to artefacts. Built from Wikidata and annotated using the CACAO ontology, the proposed system combines an RDF graph, a vector database for visual embeddings, and an object storage for digital assets, all orchestrated through a unified API gateway. This infrastructure enables hybrid and multimodal retrieval, metadata enrichment and automated text-to-graph generation, demonstrating how neuro-symbolic approaches can enhance CH data management. The paper details the architecture of the domain-specific knowledge base, demonstrating how the different components integrate to support different tasks. The resulting KB offers a scalable, interoperable, and domain-agnostic solution that can be adapted to broader CH contexts and other multimodal domains.

Keywords: Knowledge Graph · Cultural Heritage · Information Extraction · Semantic Interoperability

Resource type: Knowledge Base System
License: MIT License
DOI: https://doi.org/10.5281/zenodo.17812323
Repository: https://github.com/links-ads/eswc26-artkb
SPARQL endpoint: https://loki.linksfoundation.com/reevaluate-graphdb/sparql.

1   Introduction

In recent years, the field of cultural heritage (CH) has witnessed a growing demand for digital infrastructures capable of managing, connecting, and exposing rich heterogeneous data about artefacts, creators, and institutions [16,20]. Knowledge graphs (KGs) are graphs of data intended to accumulate and convey knowledge of the real world; their nodes represent entities of interest, while their edges represent relations between these entities [14]. They have emerged as a promising technology to represent and interlink such data, enabling enhanced search, recommendation, reasoning, and discovery. The use of knowledge graphs in the cultural heritage domain is highlighted by calls for improved interoperability [4,15], more precise semantic descriptions of artefacts [3,4], and richer contextualisation of heritage assets in digital environments [5,9].

Despite this progress, the effective deployment of knowledge graph-based systems for cultural heritage remains constrained by data heterogeneity, limited maintainability, and the absence of end-to-end operational architectures:
(i) Cultural heritage collections are inherently heterogeneous: metadata standards vary across museums and archives, artefacts are catalogued under different typologies and provenance and digitalisation details are often incomplete or inconsistent. Consequently, the retrieval efficiency and knowledge completeness remain major bottlenecks in cultural heritage knowledge graphs [34]. (ii) Many cultural heritage initiatives focus primarily on building a static knowledge graph or dataset, but fall short in supporting ongoing integration of new content (e.g., newly digitised artworks) and fail to provide mechanisms for contextualising those new entries. (iii) Although several cultural heritage knowledge graphs have been developed (e.g., the Italian cultural heritage graph [11] or ArtGraph [5]), these efforts typically address specific layers of the pipeline, like semantic modelling or data integration, while leaving aside the operational integration of ingestion pipelines and vector-based retrieval mechanisms.

A multimodal knowledge base that brings together symbolic and neural representations of cultural heritage artefacts for retrieval and discovery based on a domain-specific knowledge graph can effectively address the aforementioned challenges of fragmentation, heterogeneity, and maintainability. Motivated by these challenges, the present work proposes ArtKB: an integrated solution that unites semantic and multimodal technologies in a single deployable knowledge base for cultural heritage management.

Building upon existing efforts in semantic interoperability, such as the Europeana Data Model (EDM) [6] and CIDOC-CRM-based initiatives like ArCo [3,11], as well as recent progress in neural models for visual retrieval [5], our approach demonstrates how an end-to-end system, from data extraction to visual retrieval and user-facing exploration, can overcome fragmentation and improve the accessibility and reusability of cultural heritage data. The proposed architecture supports continuous ingestion and semantic linking of new artworks to ensure interoperability across heterogeneous datasets, enhances data discovery and retrieval through the combination of symbolic and neural representations, and promotes reuse of cultural data within and beyond institutional boundaries.

By addressing these needs, the impact of this work on the cultural heritage ecosystem is twofold. From a technological perspective, it provides museums and cultural heritage institutions with a platform that bridges descriptive metadata, images, and semantic relationships, facilitating advanced retrieval, discovery, and data integration across collections. From an organisational perspective, it lowers the entry barrier for semantic adoption by offering a solution that can be aligned with emerging European initiatives such as the *European Collaborative Cloud for Cultural Heritage (ECCCH)*. In this way, the proposed system not only advances the technical frontier of semantic applications in the arts but also supports the broader goal of enabling open, connected, and intelligent CH infrastructures.

The main contributions can be summarised as follows: (i) An ***end-to-end multimodal knowledge base*** that integrates the semantic knowledge graph with a vector database for visual embeddings and an object storage for the corresponding digital assets. This combination enables hybrid retrieval, by meaning or by visual similarity, and provides a practical example of neuro-symbolic integration, where structured knowledge and neural-based representations complement each other within a unified system. The code and documentation to instantiate a replica of ArtKB are publicly available in the project repository. (ii) A ***domain-specific knowledge graph focused on arts***, built from Wikidata [32] and structured through alignment with *Cultural Artefact Contextual Ontology* (CACAO)[1] [29]. This graph provides a structured, interoperable representation of artworks, artists, production events, materials, and places of conservation, serving as a unifying semantic backbone for the broader knowledge base. Its construction process, from data extraction and cleaning to ontology-based enrichment, offers a reproducible methodology for transforming open, heterogeneous sources into high-quality domain graphs. A snapshot of the knowledge graph, together with the list of associated multimedia entities, is accessible via the Zenodo repository. The graph can also be explored interactively through the publicly available SPARQL endpoint.

Beyond its direct application to the cultural heritage domain, the proposed architecture is domain-agnostic: the same architectural pattern can be replicated in other fields where semantic and visual information coexist, such as natural history collections or design archives. By substituting the underlying ontology and adapting the ingestion pipeline, the knowledge base can be used to manage multimodal knowledge in diverse application areas.

The paper is organised as follows: Sect. 2 discusses the related works. The proposed knowledge base architecture is presented and described in detail in Sect. 3. Section 4 describes the process and implementation of the knowledge graph construction and the use-cases supported by the whole system are presented in Sect. 5. Last, Sect. 6 provides the conclusion and future work.

---

[1] https://reevaluate.github.io/cacao-ontology/index-en.html.

2   Related Works

Research in the field of cultural heritage has increasingly embraced semantic technologies and linked data to enhance data interoperability, contextual understanding, and access to collections across institutions [5,10,15,25,34]. Early efforts such as the *Europeana Data Model (EDM)* [6] established a shared semantic framework for aggregating metadata from thousands of European institutions. Beyond EDM, the Europeana Linked Data initiative has progressively extended this framework by publishing cultural heritage data as accessible Linked Data, enabling direct integration with the broader Web of Data [18]. In the Netherlands, several institutions have pioneered semantic publishing of museum collections: the Rijksmuseum Amsterdam has made its collection available as Linked Data through the Europeana network, while initiatives such as the Dutch Digital Heritage Network (NDE) have established shared infrastructure and common vocabularies to improve interoperability across Dutch cultural heritage institutions [27]. Similarly, the *CIDOC Conceptual Reference Model (CIDOC-CRM)*, standardised as ISO 21127, has become the reference ontology for modelling cultural heritage information. CIDOC-CRM provides a formal structure for describing events, actors, places and artefacts, thereby allowing interoperability among heterogeneous cultural heritage databases and institutional repositories.

In the subsequent years, several other initiatives were launched to build knowledge graphs within the cultural heritage domain. Some of these efforts have pursued broader objectives, such as *ArCo* [11], a knowledge graph that interlinks Italian cultural heritage entities of various kinds, defined through a dedicated ontology [3], with major knowledge bases such as DBpedia [21] and the Getty Vocabulary Program (GVP) [12]. This initiative provides a large-scale semantic resource to support applications in the cultural heritage domain. Conversely, *ICH-KG* [9] is a knowledge graph automatically populated from the Chinese Intangible Cultural Heritage (ICH) database, designed to make ICH resources more accessible to the general public and to support their preservation. Other initiatives focused on more specific use cases, such as the extraction of knowledge from textual documents [25], or the modelling of artefacts within a particular collection [34] or grottoes [35].

Despite the availability of large general-purpose knowledge bases, e.g., Wikidata and DBpedia, which have proven useful for generating domain-specific knowledge graphs in fields such as biodiversity [28] and life sciences [33], most existing works in the cultural heritage domain have grounded the construction of their knowledge graphs on data collections provided by the cultural heritage institutions with which they collaborated [25,34]. A notable exception is *ArtGraph* [5], an artistic knowledge graph built upon WikiArt and DBpedia, which offers advanced capabilities for knowledge discovery. While these initiatives significantly advanced the use of semantic technologies within the cultural heritage domain, they also reveal persistent limitations that motivate the present work.

Most existing cultural heritage knowledge graphs rely on institution-specific or domain-constrained datasets, which limit their scalability and reusability. As a result, cross-institutional reasoning and federated querying across heterogeneous cultural heritage collections remain challenging. At the same time, the growing use of multimodal and AI-based approaches in art analysis, such as visual similarity search or embedding-based retrieval, has not been systematically connected to semantic representations. The present work builds upon [29], which introduced the CACAO ontology and included a preliminary mention of ArtKB. However, that work did not detail its modular architecture and design choices: the present paper provides the first comprehensive description of ArtKB as a standalone, deployable system. To the best of our knowledge, there have been no cultural heritage systems that treat knowledge graphs and AI-based visual retrieval as complementary sources of knowledge. These limitations highlight the need for more flexible, open, and integrative approaches capable of combining semantic interoperability and multimodal representation within a single cultural heritage framework.

3   Overall Architecture

To address the challenges of integrating and exploring heterogeneous CH data, we designed a knowledge base (KB) for managing, linking, and reusing data about artworks and related entities. ArtKB brings together symbolic and multimodal representations within a unified framework, enabling semantic querying, cross-domain reasoning, and neural-based retrieval across diverse collections.

The architecture of ArtKB has been designed to support both semantic interoperability and neural-based content retrieval, combining knowledge representation techniques with neural mechanisms for visual understanding. Following a modular design, the knowledge base ensures maintainability and flexibility in managing heterogeneous data and integrating new components or services as the system evolves. The architecture has been conceived to address the heterogeneity and multidimensional nature of CH data, which encompasses semantic descriptions, visual representations, and contextual information.

To effectively manage these diverse data modalities and enable both symbolic and AI-based reasoning, the knowledge base integrates four components, each addressing a specific aspect of the information life-cycle: *(i)* A ***graph database*** to store the knowledge graph expressing semantic relations and domain knowledge in a structured graph model; *(ii)* An ***object storage***, responsible for preserving and serving the digital representations of artefacts; *(iii)* A ***vector database*** to store AI-generated embeddings of visual artefacts and supports similarity-based retrieval; and *(iv)* An ***API gateway*** to provide a unified access layer to the underlying services and exposing high-level interfaces to external applications.

Together, these components establish an integrated infrastructure that bridges semantic knowledge with visual and data-driven representations of cultural artefacts. The architecture follows a component-agnostic design principle: each service was selected for its suitability for rapid prototyping and is available in a free tier, but can be replaced by any functionally equivalent alternative, including open-source solutions, by updating the corresponding communication interfaces. An overview of the knowledge base architecture is presented in Fig. 1.

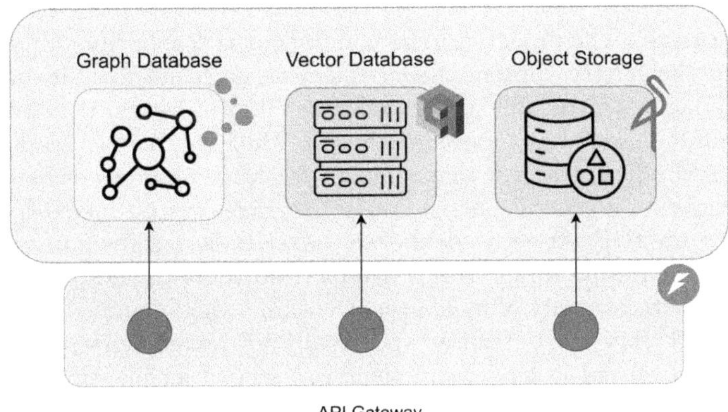

Fig. 1. The knowledge base's architecture. The API Gateway serves as the single entry point to access the Graph Database, Vector Database, and Object Storage.

Graph Database. The knowledge graph forms the semantic core of the knowledge base, modelling entities such as artefacts, artists, materials, artistic movements, and their contextual relations according to the *CACAO* [29] ontology. This structured representation enables the system to express rich, interconnected cultural heritage data in a machine-interpretable form.

To store and manage this graph, we rely on a GraphDB[2] instance, which provides robust support for RDF graphs and ensures the persistence and coherence of the semantic layer. GraphDB was selected for its mature tooling support, free tier sufficient for the needs of ArtKB, and suitability for rapid prototyping in research settings. While open-source alternatives such as QLever [2] exist, GraphDB offered the best balance between expressive querying capabilities and ease of deployment at the time of development. It offers an expressive querying interface tailored to semantic exploration via a SPARQL [13] endpoint. Through SPARQL, it becomes possible to retrieve multi-step relationships, such as chains linking an artefact to the artist who created it, the materials employed, or the broader artistic movement it belongs to, thus allowing complex reasoning and domain analysis that would be difficult to achieve through conventional database queries. Alongside the SPARQL interface, we also provide a GraphQL endpoint that serves as a complementary access layer. While SPARQL is optimised for flexible graph pattern matching, GraphQL allows clients to specify precisely the shape of the response they expect, facilitating integration with external applications and lightweight front-end services. This dual interface strategy supports

---

[2] https://graphdb.ontotext.com/.

both semantically rich queries and modern API-oriented consumption, ensuring that the graph database can operate effectively across different usage contexts and technical requirements.

Object Storage. The object storage component is responsible for storing and maintaining the binary content associated with each artefact, including high-resolution images, videos, 3D scans, and other digital assets. It is implemented using a MinIO[3]-based object storage solution, which guarantees scalability, data durability, and efficient access management for large volumes of heterogeneous files. The choice of a general-purpose object storage over an image-specific solution such as an IIIF server is deliberate: ArtKB is designed to manage any digitised representation of a cultural artefact, not only images. MinIO natively supports this heterogeneity of media types within a single, unified infrastructure, whereas an IIIF server is optimised specifically for image delivery.

Each object stored in this component is assigned an identifier that links it to its corresponding entity in the knowledge graph, enabling direct bidirectional navigation between semantic and multimedia representations. This tight coupling between semantic data and media supports multi-modal exploration and ensures consistency across the knowledge base ecosystem.

Vector Database. The vector database introduces a complementary representation layer that encodes visual features of artefacts, enabling similarity-based retrieval and linking beyond explicitly modelled semantic relations. It stores the vector embeddings of artefact images, generated through deep learning models that encode visual characteristics into numerical representations. These embeddings enable advanced content-based retrieval operations, thus supporting intelligent exploration and recommendation scenarios.

In our implementation, the vector database is implemented as an instance of Qdrant[4], which allows the creation of multiple collections of vector representations. This flexibility enables the system to manage embeddings derived from different artificial intelligence models or feature extraction strategies while maintaining a unified retrieval framework. Each vector within a collection is associated with the persistent identifier of the corresponding entity in the knowledge graph, allowing seamless navigation from perceptual representations back to their semantic counterparts. The database is optimised for similarity search according to well-defined metrics; in our case, cosine similarity is employed to measure the closeness of vectors within a single collection with respect to an external item used for the search operation. By bridging the gap between semantic and perceptual information, this component enhances the usability of the knowledge base for applications relying on computer vision and machine learning techniques.

---

[3] https://www.min.io/.
[4] https://qdrant.tech/.

API Gateway. The API gateway acts as the single entry point to the knowledge base and its underlying services. It moderates all external interactions with the system, ensuring secure and controlled access to the stored data and functionalities. As the sole component authorised to communicate with the other services, the gateway centralises all requests, while the underlying components, such as the graph database, vector database, and object storage, never interact directly with each other. This component manages authentication and authorisation procedures, and through its comprehensive set of endpoints, the gateway allows users and applications to execute SPARQL queries on the knowledge graph, retrieve the digital representation of an artefact from the object storage, or perform similarity-based searches using the vector database. Some requests involve only a single service, for instance, querying the knowledge graph or retrieving a file. Other operations, such as adding a new digital artefact, require the gateway to coordinate multiple services: the knowledge graph is updated to link the digital asset to its corresponding entity, the object storage stores the file and establishes the association, and the vector database ingests the corresponding vector representation, linked back to the knowledge graph identifier. This architecture ensures modularity and separation of concerns, while providing a unified interface for all knowledge base functionalities. By centralising service access, the API gateway simplifies integration with external applications and maintains consistency across semantic, visual, and multimedia layers of the knowledge base.

4 Knowledge Graph Construction Process

The knowledge graph constitutes the backbone of the proposed knowledge base, providing a structured and interoperable representation of cultural heritage data.

CACAO Ontology. The knowledge graph follows the CACAO[5] [29] ontology, developed as a domain-specific extension of *CIDOC-CRM* [7].

While *CIDOC-CRM* offers a robust event-centric standard for cultural heritage, its high level of abstraction can hinder direct applicability: for example, the most specific concept to represent a museum is the generic class `E74_Group`, forcing implementations to rely on external vocabularies and risking data silos. CACAO mitigates this by introducing explicit, specific subclasses for CH-related entities (e.g., `cacao:CACAO_0000020` for museums), and integrates *ODRL*[6] [17] to model structured rights expressions, a prerequisite for the legal reuse of hosted digital artefacts. The knowledge graph was constructed by extracting data from Wikidata[7] and aligning it with the CACAO ontology. Wikidata's comprehensive scope makes it a valuable source for artistic knowledge, yet its size and heterogeneity pose significant challenges in identifying domain-relevant entities. To address this, Wikidata entity classes were mapped to the CACAO concepts of

---

[5] https://reevaluate.github.io/cacao-ontology/index-en.html.
[6] http://www.w3.org/ns/odrl/2/.
[7] https://www.wikidata.org/wiki/Wikidata:Main_Page.

physical artefact, artist, museum, movement, genre, and material, and the Wikidata dump was filtered to retain only instances of the 7848 selected base classes. Each selected entity was then parsed to retain only the properties prescribed by the CACAO ontology, while preserving the original Wikidata URIs as identifiers to ensure traceability. Following subsections detail the parsing strategies applied to each entity type.

Artist. Artists are represented as instances of `cacao:CACAO_0000022` and are enriched with temporal and thematic information from Wikidata properties. Biographical details such as birth and death dates follow the CIDOC-CRM event-based pattern, linking the artist to instances of `crm:E67_Birth` and `crm:E69_Death` via `crm:P98i_was_born` and `crm:P100i_died_in`. Fields of work, artistic genres, and movements are modelled as separate entities preserving their original Wikidata identifiers, enabling reuse across multiple artists and artworks, reasoning over shared cultural contexts, and interoperability with external linked-data sources.

Physical Artefact. Each physical artefact is assigned to the appropriate subclass of `cacao:CACAO_0000023`.

Creation date and author are modelled via a `crm:E12_Production` event, linked through `crm:P4_has_time-span` and `crm:P108i_was_produced_by`. Materials are linked via `crm:P45_consists_of`; genres, movements, depicted subjects, and hosting museum are captured through the same vocabulary strategy applied to artist entities, using `cacao:CACAO_0000017`, `cacao:CACAO_0000035`, `crm:P62_depicts` and `crm:P53_has_former_or_current_location`. Physical measurements (height, width, diameter, thickness, length) are each modelled as a `crm:E54_Dimension` instance, enriched with the corresponding AAT[8] measurement type. Finally, images from Wikimedia Commons are instantiated as `cacao:CACAO_0000086` entities, assigned a unique identifier shared across the vector database and object storage, and linked to the physical artefact via `crm:P138_represents`.

Museum. Museums are represented as instances of `cacao:CACAO_0000020`, enriched with their official website (`cacao:CACAO_0000038`), hierarchical organisational relationships (`crm:P106_is_composed_of` and its inverse), curator (`crm:P109_has_current_or_former_curator`), and institutional partners (`cacao:CACAO_0000040`).

Movement, Genres and Material . Artistic movements (`cacao:CACAO_0000036`), genres (`cacao:CACAO_0000027`), and materials (`crm:E57_Material`) are modelled as distinct entities retaining their human-readable label. Materials are additionally linked to AAT entries via `skos:exactMatch` to support cross-dataset interoperability.

---

[8] http://vocab.getty.edu/aat/.

Final Result. The final knowledge graph comprises more than 8 million RDF triples, representing 1.4 million unique entities across 82 classes from CACAO, CIDOC-CRM and ODRL. The graph includes approximately $76K$ artists, $307K$ physical artefacts, $2.5K$ museums, and over $4.8K$ artistic movements, genres and materials. The graph also contains over $147K$ digital artefacts representing images, each linked to the corresponding physical artefact and consistently referenced within the object storage and vector database layers. An example of how a physical artefact and its corresponding digital versions are represented in the knowledge base is reported in Fig. 2. It shows the artefact representation in the knowledge graph, where semantic properties describe the physical artefact and associate its digitalised version. The identifier of the digital artefact in the knowledge graph can be used to retrieve the file in the object storage and the vector representation in the vector database.

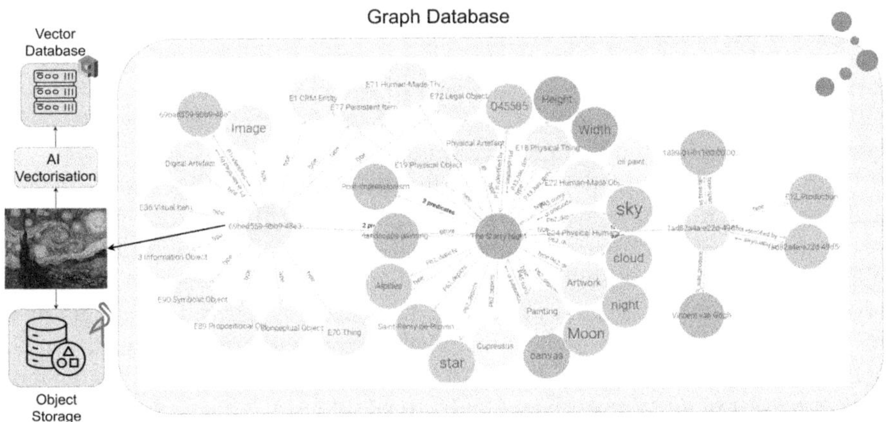

Fig. 2. Example of artefact representation in the knowledge base. The graph stores semantic information about both physical and digital artefacts, while the identifier of each digital artefact enables retrieval of its file from object storage or its vector representation from the vector database.

5   Applications and Use Cases

The proposed knowledge base is conceived not only as a repository for storing and querying cultural heritage information but also as a functional infrastructure that can actively support advanced applications, including semantic analysis, multimodal retrieval, and interactive exploration tools.

The applications of the proposed knowledge base span from curatorial research to digital cataloguing, collection management, and user-facing cultural discovery tools. At the semantic level, the SPARQL endpoint provides an expressive interface to retrieve complex relationships among entities, such as artists, materials, production events, or artistic movements. Beyond the symbolic layer, the integration of a vector database for visual embeddings enables intelligent retrieval operations based on image similarity or multimodal criteria. This allows for the identification of stylistic analogies, cross-collection visual correspondences, and clustering of artworks according to visual or semantic features.

The following subsections illustrate four representative use cases that demonstrate the flexibility and practical impact of this infrastructure. The first focuses on semantic querying and knowledge discovery through SPARQL, while the second and the third showcase, respectively, multimodal retrieval and metadata suggestion functionalities built upon the vector representation of artworks. Last, a service to insert unstructured text data into the structured KG is presented.

5.1   Semantic Knowledge Discovery

CH users often have precise information retrieval needs involving relational metadata constraints, yet lack the technical expertise to formulate SPARQL [13] queries to search in the knowledge base directly. Expressing such structured queries requires SPARQL expertise, creating a significant barrier for domain experts who have limited technical training.

To address this gap, [8] developed a retrieval system over this knowledge base, which includes an LLM-based Text2SPARQL module that automatically translates natural language queries into executable SPARQL, substantially improving knowledge base accessibility. The Text2SPARQL module operates through a multi-stage pipeline: a Mistral LLM [1] first translates natural language queries into structured JSON representations with entity placeholders, which are then resolved against the knowledge base through SPARQL lookups, leveraging entity labels (e.g., resolving "Vincent van Gogh" to `wd:Q5582`). The resolved entities are integrated into the final SPARQL query executed against the knowledge base.

As reported in [8], the module was evaluated on $4,350$ synthetic user queries derived from the knowledge base, achieving a 100% SPARQL execution success rate and a hit rate of 79.43% (i.e., the proportion of queries for which the target artefact was returned), with 98.43% precision among queries returning non-empty results. These results demonstrate the effectiveness of LLM-based Text2SPARQL in improving knowledge base accessibility, enabling CH practitioners to exploit the full expressive power of semantic querying without requiring prior expertise in query languages. This application illustrates how natural language as input over structured knowledge bases can lower the barrier to knowledge discovery in CH contexts.

5.2   Neural-Based Image Retrieval

Image retrieval is a fundamental task in computer vision and database management, increasingly approached as a dense retrieval problem using deep learning models as embedders [31]. In this context, the Vector Database introduced in Sect. 3 can be leveraged by modern Vision–Language Models (VLMs) [37] as a

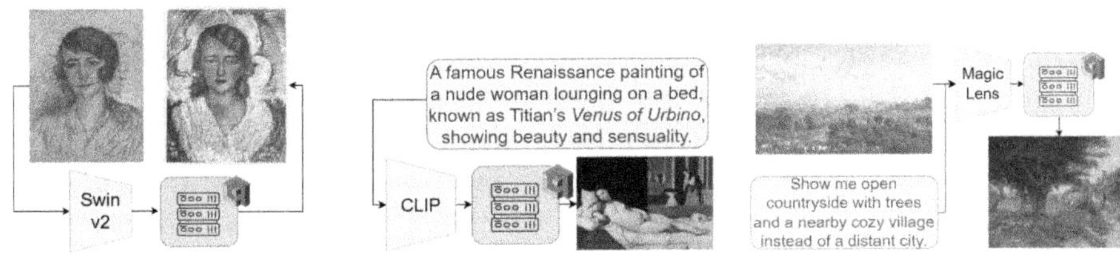

Fig. 3. Image-to-Image.   **Fig. 4.** Text-to-Image.   **Fig. 5.** Composed Image Retrieval.

unified backend for a range of image retrieval tasks, including image-to-image, text-to-image, and composed image retrieval.

The following tasks can be supported using the knowledge base as a foundation. (i) Image-to-Image (I2I) retrieval identifies artworks that are visually similar to a given input image (Fig. 3), addressable using the models introduced in Sect. 5.3. (ii) Text-to-Image (T2I) retrieval returns images that best match a textual description, bridging language and vision (Fig. 4); contrastive pre-trained VLMs such as CLIP [30] excel at this kind of cross-modal retrieval. (iii) Composed Image Retrieval (CIR) enables users to combine a reference image with a textual modification to express their search intent (Fig. 5); recent VLMs such as MagicLens [38] or LamRA [23] are well suited for this task.

Application: T2I Retrieval. The knowledge base in this paper has been used as the data foundation for a T2I retrieval module in [8]. Specifically, [8] constructs the image–text pairs from the knowledge base by combining artefact images with textual descriptions generated through metadata-filled templates instantiated from structured metadata fields (e.g., title, creator, creation time, material), complemented by visual captions (e.g., "painting of sunflowers in a yellow vase") produced by BLIP [22]. This process yielded $43,500$ image-text pairs, split into training (80%), validation (10%), and test (10%) sets, which were used to fine-tune CLIP for cross-modal retrieval over CH artefacts. As reported in [8], fine-tuning CLIP ViT-L/14 on these pairs significantly improves T2I retrieval performance over the pretrained model. *Recall@1* (the proportion of queries for which the correct artefact is ranked first) improves from 39.36 to 50.97, and *Recall@5* improves from 62.64 to 75.89. These results demonstrate that image-text pairs constructed from the knowledge base provide an effective foundation for domain-adaptive T2I retrieval over CH collections.

5.3   Neural-Aided Metadata Suggestion

The digitisation of cultural heritage artefacts is time-consuming, as adding an artwork to a digital catalogue requires manual insertion of many metadata attributes. ArtKB architecture can address this through an AI-aided metadata suggestion service, where AI-generated suggestions operate exclusively as decision-support tools: every proposed value must be manually reviewed and approved by a curator before being committed to the KB, mitigating the risks of introducing AI-derived data into a curated knowledge base. The image of an artwork is processed by a computer vision model to obtain a vector representation that can be used to query the vector database introduced in Sect. 3. This search outputs the most similar images, which are all linked to the corresponding entity in the KG. Thanks to this correspondence, the metadata of the most similar artefacts are retrieved by querying the KG. Then, a similarity-weighted majority voting scheme aggregates and ranks the metadata values. The final ranked set of suggestions for each metadata attribute is finally returned to the curator, who can now streamline the metadata input process by simply picking one or more suggestions. This process is summarised in Fig. 6.

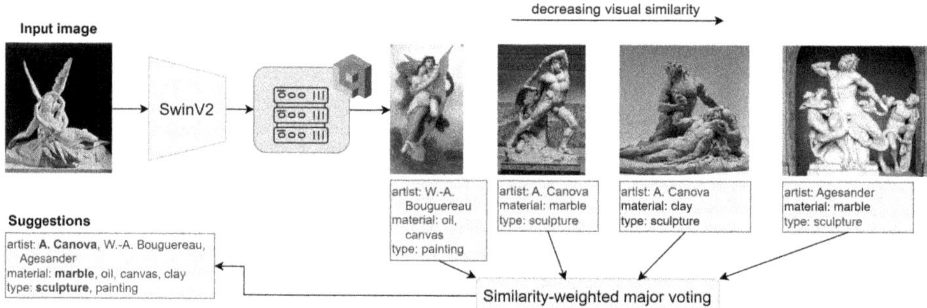

Fig. 6. Architecture of the proposed AI-aided metadata suggestion service. The input image depicts the marble sculpture "Cupid and Psyche" by the Italian sculptor Antonio Canova. Correct metadata values for the input image are highlighted in **bold**.

The vectorisation of the artefact image can be addressed using modern deep learning architectures such as Convolutional Neural Networks or Vision Transformers. In order to test the performance of the proposed metadata suggestion service, we trained a SwinV2 transformer [24] on the task of artwork instance retrieval on a large-scale, open-access dataset of artwork images, namely the MET dataset [36]. The training strategy closely follows the supervised contrastive learning framework discussed in [36]. When evaluating the image retrieval performances on the test split of the MET dataset, the trained SwinV2 model achieves strong performances (*Accuracy@1*: 66.1, *Accuracy@3*: 72.5, *mAP*: 50.9), which surpass the ones reported in the original paper for ResNet-18 and ResNet-50 architectures.

Using the SwinV2 model as a vector representation extractor, we set up an evaluation experiment for the metadata suggestion tool. Specifically, we used the validation and test splits of the MET dataset as the query set, and the training split as the database (removing training samples depicting artworks also present in the query set for a fair evaluation). For each query, we extracted the top 10 most similar images in the database, and aggregated their ground-truth metadata values for *"classification"*, *"culture"*, *"artist"*, *"medium"* and *"tags"*. A set of ranked suggestions for each query image is finally obtained by similarity-weighted majority voting, as previously explained. Suggestions are then compared to ground-truth metadata values to compute quantitative performances. Table 1 reports the results of this experiment in terms of *Recall@1* and *Recall@5*. Note that for computing *Recall@K* for a certain metadata attribute, we exclude those query images with a missing ground-truth value for that attribute; for this reason, we also report the support for the Recall metric.

Table 1. Metadata suggestion procedure results employing SwinV2 on the validation and test splits of the MET dataset.

Attribute	Recall@1	Recall@5	Support
classification	37.12	63.35	1045
culture	21.29	39.71	836
artist	5.54	12.64	451
medium	47.69	74.60	1126
tags	33.74	56.44	978

Results vary greatly depending on the attribute. Lower recall values may be explained by the extreme heterogeneity of the attribute values in the MET dataset, which exhibit long-tail distributions and poor harmonisation. Nonetheless, the results are promising and may suggest that the proposed metadata suggestion procedure can be effectively applied to other datasets and metadata attributes.

The procedure described is valid for any of the categorical attributes for which the curator seeks to obtain a suggestion (e.g., artist, material, artistic movement). The effectiveness of this service depends not only on the performance of the image retrieval model, but also on the size, heterogeneity, and completeness of the knowledge base: the more images are added to the knowledge base, the more likely it is that the images most visually similar to the input image also share its metadata attributes. In this way, the system provides a practical AI-assisted support tool that reduces curatorial workload while promoting more consistent and complete metadata acquisition across the digital collection.

5.4   Text-to-Graph Generation

Complementing the image-based metadata suggestions, the Text2RDF module enables curators to transform unstructured textual descriptions into structured data suitable for integration into the knowledge graph. This module utilises a multi-agent LLM architecture, orchestrated via the Google Agent Development Kit[9], to generate CACAO-compliant RDF triples from raw text.

---

[9] https://google.github.io/adk-docs/.

The workflow consists of two main stages. First, a set of specialised LLM agents, prompted with a summarised representation of the ontology to optimise context usage, performs information extraction tasks, including named entity recognition (NER), coreference resolution, and relation extraction. During the entity linking phase, the agents employ standard search API queries to authoritative vocabularies (e.g., Wikidata, AAT) as well as the internal KG to retrieve the top-N candidate identifiers for each entity. Second, the extracted entities and relations are transformed into RDF triples, which undergo a self-reflective validity check by a dedicated "critic" agent to ensure full adherence to the CACAO ontology before being merged into the central knowledge graph. F1, precision, recall, Ontology Conformance (OC), and hallucination metrics are evaluated on the Text2KGBench [26] benchmark following the suggestions in the Text2KGBench paper. Preliminary results show that adding a SHACL [19] tool for validation to the critic agent improves the F1 score by 9.4% (53 to 58), precision by 10.7% (56 to 62), recall by 11.3% (53 to 59) and OC by 5.8% (85 to 90) compared to the baseline approach used in [26].

By automating this conversion pipeline, the system enhances the consistency and scalability of knowledge graph enrichment, reducing manual effort and supporting more comprehensive contextualisation of cultural heritage data. Consistent with the metadata suggestion service, the RDF triples produced by this pipeline are not automatically ingested into the knowledge base. Instead, they are presented to a curator for manual validation, ensuring that only verified statements are merged into the KB and that the integrity of curated institutional data is preserved.

6   Conclusions

This paper presented an integrated semantic and multimodal knowledge base designed to address current challenges in cultural heritage data management, including fragmentation, heterogeneity, and the lack of operational end-to-end infrastructures. By combining a CIDOC-CRM–aligned knowledge graph built from Wikidata, an object storage layer for digital artefacts, and a vector database for visual embeddings, the proposed architecture enables complementary forms of retrieval and reasoning that bridge symbolic knowledge with AI-driven visual understanding. The system's API gateway ensures coherent access to all components, supporting tasks such as semantic querying, similarity-based exploration, metadata suggestion, and automatic text-to-graph generation.

The applications and use cases demonstrate how the KB can support curators, researchers, and cultural institutions by offering advanced tools for discovery, cataloguing, and knowledge enrichment. Beyond the arts domain, the architecture is inherently extensible: substituting the domain ontology and adjusting the ingestion pipeline allows it to serve as a general-purpose semantic and multimodal infrastructure for other cultural or scientific fields. Future work will address the current limitations of the proposed prototype. First, while AI-assisted services currently operate under a strict human-in-the-loop paradigm, this approach does not scale to large collections. The use of provenance tracking, will be inspected to label each triple with its source authority and allow users to distinguish between curated institutional data and AI-derived suggestions. Second, although ArtKB is already functional and publicly accessible, the current deployment should be considered a prototype and large-scale deployment has not yet been fully addressed. Future iterations will investigate the architectural trade-offs between a centralised deployment, which ensures global consistency but is challenging to manage at scale, and a distributed model in which each institution maintains its own instance, which raises non-trivial issues of inter-instance communication and coordination. Finally, further evaluation with institutional partners will guide refinements of the system's usability and scalability, supporting its evolution into a robust, open, and intelligent platform for cultural heritage data management, aligned with broader European initiatives such as the European Collaborative Cloud for Cultural Heritage.

Acknowledgments. This work has received funding from the European Union Horizon Research and Innovation programme under grant agreement No 101132389, REEVALUATE project, https://reevaluate.eu/.

References

1. Mistral AI: Mistral AI Technical Documentation. Mistral AI (2025). https://docs.mistral.ai/. Accessed 3 December 2025
2. Bast, H., Buchhold, B.: QLever: a query engine for efficient SPARQL+text search. In: Proceedings of the 2017 ACM on Conference on Information and Knowledge Management, CIKM '17, pp. 647–656. Association for Computing Machinery, New York, NY, USA (2017). https://doi.org/10.1145/3132847.3132921
3. Carriero, V.A., et al.: ArCo: the Italian Cultural Heritage knowledge graph. In: Ghidini, C., et al. (eds.) ISWC 2019. LNCS, vol. 11779, pp. 36–52. Springer, Cham (2019). https://doi.org/10.1007/978-3-030-30796-7_3
4. Casillo, M., Santo, M.D., Mosca, R., Santaniello, D.: Sharing the knowledge: exploring cultural heritage through an ontology-based platform. J. Ambient Intell. Humaniz. Comput. **14**(9), 12317–12327 (2023). https://doi.org/10.1007/S12652-023-04652-3
5. Castellano, G., Digeno, V., Sansaro, G., Vessio, G.: Leveraging knowledge graphs and deep learning for automatic art analysis. Knowl. Based Syst. **248**, 108859 (2022). https://doi.org/10.1016/j.knosys.2022.108859, https://www.sciencedirect.com/science/article/pii/S0950705122004105
6. Doerr, M., Gradmann, S., Hennicke, S., Isaac, A., Meghini, C., Sompel, H.: The Europeana Data Model (EDM), January 2010
7. Doerr, M., Ore, C.E., Stead, S.: The CIDOC conceptual reference model: a new standard for knowledge sharing. In: Tutorials, Posters, Panels and Industrial Contributions at the 26th International Conference on Conceptual Modeling, ER '07, vol. 83, pp. 51–56. Australian Computer Society, Inc., Australia (2007)
8. Duan, X., et al.: Knowledge-enhanced multimodal retrieval over cultural heritage knowledge graphs. In: The Semantic Web (2026)

9. Fan, T., Wang, H.: Research of Chinese intangible cultural heritage knowledge graph construction and attribute value extraction with graph attention network. Inf. Process. Manage. **59**(1), 102753 (2022). https://doi.org/10.1016/j.ipm.2021.102753, https://www.sciencedirect.com/science/article/pii/S030645732100234X

10. Faraj, G., Micsik, A.: Representing and validating cultural heritage knowledge graphs in CIDOC-CRM ontology. Fut. Internet **13**(11) (2021). https://doi.org/10.3390/fi13110277, https://www.mdpi.com/1999-5903/13/11/277

11. Faralli, S., Lenzi, A., Velardi, P.: A large interlinked knowledge graph of the Italian cultural heritage. In: Calzolari, N., et al. (eds.) Proceedings of the Thirteenth Language Resources and Evaluation Conference, Marseille, France, June 2022, pp. 6280–6289. European Language Resources Association (2022). https://aclanthology.org/2022.lrec-1.675/

12. Harpring, P.: Development of the Getty vocabularies: AAT, TGN, ULAN, and CONA. Art Docum. J. Art Librar. Soc. North America **29**, 67–72 (2010). https://doi.org/10.1086/adx.29.1.27949541

13. Harris, S., Seaborne, A., Prud'hommeaux, E.: SPARQL 1.1 Query Language. W3C Recommendation, March 2013. https://www.w3.org/TR/sparql11-query/

14. Hogan, A., et al.: Knowledge graphs. ACM Comput. Surv. **54**(4), 71:1–71:37 (2022). https://doi.org/10.1145/3447772

15. Huang, Y., Yu, S., Chu, J., Fan, H., Du, B.: Using knowledge graphs and deep learning algorithms to enhance digital cultural heritage management. Heritage Sci. **11** (2023). https://doi.org/10.1186/s40494-023-01042-y

16. Hyvönen, E.: Publishing and using cultural heritage linked data on the semantic web. Synth. Lect. Semant. Web Theor. Technol. **2**, 1–159 (2012). https://doi.org/10.2200/S00452ED1V01Y201210WBE003

17. Iannella, R., Rodríguez-Doncel, V., Villata, S.: ODRL information model 2.2. W3C Recommendation, February 2018. https://www.w3.org/TR/odrl-model/

18. Isaac, A., Haslhofer, B.: Europeana linked open data — data.europeana.eu. Semant. Web **4**(3), 291–297 (2013). https://doi.org/10.3233/SW-120092

19. Knublauch, H., Kontokostas, D.: Shapes Constraint Language (SHACL). Recommendation, World Wide Web Consortium (W3C), July 2017. https://www.w3.org/TR/shacl/

20. Kraev, M., Luchev, D.: Analyzing knowledge graph innovations and emerging AI technologies for cultural heritage data management. Digit. Present. Preserv. Cult. Sci. Heritage **15**, 247–258 (2025). https://doi.org/10.55630/dipp.2025.15.23

21. Lehmann, J., et al.: DBpedia - a large-scale, multilingual knowledge base extracted from Wikipedia. Semant. Web J. **6** (2014). https://doi.org/10.3233/SW-140134

22. Li, J., Li, D., Savarese, S., Hoi, S.: BLIP-2: bootstrapping language-image pre-training with frozen image encoders and large language models. In: International Conference on Machine Learning, pp. 19730–19742. PMLR (2023)

23. Liu, Y., et al.: LamRA: large multimodal model as your advanced retrieval assistant. In: Proceedings of the Computer Vision and Pattern Recognition Conference, pp. 4015–4025 (2025)

24. Liu, Z., et al.: Swin Transformer v2: scaling up capacity and resolution. In: Proceedings of the IEEE/CVF Conference on Computer Vision and Pattern Recognition (CVPR), June 2022, pp. 12009–12019 (2022)

25. Marchand, E., Gagnon, M., Zouaq, A.: Extraction of a knowledge graph from French cultural heritage documents. In: Bellatreche, L., et al. (eds.) TPDL/ADBIS-2020. CCIS, vol. 1260, pp. 23–35. Springer, Cham (2020). https://doi.org/10.1007/978-3-030-55814-7_2

26. Mihindukulasooriya, N., Tiwari, S., Enguix, C.F., Lata, K.: Text2KGBench: a benchmark for ontology-driven knowledge graph generation from text. In: Payne, T.R., et al. (eds.) The Semantic Web - ISWC 2023, pp. 247–265. Springer, Cham (2023). https://doi.org/10.1007/978-3-031-47243-5_14
27. Netwerk Digitaal Erfgoed: Dutch digital heritage network (2024). https://netwerkdigitaalerfgoed.nl/
28. Page, R.: Wikidata and the biodiversity knowledge graph. Biodivers. Inf. Sci. Stan. **3** (2019). https://doi.org/10.3897/biss.3.34742
29. Peeters, R., et al.: Rights to richness: connecting cultural artefacts with rights and context (2025)
30. Radford, A., et al.: Learning transferable visual models from natural language supervision. In: International Conference on Machine Learning, pp. 8748–8763. PMLR (2021)
31. Song, X., Lin, H., Wen, H., Hou, B., Xu, M., Nie, L.: A comprehensive survey on composed image retrieval. ACM Trans. Inf. Syst. (2025)
32. Vrandečić, D., Krötzsch, M.: Wikidata: a free collaborative knowledgebase. Commun. ACM **57**(10), 78–85 (2014). https://doi.org/10.1145/2629489, https://doi.org/10.1145/2629489
33. Waagmeester, A., et al.: Wikidata as a knowledge graph for the life sciences. eLife **9** (2020). https://doi.org/10.7554/eLife.52614
34. Wang, Y., et al.: Construction of cultural heritage knowledge graph based on graph attention neural network. Appl. Sci. **14**(18) (2024). https://doi.org/10.3390/app14188231, https://www.mdpi.com/2076-3417/14/18/8231
35. Yang, S., Hou, M.: Knowledge graph representation method for semantic 3D modeling of Chinese grottoes. Heritage Sci. **11** (2023). https://doi.org/10.1186/s40494-023-01084-2
36. Ypsilantis, N.A., Garcia, N., Han, G., Ibrahimi, S., van Noord, N., Tolias, G.: The met dataset: instance-level recognition for artworks. In: Vanschoren, J., Yeung, S. (eds.) Proceedings of the Neural Information Processing Systems Track on Datasets and Benchmarks, vol. 1. Curran (2021) https://datasets-benchmarks-proceedings.neurips.cc/paper_files/paper/2021/file/5f93f983524def3dca464469d2cf9f3e-Paper-round2.pdf
37. Zhang, J., Huang, J., Jin, S., Lu, S.: Vision-language models for vision tasks: a survey. IEEE Trans. Pattern Anal. Mach. Intell. **46**(8), 5625–5644 (2024)
38. Zhang, K., et al.: MagicLens: self-supervised image retrieval with open-ended instructions. arXiv preprint arXiv:2403.19651 (2024)

Mapping Change: A Temporal and Semantic Knowledge Base of Scottish Gazetteers

Lilin Yu[1]([✉])[iD], Angelo Salatino[2][iD], and Rosa Filgueira[1][iD]

[1] EPCC, University of Edinburgh, Edinburgh, UK
L.Yu-40@sms.ed.ac.uk, r.filgueira@epcc.ed.ac.uk
[2] Knowledge Media Institute, The Open University, Milton Keynes, UK
angelo.salatino@open.ac.uk

Abstract. We introduce MappingChange, a temporal and semantically enriched knowledge graph constructed from the "Gazetteers of Scotland" collection between 1803 and 1901. Although digitised into over 13,000 OCR-aligned ALTO/METS XML files, the collection lacks the structural information needed for effective analysis: it consists of only page-level text, lacks article-level segmentation, fails to identify places referenced within articles, and provides no semantic links connecting related entries across editions. These gaps significantly hinder reuse, especially for longitudinal studies of how historical places have evolved. To overcome these challenges, we employ edition-specific prompt strategies with large language models to extract over 45,000 article-level place descriptions. These entries are organized into a knowledge graph modelled on the Heritage Textual Ontology, capturing both bibliographic metadata and textual content. We further enriched this knowledge graph by 1) connecting semantically related entries across editions and aligning them with Wikidata and DBpedia, and 2) identifying referenced places within articles and assign modern coordinates to facilitate spatial analysis. We envision that MappingChange will serve as a foundation for downstream research in the humanities and digital humanities, with an impact extending well beyond academic settings.

Keywords: Gazetteers · Knowledge Graph · Semantic Web · Historical Texts · Ontology · Linked Data · Cultural Heritage · Geolocation

Resource Type: Dataset
License: CC BY 4.0
DOI: https://doi.org/10.5281/zenodo.15393935
URL: https://github.com/francesNLP/MappingChange
Endpoint: http://query.frances-ai.com/hto_gaz/sparql
SPARQL Query Exploration UI: http://yasgui.org/short/0M7HXSruJg.

1   Introduction

The growing availability of digitised historical collections and the exponential growth of aggregated metadata offer new opportunities for humanities research. However, these vast datasets often remain under-utilised without effective computational methods. To address this, cultural-heritage applications [17,28,30] are increasingly deploying Semantic Web technologies to improve FAIRness and interoperability. Knowledge Graphs (KGs) play a central role in this ecosystem, providing the structured representation and linking required to transform raw archives into interconnected resources suitable for a range of novel digital research methods. These infrastructures enable scholars to apply novel digital methods, allowing them to interrogate large datasets and explore an expanding array of humanities questions.

A prime candidate for such enrichment is the *Gazetteers of Scotland* (1803–1901) collection [21], which comprises the most significant descriptive gazetteers of nineteenth-century Scotland. These volumes capture towns, parishes, rivers, castles, and natural features within evolving historical, social, and economic narratives. As industrialisation, migration, land reform, and the empire reshaped the nation, these texts became critical instruments for documenting change, making them invaluable resources for studying this spatial transformation.

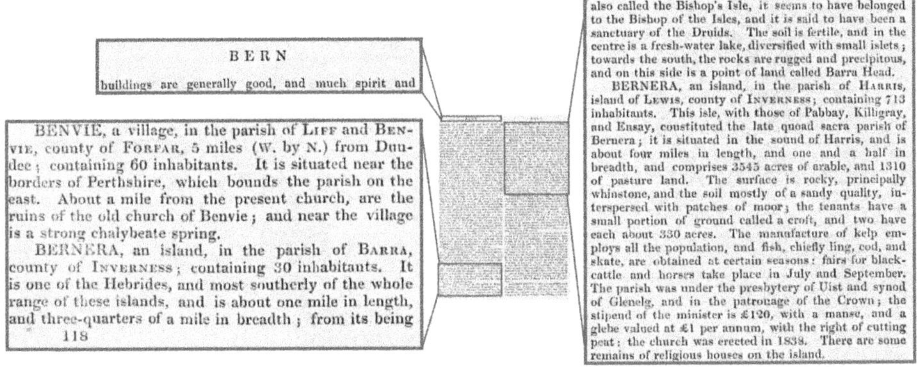

BERN

buildings are generally good, and much spirit and

BENVIE, a village, in the parish of LIFF and BENVIE, county of FORFAR, 5 miles (W. by N.) from Dundee; containing 60 inhabitants. It is situated near the borders of Perthshire, which bounds the parish on the east. About a mile from the present church, are the ruins of the old church of Benvie; and near the village is a strong chalybeate spring.

BERNERA, an island, in the parish of BARRA, county of INVERNESS; containing 30 inhabitants. It is one of the Hebrides, and most southerly of the whole range of these islands, and is about one mile in length, and three-quarters of a mile in breadth; from its being

118

also called the Bishop's Isle, it seems to have belonged to the Bishop of the Isles, and it is said to have been a sanctuary of the Druids. The soil is fertile, and in the centre is a fresh-water lake, diversified with small islets; towards the south, the rocks are rugged and precipitous, and on this side is a point of land called Barra Head.

BERNERA, an island, in the parish of HARRIS, island of LEWIS, county of INVERNESS; containing 713 inhabitants. This isle, with those of Pabbay, Killigray, and Ensay, constituted the late quoad sacra parish of Bernera; it is situated in the sound of Harris, and is about four miles in length, and one and a half in breadth, and comprises 3545 acres of arable, and 1310 of pasture land. The surface is rocky, principally whinstone, and the soil mostly of a sandy quality, interspersed with patches of moor; the tenants have a small portion of ground called a croft, and two have each about 330 acres. The manufacture of kelp employs all the population, and fish, chiefly ling, cod, and skate, are obtained at certain seasons: fairs for black-cattle and horses take place in July and September. The parish was under the presbytery of Uist and synod of Glenelg, and in the patronage of the Crown; the stipend of the minister is £120, with a manse, and a glebe valued at £1 per annum, with the right of cutting peat: the church was erected in 1838. There are some remains of religious houses on the island.

Fig. 1. Scanned image for the page 118 of the first volume, 1846 edition.

Despite their significance, the Gazetteers collection poses substantial technical challenges for computational analysis. The corpus comprises over 13,000 OCR-aligned XML files in ALTO[1] and METS[2] formats, but the digitisation encodes only page-level text. Articles frequently begin mid-page, span multiple columns or pages, and adhere to complex, edition-specific typographic conventions. While human readers can intuitively discern headings and layout hierarchy from the page images (see Fig. 1), this structural logic is absent from the

---

[1] ALTO – https://www.loc.gov/standards/alto/.
[2] METS – https://www.loc.gov/standards/mets/.

machine-readable XML. Indeed, the ALTO format captures surface-level features like text style and casing, as illustrated in Fig. 2, but fails to explicitly encode article boundaries or semantic structure, making automated reconstruction a non-trivial task.

Further complexity arises from the content itself. Many place names are polysemous: distinct locations share identical names within and across editions. Understanding historical change therefore requires disambiguating entries and link related entries across editions. Gazetteer articles also reference other places and spatial relations that, when extracted, support the reconstruction of approximate historical geographies even when coordinates are not explicitly provided.

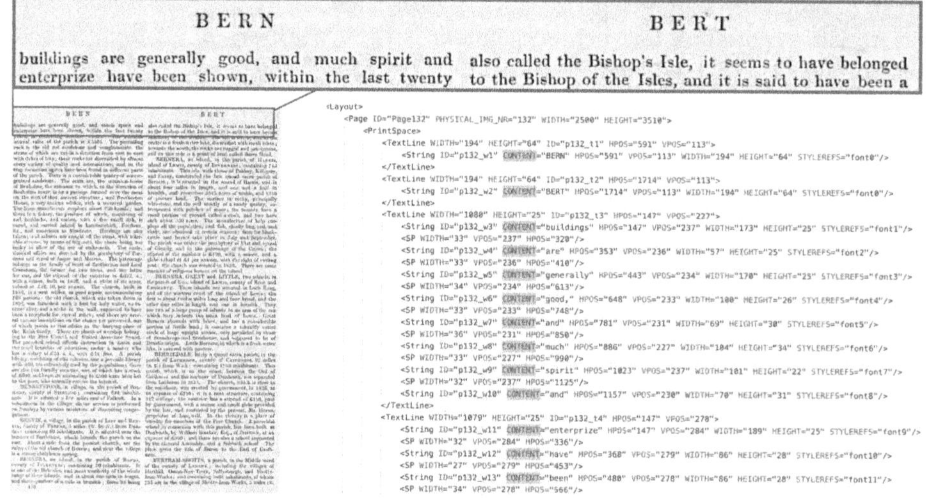


```
<Layout>
  <Page ID="Page132" PHYSICAL_IMG_NR="132" WIDTH="2500" HEIGHT="3510">
    <PrintSpace>
      <TextLine WIDTH="194" HEIGHT="64" ID="p132_t1" HPOS="591" VPOS="113">
        <String ID="p132_w1" CONTENT="BERN" HPOS="591" VPOS="113" WIDTH="194" HEIGHT="64" STYLEREFS="font0"/>
      </TextLine>
      <TextLine WIDTH="194" HEIGHT="64" ID="p132_t2" HPOS="1714" VPOS="113">
        <String ID="p132_w2" CONTENT="BERT" HPOS="1714" VPOS="113" WIDTH="194" HEIGHT="64" STYLEREFS="font0"/>
      </TextLine>
      <TextLine WIDTH="1080" HEIGHT="25" ID="p132_t3" HPOS="147" VPOS="227">
        <String ID="p132_w3" CONTENT="buildings" HPOS="147" VPOS="237" WIDTH="173" HEIGHT="25" STYLEREFS="font1"/>
        <SP WIDTH="33" VPOS="237" HPOS="320"/>
        <String ID="p132_w4" CONTENT="are" HPOS="353" VPOS="236" WIDTH="57" HEIGHT="25" STYLEREFS="font2"/>
        <SP WIDTH="33" VPOS="236" HPOS="410"/>
        <String ID="p132_w5" CONTENT="generally" HPOS="443" VPOS="234" WIDTH="170" HEIGHT="25" STYLEREFS="font3"/>
        <SP WIDTH="34" VPOS="234" HPOS="613"/>
        <String ID="p132_w6" CONTENT="good," HPOS="648" VPOS="233" WIDTH="100" HEIGHT="26" STYLEREFS="font4"/>
        <SP WIDTH="33" VPOS="233" HPOS="748"/>
        <String ID="p132_w7" CONTENT="and" HPOS="781" VPOS="231" WIDTH="69" HEIGHT="30" STYLEREFS="font5"/>
        <SP WIDTH="36" VPOS="231" HPOS="850"/>
        <String ID="p132_w8" CONTENT="much" HPOS="886" VPOS="227" WIDTH="104" HEIGHT="34" STYLEREFS="font6"/>
        <SP WIDTH="33" VPOS="227" HPOS="990"/>
        <String ID="p132_w9" CONTENT="spirit" HPOS="1023" VPOS="237" WIDTH="101" HEIGHT="22" STYLEREFS="font7"/>
        <SP WIDTH="32" VPOS="237" HPOS="1125"/>
        <String ID="p132_w10" CONTENT="and" HPOS="1157" VPOS="230" WIDTH="70" HEIGHT="31" STYLEREFS="font8"/>
      </TextLine>
      <TextLine WIDTH="1079" HEIGHT="25" ID="p132_t4" HPOS="147" VPOS="278">
        <String ID="p132_w11" CONTENT="enterprize" HPOS="147" VPOS="284" WIDTH="189" HEIGHT="25" STYLEREFS="font9"/>
        <SP WIDTH="32" VPOS="284" HPOS="336"/>
        <String ID="p132_w12" CONTENT="have" HPOS="368" VPOS="279" WIDTH="86" HEIGHT="28" STYLEREFS="font10"/>
        <SP WIDTH="27" VPOS="279" HPOS="453"/>
        <String ID="p132_w13" CONTENT="been" HPOS="480" VPOS="278" WIDTH="86" HEIGHT="28" STYLEREFS="font11"/>
        <SP WIDTH="34" VPOS="278" HPOS="566"/>
```

Fig. 2. ALTO XML for the page 118 of the first volume, 1846 edition.

To address these challenges, we present **MappingChange**, a reusable temporal and semantically enriched, article-level knowledge graph transformed from this corpus. Using edition-specific prompt strategies, we apply GPT-4 to extract over 45,000 article entries, represented using our refined Heritage Textual Ontology (HTO). We adopt our concept-linkage method from previous work [30] to connect articles across editions and align them with external sources such as Wikidata and DBpedia. For each article entry, we extract in-text place references (geotagging) using Stanza NER [26] and assign modern coordinates (georesolution) using the Edinburgh Geoparser [2], enabling temporal-spatial analysis. All datasets (DataFrames, RDF graphs), source code, and Jupyter notebooks are openly released with persistent identifiers. The data is also fully integrated into *Frances*[3], a web platform that enables temporal exploration and semantic search,

---

[3] Frances—http://www.frances-ai.com/.

removing the technical barrier of SPARQL. In summary, the main contributions
are as follows:

- we release an updated and refined version of the Heritage Textual Ontology;
- we introduce **MappingChange**, an article-level knowledge graph of the Scottish Gazetteers (1803–1901); and
- we openly release the source code along with all the datasets.

The remainder of this paper is structured as follows. Section 2 surveys related
literature, followed by an introduction to the HTO in Sect. 3. Section 4 details
the KG generation pipeline, while Sect. 5 reports on our evaluation results. We
discuss resource availability and our sustainability plan in Sect. 6, outline the
potential impact in Sect. 7, and conclude with future directions in Sect. 8.

2   Related Work

Semantic Web technologies have been widely adopted within Digital Humanities to model, enrich, and interlink diverse types of cultural data—ranging
from textual archives and historical records to musical scores and sonic heritage. This approach spans a vast array of historical contexts, including studies of conflict, slavery, and cultural evolution. Projects such as WarSampo [17],
Enslaved.org [14], and the Meetups pilot [20] from the Polifonia [6] project exemplify this by modelling and interlinking entities like people, places, and events
from curated datasets to support quantitative historical research. Additionally,
the Europeana initiative[4] aggregates metadata for millions of digital objects
using the Europeana Data Model, now serving as the core infrastructure for the
common European data space for cultural heritage.

Beyond textual data, the Polifonia project [6] applies these technologies to the
musical domain, developing a knowledge graph that connects musical heritage
across disparate European sources, enabling the analysis of musical evolution
through ontologies that capture melody, tonality, and social context. Similarly,
in the area of tangible heritage, ArCo (Architecture of Knowledge) [7] provides
a KG for Italian cultural heritage, standardising the description of architectural
and artistic assets. Moreover, the Data for History consortium [4] is establishing a
common method for modelling historical data, ensuring interoperability between
disparate historical research projects via an ontology-based approach.

The remainder of this review focuses specifically on the challenges of structuring information from historical gazetteers. In this specific domain, initiatives
like Pelagios[5] have been instrumental in linking historical resources by place
references, primarily through tools like Recogito[6], which supports manual and
semi-automated geo-annotation of texts and images [29]. Recogito provides automatic Named Entity Recognition (NER) based geotagging, and georesolution

---

[4] Europeana—https://www.europeana.eu/en.
[5] Pelagios—https://pelagios.org.
[6] Recogito—https://recogito.pelagios.org.

suggestions using historical gazetteers, but it emphasises precision and relies on manual validation. In our work, we focussed on large-scale, fully automated extraction, which prioritises recall to extract and semantically model over 45,000 entries from ten editions of the Gazetteers of Scotland. While precision can be improved through modular upgrades to subtasks, we aimed for scaled extraction and enrichment without manual effort.

PastPlace[7], a semantically enriched gazetteer developed from the Pelagios network, offers curated historical place data for Britain and supports the tracking of place references across documents. However, its manual annotation workflow and broader grouping strategy may conflate distinct entities that share a name, for example, grouping "Dundee" together with "Dundee and Perth Railway". To address this, our clustering strategy employs an alignment methodology based on semantic similarity. This ensures that only entries describing the identical place are linked across editions, enabling the precise tracking of temporal change.

In the context of article-level segmentation, traditional structuring approaches heavily rely on layout analysis and visual cues [24]. Prior work on historical text collections, including newspapers [12] and the *Encyclopaedia Britannica (EB)* [11,30], notably using the `defoe` library [11], has underscored the need for scalable, domain-adapted pipelines that can reliably structure content for downstream semantic enrichment. Recent advances in generative AI, such as GPT-4 [23], opened new possibilities for text interpretation and segmentation. These models enable extraction of structured content from historical documents without relying solely on brittle heuristics [22]. In our approach, relying on flexible prompt-based extraction enables dynamic adaptation to specific editorial styles, abbreviations, and conventions found in historical corpora.

With regard to geotagging, although rule-based systems like the Edinburgh Geoparser [2] are standard for historical georesolution [3,8,10], recent deep learning approaches offer enhanced flexibility for entity recognition [16,19]. Our evaluations on the *Encyclopaedia Britannica* [33] suggested that contextualised models (e.g., Stanza [26], Flair [27]) are particularly resilient to OCR noise. Therefore, we integrate these architectures to strengthen the recognition phase, creating a comprehensive pipeline for generating spatially enriched knowledge graphs.

3   Heritage Textual Ontology

The Heritage Textual Ontology [32] (HTO) provides the semantic foundation for modeling **MappingChange** knowledge base. Initially introduced in our Frances platform paper [30] to represent provenance-aware textual knowledge extracted from digitised heritage corpora, HTO has since been extended to support spatial and temporal aspects of historical place descriptions [33]. HTO aligns with PROV-O [13], Schema.org, CIDOC CRM [9] and its spatial-temporal extension CRMgeo [15], SKOS [18], GeoSPARQL and the Web Annotation Ontology.

---

[7] PastPlace—https://www.pastplace.org.

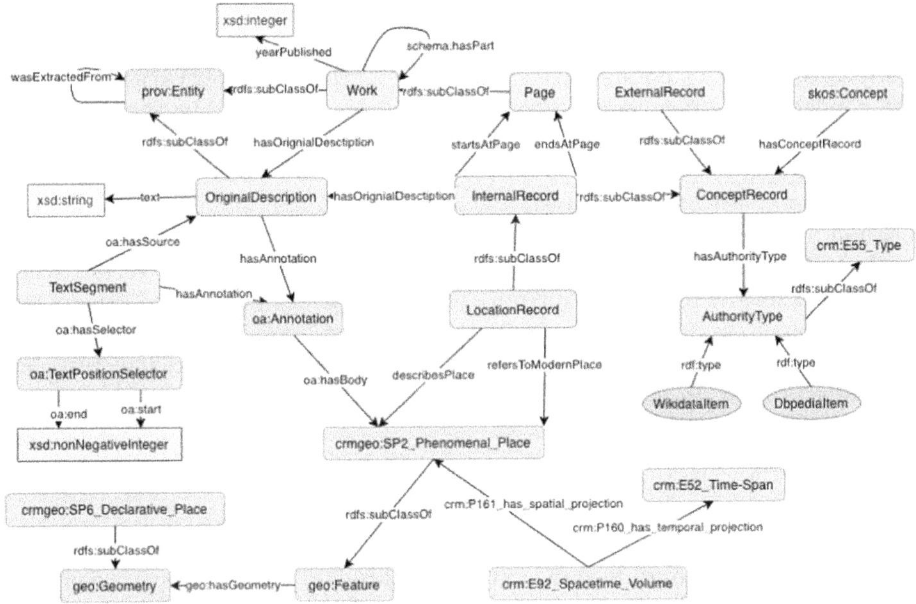

Fig. 3. Heritage Textual Ontology

Figure 3 summaries the core HTO classes and the reused ontology components that underpin the competency questions in Sect. 5.1. A place article is represented as an `hto:LocationRecord`, derived from a historical textual work (`hto:Work`). Its digitised textual content is captured as an `hto:OriginalDescription`, which is extracted from an OCR-aligned XML file modeled as a `prov:Entity`. PROV records digitization workflows, extraction activities, and provenance for the historical work and its textual content.

A work can be an edition (`hto:Series`), a volume (`hto:Volume`), or an individual page (`hto:Page`). Editions may contain multiple volumes, and volumes contain pages via `schema:hasPart`. Each place article spans one or more pages and is thus linked back to its volume and edition via this hierarchical chain. Publication year as a key biographic metadata of a work, provides temporal context for the article, indicating when a given description was widely circulated.

Each article is modeled as a distinct `hto:LocationRecord`, even when multiple entries share the same place name (e.g., "BERNERA", see Fig. 1), because they may refer to different places. To study how a place changes over time, HTO enables records to be grouped via a `skos:Concept`: a concept can be associated with multiple `hto:LocationRecord` instances through `hto:hasConceptRecord`. Additionally, place articles may be linked to modern authority records, such as Wikidata or DBpedia, via `hto:ExternalRecord`, supporting cross-edition alignment and contemporary interpretation.

To support spatial analysis, HTO uses the property `hto:describesPlace` to link an article to the real world historical place it describes, modeled as a `crmgeo:SP2_Phenomenal_Place`, together with the `crm:E52_Time-Span` representing the period of description. This historical place has its spatial geometries (such as centroids or boundaries) modeled as `crmgeo:SP6_Declarative_Place` and connected via `geo:hasGeometry`. These geometries may be stored in common GIS formats, such as WKT or GeoJSON. HTO also links a historical place article to its corresponding modern place through `hto:refersToModernPlace`, enabling temporal–spatial analysis and providing an approximate contemporary location when historical coordinates are unavailable.

To link referenced places mentioned within an article and to record where they appear in the text, HTO introduces `hto:TextSegment`, representing a specific portion of the article's `hto:OriginalDescription`. Each segment is associated with an `oa:TextPositionSelector`, which records its start and end character offsets. The segment is then connected to its identified place, historical or modern, through an `oa:Annotation`.

4   Knowledge Graph Generation Pipeline

The **MappingChange** resource is built through a modular, reproducible pipeline, illustrated in Fig. 4, that transforms OCR-aligned ALTO and METS XML files (see example Fig. 2) into structured DataFrames and semantically enriched RDF graph. To facilitate reuse, the full codebase is provided in the GitHub repository, along with step-by-step instructions in the pipeline guide[8]. The architecture is designed for reuse and adaptation across heritage corpora with similar structural complexity. Of the twelve editions digitised by the National Library of Scotland[9], ten were selected (Table 1). We excluded the 1828 edition, a town-focused summary rather than a true gazetteer, and the 1848 edition, for which only Volume II is available. The chosen editions span nearly a century (1803–1901), making them suited for diachronic analyses.

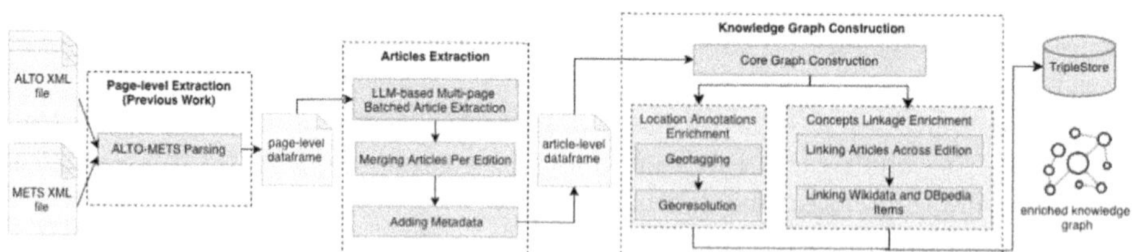

Fig. 4. Knowledge Graph Generation Pipeline.

4.1   Article Extraction

We begin with a consolidated page-level DataFrame [31] extracted from ALTO and METS XML files in our previous work [30], this dataframe contains OCR page text and associated metadata (edition, volume, page number, etc.). For article extraction, we employ GPT-4, utilising edition-specific prompts designed to handle layout complexities like alias resolution, mid-page entries, multi-volume structures, and redirects. We selected the GPT model due to its minimal hardware requirements and its ease of integration into the development workflow.

Our prompt engineering strategy addresses the unique typographic and OCR characteristics of each edition; notably, the 1883 and 1901 editions share formatting and therefore use a common prompt. Custom Python scripts tokenize the OCR pages, execute batch requests via the OpenAI API, and parse the output into structured JSON. The results are aggregated into per-edition DataFrames, containing one row per article with associated provenance metadata, where articles spanning multiple pages are merged. Additionally, we incorporate rule-based matching to capture internal references (e.g., 'See Paisley.') and alternative names (e.g., 'AVON or AVEN'), which are essential for semantic linking. Table 1 details the extraction statistics, including page counts, article totals, and average word counts per edition.

Table 1. Statistics: volume, article, and page counts, average words per article

Edition	Volumes	Pages	Articles	Words/Article
1803	1	606	1,955	182.89
1806	1	594	2,468	142.39
1825	1	222	2,141	77.07
1838	2	1,098	1,958	515.29
1842	2	1,798	2,938	404.09
1846	2	1,302	1,990	626.77
1868	2	1,908	6,382	242.85
1882	1	492	7,297	40.88
1883	6	2,038	9,115	213.02
1901	1	1,776	9,407	219.89

4.2   Core Graph Construction

Through a fully automated process, we normalise the data type and date of each article-level DataFrame and convert it into RDF triples using the Python library RDFLib [5], adhering to the `HTO` schema. Every extracted article is instantiated as an `hto:LocationRecord`, enriched with edition metadata, page references, and extraction provenance. This core graph immediately addresses specific competency questions: it satisfies CQ2 by providing extraction provenance, and partially satisfies CQ1 by exposing article descriptions and their positions within the historical documents.

4.3   Concept Linkage Enrichment

For concept linking, we group articles that describe the same place across different editions and assign them to a shared `hto:Concept` instance. This linking is performed by comparing the semantic similarity of article descriptions using sentence embeddings[10]. All articles sharing the same primary or alternative name are first grouped, and within each group every article is linked to the most semantically similar later articles that exceed a similarity threshold. This enrichment disambiguates places with identical names but different meanings or locations (e.g., the multiple instances of "BERNERA" shown in Fig. 1) and enables longitudinal analysis of place descriptions over time.

To enable comparison with modern descriptions, we further align each historical concept cluster with external resources such as Wikidata and DBpedia. For each `hto:Concept`, we compute the embedding of its latest-year article and compare it against the embeddings of candidate external entities. Matches above a similarity threshold are linked as external references. This step positions historical place records within contemporary knowledge ecosystems and facilitates cross-referencing between historical and modern contexts.

4.4   Location Annotations Enrichment

In this collection, spatial context is often conveyed through references to other places. To capture this, we apply geotagging using Stanza [26], extracting entities labeled as LOC (location), FAC (facility), and GPE (geopolitical entity) from the articles. We then perform georesolution using the Edinburgh Geoparser [2], configured with GeoNames gazetteer[11], assigning modern coordinates to both the primary place and referenced places. This process prioritises Scottish entries to align with the corpus's national scope. The resulting geospatial annotations are integrated into the core graph utilising the specific HTO constructs.

4.5   Knowledge Graph Serialization and Deployment

The final enriched knowledge graph is published in Turtle format and available on Zenodo [34]. It describes 45,651 historical place articles with over 82 million number of statements. These are grouped into 24,020 concepts representing distinct places over time. A total of 9,433 external records are linked to these concepts, with 7,037 from Wikidata and 2,396 from DBpedia. The graph includes 504,585 annotated places mentioned within articles, of which 335,249 have identified modern coordinates. Additionally, 22,018 articles contain direct references to modern locations. The knowledge graph is uploaded and spatially indexed in our cloud-hosted Apache Fuseki triple store with GeoSPARQL support, and can be queried via a public SPARQL endpoint[12].

---

[10] Sentence   Embeddings—https://huggingface.co/sentence-transformers/all-mpnet-base-v2.

[11] GeoNames gazetteer—https://www.geonames.org.

[12] SPARQL Endpoint—http://query.frances-ai.com/hto_gaz/sparql.

5   Evaluation

In this section, we evaluate the quality of the generated knowledge graph. Section 5.1 validates the graph against competency questions using SPARQL. Sections 5.2, 5.3 and 5.4 then quantify the performance of article extraction, concept record linkage, geotagging and georesolution, respectively, utilising manually annotated ground-truth datasets. Moreover, we evaluated the HTO ontology using the OOPS! OntOlogy Pitfall Scanner!) tool [25], yielding no critical pitfalls and confirming it satisfies standard design criteria regarding structural integrity, metadata completeness, and logical consistency.

5.1   Answering Competency Questions

We validate the **MappingChange** knowledge graph against five Competency Questions (CQs), demonstrating its capacity for historical and spatial analysis. Each CQ is resolved using a specific SPARQL query. For further examples and a wider range of CQs, we refer the reader to the notebook in the GitHub repository.

volume_title	s_page_num	year	text
Gazetteer of Scotland; arranged under the various descriptions of counties, parishes, islands 1825?	"38"^^xsd:integer	"1825"^^xsd:integer	a small village in the parish of Dunfermline, Fifeshire, adjoining the village of Limekilns, where there is a brewery and a quay.
gazetteer of Scotland. [With plates and maps.] 1838, Volume 1	"142"^^xsd:integer	"1838"^^xsd:integer	a small village in Fife, on the coast of the Firth of Forth, in the parish of Dunfermline.
imperial gazetteer of Scotland; or, Dictionary of Scottish topography, compiled from the most recent authorities, and forming a complete body of Scottish geography, physical, statistical, and historical 1868, Volume 1	"305"^^xsd:integer	"1868"^^xsd:integer	BRUCEHAVEN, a harbour in the parish of Inverkeithing, adjacent to the village of Limekilns, Fifeshire.
Ordnance gazetteer of Scotland 1884-1885, Volume 1	"233"^^xsd:integer	"1883"^^xsd:integer	Brucehaven, a harbour on the mutual border of Inverkeithing and Dunfermline parishes, Fife, on the Firth of Forth, adjacent to Limekilns, 3 miles S by W of Dunfermline town.
Ordnance gazetteer of Scotland 1901	"203"^^xsd:integer	"1901"^^xsd:integer	Brucehaven, a harbour in Fife, on the Firth of Forth, 3 miles S by W of Dunfermline. It was transferred by the Boundary Commissioners in 1891 from the parish of Inverkeithing to that of Dunfermline.

Fig. 5. Explore the descriptions of *Brucehaven* across editions.

CQ1: How Is a Place Described over Time? Descriptions of the same place across historical documents published in different years provide evidence of how that place changed over time. To answer this question, we produced a SPARQL query[13] that retrieves all articles describing *Brucehaven in Fife, Scotland*, as well as the volume and its page in which each description appears. The result in Fig. 5 illustrates how these temporal descriptions can reveal change: the parish affiliation of *Brucehaven* shifts from *Dunfermline* to *Inverkeithing* in the 1868 edition, and then returns to *Dunfermline* in the 1901 edition. Modern description of *Brucehaven* can be queried[14] through linked Wikidata item.

---

[13] SPARQL Query for CQ1—http://yasgui.org/short/j9IDt4egek.
[14] SPARQL Q. for linked Wikidata in CQ1—http://yasgui.org/short/uLVkr7Azlu.

CQ2: How Is an Article Extracted? This competency question targets article provenance, capturing both the underlying source files and the extraction method. We formulated a query[15] to retrieve this metadata for the *Brucehaven* article (1825 edition) introduced in the previous question. The results in Fig. 6 verify that the entry was derived from an ALTO XML file provided by the NLS Data Foundry [21] and processed via the pipeline.

text	extraction_tool	source_label	dataset
a small village in the parish of Dunfermline, Fifeshire, adjoining the village of Limekilns, where there is a brewery and a quay.	https://github.-com/francesNLP/Map-pingChange	97421702/alto/9 7422152.34.xml	https://data.nls.uk/data/digitised-collections/gazetteers-of-scotland/

Fig. 6. Explore extraction method for the article *Brucehaven* in 1825 edition.

CQ3: Where Is the Place that an Article Primarily Describes? Visualising places on a map offers an immediate spatial reference and requires either point coordinates or boundary geometries. When both historical and modern geometries are available, spatial change can be visually examined over time. To demonstrate this, we formulated a SPARQL query[16] to retrieve coordinates for the article describing Dundee in the 1803 edition, alongside those supplied by a modern gazetteer. The resulting map visualization, displayed in Fig. 7a, provides only the modern location recognised in 2025, as this work does not extract historical geometries from the text due to their complexity and irregularity. However, in contexts of collections with more stable textual patterns like the *Encyclopaedia Britannica* [33], this pipeline has successfully represented historical spatial data if extracted.

CQ4: What Places Are Mentioned in an Article? In cases when neither historical nor modern coordinates are available for the primary place, the locations mentioned within the article can provide valuable clues for approximating its spatial context. To demonstrate this, we formulated a query[17] that lists all places mentioned in the *Brucehaven* article, with their modern coordinates and the positions in the text, which allows further spatial relation examination. The result, visualised as a map in Fig. 7b, allows the user to localise *Brucehaven* as a small village in the vicinity of *Dunfermline*.

CQ5: How Do the Socio-Economic Roles of a Place Change over Time? Beyond identifying where a place is located, it is often crucial to understand how its socio-economic role evolves across editions. This question asks how economic

---

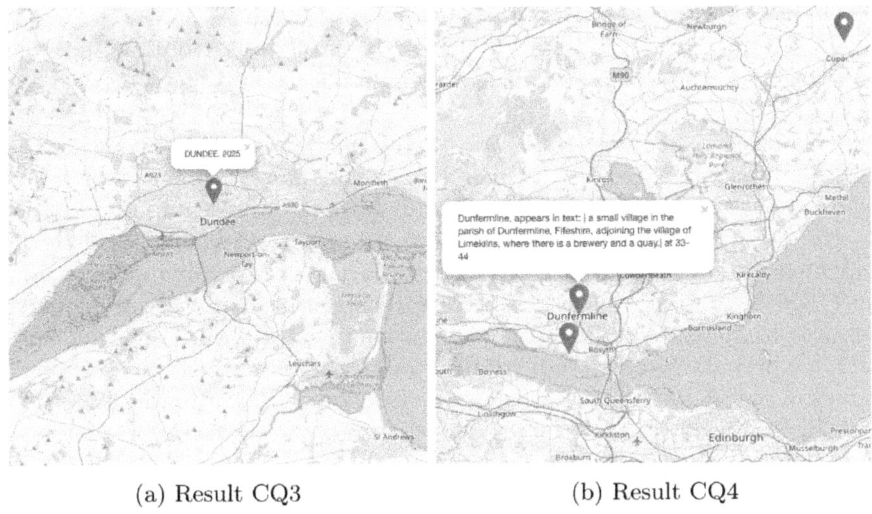

(a) Result CQ3   (b) Result CQ4

Fig. 7. Explore (a) the historical location primary described by the article *Dundee* in the 1803 edition and its modern reference; and (b) places mentioned in the article *Brucehaven* in the 1825 edition.

and infrastructural activities associated with a place (e.g., harbours, mills, factories, railways) emerge, shift, or disappear over time. Using identified articles together with the edition hierarchy and the original article description, we can follow these changes in a structured and temporally-aware manner. We answer this question in a two-step process.

Query 5.A: Retrieve Socio-Economic Descriptions Across Editions. This step identifies all editions in which socio-economic keywords[18] appear in the article for a given place. For a specified location, *Edinburgh* in this case, the query[19] retrieves the associated articles from each edition, returning the publication year, the article URI, a descriptive text snippet, and a concatenated list of all socio-economic keywords detected in that edition. This makes it possible to compare *when* socio-economic terms appear, and *which* combinations occur together in the same description.

The results show a clear and interpretable trajectory. Early editions (1806–1825) mention only harbour-related terms, reflecting Edinburgh's maritime orientation and strong connection to Leith. By 1838 and 1842, the descriptions begin to incorporate factory and mill alongside the harbour, signalling the emergence of light industry. From 1846 onwards, the articles increasingly reference railways, and by the 1868 and 1883 editions, combinations such as "harbour, rail-

---

[18] We use four representative socio-economic terms ("harbour", "mill", "factory", "railway") as examples. They capture major infrastructural transitions in 19th-century Scotland, but the approach easily extends to additional terms.

[19] SPARQL Q. and result for CQ5 step 1—http://yasgui.org/short/oFeKaGklSM.

way, factory, mill" appear. This shift from a harbour-focused town to a railway-connected industrial centre is detected directly through SPARQL, without additional NLP pipelines, demonstrating the expressiveness of the KG.

Query 5.B: Aggregate Socio-Economic Signals over Time. This second step complements the qualitative view by aggregating how many Edinburgh articles per year mention each socio-economic keyword. Instead of returning raw text snippets, the query[20] generates a compact temporal profile of terminology usage. The aggregated results reveal that terms related to harbours dominate the early editions (1806–1846), reflecting the city's historic maritime connections. In contrast, references to factories and mills appear intermittently during the mid-century, marking a period of gradual industrialisation. Most notably, railway terminology emerges in the 1840s and have a higher presents from 1883 onwards, directly mirroring historical expansions such as the consolidation of Waverley Station and the development of major branch lines.

Together, the two steps show how **MappingChange** enables both fine-grained qualitative interpretation (via text snippets and keyword combinations) and quantitative temporal analysis (via aggregated frequencies), demonstrating how a semantically structured and temporally indexed knowledge graph reveals socio-economic change across historical editions.

While all analyses from our initial dataframe notebooks[21], can now be reproduced through SPARQL, the KG provides several capabilities that dataframes alone cannot support: seamless traversal across editions and concept records; integration of modern geospatial coordinates via GeoSPARQL; direct access to provenance (e.g., extraction quality, digitisation sources); and the combination of semantic annotations, historical descriptions, and external authority links within a single query. These features enable simultaneous diachronic, semantic, and spatial analyses, workflows that would be significantly more arduous to implement with flat tables. At the same time, the KG remains compatible with downstream NLP and text-mining workflows, enabling hybrid analytical pipelines that combine symbolic and statistical methods.

5.2 Article Extraction

To evaluate the article extraction method, we manually verified 100 single-page articles with 10 articles for each edition. The result shows 95% single page articles are correctly extracted, 5% including extract text from other articles. We then selected five multi-page articles from each edition, and manually identified the number of pages for the ground truth (165) and correctly extracted (120). This evaluation yielded a high precision of 0.88 and a recall of 0.73, indicating accurate segmentation across multiple pages, though with some missed page spans.

---

[20] SPARQL Q. and result for CQ5 step 2—http://yasgui.org/short/T1JGtCd6Bi.

[21] Jupyter Notebook containing the analysis—https://github.com/francesNLP/ MappingChange/blob/main/Notebooks/Exploring_AggregatedDF.ipynb.

5.3   Concept Record Linkage

To assess the quality of concept record linking, we manually annotated a ground-truth dataset comprising 323 article-to-article links (across 175 gazetteer articles), along with 114 candidate links to Wikidata and 71 to DBpedia. These samples were drawn from 25 distinct place names, of which 10 (40%) refer to multiple distinct locations, thereby providing a rigorous test for disambiguation. We report performance using standard precision, recall, and F1-score metrics.

Table 2. Precision, Recall and F1 Score of our concept record linkage

Linkage Type	Pre.	Rec.	F1 S.
Articles Linkage	0.79	0.83	0.81
Wikidata Linkage	0.45	0.56	0.5
DBpedia Linkage	0.89	0.69	0.78

The results in Table 2 show that the embedding-based article-linking method performs very well within the Gazetteers corpus, achieving high precision and recall even in the presence of place name ambiguity. This confirms that semantic-similarity signals derived from the article descriptions are effective for identifying diachronic records referring to the same historical place.

External linking to Wikidata and DBpedia yields more varied results. Wikidata linkage shows moderate performance, with balanced precision and recall, while DBpedia linkage achieves notably high precision (0.89) but lower recall. These differences likely reflect the structural characteristics of the two knowledge bases: DBpedia tends to include fewer, less ambiguous entries per place name, making correct matches easier to identify, whereas Wikidata often contains multiple entities for the same or similar names, increasing the risk of false positives. Also, many external entities have very sparse or minimal descriptions, often only a single sentence, and in some cases none at all (e.g., the case of "Tenandry"[22] on Wikidata). Short or missing descriptions provide inadequate semantic context for embedding-based matching. Additionally, although most Gazetteer articles contain rich descriptive text, a small proportion consist only of redirect-style entries, further complicating the alignment process. Nevertheless, the ability to retrieve a substantial number of correct matches, particularly in a fully automated pipeline, supports broad coverage and enables downstream filtering or validation to improve overall quality.

5.4   Geotagging and Georesolution

To evaluate Stanza's ability to identify referenced places within article descriptions, we manually annotated 50 articles. These articles were randomly sampled and restricted to descriptions of no more than 200 tokens to facilitate efficient

---

[22] Tenandry on Wikidata—https://www.wikidata.org/wiki/Q123988596.

annotation. The resulting dataset contains 229 place mentions distributed across 2,275 tokens. We report precision, recall, and F1 score using the same evaluation metrics as in the concept linkage task. For overlapping or partial matches, we follow the method of Al-Olimat et al. [1], assigning 0.5 false positive and 0.5 false negative for each near match (e.g., "Edinburgh" vs. "Edinburgh Castle") to account for boundary ambiguity. Stanza achieves high precision (0.95), recall (0.83), and F1 score (0.89), demonstrating strong geotagging performance even in the presence of noisy OCR errors.

For georesolution, we manually assessed the predicted coordinates of both referenced and primary places of the 50 selected articles by comparing them against entries in GeoNames. Of the 159 predicted places, 125 (78.6%) were correctly resolved, demonstrating reasonable accuracy for automated coordinate assignment in historical texts.

6   Resource Availability, Reusability, Sustainability

The **MappingChange** KG adheres to the FAIR principles: it is findable (via Zenodo [34] and GitHub[23]), accessible (under open licences), interoperable (RDF/HTO-compliant), and reusable (supported by extensive documentation and a modular design). To facilitate public access and integration into diverse applications, we provide a permanent SPARQL endpoint. Furthermore, a pre-configured instance of the YASGUI interface[24] allows the community to query the KG directly without requiring local installation. Beyond the graph itself, all intermediate DataFrames generated during article extraction, concept linkage, and geospatial enrichment are archived on Zenodo. The GitHub repository also contains notebooks to ensure reproducibility and demonstrate the usage of both the DataFrames and the KG.

Complementing these technical access points, the **MappingChange** KG is also explorable via the actively maintained *Frances* semantic platform[25]. This user-friendly interface is designed for non-technical audiences, offering full-text and semantic search, RDF entity browsing, and timeline visualisations of concept evolution, all without requiring SPARQL expertise. The platform supports spatial analysis through map-based exploration, allowing users to toggle between historical[26] and modern[27] cartography for both close and distant reading.

Regarding sustainability, the resource is architected for extensibility and will be updated in tandem with related *Frances* collections. Maintenance is governed by open-source development practices, with all code and pipelines versioned in the public repository. To guarantee operational continuity and long-term access, the triple store and web services are hosted by the Edinburgh International Data Facility[28], ensuring a stable, institutionally supported environment.

---

[23] MappingChange GitHub—https://github.com/francesNLP/MappingChange.

[24] SPARQL Query Exploration UI—http://yasgui.org/short/0M7HXSruJg.

[25] Frances web platform—http://www.frances-ai.com.

[26] NLS Historical Map API—https://maps.nls.uk/projects/api/.

[27] OpenStreetMap—https://www.openstreetmap.org.

[28] EIDF—https://edinburgh-international-data-facility.ed.ac.uk.

7   Impact

We foresee **MappingChange** impacting several initiatives described below

7.1   Impact in Humanities Research

The Gazetteers of Scotland offer key textual evidence of 19th-century changes in Scottish places, yet their scale, inconsistent formatting, and page-level OCR make analysis difficult. **MappingChange** converts these unstructured volumes into a research-ready KG, making long narrative entries computable, comparable, and explorable over time.

For humanities scholars, this KG enables analyses previously impossible: tracing how places evolve across editions, tracking administrative and socio-economic changes, exploring spatial patterns like transport growth, and linking historical places to modern records. By converting narrative descriptions into queryable data, it supports reproducible humanities research.

Through its integration into the *Frances* platform, the KG becomes accessible to non-technical audiences. Historians, geographers, and community heritage groups can engage in semantic search, entity browsing, and timeline visualisation without requiring SPARQL expertise. *Frances* effectively translates the complex underlying graph into a practical research instrument, supporting both the close reading of individual entries and the distant reading of trends spanning decades.

7.2   Impact in Digital Humanities

The KG also provides a reusable foundation for future natural language processing and text-mining research on historical documents, specifically on extended named-entity recognition (e.g., persons, temporal expressions, events), relation extraction, and question answering. The KG enables the development of new analytic tools and visualizations for both close and distant reading, such as GIS-based spatial change visualisation with timeline displays, or heatmaps of demographic and infrastructural development.

7.3   Broader Applications

Beyond research, **MappingChange** offers a versatile basis for developers, educators, and cultural-heritage organisations to create new digital applications, including interactive learning environments, archival portals, museum displays, and historical visualisations. Moreover, our pipeline establishes a generalisable workflow for publishing historical texts as Linked Open Data. Its modular design allows institutions to easily adapt the method for other digitised corpora, such as the *Encyclopaedia Britannica* [33]. In this way, the resource encourages the wider uptake of Semantic Web technologies and the sustainable, standards-based publication of heritage collections.

8   Conclusions

MappingChange transforms the structurally fragmented *Gazetteers of Scotland* (1803–1901) into a semantically enriched, queryable resource for historical and spatial analysis. By combining GPT-4-driven article extraction, semantic modeling via the HTO ontology, and geospatial enrichment, the resource supports both humanistic inquiry and computational reuse. It is openly released as DataFrames, RDF graphs, and Jupyter notebooks, and is deployed through the *Frances* platform and a public SPARQL endpoint. Data and code are accessible under open licenses via Zenodo and GitHub[29].

As future we plan to improve the matching with external entities on Wikidata and DBpedia by utilising OpenStreetMap as a geospatial intermediary. Moreover, we plan to expand the knowledge graph with additional entity types (e.g., people, events), integrate cartographic metadata, and link to complementary historical resources such as the *Encyclopaedia Britannica*. We anticipate that **MappingChange** will serve as a foundation for research in humanities and digital humanities, extending its impact well beyond academic settings.

Acknowledgements. This work was supported by the Royal Society of Edinburgh Small Research Grant (RSE 2024 Autumn Small Grant, award ID: 4954).

References

1. Al-Olimat, H., Thirunarayan, K., Shalin, V., Sheth, A.: Location name extraction from targeted text streams using gazetteer-based statistical language models. In: Bender, E.M., Derczynski, L., Isabelle, P. (eds.) Proceedings of the 27th International Conference on Computational Linguistics, Santa Fe, New Mexico, USA, August 2018, pp. 1986–1997. Association for Computational Linguistics (2018). https://aclanthology.org/C18-1169/
2. Alex, B., Byrne, K., Grover, C., Tobin, R.: Adapting the Edinburgh Geoparser for historical georeferencing. Int. J. Humanit. Comput. Assist. Res. **10**(1), 69–82 (2015). https://doi.org/10.3366/ijhac.2015.0136
3. Alex, B., Grover, C., Tobin, R., Oberlander, J.: Geoparsing historical and contemporary literary text set in the city of Edinburgh. Lang. Resour. Eval. **53**(4), 651–675 (2019)
4. Beretta, F.: A challenge for historical research: making data FAIR using a collaborative ontology management environment (OntoME). Semant. Web **12**(2), 279–294 (2021)
5. Boettiger, C.: rdflib: A high level wrapper around the redland package for common rdf applications (2018). https://doi.org/10.5281/zenodo.1098478
6. Bottini, T., et al.: Polifonia: a digital harmoniser for musical heritage knowledge. H2020 (2021)
7. Carriero, V.A., et al.: ArCo: the Italian Cultural Heritage knowledge graph. In: Ghidini, C., et al. (eds.) ISWC 2019. LNCS, vol. 11779, pp. 36–52. Springer, Cham (2019). https://doi.org/10.1007/978-3-030-30796-7_3

---

[29] See first page of this manuscript.

8. Clifford, J., Alex, B., Coates, C.M., Klein, E., Watson, A.: Geoparsing history: locating commodities in ten million pages of nineteenth-century sources. Hist. Meth. J. Quant. Interdiscip. Hist. **49**(3), 115–131 (2016)
9. Doerr, M.: The CIDOC conceptual reference model: An ontological approach to semantic interoperability of metadata. AI Mag. **24**(3), 75–92 (2003)
10. Filgueira, R., Grover, C., Terras, M., Alex, B.: Geoparsing the historical gazetteers of Scotland: accurately computing location in mass digitised texts. In: Proceedings of the 8th Workshop on Challenges in the Management of Large Corpora, pp. 24–30. European Language Resources Association (2020)
11. Filgueira, R., Grover, C., et al.: Extending defoe for the efficient analysis of historical texts at scale. In: 2021 IEEE 17th International Conference on eScience (eScience), October 2021, pp. 21–29 (2021). https://doi.org/10.1109/eScience51609.2021.00012
12. Filgueira, R., et al.: defoe: a spark-based toolbox for analysing digital historical textual data. In: 2019 15th International Conference on eScience (eScience), pp. 235–242 (2019). https://doi.org/10.1109/eScience.2019.00033
13. Groth, P., et al.: PROV-overview. Working group note, W3C (2013). https://www.w3.org/TR/prov-o/
14. Hawthorne, W., Williams, D., Rehberger, D.: Enslaved: peoples of the historical slave trade (2024). https://enslaved.org/
15. Hiebel, G., Doerr, M., Eide, O.: CRMgeo: a spatiotemporal extension of CIDOC-CRM. Int. J. Digit. Libr. **18**(4), 271–279 (2017). https://doi.org/10.1007/s00799-016-0192-4
16. Hu, X., et al.: Location reference recognition from texts: a survey and comparison. ACM Comput. Surv. **56**(5) (2023). https://doi.org/10.1145/3625819
17. Hyvönen, E.: Digital humanities on the semantic web: Sampo Model and portal series. Semant. Web **14**(4), 729–744 (2023). https://doi.org/10.3233/SW-223034, https://journals.sagepub.com/doi/abs/10.3233/SW-223034
18. Miles, A., Bechhofer, S.: SKOS simple knowledge organization system reference. W3C recommendation (2009)
19. Monteiro, B.R., Davis, C.A., Fonseca, F.: A survey on the geographic scope of textual documents. Comput. Geosci. **96**, 23–34 (2016). https://doi.org/10.1016/j.cageo.2016.07.017, https://www.sciencedirect.com/science/article/pii/S0098300416301972
20. Morales Tirado, A., et al.: Musical meetups knowledge graph (MMKG): a collection of evidence for historical social network analysis. In: European Semantic Web Conference, pp. 110–127. Springer, Cham (2024). https://doi.org/10.1007/978-3-031-60635-9_7
21. National Library of Scotland: Gazetteers of Scotland (2024). https://data.nls.uk/data/digitised-collections/gazetteers-of-scotland/. Accessed 2025
22. Oberbichler, S., Mauermann, J., Tran, T.T., González-Gallardo, C.E.: Studying model design biases in LLMs for multilingual historical newspaper extraction; the Messina earthquake case study. In: Balke, W.T., Golub, K., Manolopoulos, Y., Stefanidis, K., Zhang, Z. (eds.) Linking Theory and Practice of Digital Libraries. LNCS, pp. 263–286. Springer, Cham (2026). https://doi.org/10.1007/978-3-032-05409-8_16
23. OpenAI: GPT-4 technical report. arXiv arXiv:2303.08774 [cs.CL] (2023)
24. Palfray, T., Hébert, D., Nicolas, S., Tranouez, P., Paquet, T.: Logical segmentation for article extraction in digitized old newspapers. In: Proceedings of the 2012 ACM Symposium on Document Engineering, DocEng 2012, October 2012 (2012). https://doi.org/10.1145/2361354.2361383

25. Poveda-Villalón, M., Gómez-Pérez, A., Suárez-Figueroa, M.C.: OOPS! (OntOlogy Pitfall Scanner!): an on-line tool for ontology evaluation. Int. J. Semant. Web Inf. Syst. (IJSWIS) **10**(2), 7–34 (2014)
26. Qi, P., Zhang, Y., Zhang, Y., Bolton, J., Manning, C.D.: Stanza: a Python natural language processing toolkit for many human languages. In: Proceedings of the 58th Annual Meeting of the Association for Computational Linguistics: System Demonstrations (2020)
27. Schweter, S., Akbik, A.: FLERT: document-level features for named entity recognition. CoRR abs/2011.06993 (2020). https://arxiv.org/abs/2011.06993
28. Silva, A.L., Terra, A.L.: Cultural heritage on the semantic web: the Europeana data model. IFLA J. **50**, 93–107 (2024). https://doi.org/10.1177/03400352231202506
29. Simon, R., Barker, E., Isaksen, L., CaNamares, P.D.S.: Linked data annotation without the pointy brackets: introducing Recogito 2. J. Map Geogr. Libr. **13**(1), 111–132 (2017). https://doi.org/10.1080/15420353.2017.1307303
30. Yu, L., Charlton, A., Terras, M., Filgueira, R.: Advancing frances: new heritage textual ontology, enhanced knowledge graphs, and refined search capabilities. In: 20th IEEE International Conference on e-Science, e-Science 2024, Osaka, Japan, 16–20 September 2024, pp. 1–10. IEEE (2024). https://doi.org/10.1109/e-Science62913.2024.10678663
31. Yu, L., Filgueira, R.: Gazetteer_HTO: a knowledge graph for representing the Gazetteers of Scotland (1803–1901) in page-level following Heritage Textual Ontology (2024). https://doi.org/10.5281/zenodo.17751676, data set
32. Yu, L., Filgueira, R.: The heritage textual ontology (HTO) (2024). https://github.com/frances-ai/HeritageTextOntology, https://w3id.org/hto
33. Yu, L., Filgueira, R.: Frances++: LLM-based semantic enrichment and spatial graphs for digitized historical collections. In: 2025 IEEE International Conference on eScience (eScience), pp. 269–277 (2025). https://doi.org/10.1109/eScience65000.2025.00039
34. Yu, L., Filgueira, R.: MappingChange: a temporal and semantic knowledge base of the Scottish Gazetteers (1803–1901), May 2025. https://doi.org/10.5281/zenodo.15393935

CUBE-MT: A Cultural Benchmark for Multimodal Knowledge Graph Construction with Generative Models

Albert Meroño-Peñuela[1]([envelope]), Xin Fan Guo[2], Nitisha Jain[1], Filip Birčanin[1], Timothy Neate[1], Thomas van Erven[3], Sándor Daranyi[3], and Nasrine Olson[3]

[1] King's College London, 30 Aldwych, London, UK
`albert.merono@kcl.ac.uk`
[2] Imperial College London, Exhibition Rd, London, UK
[3] University of Borås, Allégatan 1, Borås, Sweden

Abstract. Cultural heritage institutions (GLAM) have a societal role of guaranteeing access to their collections independently of location or background, as a pledge for knowledge equity. However, around one billion people (15% of the world's population) experience some form of disability that hinders such access, especially when considering the multimedia gap: differences in the use of different types of media to convey content to different audiences. Knowledge Graphs are increasingly becoming more multimodal by supporting images and text. However, the extent to which they address the multimedia gap for people with disabilities is not well understood, due to the lack of appropriate cultural evaluation frameworks. To address this, here we propose CUBE-MT, a benchmark and dataset that leverages generative models to build Multimodal Knowledge Graphs (MMKGs) with surrogate, multimedia representations, that adapt to the sensory capacities of cultural heritage collection users. We extend the CUBE (CUltural BEnchmark for Text-to-Image models) and Muse-IT datasets (MuseIT is a Horizon Europe project on innovative technologies for broadening access to cultural heritage for people with disabilities: https://www.muse-it.eu/) for paintings to encompass 6 modalities (text, images, Braille, speech, music, and 3D models); a collection of prompts to account for their cultural awareness and diversity; and a dataset with the resulting MMKG mapped to Wikidata. We show usage and evaluate the effectiveness of our approach in: (1) a quantitative assessment of cultural diversity; (2) an expert survey; and (3) a user study with people with aphasia focusing on perceptual and comprehension differences between model-generated and original MMKG objects.

Keywords: Multimodal Knowledge Graphs · Cultural Heritage · Knowledge Access

M. Acosta et al. (Eds.): ESWC 2026, LNCS 16550, pp. 95–114, 2026.
https://doi.org/10.1007/978-3-032-25159-6_6

1   Introduction

Cultural heritage is commonly preserved and promoted by GLAM (Galleries, Libraries, Archives, and Museums) institutions with a mission to provide access to knowledge [18,47]. Large digital platforms like Europeana [42] and the Cultural Heritage Cloud[1] provide this access globally on the Web, including in machine-readable form [32]. The number of cultural heritage and GLAM collection knowledge graphs (KGs) has dramatically exploded in recent years, providing a wealth of semantically enriched cultural heritage data [5,11,12,16,21,38,52].

Despite such advances, contemporary KGs face a critical **multimedia gap**. In large cross-domain KGs such as Wikidata [55], this gap manifests as under-representation of media beyond text, for instance around 95% of Wikidata items lack even basic image representations (property *P18*)[2], and coverage of audio, video, tactile, and other modalities is nearly non-existent. This creates sample biases favouring one group of readers while magnifying discrimination across gender, ethnicity, age, geography, language, and disability [2,44]. The Wikimedia Foundation has acknowledged *"the potential of different forms of media to convey content to different audiences"* and recommends *"building the necessary technology to make free knowledge content accessible in various formats and support more diverse modes of consumption and contribution"*. The multimedia gap is especially problematic for people with disabilities. Around 1 billion people (15% of the world's population)[3] experience some form of disability, of which 110-190M people experience significant disabilities. Facilitating equal access to knowledge via modality alternatives (images for those who cannot hear, sound for those who cannot see, tactile representations for those with combined impairments) is a right granted by the UN's Universal Declaration of Human Rights[4]. Yet existing KG approaches, including recent multimodal efforts, fall short to serve these populations. For instance, Multimodal Knowledge Graphs (MMKGs) in [35] cover merely a few images.

To address these gaps, we propose **CUBE-MT (CUltural BEnchmark with Multimodal Transformations)**, a systematic framework that: (1) leverages multimodal generative models to automatically synthesize six modalities (text, images, Braille, speech, music, and 3D models) for cultural heritage concepts, thereby producing multimodal knowledge graphs that close multimedia gaps and support users with different sensory capabilities; and (2) systematically evaluates both the cultural awareness and cultural diversity of the resulting multimodal representations. Unlike prior work limited to images, CUBE-MT generates representations across modalities relevant for users with diverse disabilities

---

[1] See e.g. Commission Recommendation of 10.11.2021 on a common European data space for cultural heritage https://ec.europa.eu/newsroom/dae/redirection/document/80911.

[2] https://meta.wikimedia.org/wiki/Research:Recommending_Images_to_Wikipedia _Articles.

[3] https://www.worldbank.org/en/topic/disability.

[4] https://www.un.org/en/about-us/universal-declaration-of-human-rights.

while explicitly addressing cultural competence throughout the generation process. CUBE-MT builds upon and extends two complementary resources. First, we extend CUBE [33], a text-to-image (T2I) benchmark for evaluating cultural diversity and awareness, by adding 5 additional modalities (text descriptions, Braille, speech, music, and 3D models) across three domains (cuisine, landmarks, and art-cultural attire). Second, we extend the Muse-IT dataset [53], which originally made cultural heritage paintings accessible by converting them into Multimodal Digital Objects (MMDOs) through sound (sonification with harmonic mapping) and touch (haptification with vibration mapping). In total, our benchmark covers cultural artifacts generated across these 18 countries and the six modalities described above. This combined extension allows our benchmark to support the generation and evaluation of rich multimodal representations that capture diverse cultural aspects across countries and sensory channels. In summary, our contributions are:

- A systematic benchmark for assessing cultural awareness and diversity in multimodal knowledge graph generation and completion using generative models, including an extended evaluation framework for cultural diversity that goes beyond existing approaches
- An extension to the *modalities* supported by CUBE and Muse-IT, supporting 6 modalities: images, text, Braille, speech, music, and 3D—relevant for the provision of knowledge to other human senses (hearing, touch)
- A set of multimodal *prompts* to account for the cultural awareness and cultural diversity in generating data for those modalities
- A sample *dataset* resulting from one run of the benchmark with 10,942 multimodal representations linked to 802 unique Wikidata items for cultural awareness, and 9,600 representations for cultural diversity using publicly available generative models

Our contributions are organised as follows: Sect. 3 describes the pipeline for generating cultural artifacts using generative AI models. Section 4 evaluates the cultural diversity of generated multimodal content across models and modalities. Section 5 assesses the cultural awareness and accuracy of generated artifacts through an expert survey and explores the utility of our resources for people with disabilities through a workshop with people with aphasia.

2   Related Work

Knowledge equity concerns ensuring *"inclusive and equitable quality education and promote lifelong learning opportunities for all"*[5], including people with disabilities. Cultural heritage institutions are key enablers for such learning, but biases in their collections and databases are well known and documented [60]. In the Semantic Web, Wikidata [55] follows the Wikimedia Foundation's vision of *"a world in which every single human being can freely share in the sum of all*

---

[5] UN's Sustainable Development Goal (SDG) 4: https://sdgs.un.org/goals/goal4.

knowledge" [27] and can be considered a large cultural heritage knowledge base. However, Wikidata may contain knowledge gaps, i.e. differences in reader, contributor or content coverage, which can magnify biases and favour discrimination across various dimensions (e.g. gender, recency, geographic, socio-economic [2]). Redi et al. [44] proposes a taxonomy of knowledge gaps that includes multimedia gaps as potential blockers for conveying content to different audiences. Multi-modality has been studied to address these gaps and comprehension in people with disabilities, by providing multimodal versions of podcasts (text and images) with generative models to aid understanding [10].

Generative models, thanks to the transformer architecture [54] and large-scale diffusion models [46], can generate realistic multimodal content as text, image, video, and audio [6,25,31,58], with state-of-the-art models for tasks like text-to-image, text-to-speech, text-to-audio and music. Multimodal generative models can be used for tasks such as acquisition, fusion and inference on Multimodal Knowledge Graphs [17]. However, these models are also under scrutiny due to potential harms [43]. Some of these harms can be seen as manifestations of cultural biases: the CUBE benchmark [33] aims to evaluate cultural competence of closed and open-source text-to-image models in cultural awareness and cultural diversity, and is evaluated combining different measures.

Cultural heritage knowledge graphs [11] are a primary example of the challenges around cultural biases and how more polyvocal approaches are needed [24]. [5] gives an account of existing GLAM projects that support the publication of cultural heritage semantic data as Linked Data or KGs, typically led by large digital infrastructures like Europeana [42] and models like CIDOC-CRM [15] and EDM [22]. These models are typically extended to model specific cultural heritage collections, e.g. Italian [16] and Dutch [12] cultural heritage. Knowledge graphs of music [38] and its metadata [8,20] are representatives of a modality rarely considered within KGs. More commonly, MMKGs focus on the modalities of text and literals [29] and are often used to benchmark KG embedding models [30]. Beyond literals, MMPedia [57] is an image-rich MMKG that grounds entities in KGs on images found in DBpedia [7] and the Web. In [3], the closest work in relation to ours, the authors propose a KG-based RAG architecture to derive prompts from Wikidata that are then used to generate missing images in item pages of fictional characters. Despite their relevance, none of these approaches address the issue of benchmarking cultural competence in MMKGs with multiple modalities, beyond literals and text, to address multimedia gaps.

3   The CUBE-MT Benchmark and Dataset

In this section, we describe CUBE-MT's construction process and properties.

3.1   Benchmark Construction

CUBE-MT extends CUBE [33], an existing benchmark for evaluating cultural competence of Text-to-Image (T2I) models in cultural awareness and cultural diversity, together with the Muse-IT dataset. Overall, we generate artifacts spanning six modalities, including text, images, Braille, speech, music, and 3D models. These modalities were chosen to (i) support accessibility requirements (Braille and speech for users with visual impairments; 3D models for tactile exploration), (ii) align with dissemination practices in the GLAM sector (text, images, and music), and (iii) promote reproducibility by relying on publicly available generative models (Table 1). We generate modalities to assess the two cultural dimensions described below.

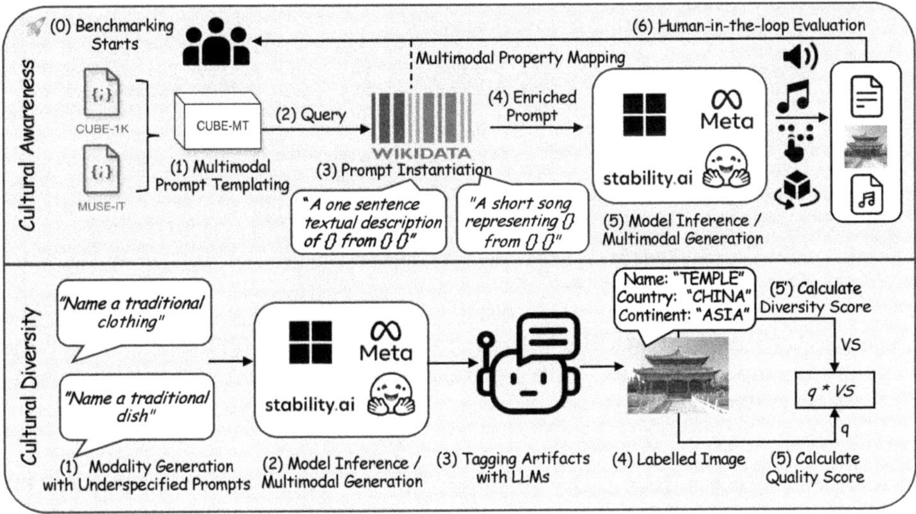

Fig. 1. The CUBE-MT Pipeline for Artifact Generation and Evaluation using Generative Models.

Cultural Awareness. Cultural awareness assesses a model's capacity to reliably and accurately portray objects and artifacts associated with specific cultures [33]. The upper panel of Fig. 1 illustrates our pipeline for collecting entities and evaluating cultural awareness across diverse modalities, extending the methodology introduced in the CUBE benchmark [33]. The pipeline integrates two primary data sources. First, we leverage CUBE-1K, a curated dataset of high-quality prompts representing widely recognized cultural artifacts linked to grounded knowledge bases (Step 1, top panel of Fig. 1). Each prompt targets a specific entity (e.g., "A high resolution image of *carne de panela* from Brazilian cuisine, realistic") and is associated with its unique Wikidata item identifier (Q-id) and relevant properties (an example entry is given in Listing 1.1). Second, we map artwork entities from the Muse-IT dataset to their corresponding Wikidata Q-ids. This alignment is achieved by traversing the Wikidata graph structure

```
1   {
2     "P31": "['Q57831']",
3     "P279": "[]",
4     "P495": "[]",
5     "P17": "['Q17']",
6     "P2012": " ",
7     "P361": "[]",
8     "id": "Q71053154",
9     "name": "Kojinyama Fortress",
10    "country": "Japan",
11    "domain": "landmarks",
12    "prompt": "A panoramic view of Kojinyama Fortress in Japan,
            realistic",
13    "prompt_text": "A one sentence textual description of Kojinyama
            Fortress from Japanese landmarks",
14    "gen_text": "txt/Q71053154.txt",
15    "gen_braille": "braille/Q71053154.txt",
16    "prompt_speech": "Kojinyama Fortress, located in Nara Prefecture,
            dates back to the Nara period and is renowned for its elegant
            hilltop design and reflections of the surrounding beauty in
            its moats and gardens.",
17    "gen_speech": "speech/Q71053154.wav",
18    "prompt_image": "A panoramic view of Kojinyama Fortress in Japan,
            realistic",
19    "gen_image": "img/Q71053154.png",
20    "prompt_music": "A short song representing Kojinyama Fortress from
            Japanese landmarks",
21    "gen_music": "music/Q71053154.wav",
22    "gen_3d": "3d/Q71053154.glb"
23  }
```

Listing 1.1. Example of a CUBE-MT item based on Wikidata Q71053154, "Kojinyama Fortress". 'prompt_' and 'gen_' attributes contain instantiated modality prompts and paths to generated objects, respectively.

to identify corresponding entity linkages (Step 2, upper section of Fig. 1). The resulting unified CUBE-MT concept space encompasses 1,300 unique cultural artifacts spanning three domains: cuisine, art, and landmarks. The generated modalities cover 18 distinct countries, expanding upon the original 8 countries in CUBE [33] to include Austria, Denmark, Germany, Ireland, Mexico, the Netherlands, Russia, Spain, Switzerland, and the United Kingdom.

Subsequently, we adopt a RAG approach and initialize each prompt with its relevant properties and submit the enriched prompt[6] for multimodal generation (Steps 3 and 4). The queried models are listed in Table 1. Finally, the generated multi-modal entities are evaluated for faithfulness and realism, with the corresponding proof of use detailed in Sect. 5.1.

Cultural Diversity. Cultural diversity is defined as a model's ability to avoid oversimplified or homogenized representations of a culture. We benchmark cultural diversity to assess a model's ability to avoid stereotypical depictions when responding to "Under-specified" prompts [33]. Under-specified prompts aim to evaluate whether generative AI models exhibit geographical or cultural biases.

---

[6] https://github.com/albertmeronyo/CUBE-MT/blob/main/resources/Examples/README.md.

These prompts omit any cultural or geographic anchor, leaving the model free to default to any culture. For instance, "Give me a traditional dish" is an under-specified prompt (see footnote 6) with no cultural anchor, whereas "Give me an image of carne de panela from Brazilian cuisine" is a fully-specified cultural artifact prompt. Under-specified prompts are generated (Step 1, lower panel of Fig. 1) and propagated to open-source models hosted on Hugging Face for multi-modal generation (Step 2). For this study, we extend the CUBE framework [33] to include text modalities but exclude audio and music, as robust automatic metrics for these modalities based on human preference scores are currently unavailable. Consequently, we benchmark only text and image outputs. Next, the generated artifacts undergo an automated tagging process (Step 3). We parse the artifacts using GPT-4, enforcing the extraction of structured metadata: *Name*, *Country*, and *Continent* (Step 4). Finally, to assess diversity both qualitatively and quantitatively, we compute the Quality-Weighted Vendi Score ($q \times VS$) (Step 5). This metric integrates a diversity score (Vendi Score) with a quality score derived from human preference proxies, following the methodology described in CUBE [33]. The evaluation process is described in   Sect. 4.

3.2   Output Dataset and Properties

With its modalities, prompt templates, and RAG approach defined, CUBE-MT can be instantiated with any generative model and produce example MMKG outputs, in this study, we publish a dataset generated using the framework described below.

Models. Table 1 shows the generative models we use for each modality. For convenience, here we select models provided by the Hugging Face Serverless API[7]. Importantly, this list of models can (and should be) changed for each instantiation of the benchmark: the purpose of CUBE-MT is to offer a framework with replaceable models that facilitates their testing and evaluation. We encourage prospective users of CUBE-MT to experiment with other models and compare their results with ours (see Sect. 5).

Execution Dataset. Executing the CUBE-MT benchmark across 14 generative models yields 10,942 multimodal representations (images, text, Braille, speech, music, and 3D models) for 802 unique Wikidata/CUBE-1k/Muse-IT items for cultural awareness and 9,600 for cultural diversity. Figure 2 shows six example items and their corresponding multimodal outputs. The dataset is publicly available via Dataverse[8] and Hugging Face[9], with permanent DOIs generated automatically per item on deposit, e.g. Kojinyama Fortress at https://dataverse.museit.eu/dataset.xhtml?persistentId=doi:10.5072/FK2/L9BLYI. Links to all resources are listed in Table 2.

---

[7] https://huggingface.co/models.
[8] https://dataverse.museit.eu/dataverse/cube-mt.
[9] https://huggingface.co/datasets/albertmeronyo/CUBE-MT.

Table 1. Models employed for modality generation and cultural competence evaluation. The final column, #Application, specifies the application scenario, where (A) denotes Cultural Awareness, (D) denotes Cultural Diversity, and (B) denotes both.

Modality	Generative model	URL	#Application
Images	Qwen Image [56]	https://huggingface.co/Qwen/Qwen-Image	B
	Stability AI Stable Diffusion 3 Medium [26]	https://huggingface.co/stabilityai/stable-diffusion-3-medium-diffusers	B
	Flux.1 Schnell [34]	https://huggingface.co/black-forest-labs/FLUX.1-schnell	B
Text	Google Gemma 2 [48]	https://huggingface.co/google/gemma-2b	D
	Meta LLama 3.1 Instruct [23]	https://huggingface.co/meta-llama/Llama-3.1-8B-Instruct	D
	Qwen 2.5 Instruct [49]	https://huggingface.co/Qwen/Qwen2.5-7B-Instruct	D
	Microsoft Phi3 Mini4K Instruct [1]	https://huggingface.co/microsoft/Phi-3-mini-4k-instruct	A
	Qwen3-Next-80B [50]	https://huggingface.co/Qwen/Qwen3-Next-80B-A3B-Instruct	A
Speech	Meta FastSpeech 2 [45]	https://huggingface.co/facebook/fastspeech2-en-ljspeech	A
	Kokoro-82M	https://huggingface.co/hexgrad/Kokoro-82M	A
	Chatterbox	https://huggingface.co/ResembleAI/chatterbox	A
Music	Meta MusicGen Small 300M [19]	https://huggingface.co/facebook/musicgen-small	A
	MusicGen Melody Large	https://huggingface.co/ylacombe/musicgen-melody-large	A
3D	Tencent Hunyuan3D 2.0 [59]	https://huggingface.co/tencent/Hunyuan3D-2	A
Braille	Algorithmic (using `PyBraille`)	https://pypi.org/project/pybraille/	A

Table 2. Links to key resources of the CUBE-MT benchmark.

Meroño-Peñuela, A. et al. (2025). *CUBE-MT: A Cultural Benchmark for Multimodal Knowledge Graph Construction with Generative Models.* doi:10.5281/zenodo.15398577

Resource	Link
CUBE-MT benchmark	https://github.com/albertmeronyo/CUBE-MT
CUBE-MT output dataset	https://dataverse.museit.eu/dataverse/cube-mt
Portal page	https://museit.eu/landing-page
Dataset generation code	https://github.com/albertmeronyo/CUBE-MT/ blob/main/cultural_diversity/ gen_diverse_modalities.ipynb
Documentation and tutorials	https://github.com/albertmeronyo/CUBE-MT/ wiki
Full dump download	https://github.com/albertmeronyo/CUBE-MT/ archive/refs/heads/main.zip
Subset of evaluation items	https://drive.google.com/drive/folders/ 1a78A7DtmHog8fACzxOEonRlojpKPn7vy? usp=sharing
Example de-referenceable resource	https://dataverse.museit.eu/dataset.xhtml? persistentId=doi:10.5072/FK2/L9BLYI
Hugging Face & dataset card	https://huggingface.co/datasets/albertmeronyo/ CUBE-MT
Zenodo	https://doi.org/10.5281/zenodo.15398577

4 Benchmarking Cultural Diversity

In this section, we present an automated framework to evaluate the cultural diversity of outputs from generative models, extending the methodology established in CUBE [33] to include text modalities.

Quality-Weighted Vendi Score. We employ the Quality-Weighted Vendi Score (qVS) to simultaneously assess the quality and diversity of generated artifacts [28,41]. While the quality measure is defined by human preference scores specific to the modality, the diversity calculation relies on a composite similarity kernel. We utilize the kernel $k(x_i, x_j)$ defined in CUBE, which is weighted by terms w_1, w_2, and w_3, corresponding to continent, country, and artifact similarity, respectively. We adopt the following weight configurations to enforce preferences for distinct cultural aspects from CUBE [33]:

- **Continent-level diversity:** ($w_1 = 1, w_2 = 0, w_3 = 0$) Evaluates the distribution of generated artifacts across continents.

- **Country-level diversity**: $(w_1 = 0, w_2 = 1, w_3 = 0)$. Evaluates the distribution of generated artifacts across countries.
- **Artifact-level diversity**: $(w_1 = 0, w_2 = 0, w_3 = 1)$. Considers only the distinctness of the artifacts, disregarding geographical origin.
- **Hierarchical geographical diversity**: $(w_1 = 1/2, w_2 = 1/2, w_3 = 0)$. Captures a hierarchical notion of diversity where both continent and country similarities are penalized equally.
- **Uniformly weighted diversity**: $(w_1 = 1/3, w_2 = 1/3, w_3 = 1/3)$. Provides equal weight to all three similarity dimensions.

The score qVS ranges from 0 to 1 (inclusive), where 0 indicates no diversity and poor quality, and 1 represents maximum diversity weighted by quality. Table 3 presents results, with quality and diversity separately, providing a diversity-focused evaluation (we also make available weighted scores that incorporate quality into the diversity metric[10]).

Text-to-Image. Following CUBE [33], we assess cultural diversity across three concepts (Art, Cuisine, Landmarks) using under-specified prompts. For each concept, we generate 400 images by combining 5 prompt template variations (see footnote 6) with 10 seed batches (seeds 0–79, 8 images per batch). GPT-4o is used to identify the artifact name and corresponding country for each generated image, after which mSigLIP-based retrieval is applied to obtain the top-5 most similar reference images from CUBE space. Cultural diversity is measured using the quality-weighted Vendi Score (qVS), which integrates artifact diversity (computed via kernel-based similarity) with HPS-v2 image quality scores. We present the mean breakdown of quality and VS component values, averaged over repetitions, in Table 3 (top).

Text-to-Text. For text generation, we follow the established workflow but exclude image generation, directly prompting models for cultural concepts (Name, Country, Origin). We employ PairRM to calculate the quality score $s(x)$ for the qVS metric. Serving as a proxy for textual quality, PairRM aligns with the visual metrics of the CUBE framework by leveraging training on human preference data. Results are shown in Table 3.

Results. We observe that contemporary T2I and T2T models exhibit limited geo-continental diversity, with T2T models showing a relative advantage in artifact diversity under under-specified prompts. The highest T2T score (≈ 0.78) nonetheless falls below the peak T2I artifact-level score, suggesting image generation encodes comparatively less artifact-level bias when prompts lack cultural specificity. Among the tested T2I models, artifact diversity is high (Flux: ≈ 0.98) but continental diversity remains poor (max: ≈ 0.39), indicating that

---

[10] https://github.com/albertmeronyo/CUBE-MT/blob/main/resources/Results/additional_tables.md.

Table 3. Breakdown of the mean quality component (q) and mean diversity component ($\overline{VS}$), averaged over repetitions. Higher scores indicate better performance. The score ranges from 0 to 1 (inclusive).

	Cuisine			Landmarks			Art (CUBE)			Art (MUSE-IT)		
	M1	M2	M3	M1	M2	M3	M1	M2	M3	M1	M2	M3
T2I (Image): M1 = Flux 1 Schnell, M2 = Qwen Image, M3 = Stable Diffusion 3 Medium												
q ($\rightarrow$)	0.2602	**0.2826**	0.2075	0.2409	**0.2876**	0.2605	0.1937	**0.2908**	0.2623	0.2152	**0.2780**	0.2623
$\overline{VS}(w_1, w_2, w_3)$												
$\overline{VS}$ (1, 0, 0)	0.2513	0.2616	**0.3852**	**0.2523**	0.2201	0.2047	0.1480	**0.1913**	0.1886	**0.3108**	0.2550	0.1965
$\overline{VS}$ (0, 1, 0)	0.6872	0.5048	**0.7079**	**0.6718**	0.4024	0.5738	**0.5857**	0.4666	0.4679	**0.7635**	0.5216	0.5737
$\overline{VS}$ (0, 0, 1)	**0.9561**	0.7385	0.9514	**0.9157**	0.7110	0.8979	**0.8414**	0.7465	0.7575	**0.9773**	0.7308	0.8224
$\overline{VS}$ ($\frac{1}{2}, \frac{1}{2}, 0$)	0.5503	0.4344	**0.6235**	**0.5309**	0.3515	0.4495	**0.4126**	0.3939	0.3725	**0.6240**	0.4427	0.4454
$\overline{VS}$ ($\frac{1}{3}, \frac{1}{3}, \frac{1}{3}$)	0.7597	0.5903	**0.7918**	**0.7243**	0.5186	0.6669	**0.6125**	0.5715	0.5570	**0.8084**	0.6002	0.6414
T2T (Text): M1 = Google Gemma 2, M2 = Meta LLama 3.1 Instruct, M3 = Qwen 2.5 Instruct												
q ($\rightarrow$)	0.0143	0.0127	**0.0786**	0.0155	0.0238	**0.1448**	0.0128	0.0190	**0.0727**	0.0120	0.0108	**0.0621**
$\overline{VS}(w_1, w_2, w_3)$												
$\overline{VS}$ (1, 0, 0)	0.3184	**0.5364**	0.1905	0.2345	**0.3369**	0.1752	0.2730	**0.2752**	0.1262	0.1626	**0.1822**	0.1326
$\overline{VS}$ (0, 1, 0)	0.4119	**0.7785**	0.2350	**0.3529**	0.3369	0.1845	0.4091	**0.4308**	0.1378	0.3477	**0.4387**	0.1821
$\overline{VS}$ (0, 0, 1)	0.5023	**0.7785**	0.2949	0.3874	**0.4172**	0.2009	0.4130	**0.5120**	0.1434	0.3546	**0.4387**	0.2164
$\overline{VS}$ ($\frac{1}{2}, \frac{1}{2}, 0$)	0.4122	**0.7255**	0.2241	0.3249	**0.3369**	0.1823	**0.3940**	0.3917	0.1348	0.2938	**0.3651**	0.1688
$\overline{VS}$ ($\frac{1}{3}, \frac{1}{3}, \frac{1}{3}$)	0.4672	**0.7551**	0.2621	0.3845	**0.3851**	0.1929	0.4371	**0.4698**	0.1404	0.3412	**0.4430**	0.1958

models recognise diverse cultural artifacts yet remain biased towards geographically homogeneous regions.

5   Use and Impact

We evaluate and show evidence of usefulness and adoption of the CUBE-MT resources in two ways: (1) by running a survey through the network of partners in a Horizon Europe project which have adopted the benchmark, asking questions about cultural awareness and accuracy of generative models; and (2) by running a workshop with people with disabilities (a diverse range of aphasia-related challenges) on strengths and weaknesses for CUBE-MT.

5.1   Survey for Evaluation of Cultural Awareness

We evaluate the accuracy and cultural awareness of the example dataset generated by the benchmark instantiated with the models of Sect. 3.2 with human experts. We collected a total of $N = 51$ responses from partners with a Semantic Web and/or Cultural Heritage background in the Horizon Europe Muse-IT project, covering a diverse set of artefacts and modalities (image, audio, 3D render). Responses were evenly distributed across the sample of 10 artefacts and their 3 modalities, ensuring a fair representation of perspectives.

Participants were asked to rate their level of agreement with a series of statements using a Likert scale from 1 (very low agreement) to 5 (very high agreement), with an additional option of 0 for 'Not relevant.' This allowed us to

Fig. 2. Multimodal transformations for the artefacts chosen for evaluation (shown here - Text, Image, 3D render and Braille; Speech and Music files available at https://huggingface.co/datasets/albertmeronyo/CUBE-MT.

capture both nuanced opinions and the relevance of each statement. A sample of results is illustrated in Fig. 3 (the complete set of results is provided at[11]). Survey respondents can be broadly characterised as holding expertise in semantic web research, cultural heritage, or related fields. We acknowledge that this constitutes a convenience sample of domain expertise whose familiarity with the project's objectives may introduce favourable disposition toward generated outputs. It was designed as a formative expert evaluation rather than an independent assessment, and results should be interpreted accordingly.

The multimodal representations were generally well received and considered useful. However, faithfulness ratings showed limited confidence in how accurately they reflected the original artefact texts, with few participants strongly agreeing. Emotional engagement was modest, with most responses indicating only low to moderate connection. Participants moderately agreed that key features were conveyed, though some important aspects may have been missed. Most also recognized the content as AI-generated, highlighting a clear distinction from human-created material. These findings suggest that while the representations are perceived as generally useful and informative, there is room for improvement in enhancing their faithfulness, emotional impact, and subtlety to make them more aligned with human expectations and experiences.

---

[11] https://github.com/albertmeronyo/CUBE-MT/blob/main/resources/workshop/survey_results.pdf.

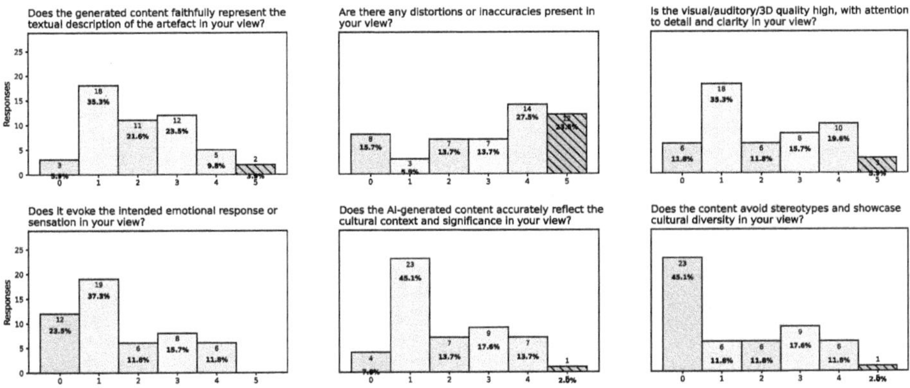

Fig. 3. Selection of questions from the online survey, where responses are expressed on a Likert scale ranging from 1 (strongly disagree) to 5 (strongly agree). The total sample size is 51 (for the full set of results). 0 indicates 'Not relevant'.

5.2   Workshop with Target Beneficiaries

Here we describe the user study held with a focus group at the APHASIA RE-CONNECT charity's premises in London, which offers peer befriending and training to support people living with aphasia. Aphasia is a language disorder frequently resulting from stroke, impacting approximately one in three stroke survivors and causing varied impairments in speaking, understanding, reading, and writing [9]. We focus on people with aphasia, for whom difficulties with recall and expression are common. Multimodal enhancements (e.g., images paired with text) can help mitigate these challenges by supporting comprehension of text-heavy documents and websites [10]. All modalities presented to workshop participants, including text descriptions, images, speech, music, and 3D-printed models were generated by the CUBE-MT pipeline.

We therefore chose CUBE-MT for its multimodality and follow a similar approach to [10] to explore (in)accuracies in its generated datasets. Three researchers (paper authors) facilitated the workshop, including two experienced Human-Computer Interaction (HCI) researchers with 8 years combined experience of supporting users with aphasia in design and testing technologies. They were responsible for guiding activities and probing participant reflections.

Participants. Five individuals with aphasia (mean age = 60.4, SD = 3.6) participated in the focus group (see Fig. 4). Their mean Aphasia Severity Rating (ASR) score was 3.2 (SD = 0.8), indicating mild to moderate aphasia. They showed relatively preserved naming abilities but had difficulties with complex sentence construction, verbal expression, and auditory comprehension. Participants relied on visual cues and context, experienced word-finding challenges, and required extra time to process and respond, but all were able to actively engage in the discussion. We deliberately used a small group size to better support the communication needs of people with aphasia. Larger groups often create additional barriers, as participants with aphasia usually need more time and focused support to express their views [36].

Fig. 4. Participants of the evaluation workshop (left); interactions with text and 3D printed models (right).

Study Method. The overall approach to this study draws from user-centred design, employing reflexive thematic analysis (TA) [13] to develop candidate topics/themes. These themes were then evaluated against three key criteria with the CUBE-MT dataset: (1) Are multimodal AI-generated formats helpful for people with aphasia? (2) Do they facilitate understanding of cultural objects? and (3) Does the presence of inaccuracies impact usefulness? This resulted in a final set of four themes, detailed in Table 4. We used a multimodal semantic guessing game, adapted from [51], to explore communication strategies in people with aphasia. Participants identified cultural artefacts (10 CUBE-MT items; Fig. 2) as modalities were gradually revealed (e.g., spoken description, images, 3D-printed object). By varying how guessable each artefact was, we observed and discussed communication breakdowns, compensatory strategies, and how different representations supported comprehension and expression in aphasia. After each round, participants reflected on which modalities helped or hindered them, highlighting how layering modalities can scaffold understanding of complex cultural content.

Analysis. The video-recorded sessions were transcribed using automated speech recognition (OpenWhisper), followed by manual review and correction. We then applied Thematic Analysis (TA) [14], an iterative method for identifying patterns and themes in qualitative data. Results are presented in Table 4. Following the six-step TA framework [14], we began by manually reading transcripts to identify recurring ideas. Initial observations were noted, generating insights relevant to the three requirements: (1), (2), and (3). Through systematic analysis,

24 codes were identified—each representing a label or phrase capturing explicit or implicit meanings tied to helpfulness, cultural understanding, or tolerance for inaccuracies. These codes were then grouped into broader patterns or themes. Four candidate themes were developed based on similarity, recurrence, and relevance. Themes were reviewed and refined for internal consistency, coherence, and clear differentiation. Final themes were explicitly defined, named, and summarized alongside corresponding codes to ensure analytical clarity and transparency.

Table 4. Initial Codes (24) Matched to Themes (4)

Theme	Codes Matched to Theme
Visual clarity and immediacy	Preference for images over text
	Immediate comprehension through visuals
	Practical understanding through 3D objects
	Easier cognitive load with images
	Initial attention to visual elements
	Images give clear cultural context
	Scale and physical attributes clear from visuals
Challenges of auditory representations	Ambiguity of auditory information
	Difficulty identifying culture from sound
	Music unclear for cultural identification
	Incorrect guesses from audio examples
	Auditory details often misleading
	Audio representations less informative than visual
	Cultural ambiguity due to unclear audio
Tolerance for inaccuracies	Visual inaccuracies better than absence of information
	Misleading visual details still somewhat useful
	Errors not affecting overall cultural understanding
	Tolerance higher for visual than textual inaccuracies
	Inaccuracies do not significantly hinder comprehension
Skepticism and trust issues	Skepticism about Wikipedia-style editing
	Concern about misinformation risk
	Trust issues limiting acceptance of AI outputs
	Questioning accuracy of textual information
	Suspicion toward AI reliability

Findings. Thematic analysis reveals clear strengths and limitations of CUBE-MT modalities for users with aphasia. Visual formats (3D objects, pictures) significantly enhance accessibility and understanding, supporting engagement with cultural objects. In contrast, auditory formats remain challenging and may require improvement or clearer integration with visual/textual context.

Accuracy, while important, did not critically affect usefulness as long as errors were minor. However, transparently acknowledging potential inaccuracies may be essential to maintain user trust.

Criteria 1: Are these AI-generated formats helpful for people with aphasia? Visual clarity and immediacy emerged as the most clearly helpful theme, indicating that images and 3D models significantly facilitate immediate understanding. Participants noted these formats reduced the cognitive load of reading textual descriptions. However, auditory formats were considerably less helpful, as they led to confusion and difficulty clearly interpreting cultural contexts.

Criteria 2: Do they make it easier to understand cultural objects? Participants clearly indicated improved understanding of cultural objects through visual representations (images, 3D) due to their clarity, tangible attributes, and immediate accessibility. Visual cues provided substantial support, especially where textual information was limited or cognitively demanding. Conversely, auditory representations did not simplify understanding; rather, they often increased ambiguity and uncertainty regarding the cultural objects discussed.

Criteria 3: Do inaccuracies in generations affect how useful these modalities are? Participants generally demonstrated tolerance for small inaccuracies, especially in visual formats, emphasizing that minor errors were not critically detrimental to overall usefulness. They preferred imperfect visuals over no information at all, suggesting pragmatic acceptance. However, they were less forgiving of inaccuracies in textual information and auditory contexts, highlighting skepticism and potential issues of misinformation and distrust. Clear accuracy expectations existed for text-based formats and cultural accuracy in sounds. Overall, minor inaccuracies in visuals were acceptable, whereas inaccuracies in textual and auditory forms significantly affected trust and perceived reliability.

6   Availability, Sustainability, and FAIRness

All CUBE-MT resources are available online[12], including the benchmark, dataset, and links to the original Wikidata items. All resources are created and made usable within the Wikidata namespace at https://www.wikidata.org/wiki/. Permanent URIs (DOIs) are created automatically when depositing to Dataverse for each individual item, and globally for the deposits in Zenodo, Datahub and Hugging Face as shown in Sect. 3.2. CUBE-MT is under version control on public GitHub repositories[13] (see Table 2). All resources are published under the CC-BY 4.0 licence and in accordance with usage of outputs of all generative models of Table 1. The storage of all resources on GitHub, Hugging Face and Dataverse guarantees their persistence beyond the project.

---

[12] https://dataverse.museit.eu/dataverse/cube-mt.
[13] https://github.com/albertmeronyo/CUBE-MT.

7 Conclusion

CUBE-MT extends existing benchmarks with 6 modalities, corresponding prompts, and a Wikidata-based RAG architecture for MMKG construction. Expert evaluation identifies cultural inaccuracies and diversity limitations in current generative models. The aphasia workshop demonstrates that image and tactile 3D representations hold considerable promise for improving cultural heritage accessibility; users showed greater tolerance for inaccuracies in visual than in auditory modalities and recognised content as AI-generated across all conditions. Future work will extend the workshop methodology to users with visual and hearing impairments, both populations under active study in the Muse-IT project. Automated evaluation metrics besides weighted Vendi scores will be investigated to complement human assessment [3], and governance over CUBE-MT outputs will be strengthened through AI-readiness standards including the MLCommons Croissant format [4,39,40]. LLM-based evaluation offers a promising direction for scaling preference simulation at the scope that MMKGs require [37].

Acknowledgements. This work was supported by the HE project Muse-IT, co-funded by the European Union under Grant Agreement No. 101061441. The views and opinions expressed in this work are those of the authors and do not necessarily reflect the views of the European Union or the European Research Executive Agency. This project has received funding from the EPSRC through the CA11y project (EP/X012395/1) and UKRI Centre for Doctoral Training in Safe and Trusted Artificial Intelligence (EP/S023356/1). We thank the anonymous reviewers for their feedback, and Barbara McGilivray for her valuable help.

References

1. Abdin, M., et al.: Phi-3 technical report: a highly capable language model locally on your phone (2024). https://arxiv.org/abs/2404.14219
2. Abián, D., Meroño-Peñuela, A., Simperl, E.: An analysis of content gaps versus user needs in the wikidata knowledge graph. In: International Semantic Web Conference, pp. 354–374. Springer (2022)
3. Ahmad, R.A., et al.: Draw me like my triples: leveraging generative AI for wikidata image completion. In: The 4th Wikidata Workshop (2023)
4. Akhtar, M., et al.: Croissant: a metadata format for ml-ready datasets. Adv. Neural. Inf. Process. Syst. **37**, 82133–82148 (2024)
5. Alexiev, V.: Museum linked open data: ontologies, datasets, projects. Digital Presentation and Preservation of Cultural and Scientific Heritage (VIII), 19–50 (2018)
6. Anil, R., et al.: Gemini: a family of highly capable multimodal models. CoRR **abs/2312.11805** (2023). https://doi.org/10.48550/ARXIV.2312.11805
7. Auer, S., Bizer, C., Kobilarov, G., Lehmann, J., Cyganiak, R., Ives, Z.: DBPEDIA: a nucleus for a web of open data. In: International Semantic Web Conference, pp. 722–735. Springer (2007)
8. de Berardinis, J., et al.: The polifonia ontology network: Building a semantic backbone for musical heritage. In: International Semantic Web Conference, pp. 302–322. Springer (2023)

9. Berthier, M.L.: Poststroke aphasia: epidemiology, pathophysiology and treatment. Drugs Aging **22**, 163–182 (2005)
10. Bircanin, F., et al.: Sounds accessible: envisioning accessible audio-media futures with people with aphasia. In: Proceedings of the 2025 CHI Conference on Human Factors in Computing Systems. CHI '25, Association for Computing Machinery, New York (2025). https://doi.org/10.1145/3706598.3714000
11. de Boer, V.: Knowledge graphs for cultural heritage and digital humanities. In: SUMAC@ ACM Multimedia, p. 3 (2023)
12. de Boer, V., et al.: Amsterdam museum linked open data. Semantic Web **4**(3), 237–243 (2013)
13. Bowman, R., Nadal, C., Morrissey, K., Thieme, A., Doherty, G.: Using thematic analysis in healthcare HCI at chi: a scoping review. In: Proceedings of the 2023 CHI Conference on Human Factors in Computing Systems, pp. 1–18 (2023)
14. Braun, V., Clarke, V.: Reflecting on reflexive thematic analysis. Qual. Res. Sport Ex. Heal. **11**(4), 589–597 (2019)
15. Bruseker, G., Carboni, N., Guillem, A.: Cultural heritage data management: the role of formal ontology and CIDOC CRM. In: Heritage and Archaeology in the Digital Age: Acquisition, Curation, and Dissemination of Spatial Cultural Heritage Data, pp. 93–131 (2017)
16. Carriero, V.A., et al.: Arco: the Italian cultural heritage knowledge graph. In: The Semantic Web–ISWC 2019: 18th International Semantic Web Conference, pp. 36–52, Auckland, New Zealand, October 26–30, 2019, Proceedings, Part II 18. Springer (2019)
17. Chen, Z., et al.: Knowledge graphs meet multi-modal learning: a comprehensive survey. CoRR **abs/2402.05391** (2024). https://doi.org/10.48550/ARXIV.2402.05391
18. Cole, J.B., Lott, L.L.: Diversity, equity, accessibility, and inclusion in museums. Rowman & Littlefield (2019)
19. Copet, J., et al.: Simple and controllable music generation (2024). https://arxiv.org/abs/2306.05284
20. Daquino, M., et al.: Characterizing the landscape of musical data on the web: state of the art and challenges (2017)
21. Daquino, M., Mambelli, F., Peroni, S., Tomasi, F., Vitali, F.: Enhancing semantic expressivity in the cultural heritage domain: exposing the zeri photo archive as linked open data. J. Comput. Cult. Heritage (JOCCH) **10**(4), 1–21 (2017)
22. Doerr, M., Gradmann, S., Hennicke, S., Isaac, A., Meghini, C., Van de Sompel, H.: The Europeana data model (EDM): object representations, context and semantics. Gothenburg; 76TH IFLA GENERAL CONFERENCE AND ASSEMBLY (2010)
23. Dubey, A., et al.: The llama 3 herd of models. arXiv e-prints pp. arXiv–2407 (2024)
24. van Erp, M., de Boer, V.: A polyvocal and contextualised semantic web. In: The Semantic Web: 18th International Conference, ESWC 2021, Virtual Event, June 6–10, 2021, Proceedings 18, pp. 506–512. Springer (2021)
25. Esser, P., et al.: Scaling rectified flow transformers for high-resolution image synthesis. In: Forty-first International Conference on Machine Learning, ICML 2024, Vienna, Austria, July 21-27, 2024. OpenReview.net (2024). https://openreview.net/forum?id=FPnUhsQJ5B
26. Esser, P., et al.: Scaling rectified flow transformers for high-resolution image synthesis (2024). https://arxiv.org/abs/2403.03206
27. Foundation, W.: Wikimedia vision. https://wikimediafoundation.org/about/vision/ (2018), retrieved 9 May 2025

28. Friedman, D., Dieng, A.B.: The vendi score: a diversity evaluation metric for machine learning. arXiv preprint arXiv:2210.02410 (2022)
29. Gesese, G.A., Alam, M., Sack, H.: Literallywikidata-a benchmark for knowledge graph completion using literals. In: The Semantic Web–ISWC 2021: 20th International Semantic Web Conference, ISWC 2021, Virtual Event, October 24–28, 2021, Proceedings 20, pp. 511–527. Springer (2021)
30. Gesese, G.A., Biswas, R., Alam, M., Sack, H.: A survey on knowledge graph embeddings with literals: which model links better literally? Semantic Web **12**(4), 617–647 (2021)
31. Girdhar, R., et al.: Imagebind one embedding space to bind them all. In: IEEE/CVF Conference on Computer Vision and Pattern Recognition, CVPR 2023, pp. 15180–15190, Vancouver, June 17-24, 2023. IEEE (2023). https://doi.org/10.1109/CVPR52729.2023.01457
32. Haslhofer, B., Isaac, A.: data.europeana.eu: the Europeana linked open data pilot. In: Proceedings of the International Conference on Dublin Core and Metadata Applications. Dublin Core Metadata Initiative (2011)
33. Kannen, N., et al.: Beyond aesthetics: cultural competence in text-to-image models (2024). https://arxiv.org/abs/2407.06863
34. Labs, B.F., et al.: Flux.1 kontext: Flow matching for in-context image generation and editing in latent space (2025). https://arxiv.org/abs/2506.15742
35. Liu, Y., Li, H., Garcia-Duran, A., Niepert, M., Onoro-Rubio, D., Rosenblum, D.S.: MMKG: multi-modal knowledge graphs (2019). https://arxiv.org/abs/1903.05485
36. Luck, A.M., Rose, M.L.: Interviewing people with aphasia: insights into method adjustments from a pilot study. Aphasiology **21**(2), 208–224 (2007)
37. Maier, B.F., et al.: LLMS reproduce human purchase intent via semantic similarity elicitation of likert ratings. arXiv preprint arXiv:2510.08338 (2025)
38. Meroño-Peñuela, A., et al.: The midi linked data cloud. In: The Semantic Web – ISWC 2017. Springer International Publishing, Cham (2017)
39. Meroño-Peñuela, A., Massey, J., Newman, A., Simperl, E.: How an AI-ready national data library would help UK science. arXiv preprint arXiv:2501.17013 (2025)
40. Meroño-Peñuela, A., Simperl, E., Kurteva, A., Reklos, I.: Kg. gov: knowledge graphs as the backbone of data governance in AI. J. Web Semantics **85**, 100847 (2025)
41. Nguyen, Q., Dieng, A.B.: Quality-weighted vendi scores and their application to diverse experimental design. arXiv preprint arXiv:2405.02449 (2024)
42. Purday, J.: Think culture: Europeana. eu from concept to construction (2009)
43. Quaye, J., et al.: Adversarial nibbler: an open red-teaming method for identifying diverse harms in text-to-image generation. In: Proceedings of the 2024 ACM Conference on Fairness, Accountability, and Transparency, pp. 388–406. FAccT '24, Association for Computing Machinery, New York (2024). https://doi.org/10.1145/3630106.3658913
44. Redi, M., Gerlach, M., Johnson, I., Morgan, J., Zia, L.: A taxonomy of knowledge gaps for wikimedia projects (second draft) (2021). https://arxiv.org/abs/2008.12314
45. Ren, Y., et al.: Fastspeech: fast, robust and controllable text to speech. In: NeurIPS 2019 (2019). https://www.microsoft.com/en-us/research/publication/fastspeech-fast-robust-and-controllable-text-to-speech/

46. Rombach, R., Blattmann, A., Lorenz, D., Esser, P., Ommer, B.: High-resolution image synthesis with latent diffusion models. In: IEEE/CVF Conference on Computer Vision and Pattern Recognition, CVPR 2022, pp. 10674–10685, New Orleans, June 18-24, 2022. IEEE (2022). https://doi.org/10.1109/CVPR52688.2022.01042
47. Sandell, R., Nightingale, E.: Museums, equality and social justice. Taylor & Francis (2012)
48. Team, G.: Gemma (2024). https://doi.org/10.34740/KAGGLE/M/3301, https://www.kaggle.com/m/3301
49. Team, Q.: Qwen2.5: a party of foundation models (2024). https://qwenlm.github.io/blog/qwen2.5/
50. Team, Q.: Qwen3 technical report (2025). https://arxiv.org/abs/2505.09388
51. Turnbull, D., Liu, R., Barrington, L., Lanckriet, G.R.: A game-based approach for collecting semantic annotations of music. In: ISMIR. vol. 7, pp. 535–538. Citeseer (2007)
52. Van Den Akker, C., et al.: Digital hermeneutics: agora and the online understanding of cultural heritage. In: Proceedings of the 3rd International Web Science Conference, pp. 1–7 (2011)
53. Van Erven, T., Darányi, S.: Turning paintings into multimodal digital objects. In: International Conference on Human-Computer Interaction, pp. 185–201. Springer (2025)
54. Vaswani, A., et al.: Attention is all you need. Adv. Neural Inf. Process. Syst. (2017)
55. Vrandečić, D., Krötzsch, M.: Wikidata: a free collaborative knowledgebase. Commun. ACM **57**(10), 78–85 (2014)
56. Wu, C., et al.: Qwen-image technical report (2025). https://arxiv.org/abs/2508.02324
57. Wu, Y., et al.: Mmpedia: a large-scale multi-modal knowledge graph. In: International Semantic Web Conference, pp. 18–37. Springer (2023)
58. Yang, Z., et al.: The dawn of LMMS: preliminary explorations with gpt-4v(ision). CoRR **abs/2309.17421** (2023). https://doi.org/10.48550/ARXIV.2309.17421
59. Zhao, Z., et al.: Hunyuan3d 2.0: scaling diffusion models for high resolution textured 3d assets generation (2025). https://arxiv.org/abs/2501.12202
60. Zhitomirsky-Geffet, M., Kizhner, I., Minster, S.: What do they make us see: a comparative study of cultural bias in online databases of two large museums. J. Document. **79**(2), 320–340 (2023)

EthereumKG: Building a Knowledge Graph of the Ethereum Blockchain

Juan Cano-Benito[1,2]($\boxtimes$) , Andrea Cimmino[2] , Sven Hertling[1] ,
Heiko Paulheim[1] , and Raúl García-Castro[2]

[1] Mannheim University, Mannheim, Baden-Württemberg, Germany
{juan.cano.de.benito,sven.hertling,heiko.paulheim}@uni-mannheim.de
[2] Universidad Politécnica de Madrid, Madrid, Spain
{juan.cano,andreajesus.cimmino,r.garcia}@upm.es

Abstract. This paper presents a method for transforming raw data
from the Ethereum blockchain (block and smart contract data) into a
knowledge graph. First, we extend current ontologies (Solidity ontology and EthOn ontology) by creating two new ontologies that model
Ethereum blocks and the Application Binary Interfaces (ABI), covering
all their properties. Based on these new and existing ontologies, we design
and implement a method to generate a knowledge graph based on semantic web standards that integrates Ethereum block, ABI and Solidity data.
The result is stored in a triple store that is organised into three interconnected knowledge graphs for blocks, ABIs, and Solidity contracts,
containing over 27 million triples. Finally, to demonstrate the applications of the Ethereum knowledge graph, three scenarios are presented.
In the first scenario, we explore the graph using SPARQL queries. In the
second scenario, we detect similar behaviour wallets that may reflect a
single entity. Finally, at the contract level, we use the ABI semantic data
to group contracts by functional similarity and detect identical contracts.

Keywords: Semantic Web · Knowledge Graph · Ontologies ·
Ethereum

1 Introduction

In the last decade, Distributed Ledger Technologies (DLT) have become a relevant technology used in industry, government, or research sectors [1]. Public DLT (known as blockchain), such as Ethereum, provides an open record of
financial transactions and smart contracts, creating a source of verifiable data.
However, the data in blockchains as Ethereum are fragmented into different
layers: block metadata and data (transactions and smart contracts). While the
metadata are in a Recursive Length Prefix (RLP), a special Ethereum format;
the ABI information is in JSON and the smart contracts are in Solidity (a programming language). This heterogeneity complicates tasks such as linking wallets

© The Author(s), under exclusive license to Springer Nature Switzerland AG 2026
M. Acosta et al. (Eds.): ESWC 2026, LNCS 16550, pp. 115–132, 2026.
https://doi.org/10.1007/978-3-032-25159-6_7

to behaviour patterns or identifying functionally related contracts [27]. Furthermore, the blockchain itself does not internally store smart contract code or ABI data, which can make Ethereum analysis even more complicated.

An approach to overcome these limitations is the combination of blockchain with semantic web technologies, allowing richer representation (by incorporating metadata or linking blockchain data to external information) or achieving semantic interoperability between blockchains [8,13,22]. The semantic web provides technologies for knowledge representation and interoperability, enabling machine-interpretable modelling of domains through ontologies and knowledge graphs. The combination of both technologies has been proposed to bridge the gap between blockchain transparency and semantic interoperability, allowing blockchain data to be accessed and analysed [7,24,25]. RDF is the standard model for describing semantic data.

In this article, we propose a method to build an Ethereum knowledge graph that addresses these limitations by aligning Ethereum data to a set of ontologies. Our knowledge graph integrates block and transaction data, smart contract ABIs, and smart contract source codes, aligning the data to new and existing Ethereum ontologies. The resulting knowledge graph is designed as a representation of a subset of Ethereum. The resulting knowledge graph allows tasks such as wallet similarity analysis and semantic clustering at the contract level, and is also stored in a triple-store that allows SPARQL queries.

The rest of the paper is organised as follows. Section 2 reviews the state of the art. In Sect. 3 two new ontologies are presented. Section 4 introduces the method for building an Ethereum Knowledge Graph. Next, Sect. 5 presents the knowledge graph and in Sect. 6 the Ethereum Knowledge Graph is evaluated. Finally, Sect. 7 presents the conclusions of this paper.

2   Related Work

One of the main challenges of combining the semantic web and blockchain lies in the lack of mechanisms that integrate both technologies [25]. Since different blockchain technologies differ in their implementation (metadata information, consensus algorithm, smart contract languages, etc.) [16], the semantic web provides a tool for achieving semantic interoperability between different implementations. Due to the potential of combining these technologies, researchers have focused on combining blockchain with the semantic web [7,23,25].

On the one hand, to build a knowledge graph, it is important that the data stored in a knowledge graph are expressed according to a semantic model, i.e., an ontology [11]. Therefore, ontologies are important when generating a knowledge graph. In [4,5,12,19] the works do not represent all data stored in an Ethereum block, since their purpose is not to replicate the structure of the blockchain, but rather its operational functioning. Some other works [3,9] represent part of the block elements, but not the entire block data. The ontology closest to this representation is EthOn [19], which is outdated, as new elements of the latest versions of Ethereum are missing. Therefore, there are no recent ontologies that cover all the properties of the Ethereum blockchain.

In the context of smart contracts, some of the main proposed contributions have been related to the representation of smart contract languages using ontologies [8,13], the auto-generation of smart contracts from domain-specific ontologies and semantic rules [10], the verification of smart contracts [18], or the improvement of smart contract discovery [2].

On the other hand, regarding knowledge graphs that represent blockchain technologies, Sopek et al. [26] introduce an approach that stores part of the Ethereum block data in RDF. However, they only store in the knowledge graph part of the block information, omitting transactions and smart contracts. Thus, our approach is the first knowledge graph that represents the blocks and smart contracts of a blockchain.

3   Extending Existing Ontologies

Building a knowledge graph of Ethereum requires a well-defined ontology to provide the underlying conceptual structure. This section aims to complement the Solidity ontology [8] with new ontologies that represent Ethereum blocks and the Application Binary Interface (ABI) code, as the knowledge graph data align with the ontology. The ontologies presented in this section were developed following the Linked Open Terms methodology [20]. This methodology, based on agile techniques, is composed of a four-stage workflow:

- **Ontology requirements specification.** The Solidity ontology needs to be extended with two new ontologies, one that represents the ABI of the contracts and another that represents the properties of Ethereum blocks. These ontologies are built for the version 1.16.6.
- **Ontology implementation.** This phase was split into three subtasks. First, the concepts of Ethereum blocks and ABI was extracted from the Ethereum documentation, and the relationships between these concepts were identified. Second, in the encoding step, the models were generated in the OWL language using Protégé. Finally, a group of experts supervised the ontologists to verify that it has no syntactic, modelling, or semantic errors and complies with all the requirements captured in the previous phase.
- **Ontology publication.** The ontologies are published in GitHub[1], a webpage[2] and Zenodo[3]. The ontologies include the code in OWL, a human-friendly documentation with a description of the classes and properties, and a graphical representation of the ontology.
- **Ontology maintenance.** The ontologies will be updated to fix possible errors or implement new requirements.

---

[1] https://github.com/oeg-upm/Ethereum-ontology.
  https://github.com/oeg-upm/Solidity-ABI-ontology.
[2] https://w3id.org/def/Ethereum.
  https://w3id.org/def/SolidityABI.
[3] https://zenodo.org/records/10719799.
  https://zenodo.org/records/10707050.

Figure 1 depicts an overview of the Ethereum ontology that models the Ethereum blockchain. The primary classes and namespaces are represented by *eth*, *abi* and *sol*. The eth namespace represents the Ethereum block ontology, explained in Sect. 3.1; the abi namespace represents the Application Binary Interface ontology, explained in Sect. 3.2; and finally, the sol namespace represents the Solidity ontology [8].

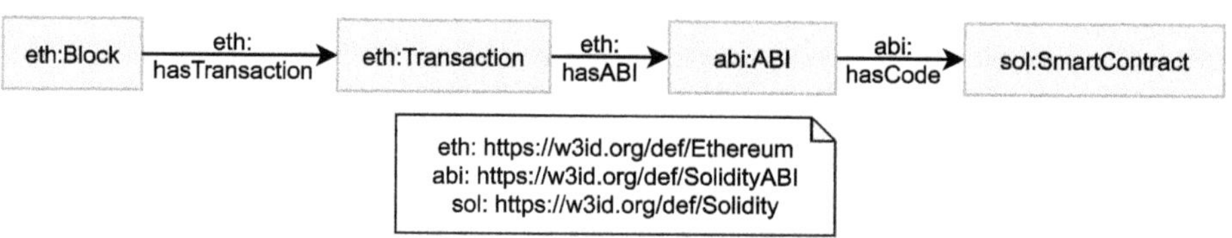

Fig. 1. Ethereum ontology overview.

The Ethereum Block, the ABI, and the Solidity ontology are linked to other ontologies. These ontologies and their elements are identified using Internationalized Resource Identifiers (IRIs). For the sake of simplicity, in the rest of this paper, these IRIs will be referenced by using the prefixes defined in Listing 1.1.

```
1  @PREFIX  sol:  <w3id.org/def/Solidity#>
2  @PREFIX  abi:  <w3id.org/def/SolidityABI#>
3  @PREFIX  eth:  <w3id.org/def/Ethereum#>
4  @PREFIX  xsd:  <www.w3.org/2001/XMLSchema#>
```

Listing 1.1. Predefined namespace prefixes

3.1   The Ethereum Block Ontology

Resource type: Ontology
License: MIT License
DOI: https://doi.org/10.5281/zenodo.10719799
URL: https://w3id.org/def/Ethereum

At the core of Ethereum technology are the blocks, which are the main element of a blockchain. Each block records recent transactions and is linked in a linear chronological order, forming a list of blocks, what is known as a blockchain. Figure 2 depicts the Ethereum block ontology.

The Ethereum block (*eth:Block*) contains the metadata and data of the block. The metadata includes the hash of the previous block, who creates the block, the time when the block was created, etc. The block stores the withdrawals produced in the block, represented by *eth:Withdrawal*. The access control system is shown through *eth:AccessList* and *eth:Key*. This system outlines the permissions and

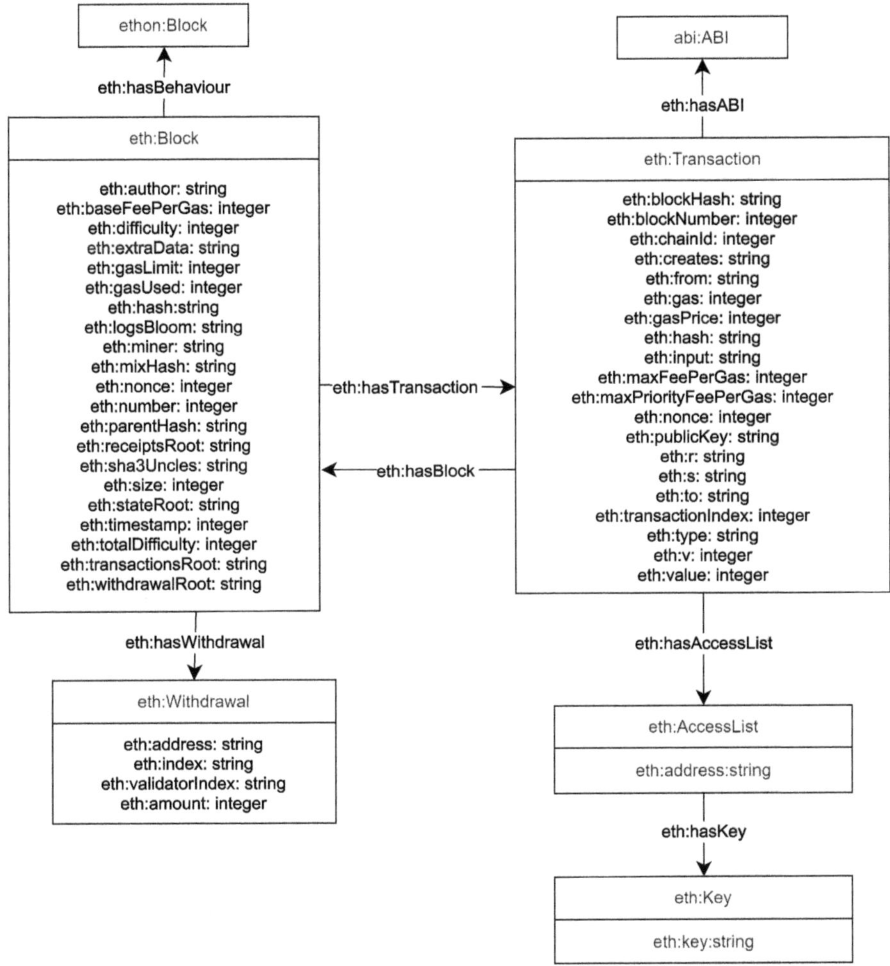

Fig. 2. Ethereum block ontology.

cryptographic keys required for specific actions and data accesses within the Ethereum ecosystem.

A transaction can be a regular transaction or a smart contract. The application binary interface (ABI) serves as an interface between smart contracts and the applications that invoke them. The ontology is related to another ontology that represents ABI code through the eth:hasABI property.

3.2 The Application Binary Interface Ontology

Resource type: Ontology
License: MIT License
DOI: https://doi.org/10.5281/zenodo.10707050

URL: https://w3id.org/def/SolidityABI

Smart contracts are written in high-level programming languages. However, in Ethereum, once these contracts are deployed, they are converted into bytecode, a form that is not directly understandable by external applications. Then, each smart contract generate its ABI. The main function of the ABI is to provide a structured way for applications to understand and interact with this bytecode. Figure 3 depicts the Ethereum ABI ontology. The presented ontology provides an overview of the components of an ABI.

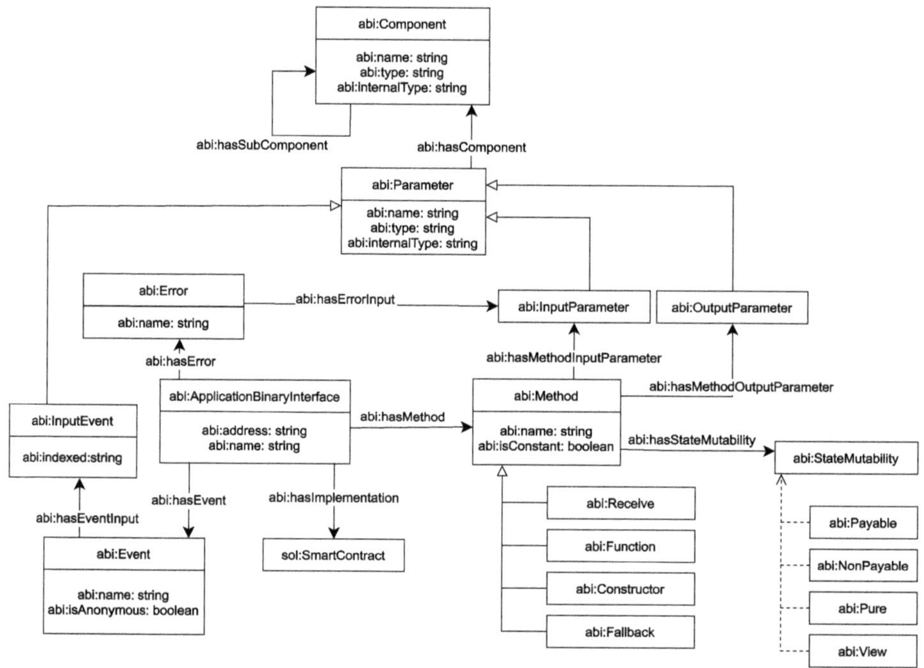

Fig. 3. Ethereum ABI ontology.

The core element of the ontology is the *abi:ApplicationBinaryInterface* class. This class is linked to a multiple of functionalities that represent the capabilities of smart contracts:

- Methods (*abi:Method*) captures specific functions that can be invoked within a smart contract. Each method has a name and is associated with certain properties, such as whether it is constant or state mutability. Methods can be functions, constructors, fallbacks, or receives.
- Events, represented by *abi:Event*, store data in the Ethereum block. Events have specific properties (the event name and a boolean flag for anonymity).
- The ABI also handles errors through *abi:Error*, providing a mechanism for transmitting error messages or conditions back to the caller.

- Components (*abi:Component*) are the representation of hashmaps (a data structure that is used to store and retrieve keys-based values). A component may encompass subcomponents, represented by the *abi:hasSubComponent* property.

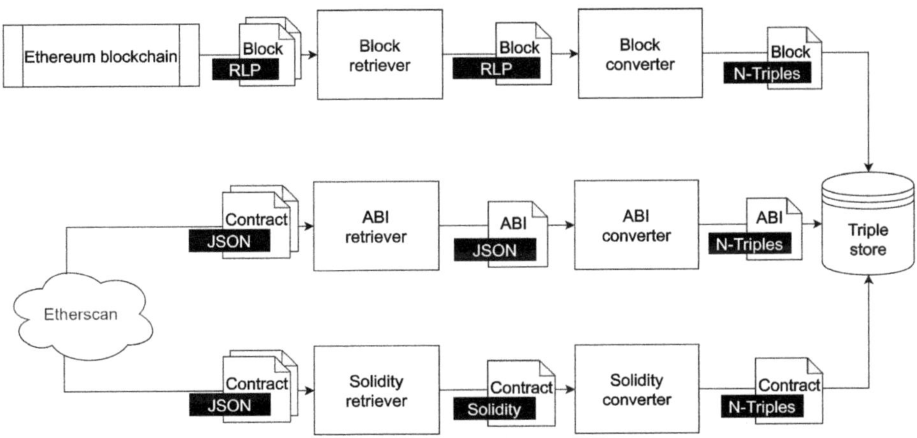

Fig. 4. Arquitecture overview.

4 Implementation of the Method

The scope of this section is to define a method to generate a knowledge graph from a corpus of valid blocks, ABIs, and Solidity smart contracts, aligning the semantic data generated with the defined ontologies. The method involves the following conceptual steps:

- **Data extraction:** A specific range of blocks is selected from the Ethereum blockchain, extracting the information from each block. Within these blocks, all smart contracts that have been deployed are identified. Since Ethereum does not store the ABI and Solidity code, an external third-party service is used to extract the ABI and Solidity code.
- **Data transformation:** Convert heterogeneous data into RDF aligned to the Ethereum ontologies. During this step, the ontologies developed in Sect. 3 are used.
- **Data storage:** The RDF generated is stored in a triple store. Although triple stores can require significant storage space compared to traditional databases [21], compression techniques are applied to reduce the storage size.

The process of building the Ethereum knowledge graph is depicted in Fig. 4 and the implementation is available online[4]. On the one hand, the first step

---

[4] https://github.com/oeg-upm/Sancus.

to implement the method is to convert the block information into RDF. These blocks are retrieved one by one using the *block retriever* and converted to RDF. After extracting the data from all the blocks and converting it to RDF, the semantic data is stored in a knowledge graph.

On the other hand, to recover contracts, as depicted in Fig. 5, once the block retriever has the block data, the software retrieved all the transactions inside a specific block; then, if the transactions is a smart contract, the ABI and the Solidity code are requested one by one to a third-party service (in this case we used Etherscan[5], an Ethereum block and smart contract explorer). Finally, this third-party service provides the ABI and the Solidity smart contract code. In our case, the third-party service returns a JSON file containing the ABI and the contract source code.

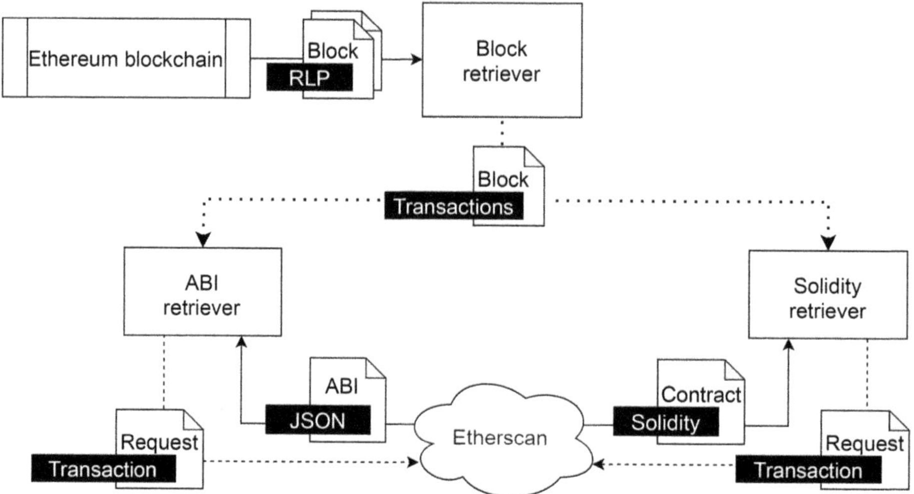

Fig. 5. Iteration to obtain block and smart contract information.

Therefore, in the second step, the ABI source code is retrieved from Etherscan using the *ABI retriever*, and then the ABI contracts are converted into RDF in the *ABI converter*. The third step is to retrieve the Solidity source code of the contracts provided by Etherscan by using the *Solidity retriever* to convert the smart contracts code into RDF with the *Solidity converter*. Finally, in the last step all the RDF generated are stored in the triple store.

---

[5] https://etherscan.io.

5   The Ethereum Knowledge Graph

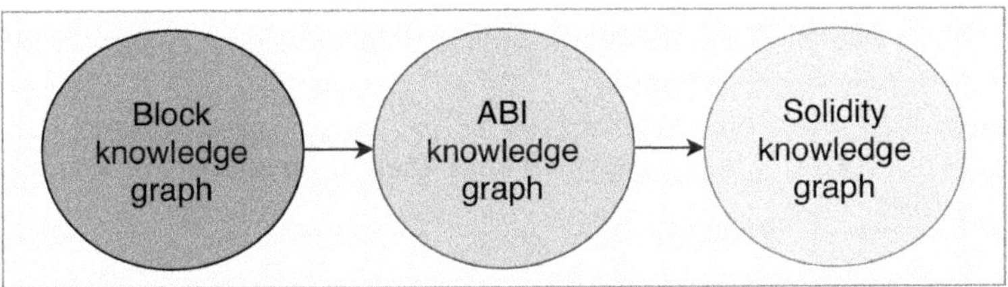

Fig. 6. Overview of the Ethereum knowledge graph.

The triple store contains three knowledge graphs, as depicted in Fig. 6 and is available in Zenodo[6] and in a triple store[7]. The knowledge graph contains block data, ABI data, and Solidity source code in RDF. The number of triples in the different knowledge graphs is shown in Table 1. The number of triples in the blocks is high, mainly due to the number of transactions in each block. While the number of triples stored in Solidity is about 15 million, the number of triples in ABI is only about 3 million. This is due to the granularity of the Solidity ontology, as opposed to the contracts stored in ABI.

Table 1. Triples comparison between the different knowledge graphs.

	Triples in blocks	Triples in ABI	Triples in Solidity
Number of triples	8,434,726	3,167,373	15,555,663

5.1   Block Retriever Implementation

First, a selected range of blocks is extracted from the Ethereum blockchain. The Ethereum block data is in a format called Recursive Length Prefix (RLP) and must be converted to RDF before storing it. This flow is depicted in Fig. 7. The main properties of the Ethereum block (such as the gas used, the hash of the block or the block number) are represented, as well as the transactions and the withdrawals.

1. **Block retriever.** The software extracts the blocks, one by one, from the Ethereum blockchain.
2. **Block converter** After receiving the block, the software extracts the data from the block, encoded in RLP. Using the Web3j library to read the RLP data, this data is extracted and converted to RDF.

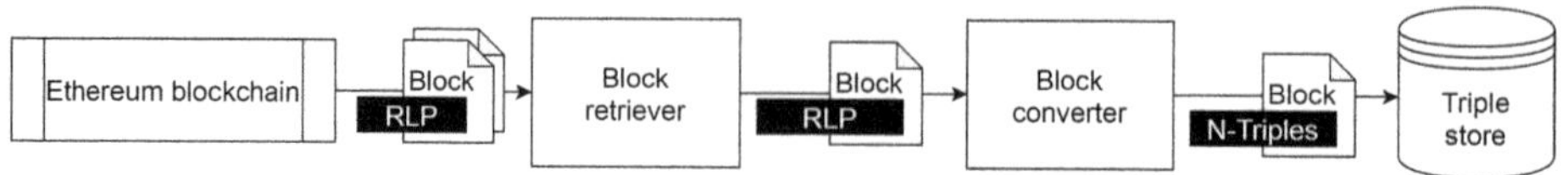

Fig. 7. Block module workflow in knowledge graphs.

The size of storing 2,608 blocks with all their information in a triple store is 8,434,726 triples.

5.2   ABI Retriever Implementation

Each block contains a list of transactions identified by a hash. As the information in Ethereum block does not indicate whether it is a smart contract, each hash is checked on a third-party platform (in this case, Etherscan was used[8]). The ABI data is in the JSON format. The information in this JSON is converted into RDF aligned with the ABI ontology. The flow is depicted in Fig. 8. Since contracts have associated licences defined in Etherscan by users, these licences have been added to the ABI knowledge graph.

1. **ABI retriever.** The software extracts the ABI in JSON from the third-party platform.
2. **ABI converter** The JSON data received is converted to RDF aligned to the ABI ontology.

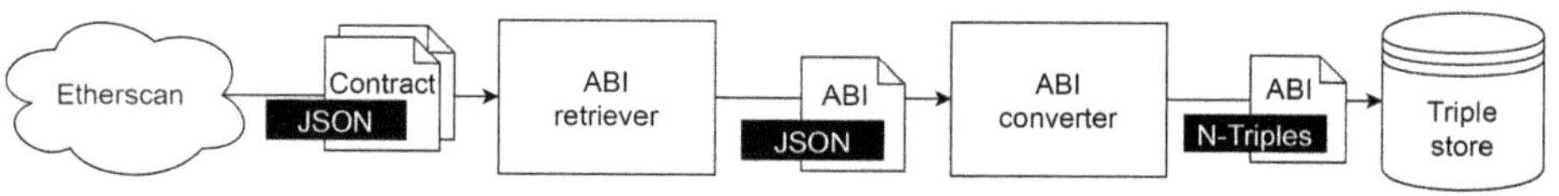

Fig. 8. ABI module workflow in Sancus.

The size of storing 8,283 ABIs with all their information in a triple store is 3,167,373 triples.

5.3   Solidity Retriever Implementation

As with the ABI, each hash is checked on a third-party platform. If the hash is a contract, the related Solidity smart contracts are retrieved (as an address may contain one or more contracts). The flow is depicted in Fig. 9.

---

[6] https://zenodo.org/records/10719604.
[7] https://ethereumkg.informatik.uni-mannheim.de/.
[8] https://etherscan.io/.

1. **Solidity retriever.** In this step, we request to the third-party service the source code from the correspondent smart contract address.
2. **Abstract syntax tree (AST) generation.** Once the Solidity source code is received, by using an LL(*) algorithm [17] and a Solidity grammar[9], the software processes the smart contract code and generates an abstract syntax tree (AST). The original AST contains all the syntactic elements of the source code, and this AST is converted to JSON.
3. **AST simplification.** The simplification stage consists of synthesising the JSON to the representative elements of a Solidity contract, eliminating superfluous nodes and retaining only those that are essential to represent the structure of the Solidity code.
4. **RDF generator.** The simplified JSON is transformed to RDF.

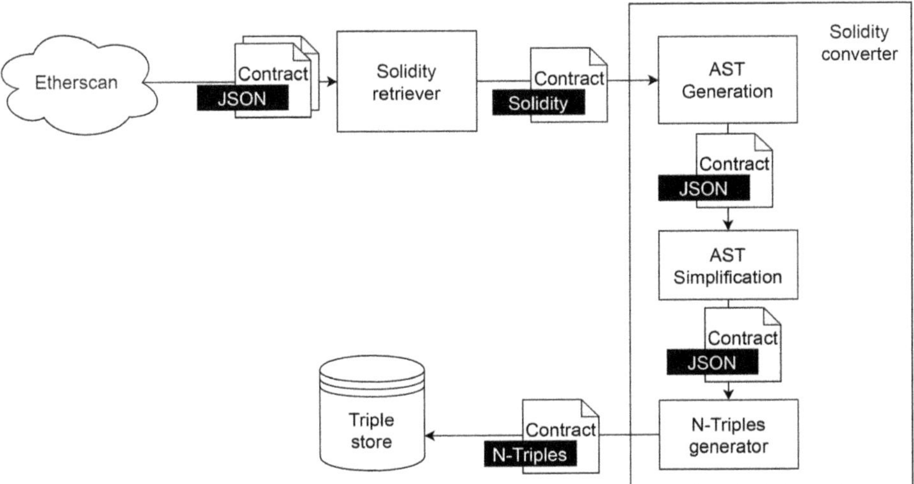

Fig. 9. Solidity module in Sancus.

Like ABI contracts, contracts have associated licences defined in Etherscan by users. The different licences stored in the extracted contracts can be found in the Table 2. The *MIT* licence is the most common explicit licence among the contracts analysed, but most of the deployed contracts do not specify a licence. This could indicate that developers are either not concerned about intellectual property or reuse of their code or are not aware of the importance of explicitly stating a licence.

Finally, the size of storing 94,364 smart contracts with all their information in a triple store is 15,555,663 triples.

---

[9] https://github.com/solidity-parser/antlr/blob/master/Solidity.g4.

Table 2. Number of licenses in all the smart contracts

Solidity license Type	Number of contracts
None	58,317
MIT	26,753
GNU GPLv3	3,533
Apache-2.0	1,725
Unlicense	1,348
GNU LGPLv3	1,331
GNU AGPLv3	813
GNU GPLv2	233
BSD-3-Clause	147
BSL 1.1	79
MPL-2.0	63
GNU LGPLv2.1	10
BSD-2-Clause	7
OSL-3.0	5

6   Usage of Knowledge Graph Data

- **Resource type:** Knowledge Graph
- **License:** CCA 4.0
- **DOI:** https://doi.org/10.5281/zenodo.17797524
- **URL:** https://zenodo.org/records/17797524

Based on the knowledge graph constructed from the Ethereum blockchain data, the analysis was divided into two parts. Firstly, we run two SPARQL queries on the knowledge graph that could not be performed by Ethereum itself. Secondly, we examine behaviour patterns at the address level to detect similar wallets (accounts). Finally, the ABI knowledge graph is used to group smart contracts by functional similarity or detect duplicated contracts.

6.1   Querying the Knowledge Graph

In the Listing 1.2, a query is performed to return the number of contracts deployed per hash. The results obtained by running this query are shown in Table 3. In this table, it is possible to see how a single user has deployed 213 contracts.

```
1   PREFIX eth: <https://w3id.org/def/Ethereum/>
2   SELECT ?creator (COUNT(DISTINCT ?abi) AS ?
        numContractsDeployed) WHERE {
3       SERVICE <http://kg.dlt.linkeddata.es/Blocks/query> {
4           GRAPH ?g {
5               ?block eth:hasTransaction ?transaction .
6               ?transaction eth:from ?creator .
7               ?transaction eth:hasABI ?abi .
8           }
9       }
10  } GROUP BY ?creator
11  ORDER BY DESC(?numContractsDeployed)
12  LIMIT 10
```

Listing 1.2. SPARQL query

Table 3. Number of contracts deployed by hash

Hash of the creator	Number of contracts deployed
0x75e89d5979e4f6fba9f97c104c2f0afb3f1dcb88	213
0x0d0707963952f2fba59dd06f2b425ace40b492fe	199
0x28c6c06298d514db089934071355e5743bf21d60	180
0xf16e9b0d03470827a95cdfd0cb8a8a3b46969b91	178
0xdfd5293d8e347dfe59e90efd55b2956a1343963d	161
0x21a31ee1afc51d94c2efccaa2092ad1028285549	157
0x6a116b531eb60d4f5e2ab31fbcaec5c962865f87	74
0x46340b20830761efd32832a74d7169b29feb9758	72
0xf89d7b9c864f589bbf53a82105107622b35eaa40	67
0x6cc5f688a315f3dc28a7781717a9a798a59fda7b	64

In the second SPARQL query, a query is made to retrieve the licence types that exist with contracts that have one of the most frequently used names by a specific user, as shown in the Listing 1.3. The results obtained in Table 4 show that different contracts with the same name have different licences, but almost half of the smart contracts do not have any type of licence attached. In conclusion, the distribution of licences suggests a possible lack of awareness or concern about the implications of software licences.

6.2   Wallets Similarity Detection

This experiment evaluates whether the knowledge graph can be used to infer relationships between accounts that have different addresses but may represent

```
1  PREFIX dc: <http://purl.org/dc/terms/>
2  PREFIX solidity: <https://w3id.org/def/Solidity#>
3  SELECT ?licenseType (COUNT(?licenseType) AS ?numLicenses)
       WHERE {
4      SERVICE <http://kg.dlt.linkeddata.es/Solidity/query> {
5          ?contract solidity:name "AdminUpgradeabilityProxy"
                   .
6          ?contract dc:license ?licenseType .
7      }
8  } GROUP BY ?licenseType
9  ORDER BY DESC(?numLicenses)
```

Listing 1.3. SPARQL query

Table 4. Different licenses and number of appearances for the contract "AdminUp-gradeabilityProxy"

Type of license	Number of appearances
MIT	47
None	42
Apache-2.0	1
GNU LGPLv3	1

the same entity. First, we generate random walks on the knowledge graph, starting only from the wallet nodes. These paths are then used as a corpus to train a skip-gram embedding model, resulting in a vector representation for each wallet that captures relations with contracts and tokens at the ABI level (methods and events). The similarity between wallets was calculated by the cosine similarity between their embedding vectors and the Jaccard similarity between their sets of destination addresses in the transaction graph. The result was a ranked list of wallet pairs. Only pairs that exceeded predefined similarity thresholds (cos_sim = 0,6 and jaccard similarity = 0,2) were used [15].

From the resulting similarity graph, connected components were extracted and interpreted as groups of potentially related wallets. The clustering process was based on the structural and semantic information embedded in the graph. Across the entire dataset, 191,146 wallets were represented, resulting in a similarity space in which approximately 4.5 million high-confidence paths were generated. After setting a threshold, the method produced 64,201 pairs of highly similar wallets and 3,492 multi-address groups. Although most of the groups contained between 2 and 5 addresses, several significantly larger components emerged, including groups with more than 100 addresses. These large components suggest patterns of shared ownership, automated activity, or highly similar activity between accounts.

The qualitative examination of these groups revealed recurring patterns of behaviour. Many groups consisted of addresses interacting with contracts that expose the same methods and events, which may suggest the presence of automated wallets. Since ABI information was included in the integration process, grouped wallets often converged not only on the same token contracts, but also on contracts that have the same methods (e.g., approve, deposit, allow, or initialise). The results of this section can be found online[10].

6.3   Contracts Semantic Analysis

We extended our analysis to smart contracts by leveraging the semantic information available in the ABI model. The experiment had two objectives: to identify families of contracts defined by a shared interface structure and to detect functionally identical or suspicious implementations that could indicate malicious behaviour, copy-and-paste contracts. Smart contracts embeddings were derived from semantics at the ABI level using gpuRDF2vec [6]. Each contract in the knowledge graph was represented by its exported interface elements, including functions, events, parameter types, and mutability attributes.

Contract embeddings were derived from ABI-level semantics. Each contract in the knowledge graph was represented by its exported interface elements, including functions, events, parameter types, and mutability attributes. This enabled a clustering approach that captures high-level functional equivalence rather than low-level syntactic identity. The resulting clusters revealed families of contracts that implement variations of the same protocol, including liquidity pools, proxy patterns, stake vaults, and ERC-based token wrappers. Some clusters contained a single contract, while others contained dozens of implementations that exposed identical or nearly identical ABI structures, suggesting the use of contracts widely used (such as OpenZeppelin contracts [14]) or copied contracts.

A second analysis was performed, focussing on duplication at the contract level by detecting interfaces that are strictly identical at the ABI level. Contracts that shared exactly the same methods, events, and mutability states were classified as implementations with identical interfaces. This technique captures copied implementations and repeated issuance of the same token or proxy template without the need to access the bytecode. A third dataset was generated by identifying the signatures of the functions that commonly appear in scam templates, such as unrestricted property transfers or externally controlled drain functions. However, this dataset is not based on execution traces, and no malicious contracts were found.

Finally, clusters of extremely similar, often identical, transactions were identified in an effort to detect Sybil candidates (addresses that behave indistinguishably from the perspective of the network) using Jaccard similarity. A total of 7,360,687 candidate Sybil pairs involving 167,580 distinct wallets were identified solely by graph-based behavioural similarity. Examples of pairs with the highest scores include accounts with perfect neighbourhood overlap (jac = 1.0) and nearly identical transaction degrees. These results should be interpreted as

---

[10] https://github.com/jucanbe/EthereumKG-ESWC26/tree/main/results/blocks.

indistinguishable behavioural identities rather than confirmed malicious actors. The results of this section can be found online[11].

7   Conclusions

In this article, we have presented a method for transforming data from the Ethereum blockchain into a knowledge graph, aligning blocks, transactions, ABIs, and Solidity source code with a set of ontologies. We have expanded existing semantic models for Ethereum and Solidity to encompass recent Ethereum features, and implemented a process that retrieves raw data from the blockchain and integrates it with off-chain data (ABI and Solidity code) using a third-party service, generating three interconnected knowledge graphs.

The experimentation demonstrates how this knowledge graph can improve the analysis of the Ethereum chain. At the block level, it is shown that combining information with semantic annotations derived from smart contracts allows the discovery of wallets that may reflect similar behaviour, automated agents, or coordinated activities. At the contract level, it is shown that ABI-level semantics can be used to group contracts according to their functional interfaces, identify families of copied or template-based implementations, and detect suspicious behaviour patterns.

The proposed method allows Ethereum to be analysed at a level that is not possible within the blockchain itself. Future work will focus on extending the approach to other blockchain technologies. In addition, we plan to extend the Ethereum knowledge graph with external data sources, such as repositories with smart contracts widely used repositories or adding a corpus of malicious smart contracts, to enrich the contextual information available for analysis.

Acknowledgments. The publication of this article was funded by the University of Mannheim and supported by MALTA PID2024-159504OB-I00 funded by MICI-U/AEI/10.13039/501100011033.

References

1. Bansod, S., Ragha, L.: Blockchain technology: applications and research challenges. In: 2020 International Conference for Emerging Technology (INCET), pp. 1–6. IEEE (2020)
2. Baqa, H., Truong, N.B., Crespi, N., Lee, G.M., Le Gall, F.: Semantic smart contracts for blockchain-based services in the Internet of Things. In: 2019 IEEE 18th International Symposium on Network Computing and Applications (NCA), pp. 1–5. IEEE (2019)
3. Bella, G., Cantone, D., Longo, C., Nicolosi Asmundo, M., Santamaria, D.F.: Blockchains through ontologies: the case study of the Ethereum ERC721 standard in oasis. In: Camacho, D., Rosaci, D., Sarné, G.M.L., Versaci, M. (eds.) Intelligent

---

[11] https://github.com/jucanbe/EthereumKG-ESWC26/tree/main/results/contracts.

Distributed Computing XIVM IDC 2021. Studies in Computational Intelligence, vol. 1026, pp. 249–259. Springer, Cham (2022). https://doi.org/10.1007/978-3-030-96627-0_23

4. Bella, G., Cantone, D., Nicolosi Asmundo, M., Santamaria, D.F.: Towards a semantic blockchain: a behaviouristic approach to modelling Ethereum. Appl. Ontol. **19**(2), 143–180 (2024)

5. Besançon, L., Da Silva, C.F., Ghodous, P., Gelas, J.P.: A blockchain ontology for DApps development. IEEE Access **10**, 49905–49933 (2022)

6. Böckling, M., Paulheim, H.: gpuRDF2vec – scalable GPU-based RDF2vec. In: International Semantic Web Conference, pp. 240–257. Springer, Switzerland (2025)

7. Cano-Benito, J., Cimmino, A., García-Castro, R.: Towards blockchain and semantic web. In: Abramowicz, W., Corchuelo, R. (eds.) BIS 2019. LNBIP, vol. 373, pp. 220–231. Springer, Cham (2019). https://doi.org/10.1007/978-3-030-36691-9_19

8. Cano-Benito, J., Cimmino, A., García-Castro, R.: Toward the ontological modeling of smart contracts: a solidity use case. IEEE Access **9**, 140156–140172 (2021)

9. Cantone, D., Longo, C.F., Nicolosi Asmundo, M., Santamaria, D.F., Santoro, C.: Ontological smart contracts in oasis: Ontology for agents, systems, and integration of services. In: Camacho, D., Rosaci, D., Sarné, G.M.L., Versaci, M. (eds.) International Symposium on Intelligent and Distributed Computing, pp. 237–247. Springer, Cham (2021). https://doi.org/10.1007/978-3-030-96627-0_22

10. Choudhury, O., Rudolph, N., Sylla, I., Fairoza, N., Das, A.: Auto-generation of smart contracts from domain-specific ontologies and semantic rules. In: 2018 IEEE International Conference on Internet of Things (iThings) and IEEE Green Computing and Communications (GreenCom) and IEEE Cyber, Physical and Social Computing (CPSCom) and IEEE Smart Data (SmartData), pp. 963–970. IEEE (2018)

11. Ehrlinger, L., Wöß, W.: Towards a definition of knowledge graphs. SEMANTiCS (Posters, Demos, SuCCESS) **48**(1–4), 2 (2016)

12. Hector, U.R., Boris, C.L.: BLONDiE: blockchain ontology with dynamic extensibility. arXiv preprint arXiv:2008.09518 (2020)

13. Hu, T., Li, B., Pan, Z., Qian, C.: Detect defects of solidity smart contract based on the knowledge graph. IEEE Trans. Reliab. (2023)

14. Kondo, M., Oliva, G.A., Jiang, Z.M., Hassan, A.E., Mizuno, O.: Code cloning in smart contracts: a case study on verified contracts from the Ethereum blockchain platform. Empir. Softw. Eng. **25**(6), 4617–4675 (2020)

15. Kopyciok, Y., Schmid, S., Victor, F.: Friend or foe? Identifying anomalous peers in Moneros P2P network. arXiv preprint arXiv:2509.10214 (2025)

16. Malsa, N., Vyas, V., Gautam, J.: Blockchain platforms and interpreting the effects of bitcoin pricing on cryptocurrencies. In: Sharma, T.K., Ahn, C.W., Verma, O.P., Panigrahi, B.K. (eds.) Soft Computing: Theories and Applications. AISC, vol. 1380, pp. 137–147. Springer, Singapore (2022). https://doi.org/10.1007/978-981-16-1740-9_13

17. Parr, T., Fisher, K.: Ll (*) the foundation of the ANTLR parser generator. ACM SIGPLAN Not. **46**(6), 425–436 (2011)

18. Petrović, N., Tošić, M.: Semantic approach to smart contract verification. Facta Univ. Ser. Autom. Control Robot. **19**(1), 021–037 (2020)

19. Pfeffer, J., et al.: EthOn-an Ethereum ontology (2016)

20. Poveda-Villalón, M., Fernández-Izquierdo, A., Fernández-López, M., García-Castro, R.: LOT: an industrial oriented ontology engineering framework. Eng. Appl. Artif. Intell. **111**, 104755 (2022)

21. Prokudin, A., Denisov, M., Sychev, O.: Disk space consumption by triple storage systems. In: Krouska, A., Troussas, C., Caro, J. (eds.) Novel & Intelligent Digital Systems Conferences, pp. 266–275. Springer, Cham (2022). https://doi.org/10.1007/978-3-031-17601-2_26

22. Ruta, M., Scioscia, F., Ieva, S., Capurso, G., Di Sciascio, E.: Semantic blockchain to improve scalability in the internet of things. Open J. IoT (OJIOT) **3**(1), 46–61 (2017)

23. Ruta, M., Scioscia, F., Ieva, S., Capurso, G., Pinto, A., Di Sciascio, E.: A blockchain infrastructure for the semantic web of things. In: SEBD (2018)

24. Seneviratne, O., McGuinness, D.L.: Web 3.0 meets web3: exploring the convergence of semantic web and blockchain technologies. In: ESWC Workshops (2023)

25. Shkembi, K., Kochovski, P., Papaioannou, T.G., Barelle, C., Stankovski, V.: Semantic web and blockchain technologies: convergence, challenges and research trends. J. Web Semant., 100809 (2023)

26. Sopek, M., Gradzki, P., Kosowski, W., Kuziski, D., Trójczak, R., Trypuz, R.: GraphChain: a distributed database with explicit semantics and chained RDF graphs. In: Companion Proceedings of the The Web Conference 2018, pp. 1171–1178 (2018)

27. Vo, H.T., Kundu, A., Mohania, M.K.: Research directions in blockchain data management and analytics. In: EDBT, pp. 445–448 (2018)

CANDI - A Semantic Framework for CAN Bus Data Modeling and System Integration

Pavle Ivanovic[✉][iD], Simon Burbach[iD], Oliver Niggemann[iD], and Maria Maleshkova[iD]

Helmut-Schmidt-University, Holstenhofweg 85, 22043 Hamburg, Germany
{ivanovip,burbachs,niggemao,maleshkm}@hsu-hh.de

Abstract. Modern automotive, maritime, railway, and airborne systems generate massive streams of operational data that challenge existing solutions for storage, security, and data management. Semantic integration techniques improve interoperability across heterogeneous sources, yet often fall short in deployment automation, scalability, and real-time operation. We present CANDI, a semantic data integration framework that enables dynamic decoding and ontological access to Controller Area Network (CAN) data. Leveraging virtual knowledge graphs, CANDI links low-level logging streams with structured semantic representations, supporting advanced diagnostics and informed decision making. The framework incorporates the DBC ontology, a CAN database extension of the W3C SSN/SOSA standards that formalizes the semantics of messages, signals, ECUs, decoding schemas, and data logging processes. Using real-world datasets, we demonstrate CANDI's contributions to end-to-end deployment automation, runtime CAN bus decoding, and secure, semantically driven analytics on streaming telemetry. The DBC ontology is rigorously evaluated for logical consistency, domain coverage, and knowledge graph instantiation, underscoring its robustness and industrial relevance.

    Ontology: https://paitools.github.io/DBCOntology/DBC.owl.
    GitHub: https://github.com/paitools/DBCOntology.
    Documentation: https://w3id.org/dbc-ontology.
    License: https://creativecommons.org/licenses/by-nc-sa/4.0.
    DOI: https://doi.org/10.5281/zenodo.17671851.

Keywords: Ontology Engineering · OBDA Framework · CAN Bus · DBC Decoding · Data Security · Deployment Automation · Real-time Analytics

1 Introduction

Driven by rapid advances in autonomous vehicles, unmanned vessels, fly-by-wire aircraft, and maglev technologies, the transportation industry is generating unprecedented volumes of data and continually pushing the frontiers of innovation. [40]. While these innovations promise greater efficiency and safety, they simultaneously introduce new challenges in data management, cybersecurity, and large-scale analytics [15]. Traditional infrastructures remain rigid, often lacking automation and optimization for advanced AI processing [42]. Digital twin technologies offer a promising paradigm, yet they struggle with heterogeneous data sources and the absence of standardized semantic integration [32].

Communication lies at the heart of modern transportation systems, requiring high speed, low latency, and robust security. The Controller Area Network (CAN) protocol has become a backbone of vehicular communication, increasingly adopted in maritime and airborne domains [1]. However, as system complexity grows, the sheer number of CAN bus signals overwhelms conventional time series databases, leading to limited retention times and inefficient storage management [27]. This creates a pressing need for solutions that can decode, integrate, and analyze CAN data streams in real-time while preserving fidelity and scalability.

Semantic integration frameworks provide a pathway toward interoperability across heterogeneous systems [34]. Yet existing approaches often fall short: they are difficult to deploy [36,37], restricted to narrow domains [5,14], dependent on centralized relational databases [17,36], or incapable of supporting real-time analytics [19,36,37]. To overcome these limitations, we present CANDI, a semantic integration framework designed for runtime decoding and dynamic analysis of raw CAN bus data. CANDI achieves full end-to-end deployment automation by directly accessing encoded CAN logs without requiring data transformation or relational database (RDB) consolidation.

At the core of CANDI is the CAN Database (DBC) ontology, an extension of W3C SSN/SOSA [20,26] standards that models CAN messages, signals, nodes, and logging processes. Together, the framework and ontology enable secure, semantically driven diagnostics and real-time analytics on live telemetry. This contribution addresses critical gaps in semantic interoperability, deployment automation, and runtime analytics, positioning CANDI as a leading virtual knowledge graph (VKG) solution for next-generation transportation systems.

The paper makes the following contributions:

- **CANDI Framework.** We present CANDI, a semantic framework for dynamic CAN bus data decoding, system integration, and deployment automation. CANDI combines VKG with ontology-based data access (OBDA) principles to bridge raw data streams with structured semantic representations, enabling runtime message decoding, semantically governed diagnostics, and real-time analytics.
- **DBC Ontology.** We introduce DBC, an ontology for CAN bus communication systems that extends the W3C SSN/SOSA standards and formalizes the semantics of signals, messages, electronic control units (ECUs), and datalogging processes. Grounded in the CAN database (DBC-file) protocol, it provides an interoperable and machine-interpretable schema for transportation and embedded-systems data.

- **Automated Deployment Pipeline.** We propose an end-to-end workflow that supports light configuration and automated deployment of VKG-based data integration components. The pipeline removes manual engineering effort, mitigates domain-expert dependency, and eliminates data migration overheads through declarative mappings and dynamic instantiation.
- **DBC Loading Tool.** We develop a tool that parses DBC-files and automatically materializes their content into a multi-sheet Knowledge Graph Matrix (KGM). The KGM representation combines ontology instances and their properties to support systematic KG population, flexible maintenance, and query-driven diagnostics.
- **Comprehensive Evaluation.** We perform an extensive evaluation of the DBC ontology with respect to logical soundness, reasoning behavior, and competency question (CQ) coverage. The framework is validated using real-world CAN bus data, multiple datasets, and two CQ suites, demonstrating efficient deployment, storage preservation, data fidelity, and real-time analytical capability.

The remainder of the paper is organized as follows. Section 2 reviews related ontologies and OBDA frameworks within the broader transportation domain. Section 3 outlines the motivation, illustrated through applied use case scenarios. Section 4 presents the DBC ontology, beginning with its design process and followed by key modeling decisions for concepts and semantic relationships. Section 5 introduces CANDI, the VKG-based integration framework for end-to-end deployment automation and real-time CAN data analysis. Section 6 describes the datasets and evaluation criteria, including CQ-based verification and validation with real-world data. Finally, Sect. 7 concludes the paper and highlights directions for future work.

2   Related Data Models and Frameworks

Although data exchange is central to communication standards such as CAN, prior work on semantic modeling of technical systems has paid limited attention to the concrete transfer of messages and signals between ECUs. Research in this area is primarily coming from the automotive domain and focuses on the high-level representation of vehicles [28]. We review two main branches of related work: 1) ontologies, ranging from general-purpose models to domain-specific ones for maritime and automotive platforms, and 2) frameworks for integrating heterogeneous and streaming sensor data using semantic technologies.

2.1   General and Domain-Specific Ontologies

A prominent example from the automotive industry is the Vehicle Signal Specification Ontology (VSSo) [28,29]. Based on the VSS standard, VSSo provides a semantically unique, extensible model for vehicle signals, integrating static properties and dynamic sensor data. However, it assumes pre-decoded sensor streams, requires access to DBC-files, and is limited to wheeled vehicles.

In our recent work, we expanded the scope to the maritime sector, contributing to enhanced vessel modeling with the AISHIP ontology [7], and providing a richer way of modeling watercraft by unifying heterogeneous data. While it supports Automatic Identification System logging, most concepts remain high-level (e.g., propulsion, motorization) and rely on aggregated measurements (e.g., typical location). We descended to the level of sensors and maritime systems with the SHIP ontology [21]. SHIP models onboard sensors, ship components, observations, thresholds, and anomalies. This paves the way for the semantic integration of sensor data, which is useful for maintenance planning and anomaly detection.

Both VSSo and the SHIP ontology build on the SSN and SOSA ontologies [20,26], which are well-established W3C standards. The industry-oriented Smart Applications REFerence ontology (SAREF) [11,31] provides a complementary model that focuses on the interoperability of IoT devices across various domains. It enables the representation of smart devices and their functions, while remaining provider-agnostic. One notable domain-specific extension is SAREF4AUTO [30], which refines these concepts for vehicle-related use cases.

Alvarez-Coello and Gómez [4] move further toward stream-level semantic enrichment. Their approach combines VSSo with the IoT-Stream model [16], which provides a lightweight layer for annotating dynamic IoT streams, enabling cross-application querying of sensor data. By doing so, they demonstrate the effectiveness of semantic technologies in deriving higher-level observations and events.

2.2 Frameworks for Semantic Data Integration

While ontologies serve as a foundational core for data integration, their embedding within a coherent framework is essential for closing the gap between heterogeneous raw data and its interpretable representation.

OBDAIR [36] addresses this challenge by providing a unified framework for accessing and integrating such sources while supporting reasoning for complex-event detection. Despite its evaluation in a maritime context, OBDAIR presents notable limitations: no real-time processing, complex deployment, and reliance on centralized RDBs, reducing flexibility in distributed or streaming settings.

Communication between industrial entities varies according to the specific use case and its implementation. To address this, Steindl and Kastner [37] developed a semantic microservice framework for digital twins that enables structured, interoperable information exchange via microservices and a knowledge graph (KG). However, real-time analytics are not yet supported.

The authors of [5] extend R2RML to build Ontop-spatial, a geospatially focused OBDA system. Its domain-specific design restricts reuse, while reliance on a RDB necessitates direct modifications and prevents real-time capabilities. Similarly, Ding et al. [14] introduce GOdIVA, another geospatial integration and analytics solution that depends on a centralized PostGIS database and likewise cannot support real-time OBDA.

The framework in [19] introduces a temporal OBDA architecture tailored to industrial diagnostic settings. The approach remains constrained by relying on static databases rather than true streaming flows. Consequently, its applicability to a real-time environment is limited.

Moving to the healthcare domain, Fareedi et al. [17] propose a framework that allows diverse data sources to be accessed as a unified KG. Although flexible, it requires complex deployment involving ontology modeling, mapping design, schema alignment, and coordination across multiple architectural layers [17].

These efforts underscore the value of semantic modeling for sensor data. Our earlier framework, MontoFlow [21], supports real-time access and analysis of maritime sensors but, like the others, assumes pre-decoded data streams. CANDI instead addresses the broader transportation domain, enables full end-to-end deployment, operates at a lower layer, focusing on CAN bus communication, and incorporates runtime automatic signal decoding together with OBDA.

3   Motivation and Use Cases

Recent advances in cyber-physical and autonomous systems have heightened the need for semantically aware access to heterogeneous, resource-constrained machine data. Although progress has been made in vehicle data standardization, most operational environments still rely on raw CAN bus logs, which inherently lack explicit structure and semantic interoperability. Conventional data processing approaches often depend on large backend infrastructures and complex ETL (Extract, Transform, and Load) pipelines [39], or proprietary decoding toolchains [25], introducing challenges of scalability, security, and vendor dependency.

Research in KGs, VKGs, and data virtualization has shown considerable promise in bridging the semantic gap between heterogeneous machine logging and high-level analytics [17,36,37]. However, existing systems often struggle in resource-constrained edge environments, lack secure mechanisms for processing encoded data, and rarely support expressive, real-time analytics. These limitations are particularly critical in mission-driven domains such as maritime SAR operations and autonomous transportation, while also impacting long-term predictive maintenance, as demonstrated in our use cases.

In contrast, CANDI extends VKG principles to encoded CAN bus data, enabling semantic access, SQL-level reasoning, and minimal exposure of sensitive decoding knowledge. This approach supports edge-only workloads (Use Case 1), secure analysis (Use Case 2), real-time diagnostics (Use Case 3), and long-term analytics (Use Case 4). By unifying these capabilities, CANDI delivers a resource-efficient, semantically transparent, and security-preserving solution to long-standing challenges in machine telemetry processing.

Use cases (motivating scenarios):

1. Edge Storage: A maritime Search and Rescue organization acquires a new vessel but lacks a backend infrastructure. Data logging is performed in the J1939 format; however, due to limited onboard space, deploying additional servers is not feasible. The crew therefore adopts the CANDI framework to conduct critical data inspection and log analysis directly on the raw data, avoiding costly data migration, relational database consolidation, and ongoing backend maintenance.

2. Data Security: An autonomous taxi company continuously records and streams critical CAN bus data to its servers for analysis and system improvement. Due to cybersecurity requirements, the data logs are retained in their encoded format and cannot be utilized without the appropriate decoding schema. To address this, the company deploys CANDI with encrypted and materialized SQL Views [9], enabling secure data analysis without decoding or exposing DBC-files online (see Sect. 5.2, security model details).

3. Real-Time Diagnostics: During high-speed cruising on a maritime rescue mission, a sensor alarm is triggered. With no time for detailed inspection, the operator must quickly identify the affected systems. The objective is to check all signals and compare them against their average values from the past ten minutes, excluding ignition and neutral readings. Using the CANDI framework, the operator promptly selects all signals that exceed the recent average by 20%, enabling rapid situational awareness and response (e.g., CQ C04_3d).

4. Advanced Analytics: To foster predictive maintenance and enhance vessel reliability, engineers examine long-term measurement trends and sensor data drifts, particularly in critical system components such as engines. Effective access to historical data is essential for these analyses, yet retention times in many time-series databases are limited, making this a persistent challenge. To overcome it, engineers deploy the CANDI framework, enabling semantically structured access to encoded system logs. Beyond evaluating seasonal heating and cooling profiles, their analyses extend to anomaly detection patterns preserved from many years ago. These insights enable condition-based rather than schedule-based maintenance, improve shipyard inspection planning, and extend engine lifecycle (e.g., domain/C03 and C04 competency questions).

4 DBC Ontology

The development of the DBC ontology was guided by the SABiOx methodology [2], an agile adaptation of the SABiO framework [3]. This methodology structures ontology engineering into a five-phase life cycle including requirements elicitation, setup, knowledge acquisition, design, and implementation. Throughout these phases, the ontology was iteratively refined to ensure methodological rigor and consistent alignment with domain requirements. Furthermore, it extends established ontologies and vocabularies, such as SSN, SOSA, and QUDT, ensuring semantic consistency.

Building on this methodological foundation, the ontology development was carried out in close collaboration with our project partner, the German Maritime Search and Rescue Service (DGzRS) [13]. This collaboration involved the design of the SmartShip system architecture for AI-based anomaly detection

[22,24] and the optimization of CAN bus data decoding [25]. The creation of the DBC ontology was therefore shaped by our partner requirements, CAN bus domain expertise, and established operational practices, which were distilled into a structured set of competency questions (CQs). In parallel, existing ontology design patterns (ODPs) were examined to identify reusable and well-established modeling approaches. Collectively, the CQs and ODPs provided the conceptual backbone of the development process, guiding iterative refinement throughout the design and implementation stages.

Finally, the DBC ontology was developed in accordance with the FAIR Guiding Principles, ensuring that it remains publicly accessible, reusable, and interoperable for the broader community. It is published under a persistent DOI and supplemented with comprehensive documentation, structured metadata, and an open license.

4.1   Core Concepts

Ontology Design Patterns (ODPs) are reusable modeling solutions that encapsulate best practices for representing common structures and relationships in ontologies [6]. They enhance consistency, reduce complexity, and accelerate development by providing validated templates that help prevent modeling errors [18]. By applying ODPs, ontology engineers can achieve semantic interoperability and scalability across diverse domains. The DBC ontology was developed on an SSN/SOSA backbone [20,26], a lightweight and widely adopted framework for modeling sensors, observations, and sample collection processes. In addition to representing physical sensing devices, SSN/SOSA supports the modeling of observable properties, enabling a natural and seamless alignment with CAN bus signaling. Wherever applicable, the DBC ontology aligns its concepts and properties with QUDT [33], ensuring consistent representation of physical quantities, measurement units, and conversions while preserving semantic coherence.

Class Specialization: Designed to model CAN bus traffic, the DBC ontology extends the core SOSA concepts of observations and observable properties through protocol-specific subclasses. These extensions are essential for the semantic representation of CAN messaging and signaling, as well as for standardized data logging (e.g., J1939 [38]). For example, we define `dbc:Message` and `dbc:Signal` as subclasses of `sosa:ObservableProperty` to represent CAN messages and their encoded signals. Likewise, `sosa:Observation` is extended to `dbc:MessageLog` and `dbc:SignalLog`, where signal logs are directly derived from the runtime decoding of message logs.

Additionally, the DBC ontology introduces two novel concepts: `dbc:Node` and `dbc:SignalEncoding`, representing ECUs and CAN bus encoding schemas, respectively. While `dbc:SignalEncoding` instances are theoretically capable of both message encoding and signal decoding, they are primarily used to facilitate signal value extraction.

Class Disjointness: To improve reasoning accuracy, query efficiency, and conceptual clarity, the DBC ontology explicitly asserts disjointness among the subclasses of the core SOSA classes, namely `sosa:Observation` and `sosa:ObservableProperty`.

4.2   Semantic Relations

A detailed depiction of the DBC semantic relations, including classes as well as object and data properties, is presented in Fig. 1. Solid lines in the diagram denote subclass relations, while dashed lines indicate combined properties. Apart from the unspecified domain of `qudt:hasUnit`, which was linked to `dbc:Signal`, all SSN/SOSA and QUDT properties retain their original signatures. This ODP design choice preserves compatibility with established query engines and enables reliable extensions for domain-specific use cases.

Regarding object properties, we introduced `dbc:isPartOf` and `dbc:hasSignal` to connect messages with their corresponding (encoded) signals. Similarly, `dbc:decodedFrom` and `dbc:encodes` capture the inverse relationship between message and signal logs, which is particularly important in use cases where both logging formats are simultaneously available. In addition, messages and signals are linked to encoding schemas through `dbc:encodedVia` and `dbc:decodedVia`, as well as to source (`dbc:hasTransmitter`) and target (`dbc:hasReceiver`) ECU nodes. Finally, we established a connection between

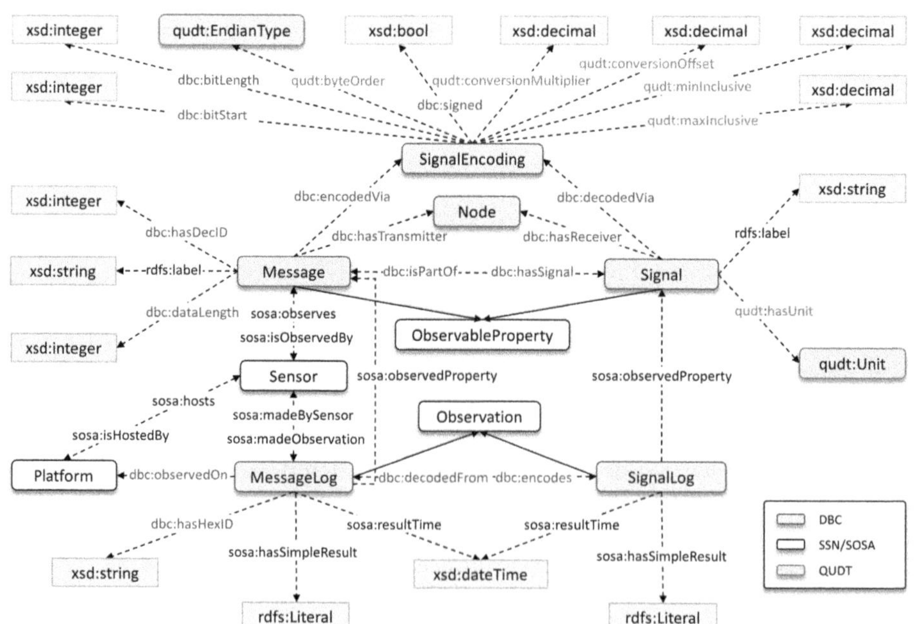

Fig. 1. DBC Ontology: The core concepts and semantic relationships.

message log observations and the corresponding CAN bus platform using
`dbc:observedOn`.

During the modeling of the `dbc:SignalEncoding` class, most data properties were mapped to QUDT equivalents. Properties without suitable matches – `dbc:bitLength`, `dbc:bitStart`, and `dbc:signed` were introduced within DBC, following established naming conventions. Together, these properties capture essential parameters for accurate and efficient decoding of CAN messages. In addition to encoding, we asserted two standard CAN identifiers, `dbc:hasDecID` and `dbc:hasHexID`, associated with CAN messages and message logs, respectively. Finally, we added the message property `dbc:dataLength`, denoting the byte length of the data payload, i.e., the maximum size of the encoded field. As illustrated in the diagram, value constraints were explicitly set to match the CAN database specification, while cardinality constraints were left flexible to accommodate diverse user logging patterns.

5   CANDI

CANDI is a system integration framework designed for dynamic and semantically structured access to raw CAN bus data. In addition to ontology-based data access (OBDA), CANDI enables runtime decoding of CAN messages, advanced diagnostics, and real-time analytics. The framework supports full deployment automation through a two-step process (Fig. 4). First, the DBC-file loading script (load_dbc.py) transforms the DBC-file content, including decoding parameters, into a KG structure called the KG Matrix (KGM). Next, the main tool (CANDI.py) integrates the KGM with the OBDA engine (Ontop), combining R2RML mappings [12] and SQL Views [9] to seamlessly connect semantic representations to raw CAN bus data. Together, these components provide a unified environment for automated data integration, semantic querying, and intelligent analysis of CAN bus systems.

5.1   Knowledge Graph Matrix

The design of the KGM was directly inspired by MontoFlow static instantiation [21], which relies on tabular individual-property mapping. However, KGM intro-

Individual	rdf:type	dbc:decodedVia	dbc:hasReceiver	dbc:isPartOf	qudt:hasUnit	sosa:isObservedBy
FuelTank01	dbc:Signal	FuelTank01Encoding	Unknown	BoeningI	L	can2_sniffer
FuelTank02	dbc:Signal	FuelTank02Encoding	Unknown	BoeningI	L	can2_sniffer
FuelTank05	dbc:Signal	FuelTank05Encoding	Unknown	BoeningI	L	can2_sniffer
FuelTank06	dbc:Signal	FuelTank06Encoding	Unknown	BoeningI	L	can2_sniffer
FuelTank07	dbc:Signal	FuelTank07Encoding	Unknown	BoeningII	L	can2_sniffer
FuelTank08	dbc:Signal	FuelTank08Encoding	Unknown	BoeningII	L	can2_sniffer
FuelTank15	dbc:Signal	FuelTank15Encoding	Unknown	BoeningII	L	can2_sniffer
BilgeWaterTank12	dbc:Signal	BilgeWaterTank12Encoding	Unknown	BoeningII	L	can2_sniffer
FreshwaterTank14	dbc:Signal	FreshwaterTank14Encoding	Unknown	BoeningIII	L	can2_sniffer
Inclination	dbc:Signal	InclinationEncoding	Unknown	BoeningIII	DEG	can2_sniffer
EngineRoomTemperature	dbc:Signal	EngineRoomTemperatureEncoding	Unknown	BoeningIII	DEG_C	can2_sniffer
FirePumpPressure	dbc:Signal	FirePumpPressureEncoding	Unknown	BoeningIII	BAR	can2_sniffer

Fig. 2. Signal representation in KGM, including individuals, type, and properties.

duces a key novelty: full process automation, eliminating the need for manual and error-prone matrix population. The developed script (load_dbc.py) automates the transformation of DBC-files into a semantically enriched KGM stored in XLSX format. It begins by loading our unit mapping file (unit_mapping.json) to align raw engineering units with QUDT standards, ensuring semantic consistency. The script then parses the DBC-file, extracts messages, signals, encodings, and nodes, and associates them with the SOSA and DBC ontology concepts (Fig. 3). Each element is structured into tabular datasets and appended to predefined sheets of the KGM template (Fig. 2), systematically capturing the relationships between ontology concepts. Through this process, CANDI enables the transformation of raw CAN bus data into semantically enhanced structures that can be leveraged within the broader framework deployment.

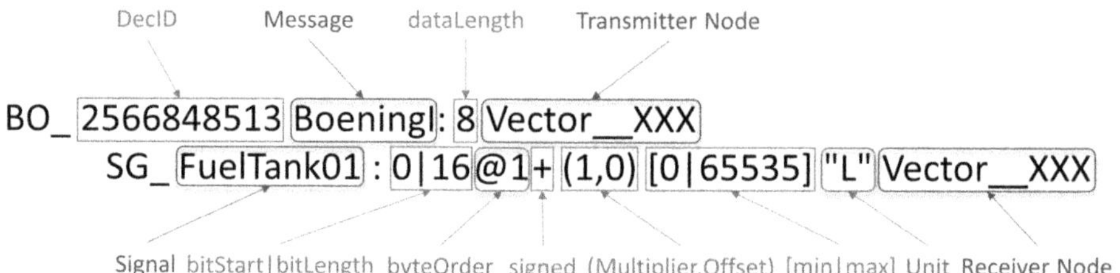

Fig. 3. An excerpt of boening.dbc file, showing encoding-to-ontology mapping.

5.2   CANDI Deployment

In the era of accelerated digitalization within the transportation industry, data security has become paramount, particularly in the deployment of autonomous vehicles [41]. While traditional communication protocols, such as the industry standard CAN, provide basic data encoding during transmission, the storage demands of decoded signals frequently exceed the capacities of local edge devices. Furthermore, transferring and maintaining sensitive data on remote servers or data centers introduces significant vulnerabilities, creating opportunities for cyber attacks that undermine automation efforts [35]. A further challenge lies in balancing immediate, real-time diagnostics with advanced, long-term analytics, which often represent bottlenecks in large end-to-end systems. OBDA architectures offer a promising direction for structured, semantically augmented data processing. However, as noted earlier, they remain difficult to deploy, are often limited to narrow domains, lack real-time support, and require manual RDB consolidation.

CANDI is a VKG framework for dynamic CAN bus decoding and real-time semantic reasoning over encoded data. It leverages the OBDA engine Ontop [8] to link R2RML mappings with SQL Views, enabling SPARQL queries to directly access CAN bus logging data without materializing RDF triples. SQL Views, or

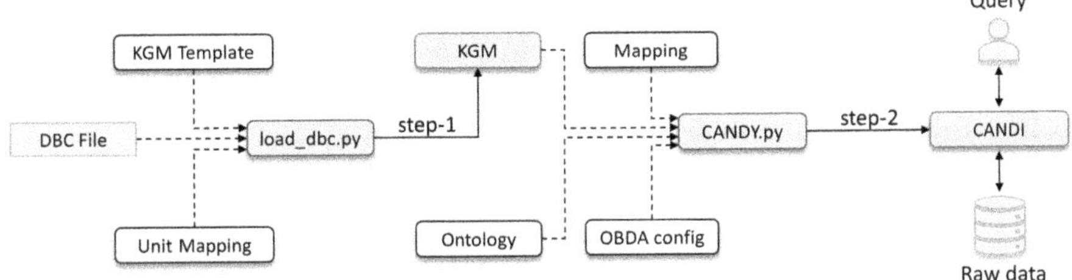

Fig. 4. Diagram of the CANDI two-step automated deployment, highlighting user inputs (gray), framework components (white), automation scripts (green), and the outputs produced at each step (blue). (Color figure online)

virtual tables, serve as a lightweight abstraction layer operating directly over streaming raw data, eliminating the need for RDB consolidation and ensuring that each SPARQL query reflects the latest logging state. All elements extracted from the DBC-file into the KGM (Fig. 4) are automatically converted by CANDI into static Views and connected via R2RML mappings to the DBC ontology. Following this deployment step, the original CAN database encoding schema is no longer required and can be removed for security reasons. Moreover, CANDI implements two dynamic SQL Views: one for accessing encoded message logs and another for decoding individual signals within each message. The decoding logic specifies bit-level extraction, scaling, sign handling, and endianness for standard 8-byte CAN messages, while delegating missing data handling to pre- or post-processing pipelines.

Once initialized (Fig. 4, step-2), the framework enables real-time diagnostics over live CAN bus telemetry, assisting operators in responding to sudden emergencies. In addition, CANDI supports the analysis of historical data archived in encoded form, ensuring both data security and storage efficiency. By embedding the decoding logic within encrypted and materialized SQL Views, the framework ensures that decoding is performed exclusively inside the database's trusted execution environment, protected by standard authentication and authorization mechanisms. This design reduces the attack surface and prevents unauthorized disclosure of decoding schemas or decoded values, assuming that encryption keys remain secure.

At present, CANDI integrates two external sources: raw message logs, which are directly accessed, and signal logs, which are dynamically decoded. The framework can be extended with additional sources through corresponding updates in the View and Mapping code sections. Furthermore, the expected CAN bus logging format (CSV) can be adapted within the message log View by modifying the file loading function or adjusting data logging fields.

6   Evaluation

In accordance with the SABiOx [2] methodology, the evaluation of the DBC ontology was conducted through two complementary perspectives: verification and validation. Verification focuses on the correctness of the ontology construction, ensuring adherence to formal design specifications, internal consistency, and methodological rigor. Validation, in contrast, addresses the ontology's capacity to achieve its intended purpose in applied contexts, demonstrating both practical relevance and adequacy within the target domain. Together, these perspectives ensure that the ontology is both methodologically sound and practically effective.

6.1   Datasets

To evaluate the DBC ontology, we employed datasets from two transportation domains: maritime and automotive. The maritime dataset originates from our project partner, DGzRS [13], and comprises DBC-files for three categories of signals: general signals, large diesel engines, and small diesel engines. These files contain 58 (15), 192 (58), and 21 (6) signals (messages), respectively. The associated logging data consists of time-series records in encoded (raw) CAN bus format, captured via the J1939 protocol [38]. As illustrated in Fig. 5 (MsglogX), the logging structure consists of four fields: timestamp, message identifier (HexID: 18ff0201), CAN bus platform (can2), and encoded data field (7b08000015000000).

The automotive dataset, provided by CSS Electronics [10], includes a demo DBC-file with 7 signals and 6 messages, complemented by logging examples from trucks, buses, and various agricultural vehicles.

During the DBC-file loading (transformation) phase, all KG triplets are automatically extracted from the input files and temporarily stored in the KGM. In the subsequent framework deployment phase, the KGM is converted into SQL Views, enabling dynamic instantiation and direct access to raw data at query runtime. An illustrative example of a semantic relation among maritime CAN bus signaling concepts is presented in Fig. 5.

6.2   Ontology Verification

Verification examines whether the ontology meets the specification requirements established at the outset of its development. To operationalize these requirements, we employ a structured set of competency questions (CQs), which define the kinds of information the ontology must be able to retrieve. These questions serve as a formal expression of the knowledge that should be captured and represented. The CQs were derived from established use cases, stakeholder expectations, and analysis of CAN bus data, ensuring that the ontology remains aligned with both domain needs and characteristics of real-world datasets. Therefore, the CQs are categorized into two complementary sets:

- **General** competency questions reflect the scope and the key concepts of the ontology.
- **Domain-specific** competency questions address operational challenges related to use cases: data security, real-time diagnostics, historical data analytics, and storage optimization.

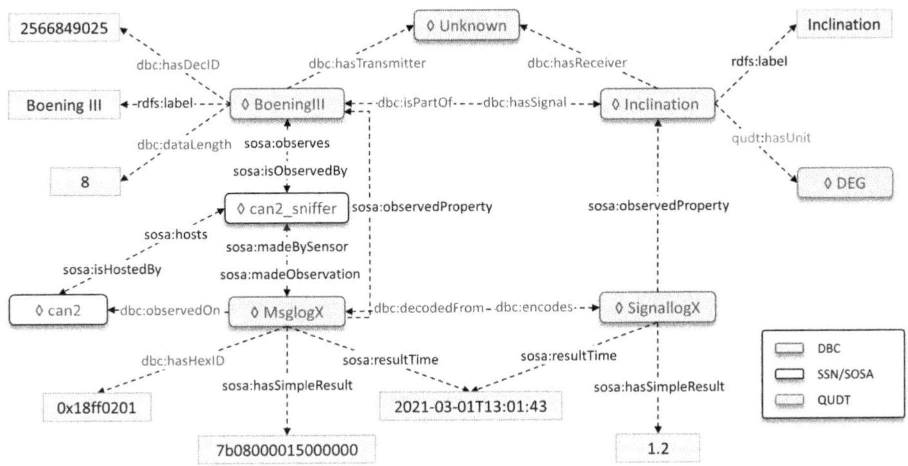

Fig. 5. An excerpt of materialized ontology for Inclination signal, encoded as part of the BoeningIII message. The other individuals include can2 platform, can2_sniffer sensor, Unknown ECU node, DEG unit, MsglogX message log, and dynamically decoded SignallogX signal log.

The evaluation of ontology design is supported through a mapping table that links competency questions to the corresponding axioms, concepts, and relations. An excerpt of the mapping table is presented in Table 1; the full version is available in the GitHub repository [23].

The demonstrated systematic alignment between CQs and the modeled ontology affirms that the ontology has been developed in accordance with its defined scope and design goals. Furthermore, the mapping of each CQ to its corresponding ontology entities confirms that the DBC ontology achieves complete (100%) domain coverage and provides comprehensive concept representation with respect to the stated functional requirements.

6.3   Ontology Validation

Validation examines the ontology's effectiveness in applied scenarios, assessing whether it can be reliably instantiated with real-world data and support the generation of meaningful, contextually appropriate insights. This perspective ensures that the ontology is not only well-constructed in a formal sense but also operationally adequate for its intended domain applications.

In the initial stage of the validation process, we loaded the DBC-files from all datasets and transformed them into a KGM. This step involves the automated extraction (via load_dbc.py) of classes, instances, properties, and literals, together with a domain–range consistency check, resulting in a multi-sheet tabular representation of triples (KGM.xlsx). In the subsequent CANDI processing phase, the KGM is converted into its final virtual-table form (SQL View). When combined with R2RML mappings and the Ontop OBDA engine, this

Table 1. Competency Questions (CQs) and corresponding Ontology entities. The CQs are categorized on core concepts: Observable Properties (C01), Sensors and Platforms (C02), Observations (C03), and Advanced Modeling (C04). They are divided into general (g) and domain-specific (d) questions.

CQ	Question	Entities
C01_3g	How are signals mapped to CAN messages?	(MessageX - type - Message) (Signal - isPartOf - MessageX)
C01_3d	Which CAN message encodes the fire pump pressure signal?	(MessageX - hasSignal - FirePumpPressure)
C02_1g	How are sniffing sensors associated with transmitter nodes?	(Message - hasTransmitter - Node) (Sensor - observes - Message)
C02_3d	Which transmitter nodes are associated with the CAN2Sniffer sensor?	(Message - hasTransmitter - Transmitter) (Message - isObservedBy - CAN2Sniffer)
C03_3g	How are signal log observations represented in the system?	(SignallogX - type - SignalLog) (SignallogX - Property - Value)
C03_2d	Which unit is assigned to the signal log of the fresh water tank?	(Signallog - observedProperty - FreshwaterTank14) (FreshwaterTank14 - hasUnit - Unit)
C04_2d	What are encoding schemas of CAN message that observes the highest number of signals?	(MessageX - type - Message) (Signal - isPartOf - MessageX) (Signal - decodedVia - Encoding)

configuration enables the ontology to dynamically instantiate not only the static individuals defined in the KGM but also the incoming streaming log data. Moreover, guided by the defined use cases, we developed a domain-specific test suite (denoted with the suffix "d" in Table 1) tailored to the maritime context. These test questions were derived from the general set of CQs but adapted to reflect complex, operationally realistic scenarios. One such example is C04_3d, which asks: Which signals exhibit measured values at least 20 percent higher than their average within the preceding ten-minute interval?

```
SELECT ?signal ?signalvalue ?meanvalue
WHERE {
  { SELECT ?signal (AVG(?val) AS ?meanvalue)
    WHERE {
      ?log sosa:observedProperty ?signal ;
           sosa:hasSimpleResult ?val ;
           sosa:resultTime ?time .
      FILTER(?time >= "2025-03-01T13:49:00"^^xsd:dateTime &&
             ?time <= "2025-03-01T13:59:00"^^xsd:dateTime)
    } GROUP BY ?signal
  }
  ?signallog sosa:observedProperty ?signal ;
             sosa:hasSimpleResult ?signalvalue ;
```

```
        sosa:resultTime ?timestamp .
  FILTER(?timestamp >= "2025-03-01T13:49:00"^^xsd:dateTime &&
         ?timestamp <= "2025-03-01T13:59:00"^^xsd:dateTime)
  FILTER(?signalvalue > 1.2 * ?meanvalue)
}
ORDER BY ?timestamp
```

To illustrate the underlying reasoning process, the corresponding SPARQL query (C04_3d) resolves the task through the following stages: 1) Dynamically decodes raw CAN log entries into their corresponding semantic signal representations. 2) Computes, for each signal, the mean measured value within the specified 10-minute window. 3) Retrieves all decoded signal observations recorded within the same interval. 4) Identifies observations whose measured values exceed 120% of their respective signal's computed mean. 5) Returns the matching signals for inspection, alongside their actual and average values.

In total, 31 CQs were defined, including 14 general and 17 domain-specific questions. All validation tests were successfully executed using SPARQL queries combined with dynamic VKG instantiation of the test datasets. The expected outputs for the sample data are provided in the project repository. In addition to validating correctness, we also evaluated the system's performance across datasets of increasing size.

Table 2. Performance evaluation across datasets of varying sizes.

Dataset	Decoded Signals	Decoding Time (s)	Throughput (signals/s)	SPARQL Lat. (ms)	Ontop Startup (s)
1 day	20,711,398	0.6816	30,386,022	738	0.726
1 week	144,979,786	4.887	29,664,632	719	0.738
2 weeks	289,959,572	9.9557	29,124,926	758	0.759
4 weeks	579,919,144	20.2616	28,621,600	789	0.741

The results (Table 2) show that the decoding pipeline consistently sustained extremely high throughput, remaining in the 28–30 million decoded signals per second range. The results demonstrate near-linear scalability: increasing the dataset from approximately 20 million to nearly 580 million decoded signals led to proportional increases in runtime without any observable degradation in per-signal performance. This stability indicates that the SQL-based decoding approach remains CPU-efficient even at multi-week data scales and provides more than three orders of magnitude ($\sim$1,800$\times$) of headroom compared to the load generated by a fully saturated 1 Mbit/s CAN bus.

When evaluating typical join operations through the Ontop CLI, SPARQL query latency remained nearly constant (720–800 ms), as queries operate primarily on the metadata layer. Similarly, Ontop startup time remained stable at approximately 0.72–0.76 s, indicating a predictable and low initialization overhead.

Overall, the evaluation demonstrates that the framework operates efficiently even on a typical laptop equipped with an i7 processor and 16 GB of RAM, with no observable performance bottlenecks.

7   Conclusion and Future Work

In this paper, we introduce CANDI, a semantic integration framework for modeling, decoding, and analyzing raw CAN bus data. Built on VKGs, CANDI supports end-to-end deployment automation through our KGM structure. We also present the DBC ontology, an extension of W3C SSN/SOSA, which models signals, messages, ECUs, and data logging as a core component of the framework. Together, these contributions enable runtime decoding of raw CAN bus messages, secure, semantically governed diagnostics, and real-time analytics on streaming telemetry. The evaluation of maritime and automotive datasets demonstrates semantic coherence, efficient storage, and advanced analytical capability, meeting all design requirements.

Future work will focus on two main directions: user interaction and performance optimization. Because most framework users, such as ship crew members, are not trained in semantic querying, we plan to develop a large language model (LLM) interface to provide real-time, factual system knowledge in a natural way. In parallel, we will pursue performance improvements in CAN bus decoding, with particular attention to edge devices and embedded systems. These efforts will strengthen CANDI as a leading framework for semantic analytics and real-time diagnostics of CAN bus data.

Resource Availability Statement The CANDI-code[1], the DBC ontology[2], the CQs [3] with their mapping to SPARQL [4] are available on GitHub [23] and Zenodo (DOI: 10.5281/zenodo.17671851). Documentation is available via a persistent W3ID identifier[5]. All resources are licensed under Creative Commons Attribution-NonCommercial-ShareAlike 4.0 International.

Acknowledgment. This research, as part of the project SmartShip, is funded by dtec.bw – Digitalization and Technology Research Center of the Bundeswehr, which we gratefully acknowledge. dtec.bw is funded by the European Union – NextGenerationEU.

Disclosure of Interests. The authors have no competing interests to declare.

---

[1] https://github.com/paitools/DBCOntology/blob/c2857a5d0c7d915ec3c00d1aeafd5
bd4d1631cf3/CANDI.py.

[2] https://paitools.github.io/DBCOntology/DBC.owl.

[3] https://github.com/paitools/DBCOntology/blob/c2857a5d0c7d915ec3c00d1aeafd5
bd4d1631cf3/CQs/Competency\%20Questions.txt.

[4] https://github.com/paitools/DBCOntology/tree/c2857a5d0c7d915ec3c00d1aeafd5
bd4d1631cf3/CQs.

[5] https://w3id.org/dbc-ontology.

References

1. Road vehicles — Controller area network (CAN) — Part 1: Data link layer and physical coding sublayer (2024). https://www.iso.org/standard/86384.html. Accessed 02 Dec 2025
2. Aguiar, C.Z., Souza, V.E.S.: SABiOx: the extended systematic approach for building ontologies. In: Proceedings of the 17th Seminar on Ontology Research in Brazil (ONTOBRAS 2024) and 8th Doctoral and Masters Consortium on Ontologies, WTDO 2024, Vitória, Brazil, October 2024. CEUR Workshop Proceedings (2024). Available under Creative Commons License Attribution 4.0 International (CC BY 4.0). https://nemo.inf.ufes.br/projetos/sabiox
3. de Almeida Falbo, R.: SABiO: systematic approach for building ontologies. In: Ontology and Conceptual Modeling - Onto. Com/ODISE@FOIS (2014)
4. Alvarez-Coello, D., Gomez, J.M.: Ontology-based integration of vehicle-related data. In: 2021 IEEE 15th International Conference on Semantic Computing (ICSC), January 2021. IEEE (2021). https://doi.org/10.1109/icsc50631.2021.00078
5. Bereta, K., Xiao, G., Koubarakis, M.: Ontop-spatial: ontop of geospatial databases. J. Web Semant. **58**, 100514 (2019). https://doi.org/10.1016/j.websem.2019.100514
6. Blomqvist, E., Gangemi, A., Presutti, V.: Experiments on pattern-based ontology design. In: Proceedings of the Fifth International Conference on Knowledge Capture, K-CAP '09, September 2009, pp. 41–48. ACM (2009). https://doi.org/10.1145/1597735.1597743
7. Burbach, S., Mackert, L., Maleshkova, M.: AISHIP: an ontology for extended vessel representation and multimodal data integration. In: ISWC 2025 Companion Volume. CEUR Workshop Proceedings, Nara, Japan (2025). https://ceur-ws.org/Vol-4093/paper2.pdf. Accessed 01 Dec 2025
8. Calvanese, D., et al.: Ontop: answering SPARQL queries over relational databases. Semant. Web **8**(3), 471–487 (2016). https://doi.org/10.3233/sw-160217
9. Chirkova, R., Yang, J.: Materialized views. Found. Trends® Databases **4**(4), 295–405 (2012). https://doi.org/10.1561/1900000020
10. CSS Electronics: J1939 Data Pack [PDF, Demo DBC & Sample Data] (n.d.). https://www.csselectronics.com/pages/j1939-data-pack-heavy-duty. Accessed 03 Dec 2025
11. Daniele, L., den Hartog, F., Roes, J.: Created in close interaction with the industry: the Smart Appliances REFerence (SAREF) ontology. In: Cuel, R., Young, R. (eds.) FOMI 2015. LNBIP, vol. 225, pp. 100–112. Springer, Cham (2015). https://doi.org/10.1007/978-3-319-21545-7_9
12. Das, S., Sundara, S., Cyganiak, R.: R2RML: RDB to RDF mapping language. In: W3c Recommendation, World Wide Web Consortium (W3C) (2012). https://www.w3.org/TR/r2rml/. Accessed 01 Dec 2025
13. Die Seenotretter – DGzRS: ABOUT THE DGzRS (2024). https://www.seenotretter.de. Accessed 01 Dec 2025
14. Ding, L., Xiao, G., Calvanese, D., Meng, L.: A framework uniting ontology-based geodata integration and geovisual analytics. ISPRS Int. J. Geo Inf. **9**(8), 474 (2020). https://doi.org/10.3390/ijgi9080474
15. Durlik, I., Miller, T., Kostecka, E., Zwierzewicz, Z., Łobodzińska, A.: Cybersecurity in autonomous vehicles—are we ready for the challenge? Electronics **13**(13) (2024). https://doi.org/10.3390/electronics13132654

16. Elsaleh, T., Enshaeifar, S., Rezvani, R., Acton, S.T., Janeiko, V., Bermudez-Edo, M.: IoT-stream: a lightweight ontology for internet of things data streams and its use with data analytics and event detection services. Sensors **20**(4), 953 (2020). https://doi.org/10.3390/s20040953
17. Fareedi, A.A., Gagnon, S., Ghazawneh, A., Valverde, R.: Semantic fusion of health data: implementing a federated virtualized knowledge graph framework leveraging ontop system. Fut. Internet **17**(6), 245 (2025). https://doi.org/10.3390/fi17060245
18. Gangemi, A., Presutti, V.: Ontology design patterns. In: Staab, S., Studer, R. (eds.) Handbook on Ontologies. IHIS, pp. 221–243. Springer, Heidelberg (2009). https://doi.org/10.1007/978-3-540-92673-3_10
19. Güzel Kalayci, E., Brandt, S., Calvanese, D., Ryzhikov, V., Xiao, G., Zakharyaschev, M.: Ontology–based access to temporal data with Ontop: a framework proposal. Int. J. Appl. Math. Comput. Sci. **29**(1), 17–30 (2019). https://doi.org/10.2478/amcs-2019-0002
20. Haller, A., Janowicz, K., Cox, S., Le Phuoc, D., Taylor, K., Lefrançois, M.: Semantic sensor network ontology. In: W3C Recommendation, World Wide Web Consortium (W3C) (2017). https://www.w3.org/TR/vocab-ssn/. Accessed 02 Dec 2025
21. Ivanovic, P., Burbach, S., Maleshkova, M.: MontoFlow – a maritime ontology framework for modeling ship sensory systems. In: Garijo, D., et al. (eds.) The Semantic Web, ISWC 2025. LNCS, vol. 16141, pp. 276–294. Springer, Cham (2025). https://doi.org/10.1007/978-3-032-09530-5_16
22. Ivanovic, P., Ganbarov, A., Neumann, P.: System architecture for real-time condition monitoring and anomaly detection on ships. In: 2023 22nd International Symposium on Parallel and Distributed Computing (ISPDC), July 2023, pp. 45–52. IEEE (2023). https://doi.org/10.1109/ispdc59212.2023.00019
23. Ivanovic, P., Maleshkova, M.: paitools/DBCOntology:CANDI/DBC v1.0.0: pre-publication release, November 2025. https://doi.org/10.5281/zenodo.17671851, https://paitools.github.io/DBCOntology
24. Ivanovic, P., Schöttler, J., Windmann, A., Neumann, P.: SmartShip - AI-based assistance systems for the maritime sector. In: Forschungsaktivitäten im Zentrum für Digitalisierungs- und Technologieforschung der Bundeswehr dtec.bw: Band 1, 2022, pp. 292–296. Universitätsbibliothek der HSU/UniBw H (December 2022). https://doi.org/10.24405/14567, project: Digitale Zwillinge für Intelligente Schiffe und für Schiffsflotten. Part of: https://doi.org/10.24405/14522
25. Ivanovic, P., Windmann, A., Neumann, P.: SmartShip: data decoding optimization for onboard AI anomaly detection. In: 2024 23rd International Symposium on Parallel and Distributed Computing (ISPDC), pp. 1–5 (2024). https://doi.org/10.1109/ISPDC62236.2024.10705391
26. Janowicz, K., Haller, A., Cox, S.J., Le Phuoc, D., Lefrançois, M.: SOSA: a lightweight ontology for sensors, observations, samples, and actuators. J. Web Semant. **56**, 1–10 (2019). https://doi.org/10.1016/j.websem.2018.06.003
27. Jensen, S., Pedersen, T., Thomsen, C.: Time series management systems: a survey. IEEE Trans. Knowl. Data Eng. (2017). https://doi.org/10.1109/TKDE.2017.2740932
28. Klotz, B., Troncy, R., Wilms, D., Bonnet, C.: VSSo: a vehicle signal and attribute ontology. In: Proceedings of the 9th International Semantic Sensor Networks Workshop, SSN 2018, Monterey, CA, USA (2018). https://ssn2018.github.io/submissions/SSN2018_paper_4_submitted.pdf. Accessed 01 Dec 2025
29. Klotz, B., Troncy, R., Wilms, D.: VSSo: vehicle signal specification ontology. In: W3C Recommendation, World Wide Web Consortium (W3C) (2017). https://www.w3.org/TR/vsso/. Accessed 01 Dec 2025

30. Lefrançois, M.: SAREF4AUTO: an extension of SAREF for the automotive domain (2025). https://saref.etsi.org/saref4auto/v2.1.1/. Accessed 02 Dec 2025
31. Lefrançois, M., et al.: SAREF: the smart applications REFerence ontology (Version 4.1.1) (2024). https://saref.etsi.org/core/v4.1.1/. Accessed 02 Dec 2025
32. Lu, Y., Liu, C., Wang, K.I.K., Huang, H., Xu, X.: Digital Twin-driven smart manufacturing: connotation, reference model, applications and research issues. Robot. Comput. Integr. Manuf. **61**, 101837 (2020). https://doi.org/10.1016/j.rcim.2019.101837
33. QUDT.org: QUDT Ontologies Overview (2025). https://www.qudt.org/pages/QUDToverviewPage.html. Accessed 01 Dec 2025
34. Ranpara, R.: A semantic and ontology-based framework for enhancing interoperability and automation in IoT systems. Discov. IoT **5**(22) (2025). https://doi.org/10.1007/s43926-025-00122-8
35. Salek, M.S., et al.: A review on cybersecurity of cloud computing for supporting connected vehicle applications. IEEE Internet Things J. **9**(11), 8250–8268 (2022). https://doi.org/10.1109/JIOT.2022.3152477
36. Santipantakis, G., Kotis, K., Vouros, G.A.: OBDAIR: ontology-based distributed framework for accessing, integrating and reasoning with data in disparate data sources. Exp. Syst. Appl. **90**, 464–483 (2017). https://doi.org/10.1016/j.eswa.2017.08.031
37. Steindl, G., Kastner, W.: Semantic microservice framework for digital twins. Appl. Sci. **11**(12), 5633 (2021). https://doi.org/10.3390/app11125633
38. Voss, W.: A Comprehensible Guide to J1939. Copperhill Technologies Corporation (2008). Paperback, perfect binding. SKU: 978-09765116-3-2
39. Walha, A., Ghozzi, F., Gargouri, F.: Data integration from traditional to big data: main features and comparisons of ETL approaches. J. Supercomput. **80**(19), 26687–26725 (2024). https://doi.org/10.1007/s11227-024-06413-1
40. World Intellectual Property Organization: WIPO Technology Trends: Future of Transportation (2025). https://www.wipo.int/pressroom/en/articles/2025/article_0002.html. Accessed 02 Dec 2025
41. Xu, Y., Wei, J., Mi, T., Chen, Z.: Data security in autonomous driving: multifaceted challenges of technology, law, and social ethics. World Electr. Veh. J. **16**(1) (2025). https://doi.org/10.3390/wevj16010006
42. Zhang, Q., Yang, L.T., Chen, Z., Li, P.: A survey on deep learning for big data. Inf. Fus. **42**, 146–157 (2018). https://doi.org/10.1016/j.inffus.2017.10.006

CR-TKGQA: A Temporal Knowledge Graph Question Answering Dataset Involving Complex Reasoning

Yuheng Bao[ID], Dingkun Xu[ID], E. Longfei[ID], Jiayu Shen[ID], Gong Cheng[ID], and Yuzhong Qu[(✉)][ID]

State Key Lab for Novel Software Technology, Nanjing University, Nanjing, China
{yhbao,dkxu,lfe,jyshen}@smail.nju.edu.cn, {gcheng,yzqu}@nju.edu.cn

Abstract. Complex reasoning in Temporal Knowledge Graph Question Answering (TKGQA), which requires both Knowledge Graph (KG) navigation and temporal computation, has received limited attention because current datasets often have limited coverage of multi-hop retrieval and computations. To address this gap, we propose a taxonomy of complex reasoning in TKGQA, which consists of two dimensions: structural complexity and computational complexity. Guided by this taxonomy, we construct **CR-TKGQA**, a new TKGQA dataset for Complex Reasoning. CR-TKGQA consists of more than 30,000 natural language questions, each paired with an executable SPARQL query on Wikidata. Its intent-driven construction ensures broad coverage across both complexity dimensions, achieved by first seeding computational complexity from human intent and then automatically augmenting structural complexity. Experiments show CR-TKGQA poses challenges to current baselines.

Keywords: Temporal Knowledge Graph Question Answering · Dataset · Knowledge Graph

Resource type: Dataset
License: Apache License 2.0
DOI: 10.5281/zenodo.17706444
URL: https://doi.org/10.5281/zenodo.17706444.

1 Introduction

Temporal Knowledge Graph Question Answering (TKGQA) focuses on utilizing Knowledge Graphs (KGs) to answer *temporal questions* [11]. As the field advances, the frontier of TKGQA is moving beyond simple fact lookups to complex questions that require the integration of multiple facts or contain complex computations [13]. Such questions demand what we term *complex reasoning* in TKGQA, which we define as the combination of two challenges: i) navigating

M. Acosta et al. (Eds.): ESWC 2026, LNCS 16550, pp. 152–171, 2026.
https://doi.org/10.1007/978-3-032-25159-6_9

the KG to retrieve multiple facts, ii) applying temporal computations over the retrieved facts. For instance, answering *Who is the youngest winner of the prize that Anne Marie Ottersen received before 2001* requires multi-hop fact retrieval (Anne → the prize she received → other winners), with computations to compare each winner's award time with *2001*, derive each winner's age from their birth date and the award time, and an ordinal on the winner's age. This kind of reasoning is different from *deductive reasoning*, which infers new knowledge from existing rules via logical entailment. Instead, our focus is on the multi-step retrieval and computation over temporal facts.

Despite the importance of complex reasoning, current datasets have not yet fully covered its spectrum. In terms of KG navigation, many datasets provide limited coverage of multi-hop retrieval, where answers cannot be found directly but require navigating through intermediate facts. The range of temporal computations is likewise restricted; for instance, computations on duration, such as computing an *age*, are often absent from current datasets.

These limitations, which constrain the development of complex reasoning capabilities, largely stem from the absence of a taxonomy to measure the challenge of complex reasoning in TKGQA demanded by the questions. To fill this gap, we first introduce a two-dimensional taxonomy of complex reasoning in TKGQA. This taxonomy consists of two dimensions: *structural complexity* and *computational complexity*. By parsing the SPARQL queries [8] or analyzing the question templates of existing datasets to infer their implicit query structures, we find significant gaps across their coverages, as detailed in Table 2.

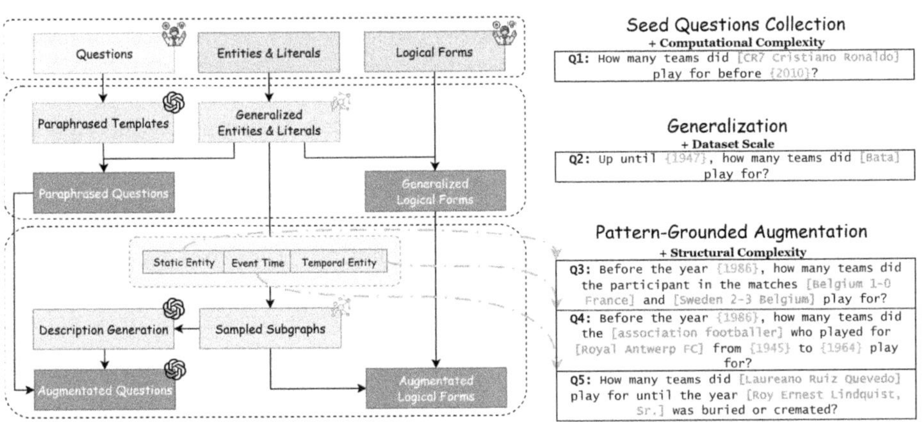

Fig. 1. The overview of construction of the CR-TKGQA dataset.

Guided by this taxonomy, we construct **CR-TKGQA**, a TKGQA dataset involving Complex Reasoning on Wikidata [22]. Its intent-driven construction directly follows our taxonomy, ensuring broad coverage of both structural complexity and computational complexity. The construction process, as shown in

Fig. 1, contains three stages. We start by seeding a wide variety of computational complexities from human intent by manually authoring diverse questions and their corresponding SPARQL queries, from simple comparisons (*before*) to complex aggregations (*average age*). We then scale the dataset's size and diversity through automatic generalization. We subsequently design augmentation strategies to enhance structural complexity, effectively introducing diverse graph pattern structures.

Ultimately, we produce over 30,000 complex temporal questions, each paired with an executable SPARQL query on Wikidata, providing a large-scale resource for training and evaluation.

The rest of the paper is organized as follows. Section 2 introduces our taxonomy and reviews related works. Section 3 describes the construction and analysis of the CR-TKGQA dataset. Section 4 evaluates baselines on the CR-TKGQA dataset. Section 5 draws a conclusion.

2   Complex Reasoning in TKGQA

In this section, we first define the core challenges of complex reasoning in TKGQA through our two-dimensional taxonomy (Subsect. 2.1). We then use this taxonomy to analyze existing TKGQA methods (Subsect. 2.3), and to examine the characteristics of current benchmarks (Subsect. 2.2).

2.1   A Taxonomy of Complex Reasoning for TKGQA

We begin by elucidating the challenges of the TKGQA task, which is characterized by two underlying dimensions of complexity: *structural complexity* and *computational complexity*.

The process of TKGQA can be conceptualized as parsing the question into a SPARQL query [8]. To make this concrete, let us consider the structure of the data these queries operate on. On KGs, specifically Wikidata, *facts* are stored as triples. These facts can be binary (a simple relation) or n-ary, where additional *qualifiers* provide additional information. *Temporal facts* are those where the object or a qualifier value is a timestamp. For example, the 3-ary fact *<Anne Marie Ottersen, award_received, Amanda Award for Best Actress, point_in_time, 1986>* contains a *temporal value 1986* of the type *time point*. Another type of temporal value is *duration*, which is usually derived from intervals between time points. In this paper, we use the term *entity* to refer to items in the KG (e.g., *Anne Marie Ottersen*), and distinguish them from *literals*, which are values like timestamps (e.g., *1986*) or strings.

SPARQL queries consist of two fundamental components: a *graph pattern* that identifies the related facts to be retrieved from the KG; a series of *temporal computations* applied to these retrieved facts. These two components directly correspond to the aforementioned challenges of TKGQA: The challenge of KG navigation is embodied in the complexity of graph patterns; the challenge of applying temporal computations is reflected in the complexity of these temporal computations. Inspired by this intuition, our two-dimensional taxonomy of complex reasoning in TKGQA, as shown in Table 1, is built upon two dimensions: structural complexity and computational complexity, and derived from the question's underlying SPARQL query structure.

Crucially, this taxonomy based on query structure provides a unifying framework to align taxonomies that are based on linguistic phenomena from existing works. The key insight is that the mapping from linguistic surface forms to underlying queries is fundamentally many-to-one. For example, consider *implicit temporal questions* [11], which identifies questions with free-text temporal expression such as *Obama's presidency*, and *time join* [17], which considers the feature of an event during the other event. Despite their distinct origins, both can be unified under the same query structure: first retrieving multiple temporal facts, then performing computations on the temporal values from these facts.

Our taxonomy serves two purposes. First, for datasets where SPARQL queries are available, it permits precise, fine-grained analysis of question complexity. Second, it allows us to project template-based datasets onto a standardized complexity taxonomy by projecting their taxonomy to ours.

Structural Complexity. Structural complexity measures the intricacy of the graph pattern required to answer a question. To analyze it systematically, we first decompose the SPARQL query based on Wikidata's data model of statements. Specifically, we group triple patterns that connect to a common statement node, as this structure corresponds to retrieving a single fact. We then identify temporal variables (i.e., the variables whose values will be temporal values) within these triple patterns by inspecting their property datatypes.[1].

Based on this formalization, we identify two fundamental structures, which we refer to as the two *aspects* of structural complexity: i) *Temporal fact fusion*: the graph pattern contains temporal variables from two or more distinct facts (i.e., from separate groups of statement-centric triples). ii) *Multi-hop reasoning*: a path spans multiple facts linked by shared variables.

Temporal fact fusion involves retrieving temporal values by temporal variables in multiple, distinct facts. This structure is inherently complex, independent of any subsequent computations. For instance, the question about *Grover Cleveland's terms* in Table 1 requires matching the triple patterns for two separate facts merely to list their start times. While this example demonstrates pure structural complexity, such patterns are often combined with computations, as in the question *How long is between Grover Cleveland's two presidential terms*, where a duration is derived from the two retrieved time points.

Multi-hop reasoning uses entities as intermediate nodes for KG navigation, as in the example (Table 1), the reference *the award that Anne Marie Ottersen*

---

[1] In this paper, we distinguish two types of properties on Wikidata: predicates (prefixes wdt:, p:/ps:) define core relations, while qualifiers (prefix pq:) provide contextual details attached to a statement.

Table 1. An overview of our taxonomy of complex reasoning in TKGQA, and its two dimensions: structural complexity and computational complexity.

(a) Structural Complexity

Aspect	Definition	Example
Temporal Fusion	Fact Integrating temporal values from multiple facts.	*When did Grover Cleveland's two presidential terms begin*
Multi-hop Reasoning	Navigating a chain of facts.	*Who were the winners of the award that Anne Marie Ottersen received*

(b) Computational Complexity

Aspect	Definition	Example
Comparison		
Time Point Comparison	Comparing two time points.	*before 2001*
Duration Comparison	Comparing two durations.	*longer than 10 years*
Arithmetic		
Duration Derivation	Calculating duration between time points.	*the presidency length of Obama*
Duration Calculation	Adding/Subtracting durations.	*the total length of two presidents' terms*
Time Point Shift	Adding a duration to a time point.	*5 years after WWII*
Granularity Conversion	Converting time units.	*3 years (in days)*
Aggregation		
Time Point Ordinal	Ordering time points.	*the earliest event*
Duration Ordinal	Ordering durations by length.	*the youngest winner*
Statistics	Frequency and AVG/SUM of durations.	*the average age*

received must first be resolved to the entity *Amanda Award for Best Actress* via her award history. Subsequent reasoning then proceeds from this inferred entity.

Computational Complexity. Computational complexity comes from the set of temporal computations demanded by the question. In TKGQA, these computations are predominantly temporal, as they primarily act upon temporal values like time points and durations.

We classify these temporal computations into three categories, and refine these categories into 9 kinds of computations, which we refer to as the 9 *aspects* of computational complexity, as in Table 1: i) *Comparison* ($>$, $=$, and $<$) accesses the relational order of time points or the length of durations. ii) *Arithmetic* ($+$,

−, and granularity conversion) derives new temporal values by applying calculations to time points and durations, including deriving a duration from time points, shifting a time point by a duration, and converting between granularities. iii) *Aggregation* calculates a summary value over a set of temporal values. In this paper, we consider: 1) ordinal over time points and durations; 2) statistics, including average (AVG) and sum (SUM) of durations, and frequency, defined as the number of events divided by a duration.

For illustration, answering the question *Who was the youngest winner of the award that Anne Marie Ottersen received before 2001* involves all three categories of computations. First, a granularity conversion is needed, as the question specifies the year *2001* while award times in the KG are often stored as dates. After aligning the granularities, the process involves: i) a time point comparison, as the award time must be before *2001*; ii) a duration derivation to compute each winner's age upon the award; and iii) a duration ordinal to find the winner with the minimum age. In practice, these computations can be identified by analyzing the operators and the type of operands in FILTERs, BINDs, and solution modifiers in the SPARQL query.

2.2 Existing TKGQA Datasets

After establishing our taxonomy, we examine the existing TKGQA datasets from the perspective of this taxonomy. This examination substantiates the gaps in current datasets highlighted in the introduction and further demonstrates the necessity of a more comprehensive benchmark.

Coverage of Complexity. Early TKGQA datasets, such as TempQuestions [11] and TimeQuestions [13], extract temporal questions from QA datasets like CWQ [20]. Regarding structural complexity, they include multi-hop reasoning and temporal fact fusion; regarding computational complexity, they capture ordering relationships, including time point comparison and time point ordinals.

Recent template-based TKGQA datasets [3,6,15,17,19] populate predefined templates to generate a large-scale questions, which deepens the coverage of computational complexity. CronQuestions [17] first utilizes templates to produce large-scale TKGQA dataset, considering mainly time point comparison (*before/after*) and ordinal (*First/Last*), and temporal fact fusion (*Time join*); MultiTQ [3] largely covers this scope while introducing granularity conversion; TimelineKGQA [19] incorporates duration calculation/ordinal; ETRQA [15] achieves a relatively broad coverage of computational complexity, lacking only statistics. However, these datasets often neglect structural complexity. Among template-based datasets, only ComplexTempQA [6] considers multi-hop reasoning.

Question Taxonomies. One potential reason for this limited coverage may be their question taxonomies that focus on the *phenomenon*, i.e., the surface form

Table 2. The comparison across TKGQA datasets based on our taxonomy (See Table 1 for details). While ETRQA and Complex-TempQA are included, they are not standard TKGQA datasets as they adopt open-book/open-domain settings. TempQA-WD's coverage is identified by parsing its SPARQL queries; Other datasets are all template-based, whose coverage is identified by checking their query templates and question taxonomies.

	TempQA-WD [16]	Multi-TQ [3]	Timeline-KGQA [19]	ETR-QA [15]	Complex-TempQA [6]	**CR-TKGQA**
Structural Complexity						
Temporal Fact Fusion	✓	✓	✓	✓	✓	✓
Multi-hop Reasoning	✓				✓	✓
Computational Complexity						
Comparison						
Time Point Comparison	✓	✓	✓	✓	✓	✓
Duration Comparison				✓	✓	✓
Arithmetic						
Duration Derivation			✓	✓		✓
Duration Calculation			✓	✓		✓
Time Point Shift	✓			✓		✓
Granularity Conversion	✓	✓		✓		✓
Aggregation						
Time Point Ordinal	✓	✓	✓	✓	✓	✓
Duration Ordinal			✓	✓		✓
Statistics						✓

of the question, while neglecting their underlying query structures. TempQuestions first identifies implicit temporal questions by implicit temporal expressions. MultiTQ considers multiple computations (*Before Last/After First*). ETRQA proposes a taxonomy based on question composition and answer type, identifying compound questions that combine various temporal constraints (e.g., explicit, implicit) with temporal comparisons (e.g., order, duration).

While the phenomenon can correlate with underlying query structures, the mapping, relying on the surface forms, is coarse. By proposing a taxonomy based on query structure, we transform the taxonomy based on fragmented surface form into a rigorous, unified taxonomy for analysis.

Lack of SPARQL Queries. Another noteworthy characteristic is the widespread absence of ground-truth SPARQL queries: Among major TKGQA datasets, only TempQA-WD [16] (a migration of TempQuestions to Wikidata) provides SPARQL queries for questions. This is often a consequence of task design: many recent datasets are built upon smaller Temporal Knowledge Graphs (TKGs) [3,18]. ETRQA [15] further adopts an *open-book* setting, providing necessary facts as context. Similarly, ComplexTempQA primarily evaluates a model's parametric knowledge in an open-domain setting. Although these approaches obviate the need for SPARQL queries, they also prevent query-based analysis.

Comparison to Our Dataset . In light of this analysis, CR-TKGQA is designed to complement existing resources in two key ways: i) CR-TKGQA achieves a broader coverage of structural complexity and computational complexity; ii) CR-TKGQA provides executable SPARQL queries on Wikidata for each question, serving as a valuable resource for analysis and training.

2.3   Methods for TKGQA

Having reviewed existing TKGQA datasets, we analyze existing TKGQA methods, focusing on how they address the challenge of complex reasoning. We categorize these methods into two classes based on training requirements.

We first discuss methods that require training. TKGQA methods before the Large Language Model (LLM) era often utilize question-answer pairs as the training data. They address the challenge by adopting specialized designs, such as time-aware embeddings (e.g., EXAQT [13]) or heuristic-based decomposition or structure (e.g., Tequila [12] and SF-TQA [4]). While effective on specific benchmarks, these specialized designs may not generalize to the structural complexities and computational complexities in CR-TKGQA. Apart from these dedicated TKGQA methods, a mainstream KGQA paradigm before the LLM era is *sequence-to-sequence (seq2seq) fine-tuning* using question-SPARQL query pairs [23]. This approach directly learns a mapping from question to query, but often struggles to generalize to unseen query patterns.

We then turn to the few-shot LLM-based methods free of training data. The internal knowledge and reasoning ability of LLMs allows *few-shot LLM-based KGQA methods* to directly answer the question, perform Retrieval-Augmented Generation (RAG) [1], or act as agents that compose SPARQL queries to interact with the KG [14]. *Few-shot LLM-based TKGQA methods* augment these KGQA methods with specialized temporal modules or prompts (e.g., ARI [2] and RTQA [5]). However, whether these LLM-based methods can handle the challenges in CR-TKGQA remains a question.

3   CR-TKGQA Dataset

In this section, we present CR-TKGQA, a TKGQA dataset involving complex reasoning, on Wikidata [22]. We first describe the construction of the dataset in Subsect. 3.1, then conduct an in-depth analysis in Subsect. 3.2.

3.1   Construction of CR-TKGQA Dataset

The goal of CR-TKGQA dataset is to generate temporal questions containing both structural complexity and computational complexity at scale, paired with SPARQL queries. Since LLMs struggle with accurately generating such questions and mapping them to SPARQL queries due to common-sense knowledge requirements, We employ an *intent-driven* approach, adapting the paradigm from MarkQA [10] with time-specific extensions. We first collect human-authored questions and their corresponding SPARQL queries, which cover a variety of computational complexities. We then perform time-specific automatic generalization for scaling and augmentation to enrich structural complexity.

Transferability . Our approach is KG-agnostic and can be transferred to any KG that supports SPARQL endpoints. To migrate this pipeline, the experts would need to i) annotate seed question-SPARQL query pairs, and ii) design patterns for sampling strategies regarding the KG schema. Nevertheless, we note that Wikidata's user-friendly UI for browsing facilitates easier annotation.

Seed Questions Collection. We begin by collecting a set of manually authored temporal questions to serve as the foundation of the CR-TKGQA dataset, namely Seed Questions (SQs). We hope that these SQs can cover diverse temporal phenomena from simple queries about temporal facts to complex temporal operations. These SQs should also be expressed concisely and naturally, distinct from rigid templates. These characteristics are then inherited by the subsequent questions generated in subsequent automated generalization and augmentation.

We select 40 common predicates of temporal facts on Wikidata to define the core topics for the seed questions. We invite three KGQA experts to author seed questions and their corresponding SPARQL queries, with access to the Wikidata official website as the knowledge source. This process contains three principles: i) Questions must be temporal in nature and center on a topic property, i.e., the question should originate from temporal facts where this property acts as the predicate. ii) Questions should span a spectrum of temporal phenomena, from simple fact retrieval to complex reasoning (e.g., duration ordinal and fact counting). The language must be both natural and concise. iii) All entities and literals are marked for subsequent automated generalization and augmentation. Upon completion, each question is cross-validated by all three experts.

Generalization. We then generalize the seed questions to increase the scale of the dataset, producing Generalized Questions (GQs). The generalization replaces entities and literals in both the questions and their corresponding SPARQL queries with new values. Generalization contains two parallel processes: literal perturbation and entity substitution.

For literal perturbation, we extend MarkQA's approach by designing diverse perturbation strategies specifically for literals according to their types. For non-periodic literals such as years, ordinals, and duration, we apply Gaussian-distributed offsets. For instance, a year value is perturbed by adding a random offset $X \sim N(\mu = 0, \sigma^2 = 50)$, where *2014* is perturbed into *1947* in *Q2* of Fig. 1. For periodic literals (months and days), we perform random sampling within their valid ranges. All perturbed literals must constitute legal temporal values (e.g., years must fall between -2020 and 2024). For entity substitution, we follow MarkQA to randomly select entities in the SPARQL query to be replaced by variables, and impose type constraints on these variables (they must have common types with the original entities). Execution of these modified queries against the KG retrieves valid entity substitutions. In *Q2* of Fig. 1, we find a substitution *Jan de Kreek* for Ronaldo, who also has the type of *human*.

After obtaining literal perturbations and entity substitutions, we randomly select the combination of both and retain only those combinations that yield a manageable number of answers (1 to 10) after being applied to the SPARQL query. The selected substitutions are then used to replace the corresponding literals and entities in the question and the query. To maintain a balanced distribution in dataset, we limit each seed question to generate at most 15 generalized questions. To ensure linguistic diversity, we employ DeepSeek-v3 to paraphrase the generalized questions derived from the same seed question.

Pattern-Grounded Augmentation. We further augment the dataset by introducing diverse graph pattern structures through a Pattern-Grounded Augmentation (PGA) process, producing Augmented Questions (AQs). This process involves replacing entities or literals in questions with query snippets, whose execution returns precisely the original entity or literal to be replaced.

The query snippets are instantiated by grounding the predefined topological patterns containing variable slots. We design three PGA strategies: i) *Static Entity Augmentation (SEA)* (*Q3* in Fig. 1) adopts the composition concept from MarkQA and replaces entities with non-temporal query snippets, augmenting multi-hop reasoning; ii) *Temporal Entity Augmentation (TEA)* (*Q4* in Fig. 1) replaces entities with time-related snippets, augmenting multi-hop reasoning while adding temporal fact fusion; iii) *Event Time Augmentation (ETA)* (*Q5* in Fig. 1) extends augmentation to literals by replacing them with temporal facts, augmenting temporal fact fusion. Consider the SEA example *Q3* in Fig. 1: the topological pattern `(?e1,?p1,?x)`, `(?e2,?p2,?x)` is grounded by setting `?p1` and `?p2` to property *participant*, and `?e1` and `?e2` to two specific matches. The resulting query snippet uniquely identifies a player *Victor Mees*.

The process for grounding these patterns differs by strategy. For SEA and TEA, we execute the patterns on the KG, constraining the result to be exactly the entity to be replaced. For ETA, we only consider time points with year granularity in this paper, and require the year of the resulting temporal fact to match the literal to be replaced. Once the query snippets are obtained, we generate natural language description for each query snippet using `DeepSeek-v3`. The mention of the entity or year literal in the question is replaced with this description; The entity or the year literal in the SPARQL query is replaced by the query snippets. We ensure the original and augmented SPARQL queries return identical results on the KG.

To control the scale of the dataset and the length of questions, we impose two constraints during PGA. i) Each seed or generalized question undergoes augmentation at most twice. ii) We randomly select a maximum of two entities or year literals for replacement within a single augmentation.

Alignment, Data Partitioning, and Quality Checking. To ensure the answers in our dataset reflect the most current state of Wikidata, we executed all SPARQL queries against the official Wikidata endpoint and updated the answers accordingly[2]. Any SPARQL queries (and their corresponding questions) that return empty result sets are discarded.

Following the dataset splitting methodology of GrailQA [7], we partition our dataset into training (train), development (dev), and test sets with a ratio of 7:2:1. The dev and test sets comprise three distinct levels of generalization: i) I.I.D. (25%): Instances are randomly sampled from the train split. ii) Compositional (25%): Instances contain properties that can be found in the train split, but exhibit new query structures absent from the train split. iii) Zero-shot (50%): Instances contain properties that are absent from the train split.

To validate dataset quality, three experts randomly selected 100 question-SPARQL query pair samples and manually examine them. For each question, we check i) the linguistic fluency of the questions, and ii) the logical alignment between the question and its SPARQL query. A question is accepted only if at least two experts agree on both criteria. In total, 97 samples pass this validation.

Taking the *Cristiano Ronaldo* seed in Fig. 1 for an example, we reject augmentations such as: i) replacing *Ronaldo* with *The football player who has the father of Jose Dinis Aveiro* for linguistic awkwardness; and ii) replacing *Ronaldo* with *the football player who is the father of Jose Dinis Aveiro* because it reverses the relation polarity, causing a logical misalignment with the SPARQL query.

Table 3. The distribution of different types of questions in CR-TKGQA dataset. # denotes the number of corresponding types of questions.

Split	#Questions	#SQ	#GQ	#AQ		
				#SEA	#TEA	#ETA
Train	21,701	807	9,497	4,928	4,981	1,488
Dev	6,327	165	1,682	1,998	2,029	453
Test	3,050	104	989	667	875	415
Total	31,078	1,076	12,168	7,593	7,885	2,356

3.2 Analysis of CR-TKGQA Dataset

Scale of Dataset. First, we present the distribution of questions in each partition of CR-TKGQA in Table 3. In total, CR-TKGQA contains 31,078 questions. Although this number is smaller than recent template-based TKGQA datasets such as MultiTQ (500K), it is substantially larger than early extraction-based datasets like TempQA-WD (839) and TimeQuestions (16,859). This positions

---

[2] The time of alignment is November 2025. We provide a Wikidata dump on 2025-11-14 in https://box.nju.edu.cn/d/53ecfc574a0d499384ce/ for local deployment.

CR-TKGQA as a moderate-scale, high-quality resource. Moreover, CR-TKGQA includes questions with multi-field answers. It also encompasses comprehensive answer types (entities, time points, numbers, Boolean values).

Table 4. Fine-grained analysis of complexity aspects in TempQA-WD and CR-TKGQA, measured by: i) the proportion of questions exhibiting each complexity aspect, and ii) the number of canonicalized SPARQL queries (i.e., query structures with entities/literals masked).

	TempQA-WD	CR-TKGQA (Ours)		
		Overall	SQ	GQ	AQ
Structural Complexity					
Temporal Fact Fusion	18.0%	76.0%	58.1%	64.6%	84.8%
Multi-hop Reasoning	29.8%	38.5%	27.4%	19.0%	52.4%
Computational Complexity					
Comparison					
Time Point Comparison	53.2%	69.8%	51.8%	54.9%	81.0%
Duration Comparison	–	4.8%	9.4%	5.0%	4.4%
Arithmetic					
Duration Derivation	–	26.8%	36.0%	32.3%	22.5%
Duration Calculation	–	0.8%	0.7%	1.0%	0.6%
Time Point Shift	4.6%	0.6%	0.3%	0.4%	0.7%
Granularity Conversion	4.1%	39.6%	30.8%	22.6%	51.8%
Aggregation					
Time Point Ordinal	12.6%	20.8%	24.0%	22.9%	19.2%
Duration Ordinal	–	6.4%	13.3%	10.0 %	3.5%
Statistics	–	4.0%	3.7%	4.6 %	3.7%
Other Features					
#Canonicalized SPARQL queries	286	10,252	636	524	9,626

Complexity Analysis. Having established its scale, we now delve into the complexity of CR-TKGQA. As our taxonomy enables fine-grained analysis when SPARQL queries are available, we conduct a fine-grained analysis based on our taxonomy with a direct comparison to TempQA-WD, the only existing TKGQA dataset that provides SPARQL queries on Wikidata. The results, as detailed in Table 4, reveal two crucial findings.

First, CR-TKGQA is significantly more complex than TempQA-WD. Regarding the structural complexity, the proportion of questions involving temporal fact fusion or multi-hop reasoning is substantially larger than that of TempQA-WD. Specifically, 33.9% of questions in CR-TKGQA involve both temporal fact fusion and multi-hop reasoning, whereas in TempQA-WD, there are only 1.3% (not tabulated). Regarding computational complexity, CR-TKGQA covers all nine categories of computations. CR-TKGQA also substantially exceeds TempQA-WD in the number of canonicalized SPARQL queries.

Second, our intent-driven approach systematically introduces complexity as designed. The SQs have already captured the full range of computational complexity. Subsequently, our pattern-grounded augmentation significantly strengthens structural complexity: The AQs exhibit a high proportion of temporal fact fusion and multi-hop reasoning (84.8% and 52.4%, respectively) and a more than tenfold increase in the number of canonicalized SPARQL queries compared to the SQs.

4   Experiments

In this section, we conduct a comprehensive evaluation of representative baselines on CR-TKGQA. Our goal is two-fold: first, to establish CR-TKGQA as a challenging benchmark; second, to use our taxonomy to conduct a fine-grained analysis of baseline performance.

4.1   Experiment Setups

We use the Wikidata official SPARQL endpoint[3] for evaluation. To ensure a comprehensive evaluation, we select the following representative baselines identified in our analysis (Sect. 2.3):

- Pre-LLM era TKGQA method: EXAQT [13] and SF-TQA [4] are chosen to evaluate traditional TKGQA systems.
- Sequence-to-sequence (seq2seq) fine-tuning: We fine-tune Llama-3-8B-Instruct (abbreviated as LLaMA) on the question-SPARQL query pairs from our training set using Low-Rank Adaptation (LoRA) [9] to establish performance when full training data is available.
- Few-shot LLM-based KGQA: We prompt the LLM to answer the question using Chain-of-Thought (DirectQA), and select SPINACH [14], a state-of-the-art KG exploration agent on Wikidata that adopts an iterative refinement paradigm to construct SPARQL queries.
- Few-shot LLM-based TKGQA: RTQA [5] is a TKGQA method that first decomposes the question, then recursively answers the question. Since RTQA is designed for TKGs and relies on a textualized KG for retrieval, we retrieve the one-hop subgraph on Wikidata for every entity in the queries of the full dataset to form a file as the textualized KG.

Evaluation Metric. We use the F1 score on answers as the evaluation metric. Since some questions in our dataset involve multi-field answers, we adopt a flexible evaluation strategy similar to [14]: a model's answer is considered correct as long as it contains all correct tuples, even if it includes additional columns. We evaluate EXAQT, SF-TQA, LLaMA, and SPINACH based on the execution results of their generated queries. For RTQA and DirectQA, which

---

[3] https://query.wikidata.org/.

provide answers directly in natural language, we use string matching against the labels and aliases. To reduce computational cost, we evaluate all baselines on a random sample of 1,000 questions from the test set.

Settings. *Linkings.* Due to the complexity of questions in CR-TKGQA, which makes entity and relation linking challenging, we evaluate all models using gold-standard linking results. For EXAQT and SF-TQA, we replace their linking results with gold-standard relations/entities. For LLaMA, we append entities/relations to the question during training and inference. For LLM-based methods, we concatenate gold-standard entities after the question.

Models and Training. We use DeepSeek-v3.2-Exp as the LLM backbone for few-shot methods. Other methods are trained and evaluated on an NVIDIA RTX A6000 GPU.

4.2   Results

Table 5. Baseline performance (F1 score in percentage) on CR-TKGQA dataset. * denotes that the method is designed for Temporal KGs (TKGs) and is adapted for Wikidata by our modification, which may not reflect its original performance.

Method	Overall	I.I.D	Compositional	Zero-shot
Pre-LLM TKGQA				
SF-TQA	7.5	8.3	9.6	6.1
EXAQT	9.2	12.3	14.2	4.9
Seq2seq Fine-Tuning				
LLaMA	43.2	74.3	74.6	10.1
Few-shot LLM-Based KGQA				
DirectQA	29.7	28.7	27.2	31.6
SPINACH	43.9	46.5	44.8	42.1
Few-shot LLM-Based TKGQA				
RTQA*	27.6	28.7	23.4	29.3

We report the main results in Table 5. These results reveal two key findings.

First, CR-TKGQA is more challenging than existing TKGQA datasets. Baselines achieve strong results on previous benchmarks, both in their original reports (SPINACH:F1 score = 69.5 in QALD-10 [21], RTQA: Hits@1 = 76.5% on MultiTQ [3], EXAQT/SF-TQA: Hits@1 = 56.5%/53.9% on TimeQuestions [13]) and under our settings on TempQA-WD [16] (F1 scores of 71.5, 59.5, and 52.5 for SPINACH, RTQA, and DirectQA, respectively). The sharp performance drop of these baselines on CR-TKGQA clearly underscores its challenge.

Second, CR-TKGQA reveals the characteristics of different LLM-based baselines. While fine-tuned LLaMA model and few-shot LLM-based baselines outperform pre-LLM era TKGQA methods, their performance varies. SPINACH achieves the best overall performance, slightly surpassing the fine-tuned LLaMA. Its decisive advantage lies in the zero-shot split, where it outperforms LLaMA by a large margin (42.1 vs. 10.1), highlighting a known limitation of the fine-tuning approach. Among few-shot LLM-based baselines, the performance of RAG-based RTQA is suboptimal. This is likely because its text-based retrieval, limited to one-hop neighborhoods in our setup, is insufficient for retrieving multiple facts required by our dataset. However, expanding the context is impractical due to exponential growth. The varied performance across both the fine-tuned model and the few-shot LLM-based baselines indicates that complex reasoning in TKGQA remains challenging, even for the fine-tuning paradigm and powerful LLMs (Table 6).

Table 6. Fine-grained performance of few-shot LLM-based baselines based on our taxonomy. **F1** is the average F1 score on the subset containing the complexity aspect, and Δ is the difference of F1 score between the subset and the overall performance in Table 5. † denotes subsets with small size (n ≤ 10).

	Subset Size	SPINACH		DirectQA		RTQA	
		F1	Δ	F1	Δ	F1	Δ
Structural Complexity							
Temporal Fact Fusion	741	42.7	1.2 ↓	28.8	0.9 ↓	25.4	2.2 ↓
Multi-hop Reasoning	590	32.2	11.7 ↓	22.2	7.5 ↓	15.9	11.7 ↓
Computational Complexity							
Comparison							
Time Point Comparison	767	43.8	0.1 ↓	28.1	1.6 ↓	24.9	2.7 ↓
Duration Comparison	66	33.3	10.6 ↓	45.9	16.2 ↑	25.5	2.1 ↓
Arithmetic							
Duration Derivation	241	40.6	3.3 ↓	31.0	1.3 ↑	29.8	2.2 ↑
Duration Calculation†	2	100.0	56.1 ↑	50.0	20.3 ↑	0.0	27.6 ↓
Time Point Shift†	5	20.0	23.9 ↓	20.0	9.7 ↓	20.0	7.6 ↓
Granularity Conversion	52	45.5	1.6 ↑	11.5	18.2 ↓	21.2	6.4 ↓
Aggregation							
Time Point Ordinal	191	35.7	8.2 ↓	19.9	9.8 ↓	18.3	9.3 ↓
Duration Ordinal	32	22.4	21.5 ↓	21.9	7.8 ↓	21.9	5.7 ↓
Statistics	35	20.0	23.9 ↓	24.3	5.4 ↓	20.0	7.6 ↓

We then turn to the fine-grained analysis of the performance of baselines across the complexity aspects of our taxonomy. For this analysis, we focus on the few-shot LLM-based baselines. SF-TQA and EXAQT are excluded due to their suboptimal performance, and LLaMA is omitted because its results are skewed by its performance on the zero-shot split. For each complexity aspect in our taxonomy, we calculate the baselines' average F1 score on the subset of questions that contains this aspect. We then compare this subset-specific F1 score against the model's overall performance (as reported in Table 5) to analyze the performance change (i.e., drop ↓ or increase ↑). A significant performance degradation on a subset indicates a specific challenge in handling that aspect.

The primary finding is that the complex aspects defined in our taxonomy effectively represent the challenges for baselines. For the vast majority of aspects of both structural and computational complexity, baselines generally exhibit a performance degradation, with only a few exceptions in arithmetic and comparison tasks. This consistent performance degradation validates our taxonomy's core premise: the identified structural patterns and computations reflect the challenges of complex reasoning in TKGQA.

Furthermore, a clear hierarchy of difficulty exists within each dimension. In structural complexity, multi-hop reasoning is significantly more challenging than temporal fact fusion, with a substantial performance degradation on all baselines. In computational complexity, for all baselines, aggregation consistently leads to a performance degradation, identifying aggregation as the most challenging category. Specifically, Duration Ordinal and Statistics stand out as the most challenging, with consistently low F1 scores (around 20%) across all baselines.

Finally, the fine-grained analysis highlights distinct baseline characteristics. An observation is that, while SPINACH generally outperforms DirectQA and RTQA across most aspects, its degree of performance degradation on duration-related computations is noticeably more significant. A possible explanation is that DirectQA and RTQA may leverage parametric knowledge where durations are implicitly stored, whereas SPINACH must first retrieve time points before deriving durations, since most temporal values stored in Wikidata are time points. As a case study, we investigate the duration-related questions in the results from Table 5. Out of 241 duration-related questions, there are 34 questions that SPINACH fails due to errors in SPARQL, whereas DirectQA succeeds.

Performance Without Gold-Standard Linking. To evaluate the impact of entity linking on performance, we first employ GPT-4o-mini to extract entity mentions from the question, retrieve the top-5 candidate entities via the Wikidata API for each mention, and use GPT-4o-mini to select the most plausible entity for each mention. We then use these linked entities to evaluate SPINACH on the same 1,000 samples as Table 5, achieving an F1 score of 41.8. Compared to the performance in Table 5, there is a slight drop, confirming that entity linking in CR-TKGQA remains a challenge. Crucially, the performance with gold-standard linking indicates that complex reasoning remains the core bottleneck.

Summary of Experiments. Our experiments highlight the significant challenges CR-TKGQA presents to current baselines. The fine-grained analysis of complexity aspects identifies the challenge from each complexity aspect and the characteristics of different baselines. To sum up, the experiments suggest that CR-TKGQA is a valuable benchmark for the development of TKGQA models, and our two-dimensional taxonomy serves as a useful taxonomy for analysis.

5   Conclusion

In this paper, to address the limited complexity coverage in current TKGQA datasets, we first introduce a two-dimensional taxonomy of complex reasoning in TKGQA that consists of structural complexity and operational dimensions, enabling fine-grained analysis. We then construct CR-TKGQA, a TKGQA dataset involving complex reasoning. Through an intent-driven construction process involving manual seeding for operational complexity and automatic scaling for structural complexity, CR-TKGQA offers a broad coverage of our complexity taxonomy. The inclusion of executable SPARQL queries for every question further enhances its value as a resource for both training and analysis.

Our experiments highlight that CR-TKGQA poses a significant challenge to current baselines, revealing their limitations in handling complex KG navigation and temporal computations. The detailed analysis guided by our taxonomy validates the value of our taxonomy. By providing a more challenging and systematically constructed dataset, CR-TKGQA can serve as a useful benchmark for the development of TKGQA systems.

Despite these contributions, we acknowledge the following limitations for future work: i) Our current taxonomy is primarily categorical, classifying questions based on the presence of discrete structural and computational patterns. A promising direction for future work is to evolve our taxonomy into a quantitative framework. This may involve assigning complexity scores by quantified measurements, e.g., the number of fused temporal facts and the number of computations. ii) Our current analysis focuses on the impact of individual complexity aspects. A valuable direction for future work is to investigate the interaction effects between the two dimensions when they intertwine.

Acknowledgement. This work is supported by the National Science and Technology Innovation 2030 Major Program (2025ZD0544900). We also would like to thank the anonymous reviewers for their insightful comments and constructive suggestions, which help to improve the quality of this paper.

Supplemental Material Statement. The source code is available in the GitHub repository https://github.com/nju-websoft/CR-TKGQA.

Use of Generative AI. Gemini-2.5pro and GPT-5 are utilized to polish the language of this paper. DeepSeek-v3 is used for rephrasing our seed questions and generating natural language descriptions of query snippets in the construction of our dataset. DeepSeek-v3.2-exp serves as the LLM backbone of LLM-based methods in the experiments in this paper.

References

1. Chen, L., Tong, P., Jin, Z., Sun, Y., Ye, J., Xiong, H.: Plan-on-graph: self-correcting adaptive planning of large language model on knowledge graphs. In: Globersons, A., et al. (eds.) Advances in Neural Information Processing Systems 38: Annual Conference on Neural Information Processing Systems 2024, NeurIPS 2024, Vancouver, BC, Canada, 10–15 December 2024 (2024). http://papers.nips.cc/paper_files/paper/2024/hash/4254e856d01a5e7b7ea050477c3ef9b9-Abstract-Conference.html

2. Chen, Z., Li, D., Zhao, X., Hu, B., Zhang, M.: Temporal knowledge question answering via abstract reasoning induction. In: Ku, L., Martins, A., Srikumar, V. (eds.) Proceedings of the 62nd Annual Meeting of the Association for Computational Linguistics (Volume 1: Long Papers), ACL 2024, Bangkok, Thailand, 11–16 August 2024, pp. 4872–4889. Association for Computational Linguistics (2024). https://doi.org/10.18653/V1/2024.ACL-LONG.267

3. Chen, Z., Liao, J., Zhao, X.: Multi-granularity temporal question answering over knowledge graphs. In: Rogers, A., Boyd-Graber, J.L., Okazaki, N. (eds.) Proceedings of the 61st Annual Meeting of the Association for Computational Linguistics (Volume 1: Long Papers), ACL 2023, Toronto, Canada, 9–14 July 2023, pp. 11378–11392. Association for Computational Linguistics (2023). https://doi.org/10.18653/V1/2023.ACL-LONG.637

4. Ding, W., Chen, H., Li, H., Qu, Y.: Semantic framework based query generation for temporal question answering over knowledge graphs. In: Goldberg, Y., Kozareva, Z., Zhang, Y. (eds.) Proceedings of the 2022 Conference on Empirical Methods in Natural Language Processing, EMNLP 2022, Abu Dhabi, United Arab Emirates, 7–11 December 2022, pp. 1867–1877. Association for Computational Linguistics (2022). https://doi.org/10.18653/V1/2022.EMNLP-MAIN.122

5. Gong, Z., Li, J., Liu, Z., Liang, L., Chen, H., Zhang, W.: RTQA: recursive thinking for complex temporal knowledge graph question answering with large language models. In: Christodoulopoulos, C., Chakraborty, T., Rose, C., Peng, V. (eds.) Proceedings of the 2025 Conference on Empirical Methods in Natural Language Processing, EMNLP 2025, Suzhou, China, 4–9 November 2025, pp. 9853–9870. Association for Computational Linguistics (2025). https://doi.org/10.18653/V1/2025.EMNLP-MAIN.499

6. Gruber, R., Abdallah, A., Färber, M., Jatowt, A.: ComplexTempQA: a 100m dataset for complex temporal question answering. In: Christodoulopoulos, C., Chakraborty, T., Rose, C., Peng, V. (eds.) Proceedings of the 2025 Conference on Empirical Methods in Natural Language Processing, EMNLP 2025, Suzhou, China, 4–9 November 2025, pp. 9100–9112. Association for Computational Linguistics (2025). https://doi.org/10.18653/V1/2025.EMNLP-MAIN.463

7. Gu, Y., et al.: Beyond I.I.D.: three levels of generalization for question answering on knowledge bases. In: Leskovec, J., Grobelnik, M., Najork, M., Tang, J., Zia, L. (eds.) WWW '21: The Web Conference 2021, Virtual Event/Ljubljana, Slovenia, 19–23 April 2021, pp. 3477–3488. ACM/IW3C2 (2021). https://doi.org/10.1145/3442381.3449992

8. Harris, S., Seaborne, A., Prud'hommeaux, E.: SPARQL 1.1 query language. w3c recommendation (2013). https://www.w3.org/TR/sparql11-query (2013)

9. Hu, E.J., et al.: LoRA: low-rank adaptation of large language models. In: The Tenth International Conference on Learning Representations, ICLR 2022, Virtual Event, 25–29 April 2022. OpenReview.net (2022). https://openreview.net/forum?id=nZeVKeeFYf9

10. Huang, X., Cheng, S., Bao, Y., Huang, S., Qu, Y.: MarkQA: a large scale KBQA dataset with numerical reasoning. In: Bouamor, H., Pino, J., Bali, K. (eds.) Proceedings of the 2023 Conference on Empirical Methods in Natural Language Processing, EMNLP 2023, Singapore, 6–10 December 2023, pp. 10241–10259. Association for Computational Linguistics (2023). https://doi.org/10.18653/V1/2023.EMNLP-MAIN.633

11. Jia, Z., Abujabal, A., Roy, R.S., Strötgen, J., Weikum, G.: TempQuestions: a benchmark for temporal question answering. In: Champin, P., Gandon, F., Lalmas, M., Ipeirotis, P.G. (eds.) Companion of the The Web Conference 2018 on The Web Conference 2018, WWW 2018, Lyon, France, 23–27 April 2018, pp. 1057–1062. ACM (2018). https://doi.org/10.1145/3184558.3191536

12. Jia, Z., Abujabal, A., Roy, R.S., Strötgen, J., Weikum, G.: TEQUILA: temporal question answering over knowledge bases. In: Cuzzocrea, A., et al. (eds.) Proceedings of the 27th ACM International Conference on Information and Knowledge Management, CIKM 2018, Torino, Italy, 22–26 October 2018, pp. 1807–1810. ACM (2018). https://doi.org/10.1145/3269206.3269247

13. Jia, Z., Pramanik, S., Roy, R.S., Weikum, G.: Complex temporal question answering on knowledge graphs. In: Demartini, G., Zuccon, G., Culpepper, J.S., Huang, Z., Tong, H. (eds.) The 30th ACM International Conference on Information and Knowledge Management, CIKM '21, Virtual Event, Queensland, Australia, 1–5 November 2021, pp. 792–802. ACM (2021). https://doi.org/10.1145/3459637.3482416

14. Liu, S., Semnani, S.J., Triedman, H., Xu, J., Zhao, I.D., Lam, M.S.: SPINACH: SPARQL-based information navigation for challenging real-world questions. In: Al-Onaizan, Y., Bansal, M., Chen, Y. (eds.) Findings of the Association for Computational Linguistics, EMNLP 2024. Findings of ACL, Miami, Florida, USA, 12–16 November 2024, pp. 15977–16001. Association for Computational Linguistics (2024). https://doi.org/10.18653/V1/2024.FINDINGS-EMNLP.938

15. Luo, S., et al.: ETRQA: a comprehensive benchmark for evaluating event temporal reasoning abilities of large language models. In: Che, W., Nabende, J., Shutova, E., Pilehvar, M.T. (eds.) Findings of the Association for Computational Linguistics, ACL 2025. Findings of ACL, Vienna, Austria, 27 July–1 August 2025, pp. 23321–23339. Association for Computational Linguistics (2025). https://aclanthology.org/2025.findings-acl.1198/

16. Neelam, S., et al.: SYGMA: a system for generalizable and modular question answering over knowledge bases. In: Goldberg, Y., Kozareva, Z., Zhang, Y. (eds.) Findings of the Association for Computational Linguistics: EMNLP 2022, Abu Dhabi, United Arab Emirates, 7–11 December 2022. Findings of ACL, pp. 3866–3879. Association for Computational Linguistics (2022). https://doi.org/10.18653/V1/2022.FINDINGS-EMNLP.284

17. Saxena, A., Chakrabarti, S., Talukdar, P.P.: Question answering over temporal knowledge graphs. In: Zong, C., Xia, F., Li, W., Navigli, R. (eds.) Proceedings of the 59th Annual Meeting of the Association for Computational Linguistics and the 11th International Joint Conference on Natural Language Processing, ACL/IJCNLP 2021, (Volume 1: Long Papers), Virtual Event, 1–6 August 2021, pp. 6663–6676. Association for Computational Linguistics (2021). https://doi.org/10.18653/V1/2021.ACL-LONG.520

18. Saxena, A., Tripathi, A., Talukdar, P.P.: Improving multi-hop question answering over knowledge graphs using knowledge base embeddings. In: Jurafsky, D., Chai, J., Schluter, N., Tetreault, J.R. (eds.) Proceedings of the 58th Annual Meeting of the Association for Computational Linguistics, ACL 2020, Online, 5–10 July 2020, pp. 4498–4507. Association for Computational Linguistics (2020). https://doi.org/10.18653/V1/2020.ACL-MAIN.412
19. Sun, Q., Li, S., Huynh, D., Reynolds, M., Liu, W.: TimelineKGQA: a comprehensive question-answer pair generator for temporal knowledge graphs. In: Long, G., Blumestein, M., Chang, Y., Lewin-Eytan, L., Huang, Z.H., Yom-Tov, E. (eds.) Companion Proceedings of the ACM on Web Conference 2025, WWW 2025, Sydney, NSW, Australia, 28 April 2025–2 May 2025, pp. 797–800. ACM (2025). https://doi.org/10.1145/3701716.3715308
20. Talmor, A., Berant, J.: The web as a knowledge-base for answering complex questions. In: Walker, M.A., Ji, H., Stent, A. (eds.) Proceedings of the 2018 Conference of the North American Chapter of the Association for Computational Linguistics: Human Language Technologies, NAACL-HLT 2018, New Orleans, Louisiana, USA, 1–6 June 2018, Volume 1 (Long Papers), pp. 641–651. Association for Computational Linguistics (2018). https://doi.org/10.18653/V1/N18-1059
21. Usbeck, R., et al.: QALD-10 - the 10th challenge on question answering over linked data: shifting from DBpedia to Wikidata as a KG for KGQA. Semant. Web **15**(6), 2193–2207 (2024). https://doi.org/10.3233/SW-233471
22. Vrandecic, D., Krötzsch, M.: Wikidata: a free collaborative knowledgebase. Commun. ACM **57**(10), 78–85 (2014). https://doi.org/10.1145/2629489
23. Ye, X., Yavuz, S., Hashimoto, K., Zhou, Y., Xiong, C.: RNG-KBQA: generation augmented iterative ranking for knowledge base question answering. In: Muresan, S., Nakov, P., Villavicencio, A. (eds.) Proceedings of the 60th Annual Meeting of the Association for Computational Linguistics (Volume 1: Long Papers), ACL 2022, Dublin, Ireland, 22–27 May 2022, pp. 6032–6043. Association for Computational Linguistics (2022). https://doi.org/10.18653/V1/2022.ACL-LONG.417

SemTS: Ontology and Vocabularies for the Semantic Categorization of Time Series Knowledge

Alexander Graß[1,2(✉)], Rohit A. Deshmukh[1,2], Christoph Lange[1,2], Diego Collarana[1], Christian Beecks[1,3], and Stefan Decker[1,2]

[1] Fraunhofer Institute for Applied Information Technology FIT, Sankt Augustin, Germany
{alexander.grass,rohit.deshmukh,christoph.lange,diego.collarana, christian.beecks,stefan.decker}@fit.fraunhofer.de
[2] RWTH Aachen University, Aachen, Germany
[3] FernUniversität in Hagen, Hagen, Germany

Abstract. Although time series analytics plays an important role across diverse application domains, efficiently managing the resulting insights remains a significant challenge. While specialized ontologies structure information of analytical models, they provide limited support for standardizing and arranging inferred knowledge. The absence of a unified data model to categorize findings such as anomalies, trends, or patterns complicates reuse and inhibits synergy effects between subsequent utilization stages. This paper presents the *Semantic Time Series Ontology - SemTS*, an ontology designed to classify time series characteristics as explicit knowledge entities, facilitating their consistent and semantic representation. The associated integration of related concepts through specially created vocabularies improves the dissemination of insights across various abstraction levels. SemTS further enables the description and incorporation of scenario-specific information, including domain expertise, to enrich analytical contexts. To demonstrate its practical utility, we showcase various aspects of SemTS through competency questions that illustrate how the ontology can be employed to efficiently query and validate semantic time series information. By systematically combining inferred knowledge and predefined facts, SemTS provides a comprehensive framework for improving the reusability and integration of insights affiliated with time series data.

Keywords: Ontology · Vocabulary · Time Series · Data Analytics · Knowledge Graph

1 Introduction and Motivation

The increasing importance of time series analytics in various scenarios such as industrial automation [49], financial market analysis [41], and healthcare monitoring [33] is a major reason for the prevailing popularity of time series database management systems[1]. By enabling optimized data provisioning through efficient querying and retrieval, these systems form the foundation for various data analysis pipelines [22]. However, as multiple analyses are carried out to gain valuable and exploitable insights from large time series databases, there is a lack of structuring, persisting, and maintaining the revealed insights at the end of each data pipeline. This results in a missed opportunity to optimize subsequent analyses and decision-making processes via knowledge transfer [39]. Performing analysis tasks independently, due to a missing integration of former results or domain expertise, can lead to overlooked synergies or even unintended outcomes.

One specific example is forecasting time series data without addressing previously identified anomalies. Figure 1 illustrates this particular problem by a regression conducted on a periodic time series, both with (blue) and without (green) accounting for a previously identified anomaly. When neglecting the anomaly, the modeling process is affected by the uncharacteristic behavior, leading to a biased representation of the underlying wave signal (green).

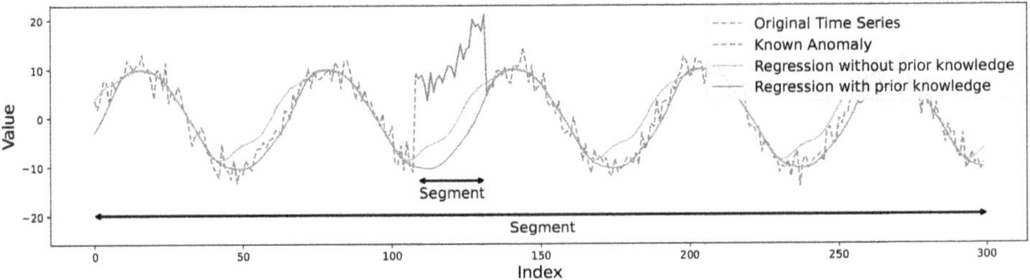

Fig. 1. Impact of knowledge reuse: Example of two Fourier regression models, representing outcomes with knowledge integration (blue) and without it (green). Excluding the anomaly reduces the Mean Squared Error (MSE) from 20.74 to 7.73. Both, the whole time series and the interval defining the anomaly represent knowledge segments. (Color figure online)

This simplified application scenario emphasizes the benefits of reusing insights gained from previous data analyses. Already discarding actionable insights like anomalies, patterns, or cluster assignments after a single exploitation consequently limits reusability and prevents the inclusion of inferred knowledge in future analyses. Moreover, repeatedly answering identical or similar questions over time leads to redundancy in analyses. After process experts and data analysts assess results, insights are frequently lost due to insufficient retention and documentation, further amplified by personnel turnover. Although preserved analysis artifacts usually comprise trained AI models and corresponding evaluation metrics [20], these artifacts do not account for the storage of identified insights in a FAIR [50] way. Therefore, a repeated generation of insights is an avoidable consumption of resources and time. The mentioned issues thus con-

---

[1] https://db-engines.com/en/ranking_categories, Accessed: 2025-12-03.

verge to the following research question: *How to semantically represent analytical and scenario-specific insights associated with time series data for reusability and future accessibility?* As results from individual analysis tasks can differ in representation and interpretation, a rudimentary storage of heterogeneous metadata with ambiguous meaning is unsuitable. While semantic concepts in the scope of data analysis exist (cf. Sect. 2), they rather focus on dataset definitions and methodical information than on a structured classification of related facts and generated knowledge.

In this paper, we propose an ontology, namely the ***Semantic Time Series Ontology***, to enable a semantic categorization and integration of relevant time series information. SemTS provides a framework to represent insights derived from time series analyses together with preexisting domain information to facilitate advanced querying, inference, and reasoning capabilities for the retrieval of data characteristics and facts. Furthermore, we introduce associated vocabularies to enable a detailed and expandable specification of concepts for both data-driven and scenario-specific knowledge. Although a more precise definition depends on the application context, throughout the rest of this paper, we refer to knowledge as condensed and valuable information, aligned with the DIKW information hierarchy model [30]. The subsequent sections of this paper are organized as follows: Sect. 2 provides an overview of related semantic data models. In Sect. 3, we introduce and elaborate on SemTS, covering identified requirements, its conceptual structure, and related vocabularies. Following Sect. 4, where we assess the applicability of SemTS, in Sect. 5 we discuss its limitations and impact. Lastly, we wrap up the presentation of our work in Sect. 6.

2   Related Work

To reflect analytic lifecycle phases in process models [26], several ontologies and vocabularies have been developed, which semantically organize entities associated with data management and analysis. A prominent contribution in the context of data interoperability and data management is the Data Catalog (DCAT) vocabulary [2], which resembles an approach to standardize dataset information across multiple platforms. PROV-O [27] offers transparency and accountability through the representation of provenance information. Shifting focus from data management to data analysis, the Data Cube Vocabulary [38] structures multidimensional data including characteristics that can be consumed for analytical purposes. The necessity of a definition of entities in the field of data science and adjacent domains has led to the development of more comprehensive and customized solutions. OntoDM [32] is an ontology to standardize the description of data mining knowledge and processes, whereas DMOP [23] aims at the optimization and performance aspects of these processes. Particularly focusing on machine learning experiments and associated metadata, Exposé [47] offers descriptions of experimental designs and their execution. The MEX vocabulary [12] and PROV-ML [43], which extends the PROV-O standard specifically for machine learning, both capture and represent the provenance of machine learning models and related processes. Since the listed approaches vary in their expressiveness due to a different weighting of common contributions, ML-Schema [36] compares the previous approaches and integrates them into a single solution. A recent approach, named the Machine Learning Sailor Ontology, short MLSO [7], further extends ML-Schema by an integration of the previous data management approaches and taxonomies, covering data science pipelines from data management up to data analysis and evaluation. Although the process of analytical modeling and preceding stages can be formalized by these means, it does not account for a structured cataloging of produced analysis outcomes, including cross-correlations, patterns and other comprehensive structural time series characteristics. While concepts like SOSA [21] or IoT-Stream [11] exist to annotate events or interesting time series aspects, they do not provide a detailed classification of analytical knowledge or emphasize the inclusion of domain information. SemML [52,53], in contrast, enables a structured integration of prior domain information into predefined ML workflows. However, the latter contributions prioritize the execution of semantically enriched analysis pipelines over providing a detailed classification of insights. To address the aforementioned issues, in the next section we present SemTS, an ontology designed to represent knowledge derived from time series analytics and associated scenario specifics. It draws upon and extends the preliminary, conceptual elaboration [17] and is motivated by the idea of knowledge graphs for query-induced analyses of time series information [15].

3   The Semantic Time Series Ontology

The main concept of SemTS constitutes a specification of informative intervals of arbitrary length within time series data, further referred to as segments. Any segment comprises characteristic knowledge linked to the covered time interval. Examples for characteristic knowledge range from common time series features [6], and structural particularities like anomalies [40] or motifs [29], to apriori information provided by domain experts. Apart from annotations, this predefined information, for instance, can correspond to scenario-specific rules or facts. With respect to the previous example from Fig. 1, the segment colored in red defines a time series interval together with data-related information and details about the anomaly. SemTS can thus be summarized as *enabler of a semantic and queryable index structure for valuable facts associated with time series data.*

For the design and implementation of SemTS, we followed an agile and iterative approach that emphasized close collaboration between Semantic Web experts and domain experts throughout the development lifecycle. Moreover, we used specific techniques from various ontology development methodologies [44], including the use of: (i) competency questions for scoping and validating SemTS from Ontology Development 101 (OD101), (ii) pathways such as ontological and non-ontological resource reengineering and reuse from NeOn, (iii) iterative pair design from eXtreme Design (XD), and (iv) tooling for visual representation, validation (OOPS! [35] and FOOPS! [14]), and publication (WIDOCO) as recommended by LOT [34]. While technical details and documentation are elaborated in Sect. 3.4, the use of validation tools and competency questions are elaborated in Sect. 4.

Starting with an overview of ontology requirements, this section proceeds with a discussion of components, underlying concepts, vocabularies, and a description of technical specifications.

3.1 Requirements

The SemTS ontology is developed to meet several key requirements, particularly driven by industrial settings [42] with an immense generation of time-sensitive data. An initial set of core requirements was derived from interviews with different expert groups. Associated interview templates as well as inferred requirements are available in the online repository (cf. Sect. 3.4). As an application-agnostic ontology, SemTS should facilitate a detailed, structured classification and description of time series specifics, including associated knowledge that is independent of particular setups and platforms to ensure wide applicability and flexibility across various environments. SemTS should further enable the representation of knowledge values regardless of the complexity of data types, making the ontology adaptable to diverse data structures and formats.

Defined knowledge should be organized using hierarchical concepts, capturing characteristics from data analyses along with methodical presets and domain- or scenario-specific information. It should allow for a mapping of knowledge to entire time series or subsequences of potentially reduced dimensionality. Detailed coverage of segment data information, including dimensionality and time information, is essential to accommodate diverse time series configurations. This includes descriptions of data not already in the dataset, such as information from forecasts. The ontology should further enable nested groupings of knowledge to infer complex hierarchies of structural and semantic relationships and support the description of knowledge qualities, offering insights into the reliability and validity of information.

One primary requirement for the representation and combination of knowledge is the integration of generated time series insights and domain information. This is important because of the potentially varying data formats originating from diverse database systems, each designed to accommodate specific contextual information, and the necessity to support informed analytical methods. Provenance is essential to support backtracking of propagated knowledge, allowing users to trace the origin and evolution of insights. In this context, SemTS should also allow for references to more sophisticated descriptions associated with data analysis runs, supporting complex analytical processes. Moreover, it should support advanced querying, inference, and reasoning operations to efficiently reuse time series knowledge. Regarding technical specifications, SemTS should adhere to best practices within the Semantic Web community for the reuse, documentation, and publication of existing ontologies.

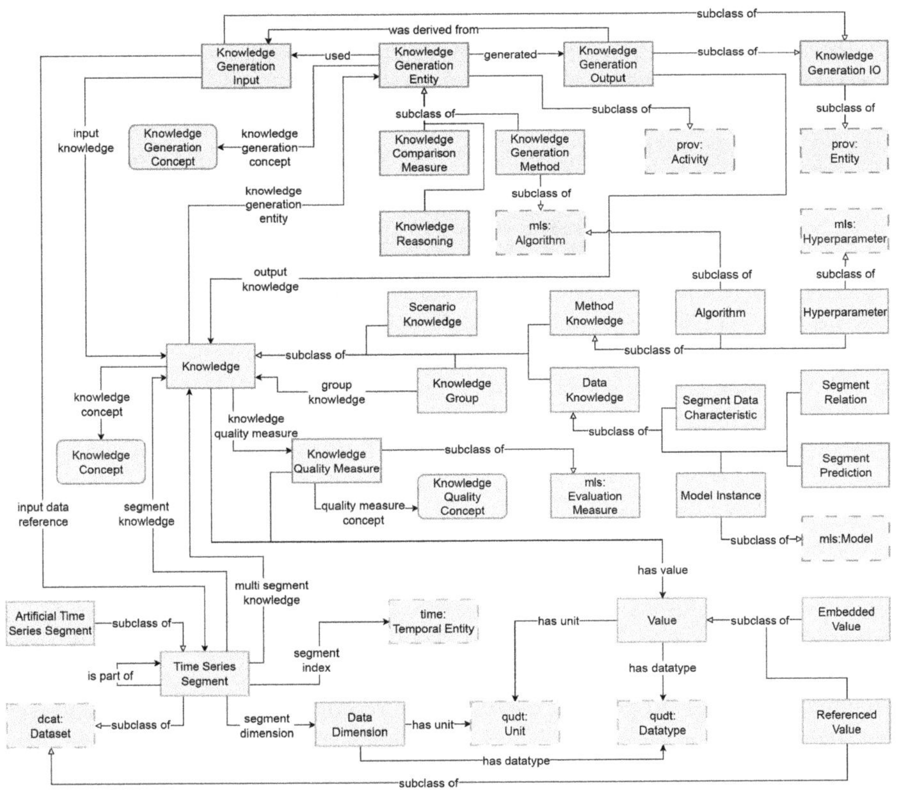

Fig. 2. Layered abstraction of the SemTS ontology incorporating external, reused schemas: Data Management (green), Knowledge Classification (blue), and Knowledge Generation (red). Classes from external schemas are labeled by a prefix and shown with dotted borders, while concept classes that extend SKOS concepts are illustrated with rounded boxes. (Color figure online)

3.2 Ontology Description

SemTS pursues the idea of a segment-based specification of time series insights. These segments capture knowledge tied to their covered timeframe and dimensionality, such as basic time series characteristics, prediction details, or insights provided by domain experts. This allows for an assignment of knowledge to any type of structural time series complexity, ranging from segments describing single-points of univariate data, up to multiple segments of entire multivariate time series from different collections.

At a conceptual level, SemTS can be divided into three interconnected layers of semantic entities. The most fundamental layer is the Data Management layer, which is essential for organizing data, including initial time series inputs as well as concrete knowledge values. The second layer is the Knowledge Classification layer, which outlines a coarse taxonomy of classes to represent various types of knowledge. The final Knowledge Generation layer addresses the creation of knowledge from processed time series data or the dissemination of existing knowledge, optionally accompanied by raw time series data.

An overview of involved classes, grouped by conceptual layer, is illustrated in Fig. 2. Aligned with this visualization, the following subsections provides a detailed explanation of each layer.

Time Series Data Management

Any *Time Series Segment* is essentially characterized by its dataset reference, a time interval, dimensions and the knowledge corresponding to inferred or predefined insights. While the segment metadata is specified via the inheritance from *dcat: Dataset*, index information is defined by referencing *time: Temporal Entity* from OWL-Time [1] and might comprise multiple adjacent time steps or a single point in time. Dimensions are represented by the class *Data Dimension*, which details the data specifics unit and datatype, both defined through QUDT [37] using *qudt: Unit* and *qudt: Datatype* respectively. In cases where knowledge linked to a time series segment does not address already existing time frames, e.g., potentially forecasts, SemTS provides a subclass named *Artificial Time Series Segment*. A hierarchy of nested segments can further be described by the transitive self-relation *is part of*. This use of axioms in SemTS allows for the inference of additional relationships, such that if for instance Segment A is part of Segment B and Segment B is part of Segment C, then Segment A is also part of Segment C. This property is useful for representing hierarchical relationships between parent time series and their subsegments, such as identifying that a daily segment is part of a larger weekly segment. The additional definition of relations from Allen's algebra [3] in combination with OWL-Time allows for the inference of several properties defining interval relations. By modeling both, entire time series and particular temporal slices, SemTS allows for flexibility in accommodating varying levels of data granularity, thus supporting diverse analytical perspectives. Similar to dimensional metadata, values associated with knowledge instances come with a unit and a datatype specified by QUDT. Primitive value representations, such as a single floating point number or a string, are defined by *Embedded Value*, while complex data representations use *Referenced Value*. The latter class also inherits from *dcat: Dataset*, allowing for stored, arbitrarily nested data structures, such as annotations for tracked objects in video data.

Classification of Time Series Knowledge

At a broad level, SemTS categorizes time series knowledge into three types: (i) *Data Knowledge*, (ii) *Scenario Knowledge*, and (iii) *Method Knowledge*. As the class *Knowledge Group* represents collections of individual knowledge entities and extends *Knowledge* it is defined in a recursive manner. To characterize the reliability and accuracy of any knowledge, SemTS introduces the class *Knowledge Quality Measure*. This class can be attributed with either a qualitative or quantitative value, along with a concept that specifies the type of quality measure.

Data Knowledge encompasses insights obtained directly from the data or through analytical processes. It is partitioned into further subclasses to specify the general type of derived knowledge. In SemTS, knowledge classes do not focus on specific details but represent a high-level taxonomy for varying granularities and categories of data insights. For instance, *Segment Data Characteristic* identifies features such as the median of a single segment, while *Segment Relation* identifies relationships between multiple segments, including distances or correlations. This distinction is further reflected by the object properties connecting *Time Series Segment* to *Knowledge*. The *segment knowledge* property describes knowledge directly tied to a particular time series segment, whereas *multi segment knowledge* encompasses knowledge related to groups of segments. Other classes of type *Data Knowledge* are *Segment Prediction*, which defines knowledge related to the result or specifics of a prediction, whereas *Model Instance* specifies instances of the analysis model.

Scenario Knowledge mainly describes verified content, such as data annotations or domain-specific process knowledge. This type commonly equals prior information from domain experts, but can also be used to define facts associated with inferred knowledge. Examples range from class labels and comments to event information or mathematical equations describing a certain signal behavior.

Method Knowledge encompasses data-driven specifics of methodological information, however, with a certain rationale. It outlines analytical method configurations, including algorithmic representations of established process details. It describes methodological recommendations or setups that have proven successful for the analysis of particular segments. Both the subclasses *Algorithm* and *Hyperparameter* extend the classes named identically in ML-Schema, to allow for references to more comprehensive metainformation.

During knowledge generation and knowledge propagation, both input knowledge and generated output knowledge can form intricate ensembles comprising various types of knowledge. These ensembles may even span multiple hierarchical levels. To address this complexity, SemTS incorporates the class *Knowledge Group* to represent collections of knowledge. This class, which itself is also a subclass of *Knowledge*, allows for the definition of complex hierarchies due to its recursive design. For instance, consider the clustering of time series data to automatically deduce groupings. Although performing a single analysis task, the result generally consists of multiple clusters, which again comprises various components to define each cluster. Each of these components, such as cluster representatives or cluster assignments, can be regarded as knowledge on its own.

Generation of Time Series Knowledge

While a classification of knowledge by SemTS is beneficial for making specific inquiries about analytical results and related facts, one primary benefit lies in enabling knowledge propagation and informed analyses. The reason is that classified knowledge not only facilitates semantic data retrieval, but also serves as valuable input to subsequent analyses (cf. Fig. 1) and semantic reasoning. It is important to note that by design, knowledge propagation in SemTS does not include dataset transformations, as these transformations are considered building blocks in data engineering pipelines, which are already specified

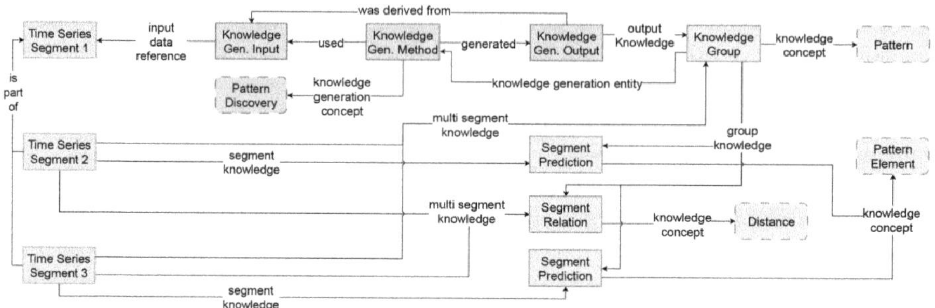

Fig. 3. Example interaction of core entities from Fig. 2. It emphasizes the difference between knowledge linked to individual segments (*segmentKnowledge*) and knowledge that is assigned to multiple segments (*multiSegmentKnowledge*), with knowledge concepts being used from the corresponding vocabularies (cf. Sect. 3.3).

in other semantic schemas such as DMOP. Instead, the concept of SemTS aims at enhancing machine-readable insights at the end of these pipelines, allowing for the inference of new knowledge from existing information. However, to enable provenance and details about the generation process, it is important to define knowledge transformations. In SemTS, this is achieved through a specification of input-output relationships and the corresponding generation entity. The class *Knowledge Generation Input* determines the input for knowledge generation processes, comprising insights from raw data, existing knowledge, or a combination of both. Conversely, *Knowledge Generation Output* specifies the output, focusing only on the representation of generated knowledge, due to the intentional limitation of not representing the creation of new time series data. Each output is produced by an independent *Knowledge Generation Entity*. By defining the generation of knowledge through comparison, reasoning, or algorithmic processes, the *Knowledge Generation Entity* is divided into three subclasses. The subclass algorithm-based knowledge generation, defined by the *Knowledge Generation Method*, further derives from *mls: Algorithm*, allowing for detailed descriptions of methodological aspects via ML-Schema (e.g., details on different runs). To maintain provenance during knowledge propagation, SemTS adopts and utilizes the PROV ontology. Consequently, any *Knowledge Generation IO* class is a subclass of *prov: Entity*, while classes categorized as *Knowledge Generation Entity* inherit from *prov: Activity*. This entails reusing essential object properties among these classes, such as *used, was derived from*, and *generated*. Equal to the conceptual definition of knowledge, SemTS also defines concepts for knowledge generation entities. Figure 3 depicts an abstract, exemplary illustration of the interaction between class instances from central entities of the three mentioned layers: Data Management, Knowledge Classification, and Knowledge Generation. The example illustrates knowledge extracted via pattern discovery conducted on a single time series segment, resulting in two additional segments, each defining an instance of the pattern.

3.3   Vocabularies

Alongside the presented ontology, SemTS offers two standardized vocabularies for representing concepts associated with data and scenario knowledge. Both vocabularies describe instances of type *Knowledge Concept* and extend the SKOS [31] framework for consistent, structured knowledge representation in order to enhance the semantic richness of related ontology classes via detailed taxonomies. The `Data Knowledge vocabulary` (`semts-dv`) is focused on defining insights inferred from time series data, detailing *Data Knowledge*. Examples of concepts include anomalies, distance measures, and structural properties. Accordingly, the `Scenario Data vocabulary` (`semts-sv`) provides concepts to enhance classes of type *Scenario Knowledge*, encompassing instances like annotations, mathematical expressions, and notes. The additional use of vocabularies for knowledge representation is beneficial to manage and categorize the diverse and complex types of time series knowledge with a higher level of granularity, avoiding a dependency only on ambiguous, string-based descriptions. A major reason for utilizing vocabularies in SemTS is their dynamic contribution to the extensibility of established *Knowledge* subclasses. As new types of knowledge emerge while methodologies evolve, vocabularies can adapt to incorporate these changes, ensuring that the ontology remains relevant and comprehensive. This adaptability is vital for accommodating advancements in data analysis techniques and changes in domain-specific conditions. Both vocabularies make use of *skos:Concept* and *skos:Collection* and related properties like *skos:broader* and *skos:member* to characterize relationships and hierarchies of time series knowledge. For analysis results, the definition of hierarchies is essential to represent the occurrence of nested knowledge compositions. A core idea in SemTS is thus to align the class of recursively defined knowledge groups with cascading structures of the SKOS framework. This facilitates a mapping of static yet generic structures from the ontology to dynamically definable knowledge types from the vocabulary. An example of such a mapping in the case of time series clustering is presented in Fig. 4.

Further concepts classified as *Knowledge Generation Concept* or *Knowledge Quality Concept* are made compatible to SKOS-based taxonomies like MLSO.

Although ontology instances may differ in their property values, their knowledge type is consistently defined by the associated vocabulary concepts. The supplementary use of *skos:Collection* complements the organization of knowledge in accordance with knowledge groups.

3.4   Technical Specification and Documentation

The SemTS ontology and its vocabularies have been formalized in multiple serialization formats, including RDF/TTL, JSON-LD, OWL, and N-Triples.

SemTS assets and documentation follow the Linked Data Principles and WIDOCO [13] best practices. Persistent identifiers based on W3ID are used, with the base URI `https://w3id.org/semts`. Distinct URI paths for the ontology, vocabularies and further assets ensure clear separation:

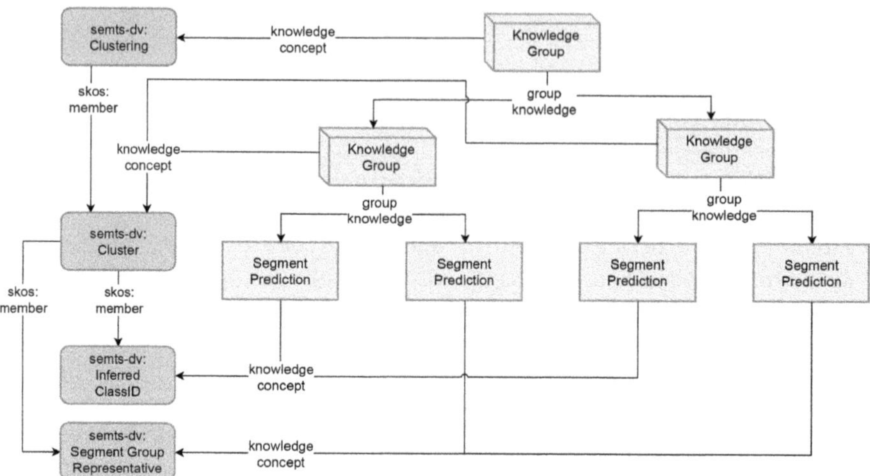

Fig. 4. Alignment of ontology and vocabulary hierarchies: SemTS facilitates the representation of arbitrarily nested groupings, supported by semantic structuring through SKOS-based concepts associated with each knowledge entity.

Ontology:
https://w3id.org/semts/ontology/{version}
Data Knowledge Vocabulary:
https://w3id.org/semts/vocabulary/data-knowledge
Scenario Knowledge Vocabulary:
https://w3id.org/semts/vocabulary/scenario-knowledge
Online Github Repository (Code & Further Assets):
https://w3id.org/semts/code

where the optional path {version} defines a particular version of SemTS. Human-readable HTML documentation and machine-readable representations are generated using WIDOCO and served with content negotiation via standard HTTP Accept headers and media types. SemTS is indexed in the Linked Open Vocabularies (LOV) [46] and listed on *prefix.cc*.

SemTS is maintained in a public GitHub repository under a CC BY 4.0 license. GitHub's collaborative features, such as version control, issue tracking, and pull requests, are planned to be leveraged to support community-driven development and continuous improvement. Specifically, the SemTS vocabularies are intended as living documentation, regularly expanded with new terms through community input. Each release snapshot is published on GitHub and followed by an update to the WIDOCO-based documentation.

3.5   Accessibility and Tooling Support

While providing semantic specifications and documentation, we additionally address practical accessibility through dedicated software solutions. These solutions supports users in populating the ontology without extensive expertise in RDF serialization formats or ontology engineering. As the information for SemTS typically stems from AI/ML program code, since it captures processes and knowledge extracted from conducted analytical pipelines, two complementary approaches have been developed that support different integration scenarios.

The SemTS Population Library is a Python library built on RDFLib [25] that provides a programmatic interface for constructing SemTS-compliant knowledge graphs. The library exposes Python classes and methods corresponding to SemTS concepts, allowing developers to create time series segments, instantiate knowledge-related entities, and establish relationships between these entities. Temporal modeling through OWL-Time, provenance tracking via PROV-O, and machine learning metadata through ML-Schema are handled through the library's API without requiring direct RDF triple manipulation. The library includes SHACL [24] validation against SemTS shapes to verify structural correctness of generated graphs and supports optional reasoning capabilities. This solution addresses scenarios where developers explicitly construct knowledge graphs within their application code and require fine-grained control over the instantiated entities and their properties.

The Code2Onto Framework [18] addresses the population of SemTS from analytical source code and its associated runtime context. Since time series knowledge in SemTS often originates from data analysis workflows, relevant information such as computed features, detected anomalies, model configurations, or method parameters exists within the execution environment of analytical scripts. Code2Onto employs Large Language Models within a multi-agent architecture to analyze source code, extract values from runtime variables, and generate corresponding SemTS instances. The framework separates local agents that access in-process variables from remote agents that provide ontology resources and validation services. Integration requires minimal code modification, and generated instances undergo OWL reasoning and SHACL validation before persistence. While developed to simplify SemTS population from time series analytics code, the framework architecture is ontology-agnostic through configurable prompt templates and semantic resources.

The two solutions serve complementary purposes in making SemTS accessible. The population library supports explicit, developer-controlled knowledge graph construction, whereas Code2Onto enables automated extraction of implicit knowledge from existing analytical workflows. Both solutions are publicly available under: https://github.com/semts-ontology.

4   Evaluation

In addition to a verification of SemTS by means of corresponding tools and techniques, this section provides an outline of the conducted data-driven evaluation of SemTS by competency questions to ensure its practical applicability and semantic coherence.

Table 1. Evaluation of SemTS through competency questions.

Competency Question: Is it possible to retrieve segments representing knowledge with a concrete specification?

Example: Retrieve two segments having a positive linear trend with slope ≥ 1

SPARQL Query

```
PREFIX semts: <https://w3id.org/semts/ontology#>
PREFIX semts-dv: <https://w3id.org/semts/vocabulary/data-knowledge#>

SELECT ?segment ?valueString
WHERE {
    ?segment semts:segmentKnowledge ?knowledge .
    ?knowledge semts:knowledgeConcept semts-dv:LinearSlope .
    ?knowledge semts:hasValue ?value .
    ?value semts:valueString ?valueString .
    FILTER(xsd:float(?valueString) > 1)
} LIMIT 2
```

Formatted Response

```
Segment: TimeSeriesSegment_026607a659, Slope: 1.304
Segment: TimeSeriesSegment_63126db1dc, Slope: 4.606
```

Visualization

Structure and Quality Assessment. To evaluate the ontology's structure, we used the OntOlogy Pitfall Scanner (OOPS!) [35] to detect important and critical modeling pitfalls such as missing disjointness or incomplete relations. According to OOPS, any critical and important pitfalls could be addressed with results being linked in the available online documentation of SemTS. Additionally we evaluated compliance with the FAIR principles using FOOPS! [14], and resolved any essential requirements to improve FAIRness. The result can be validated online using the FOOPS! scanner tool with the ontology IRI.

Data-driven Ontology Validation. SemTS was primarily evaluated using a catalog of competency questions applied to several real-world datasets from both industry and research projects covering multiple concrete application scenarios [4,5,16]. Information about the corresponding projects, along with examples, is available in the SemTS repository within the *application scenarios* folder. To support a traceable evaluation of the ontology using publicly accessible data, we further applied various analysis methods [28] to samples from the UCR Time Series Archive [8]. These samples include data from the *Trace* dataset, which simulates instrumentation failures in industrial plants and the *GestureMidAirD1* dataset, which captures a single dimension of hand trajectories. By using a competency question-based evaluation approach, we assessed key aspects of the ontology to demonstrate its practical applicability. More than 20 competency questions were derived from the requirements (cf. Sect. 3.1) established during

Listing 1.1. Example of a SHACL shape to validate instances of SemTS

```
@prefix sh: <http://www.w3.org/ns/shacl#> .
@prefix semts: <https://w3id.org/semts/ontology#> .
semts:SegmentPredictionGenerationMethodShape
  a sh:NodeShape ;
  sh:targetClass semts:SegmentPrediction ;
    sh:property [
      sh:path semts:knowledgeGenerationEntity ;
      sh:class semts:KnowledgeGenerationMethod ;
      sh:minCount 1 ;
      sh:maxCount 1 ;
      sh:name "Unique Knowledge Generation Method" ;
      sh:description "Every SegmentPrediction must be generated by exactly
          one KnowledgeGenerationMethod" ;
      sh:message "SegmentPrediction must have exactly one semts:
          knowledgeGenerationEntity of type semts:KnowledgeGenerationMethod "
    ] .
```

the ontology design process, addressing different categories such as data management, knowledge classification and knowledge generation. Examples include the validation of segment information, methodical details and concept-driven retrieval of specific knowledge. The list of competency questions, alongside a dedicated script to execute these questions as SPARQL queries on the *UCR knowledge graph* and other available resources is also available in the SemTS repository.

Performing the evaluation both on the project datasets and the synthetical dataset, confirmed the expected behavior in nearly all cases and resulted in the accurate retrieval of requested instances with respect to semantic and syntactic correctness. Inconsistencies occurring during the evaluation were resolved by further discussions and updates to the ontology. Table 1 illustrates the evaluation by a concrete example. A next step is to conduct a more comprehensive evaluation that incorporates reasoning capabilities. This evaluation is planned to be carried out as part of an agentic data science framework within the Jupiter AI Factory (JAIF) project[2].

Instance Validation. To validate SemTS instance (ABox) data, we used SHACL [24] shapes to enforce structural and syntactic constraints (e.g., cardinalities and datatypes). The corresponding collection of SHACL shapes has been made available in the online repository and was already verified in the scope of Code2Onto with SemTS being used for evaluation. An example is provided in Listing 1.1.

5   Discussion and Future Work

This section explores the challenges and limitations encountered in developing an ontology for classified time series knowledge, as well as potential future advancements. We demonstrate the impact of SemTS in and outline future directions for expanding the applicability of SemTS in industrial and data science contexts.

---

[2] https://w3id.org/jupiter, Accessed: 2026-03-09.

Challenges and Limitations. The field of data analytics poses challenges in creating an ontology for classified time series knowledge and its generation process due to evolving analytical methods. Results of new methods may deviate from established knowledge structures, complicating the development of a consistent framework. Additionally, defining what constitutes knowledge lacks standardization [9], requiring ontology classes to be broad enough for diverse contexts yet semantically rich for detailed taxonomies. This complexity is evident in nested knowledge hierarchies, such as identifying kernel compositions in time series regression. While kernels capture overall structure, hyperparameters define characteristics, leading to varying levels of granularity in data representations. This problem further transitions to related value representations, which potentially consist of multiple intricate data structures.

Impact and Future Work. SemTS was primarily developed for industrial applications like the knowlEdge [4] project, due to the extensive availability of temporal process data and associated scenario-related facts. It is currently being taken into account for further development and evaluation in the Next Generation Dataspaces Initiative (NGDI)[3]. The project aims to enhance the interoperability, quality, and usability among dataspaces in order to facilitate the next generation of a cross-domain dataspace infrastructure. Within the project, the purpose of SemTS is to enable a standardized semantic description and assessment of time series knowledge in multiple dataspace applications.

With the recent success and influence of Large Language Models (LLMs) and related techniques like Graph-based Retrieval Augmented Generation (Graph RAG) [48], the scope of application has expanded. Therefore, SemTS has additionally been incorporated into the development of LLM-based data science agents[4] [19,48], where it serves as an ontology for managing, retrieving, and utilizing time series knowledge from conducted analyses. By offering a semantically enriched representation of insights, modern methods like Graph RAG can facilitate validated and more detailed reasoning processes.

Beyond developing user-friendly methods for populating SemTS, we are exploring the automatic generation and integration of time series insights within external solutions such as MLflow [51]. We are pursuing two potential strategies: (i) mapping knowledge derived from external tools to SemTS instances, and (ii) providing SemTS instances as logged artifacts for these tools.

Furthermore, apart from possible vocabulary extensions, we are currently developing application profiles to enable a more precise and domain-specific adaptation of SemTS across various application areas. For this purpose, we make use of AP-first [10], a methodology originating from the Culture Dataspace project [45].

---

[3] https://tinyurl.com/ngdinitiative, Accessed: 2025-12-03.
[4] https://tinyurl.com/dsagents-infos, (pdf), Accessed: 2025-12-03.

6   Conclusion

In this paper, we introduced SemTS, an ontology to address the need for a structured representation and classification of time series knowledge. Complementing the SemTS ontology, we provide two standardized, yet extendable controlled vocabularies, defining concepts for time series insights as well as contextual knowledge. By offering a solution for reusing analytical insights and scenario-specific facts, SemTS supports semantic retrieval and exploitation of time series knowledge with respect to informed data analytics. Apart from ontology design, we discussed associated requirements and challenges of SemTS and demonstrated its impact in different projects and initiatives. Additionally, we provided a technical specification and discussed the evaluation of SemTS, ensuring the structural and semantic correctness of SemTS utilizing concrete data from industry and research projects. We concluded with a brief discussion on planned future work, emphasizing how SemTS can be optimized to drive further innovations like the development of automated data science agents.

Acknowledgments. This work was supported by a Fraunhofer ICON grant through the Next Generation Dataspaces Initiative (NGDI), and the German Ministry for Research and Education (BMBF) project WestAI (Grant no. 01IS22094D). It was also supported by the JUPITER Al Factory (JAIF) Project, funded via the EuroHPC Joint Understaking (JU) under Grant ID 101250682 (https://cordis.europa.eu/project/id/101250682); and in part by the German Federal Government and the German States of North Rhine-Westphalia and Hesse, coordinated by the Forschungszentrum Jülich and the involving consortium partners, including RWTH Aachen University, Fraunhofer, and the Hessian AI Ecosystem.

References

1. Time Ontology in OWL (2022). https://www.w3.org/TR/owl-time/
2. Albertoni, R., Browning, D., Cox, S.J.D., Gonzalez Beltran, A., Perego, A., Winstanley, P.: Data catalog vocabulary (DCAT) - version 3. Recommendation, W3C (2024). https://www.w3.org/TR/vocab-dcat-3/
3. Allen, J.F., Ferguson, G.: Actions and events in interval temporal logic. J. Log. Comput. **4**(5), 531–579 (1994)
4. Alvarez-Napagao, S., Ashmore, B., Barroso, M., et al.: Knowledge project –concept, methodology and innovations for artificial intelligence in industry 4.0. In: 2021 IEEE 19th International Conference on Industrial Informatics (INDIN), pp. 1–7 (2021). https://doi.org/10.1109/INDIN45523.2021.9557410
5. Beecks, C., Amalraj, A., Graß, A., et al.: Leveraging yolo for real-time video analysis of animal welfare in pig slaughtering processes. In: German Conference on Artificial Intelligence (Künstliche Intelligenz), pp. 275–281. Springer, Cham (2024)
6. Christ, M., Braun, N., Neuffer, J., et al.: Time series feature extraction on basis of scalable hypothesis tests (tsfresh-a python package). Neurocomputing **307**, 72–77 (2018)
7. Dasoulas, I., Yang, D., Dimou, A.: Mlsea: a semantic layer for discoverable machine learning. In: European Semantic Web Conference, pp. 178–198. Springer, Cham (2024)

8. Dau, H.A., Keogh, E., Kamgar, K., et al.: The UCR time series classification archive (2018). https://www.cs.ucr.edu/~eamonn/time_series_data_2018/
9. Davenport, T., Prusak, L.: Working Knowledge: How Organizations Manage What They Know, vol. 1 (1998). https://doi.org/10.1145/348772.348775
10. Deshmukh, R.A., Mustaf, D.M., Toubekis, G., et al.: Semantic data modeling for dataspaces: the extensible culture information model and the AP-first methodology for application profile development. Semant. Web J. (2025, under review)
11. Elsaleh, T., Enshaeifar, S., Rezvani, R., et al.: IoT-stream: a lightweight ontology for internet of things data streams and its use with data analytics and event detection services. Sensors 20(4), 953 (2020)
12. Esteves, D., Moussallem, D., Neto, C.B., et al.: Mex vocabulary: a lightweight interchange format for machine learning experiments. In: Proceedings of the 11th International Conference on Semantic Systems, pp. 169–176 (2015)
13. Garijo, D.: Widoco: a wizard for documenting ontologies. In: International Semantic Web Conference, pp. 94–102. Springer, Cham (2017). http://dgarijo.com/papers/widoco-iswc2017.pdf
14. Garijo, D., Corcho, O., Poveda-Villalón, M.: Foops!: an ontology pitfall scanner for the fair principles. 2980 (2021). http://ceur-ws.org/Vol-2980/paper321.pdf
15. Graß, A., Beecks, C., Chala, S.A., et al.: A knowledge graph for query-induced analyses of hierarchically structured time series information. In: European Conference on Advances in Databases and Information Systems, pp. 174–184. Springer, Cham (2023)
16. Graß, A., Beecks, C., Soto, J.A.C.: Unsupervised anomaly detection in production lines. In: Machine Learning for Cyber Physical Systems: Selected papers from the International Conference ML4CPS 2018, pp. 18–25. Springer, Cham (2018)
17. Graß, A., Deshmukh, R.A., Beecks, C., et al.: Towards an ontology for representing time series knowledge: motivation, requirements and concept. In: International Conference on Advanced Information Systems Engineering, pp. 103–110. Springer, Cham (2025)
18. Graß, A., Lehmkuhl, J., Collarana, D., Decker, S., Beecks, C.: Code2onto: multi-agent system for code-driven ontology population. In: 2025 IEEE International Conference on Big Data (BigData), pp. 5642–5651. IEEE (2025)
19. Graß, A., Pack, C.I., Collarana, D., Decker, S., Beecks, C.: Semantic intelligence: graph rag-driven agents for time series analytics. In: International Conference on Intelligent Data Engineering and Automated Learning, pp. 226–231. Springer, Cham (2025)
20. Idowu, S., Strüber, D., Berger, T.: Asset management in machine learning: state-of-research and state-of-practice. ACM Comput. Surv. 55(7), 1–35 (2022)
21. Janowicz, K., Haller, A., Cox, S.J., et al.: Sosa: a lightweight ontology for sensors, observations, samples, and actuators. J. Web Semant. 56, 1–10 (2019)
22. Jensen, S.K., Pedersen, T.B., Thomsen, C.: Time series management systems: a survey. IEEE Trans. Knowl. Data Eng. 29(11), 2581–2600 (2017)
23. Keet, C.M., d'Amato, C., Khan, Z.C., et al.: Exploring reasoning with the dmop ontology (2014)
24. Knublauch, H., Kontokostas, D.: Shapes constraint language (shacl) (2017). https://www.w3.org/TR/shacl/, w3C Recommendation
25. Krech, D., et al.: Rdflib (2023). https://doi.org/10.5281/zenodo.6845245
26. Kutzias, D., Dukino, C., Kötter, F., Kett, H.: Comparative analysis of process models for data science projects. In: ICAART (3), pp. 1052–1062 (2023)
27. Lebo, T., Sahoo, S., McGuinness, D., et al.: Prov-o: the prov ontology. W3C Recommendation 30 (2013)

28. Löning, M., Bagnall, A., Ganesh, S., et al.: sktime: a unified interface for machine learning with time series. arXiv preprint arXiv:1909.07872 (2019)
29. Madrid, F., Imani, S., Mercer, R., et al.: Matrix profile xx: finding and visualizing time series motifs of all lengths using the matrix profile. In: 2019 IEEE International Conference on Big Knowledge (ICBK), pp. 175–182. IEEE (2019)
30. McDowell, K.: Storytelling wisdom: story, information, and dikw. J. Assoc. Inf. Sci. Technol. **72**(10) (2021)
31. Miles, A., Pérez-Agüera, J.R.: Skos: simple knowledge organisation for the web. Cataloging Classification Q. **43**(3–4), 69–83 (2007)
32. Panov, P., Džeroski, S., Soldatova, L.: Ontodm: an ontology of data mining. In: 2008 IEEE International Conference on Data Mining Workshops, pp. 752–760. IEEE (2008)
33. Piccialli, F., Giampaolo, F., Prezioso, E., et al.: Artificial intelligence and healthcare: forecasting of medical bookings through multi-source time-series fusion. Inf. Fusion **74**, 1–16 (2021)
34. Poveda-Villalón, M., Fernández-Izquierdo, A., Fernández-López, M., García-Castro, R.: Lot: an industrial oriented ontology engineering framework. Eng. Appl. Artif. Intell. **111**, 104755 (2022)
35. Poveda-Villalón, M., Gómez-Pérez, A., Suárez-Figueroa, M.C.: OOPS! (OntOlogy Pitfall Scanner!): an on-line tool for ontology evaluation. Int. J. Semant. Web Inf. Syst. (IJSWIS) **10**(2), 7–34 (2014)
36. Publio, G.C., Ławrynowicz, A., Soldatova, L., et al.: ML-schema: an interchangeable format for description of machine learning experiments. Semant. Web 1–11 (2020)
37. QUDT Organization: QUDT; quantities, units, dimensions and types (2011). https://fairsharing.org/10.25504/FAIRsharing.d3pqw7
38. Cyganiak, R., Reynolds, D.: The RDF Data Cube Vocabulary. World Wide Web Consortium (W3C) Recommendation (2014). https://www.w3.org/TR/vocab-data-cube/
39. von Rueden, L., Mayer, S., Beckh, K., et al.: Informed machine learning – a taxonomy and survey of integrating prior knowledge into learning systems. IEEE Trans. Knowl. Data Eng. **35**(1), 614–633 (2023). https://doi.org/10.1109/TKDE.2021.3079836
40. Schmidl, S., Wenig, P., Papenbrock, T.: Anomaly detection in time series: a comprehensive evaluation. Proc. VLDB Endow. **15**(9), 1779–1797 (2022)
41. Sezer, O.B., Gudelek, M.U., Ozbayoglu, A.M.: Financial time series forecasting with deep learning: a systematic literature review: 2005–2019. Appl. Soft Comput. **90**, 106181 (2020)
42. Soldatos, J.: Artificial Intelligence in Manufacturing: Enabling Intelligent. Flexible and Cost-Effective Production Through AI, Springer Nature (2024)
43. Souza, R., Azevedo, L., Lourenço, V., et al.: Provenance data in the machine learning lifecycle in computational science and engineering. In: 2019 IEEE/ACM Workflows in Support of Large-Scale Science (WORKS), pp. 1–10. IEEE (2019)
44. Theissen-Lipp, J., Decker, S.J., et al.: Semantic foundations of dataspaces. Technical report, Fachgruppe Informatik (2024)
45. Toubekis, G., Decker, S.: The culture dataspace (datenraum kultur)-a data-sovereign open-source digital infrastructure based on the eclipse dataspace components (EDC) framework. Int. Arch. Photogramm. Remote. Sens. Spat. Inf. Sci. **48**, 1515–1523 (2025)

46. Vandenbussche, P.Y., Atemezing, G.A., Poveda-Villalón, M., Vatant, B.: Linked open vocabularies (LOV): a gateway to reusable semantic vocabularies on the web. Semant. Web **8**(3), 437–452 (2016)
47. Vanschoren, J., Soldatova, L.: Exposé: an ontology for data mining experiments. In: International Workshop on Third Generation Data Mining: Towards Service-Oriented Knowledge Discovery (SoKD-2010), pp. 31–46 (2010)
48. Vargas, D.C., Pack, C.I., Liao, Y.Y., et al.: Graph rag in the wild: insights and best practices from real-world applications. Semantic Web – Interoperability, Usability, Applicability (2025, under review)
49. Wen, Q., Yang, L., Zhou, T., et al.: Robust time series analysis and applications: an industrial perspective. In: Proceedings of the 28th ACM SIGKDD Conference on Knowledge Discovery and Data Mining, pp. 4836–4837 (2022)
50. Wilkinson, M., Dumontier, M., Aalbersberg, I., et al.: The FAIR guiding principles for scientific data management and stewardship. Sci. Data **3** (2016). https://doi.org/10.1038/sdata.2016.18
51. Zaharia, M.A., Chen, A., Davidson, A., et al.: Accelerating the machine learning lifecycle with MLflow. IEEE Data Eng. Bull. **41**, 39–45 (2018). https://api.semanticscholar.org/CorpusID:83459546
52. Zhou, B., Svetashova, Y., Gusmao, A., et al.: Semml: facilitating development of ml models for condition monitoring with semantics. J. Web Semant. **71**, 100664 (2021)
53. Zhou, D., Zhou, B., Zheng, Z., et al.: Ontology reshaping for knowledge graph construction: applied on Bosch welding case. In: International Semantic Web Conference, pp. 770–790. Springer, Cham (2022)

Tool4Boxology: A Semantic Toolbox for Constructing and Analysing Neuro-Symbolic Architectures

Johannes E. Bendler[1], Yashrajsinh Chudasama[2,3]([✉]), Mahsa Forghani[2,3],
Enrique Iglesias[2,4], Disha Purohit[2,3], Jacquiline Roney[1], Annette ten Teije[1],
Frank van Harmelen[1], and Maria-Esther Vidal[2,3]

[1] Department of Computer Science, Vrije Universiteit Amsterdam,
Amsterdam, The Netherlands
[2] TIB-Leibniz Information Centre for Science and Technology, Hannover, Germany
`yashrajsinh.chudasama@tib.eu`
[3] Leibniz University Hannover, Hannover, Germany
[4] L3S Research Center Germany, Hannover, Germany

Abstract. Neuro-symbolic (NeSy) AI systems are increasingly used in domains such as healthcare and finance, yet their architectures are often described only in prose or diagrams, making them difficult to compare, reproduce, or integrate. Existing modelling tools, such as UML or ML workflow editors, do not capture the semantics of NeSy components or enforce syntactic and semantic constraints on architectural validity. To address this gap, we provide a semantic, valid, and interoperable representation of NeSy architectures and present Tool4Boxology, a toolbox for constructing, validating, and analysing such architectures. The resource provides (i) a graphical editor for architecture diagrams with syntax validation, (ii) an ontology that describes architectural components and their interconnections, (iii) a pipeline that uses the RDF Mapping Language to generate ontology-compliant knowledge graphs from diagrams, (iv) SHACL constraints that enforce semantic coherence conditions, and (v) querying and validation mechanisms on these knowledge graphs for pattern detection, correctness checking, and reuse. We demonstrate the utility of Tool4Boxology by modelling a corpus of 61 NeSy systems from the biomedical literature and 7 from other domains, enforcing correctness constraints on these models, and querying them for recurring sub-patterns. Tool4Boxology offers researchers and practitioners a transparent, comparable, and FAIR-aligned foundation for documenting and analysing NeSy AI systems. The application, source code, ontology, and a corpus of neuro-symbolic architectures are available under a permissive license, along with an instruction video. A Jupyter Notebook provides runnable examples of KG exploration and analysis.

Keywords: Neuro-symbolic AI · Design Patterns · Knowledge Graphs

---

Alphabetical order, all authors (Bendler, Chudasama, Forghani, Iglesias, Purohit, and Roney) are joint first author.

© The Author(s), under exclusive license to Springer Nature Switzerland AG 2026
M. Acosta et al. (Eds.): ESWC 2026, LNCS 16550, pp. 191–211, 2026.
https://doi.org/10.1007/978-3-032-25159-6_11

Resource type: Software
License: Apache Licence 2.0
DOI: https://zenodo.org/records/17711495
URL: https://github.com/SDM-TIB/Tool4Boxology

1   Introduction

Neuro-symbolic (NeSy) systems are becoming increasingly prevalent in Artificial Intelligence [21]. As these systems grow in number and complexity, the absence of formalised and interoperable architectural representations becomes more apparent, highlighting a pressing need for standardised, machine-readable formats to support transparency, reusability, and comparative analysis. Despite the increasing complexity of AI systems, their designs are typically presented in unstructured formats buried in prose, figures, or disparate diagrams in research papers without a framework for machine-processable representations of the architectures [27]. This significantly hinders the ability to analyse, replicate and build on existing systems [4].

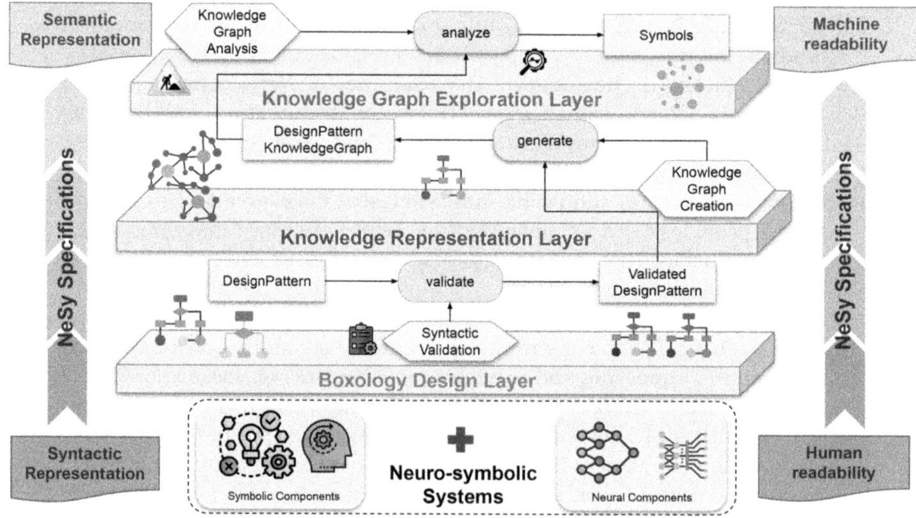

Fig. 1. Graphical abstract. Tool4Boxology is a layered framework for constructing and analysing the architecture of neuro-symbolic systems. The *Boxology Design Layer* ensures syntactic correctness of architectures, the *Knowledge Representation Layer* transforms validated designs into Knowledge Graphs, and the *Knowledge Graph Exploration Layer* enables their querying and analysis.

Additionally, traditional research platforms such as Google Scholar and academic databases that offer access to vast amounts of research operate primarily through keyword-based retrieval and not semantic understanding [3]. These platforms do not use structured semantic models or inference to interpret, compare, or retrieve articles based on system architecture, methodology, or design components [16]. Consequently, researchers are often required to manually sort through unstructured documents to identify relevant or comparable systems, which is time-consuming and inefficient. This issue is especially urgent as AI systems become more integrated and complex and are adopted in critical sectors such as healthcare, finance, and policy [34]. As a result, valuable system-level knowledge from independent researchers is frequently buried in informal sources such as code repositories or isolated implementations, with no formal mechanism for integration into a shared research infrastructure. Collectively, these conditions hinder the reproducibility and scalability of AI research [27].

Problem Statement and Proposed Solution. We address the problem of *constructing machine-readable, syntactically valid, and semantically interoperable representations of neuro-symbolic architectures.* We present Tool4Boxology, a semantic toolbox that provides an end-to-end workflow for constructing, validating, and analysing neuro-symbolic architectures. As illustrated in Fig. 1, our approach is organised into three layers: (i) the *Boxology Design Layer*, where users create neuro-symbolic design patterns using a graphical editor equipped with syntactic validation rules; (ii) the *Knowledge Representation Layer*, where validated diagrams are transformed into a knowledge graph through a unified ontology and mappings in the RDF Mapping Language (RML); (iii) the *Knowledge Graph Exploration Layer*, where the resulting semantic representation supports SPARQL querying, pattern discovery, and formal analysis. Together, these layers bridge human-readable architectural diagrams and machine-readable semantic models, enabling both consistency checking and machine-supported exploration of neuro-symbolic design patterns.

Results and Validation. We demonstrate the capabilities of Tool4Boxology by modelling 61 neuro-symbolic systems from the biomedical literature, and 7 systems from other domains. Through the toolbox, we detect structural inconsistencies via syntactic rules and generate knowledge graphs suitable for querying and comparing architectural motifs. The toolbox provides transparent, comparable, and FAIR-aligned representations of complex neuro-symbolic systems that were previously available only as informal diagrams and text.

Contributions. (i) a graphical language (extending the boxology language by Bekkum et al. [36]) and syntactic validator for neuro-symbolic architectures expressed in this language (*Boxology Design Layer*); (ii) an ontology and an RDF mapping for generating machine-readable knowledge graphs (*Knowledge Representation Layer*); (iii) a resource comprising a graphical editor, an ontology, RDF mappings, rules for syntactic and semantic validation, and a corpus of 61 neuro-symbolic architectures from the literature represented as queryable knowledge graphs (*Knowledge Graph Exploration Layer*).

The paper is structured as follows. Section 2 introduces the Boxology language. Section 3 presents the Tool4Boxology. Section 4 demonstrates its application to biomedical neuro-symbolic systems. Section 5 presents the resource. Section 6 discusses related work, and Sect. 7 concludes with outlook and extensions.

2   The Boxology Language

Boxology is a typed visual language that represents the computational structure of neuro-symbolic systems, developed in [37] and later extended in [36]. The main idea is that most systems, despite their different implementations, can be described using a few basic patterns. These patterns define how inputs, models, processes, and outputs relate to each other. Using these patterns, Boxology diagrams are semantically clear, structurally correct, and easier to compare between different works. Figure 2 shows the basic building blocks and how they combine into a structured computational workflow.

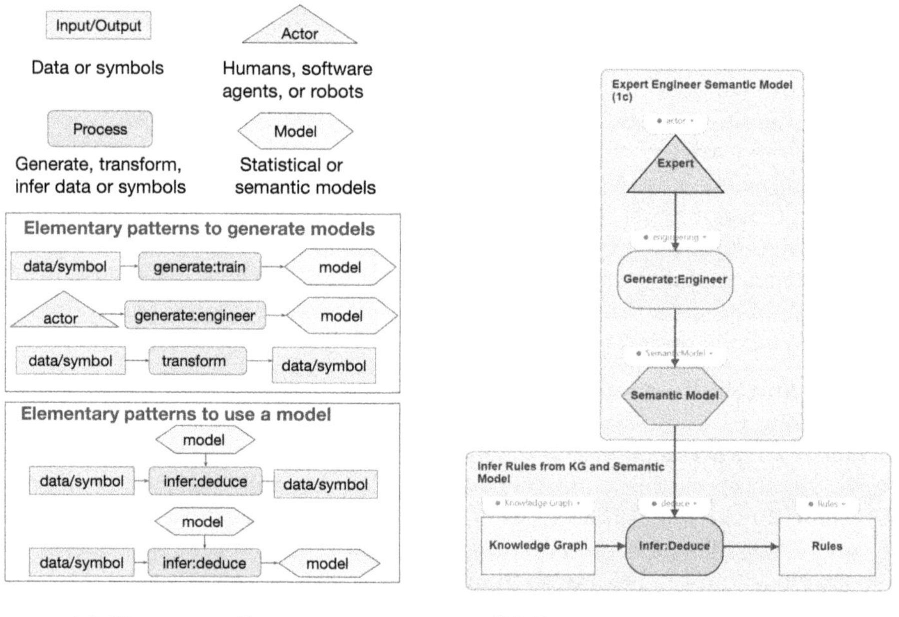

(a) Elementary Patterns.   (b) Illustrative Boxology diagram.

Fig. 2. Boxology Design Patterns. The left diagram shows the core Boxology components and some elementary patterns, which define the typed grammar for valid input-processoutput combinations. The right diagram shows how these patterns connect to form a complete, well-typed neuro-symbolic workflow.

Boxology Artefacts. The boxology defines five artefact types: (i) data, e.g., audio and numeric datasets; (ii) symbols, such as labels or relations; (iii) actors, which are human or software agents; (iv) processes, which denote computational operations; and (v) models, which are semantic, symbolic, statistical, or neural models. Interactions between these components are governed by elementary patterns such as using data to train a model, enabling an actor to engineer a model, transforming data, or combining a model with data to deduce an output. Every process in a Boxology diagram must instantiate exactly one such pattern, and all connections must satisfy the required inputoutput types. Larger computational flows are constructed by composing these elementary patterns, where the output of one pattern (data, symbols, or models) becomes the input of the next and is further processed to generate new symbols, data, and models within the architecture.

Artefacts are shown as boxes in diagrams; they include Data (e.g., text, images, embeddings), Symbols (e.g., labels, relations, triples, KGs), Actors (humans or agents), Processes (infer, transform, generate), and Models (semantic, symbolic, statistical, or neuronal). Elementary patterns specify which types are allowed to be combined. For example, `transform` always consumes and produces Data or Symbol artefacts, while `train` must output a Model. Table 1 summarises the artefact types and their roles, forming the vocabulary later formalised in the T4B ontology (Subsect. 3.2).

Table 1. Summary of Boxology language elements.

Element	Description
Data	Input/output artefacts such as text, images, embeddings, or traces
Symbol	Input/output artefacts such as labels, relations, RDF triples, knowledge graph fragments
Actor	Human experts, software agents, or systems interacting with or modifying the model
Model	Semantic or neural models used for inference or prediction
Generate/Engineer	Processes in which an actor creates or modifies artefacts
Generate/Train	Processes that create or induce models
Transform	Processes that modify data or symbols
Infer/Deduce	Processes that apply models to derive new outputs
Elementary Pattern	Typed minimal patterns
Composite Pattern	Larger workflow formed by composing elementary patterns

Elementary Patterns. The elementary patterns shown in Fig. 2 are used to define the typed grammar rules for the Boxology language. Each pattern specifies a minimal, valid computational step by constraining which type and amount of artefacts may serve as inputs, which process type may operate on them, and which outputs are permitted. As an example, the grammar rule for the `generate:train` pattern states that it produces a single model from 1 or more `data` or `symbol` inputs, plus an optional input `model`:

$$\{\texttt{model[0..1]},(\texttt{data} \mid \texttt{symbol})\texttt{[1..]}\} \rightarrow \texttt{generate:train} \rightarrow \{\texttt{model[1]}\}.$$

Rules for all other patterns can be found in Appendix A. Through these rules, elementary patterns encode not only the correctness conditions of individual steps but also the permitted interfaces through which elementary patterns can be composed, enabling the construction of complex neuro-symbolic workflows.

Boxology Models. Models are the knowledge-bearing artefacts in the boxology language and encode either declarative structure or learnt behaviour. The boxology distinguishes four main families. *Semantic models* capture explicit knowledge, including RDF graphs, OWL/RDFS ontologies, SHACL shapes, and rule sets. *Inductive learning models* learn patterns from data and follow established KG learning taxonomies ([12, Chapter 5]), covering symbolic approaches (e.g., community detection, rule and axiom mining) and numerical methods (e.g., graph analytics, embeddings, supervised GNNs). *Neural models* comprise deep architectures such as CNNs, RNNs, LSTMs, Transformers, and GNNs, while *statistical models* include classification and regression models. Models are produced by `train` or `engineer`, consumed by `deduce`, and constrained by the typed grammar of the language to ensure consistent and semantically valid use within Boxology architectures.

Boxology Processes. Boxology processes operate on artefacts according to well-typed elementary patterns. The three core process families are: `generate` (for training or engineering models), `transform` (for embedding, augmenting, or cleaning data), and `infer` (for deducing new outputs). Each process type enforces strict typing rules: `train` consumes data or symbols and produces a model, `deduce` requires at least one model plus data or symbols, and `transform` only produces data or symbolic artefacts. These constraints correspond directly to the formal validation rules listed in Appendix A, allowing misconfigured flows to be detected as type-constraint violations.

Boxology Expressiveness. The expressiveness of the Boxology comes from composing its elementary patterns into larger computational flows. By connecting (elementary) patterns, a diagram can represent complete neuro-symbolic workflows, including multi-stage preprocessing pipelines, model training and refinement loops, hybrid symbolicneural inference, and human-in-the-loop interactions. The language remains extensible while preserving its typing discipline: for example, LLM-based systems can be captured by treating prompts as data artefacts and LLMs as neural models used in `generate` or `infer` patterns. Extensions proposed in the literature (e.g., interaction processes between multiple actors [26]) extend the conceptual Boxology framework but are not yet supported in the current Tool4Boxology implementation. Empirical studies show that many published neuro-symbolic systems can be decomposed into combinations of elementary patterns [5], demonstrating that the Boxology captures the core structural motifs of these hybrid systems.

3   Tool4Boxology - A Toolbox for Neuro-Symbolic Systems

The toolbox presented in this paper consists of several components. An overview of these components is provided in Fig. 3, with an explicit focus on the conversion of visual diagrams to a KG representation. The following section will explain this figure and how the elements relate and interact with each other.

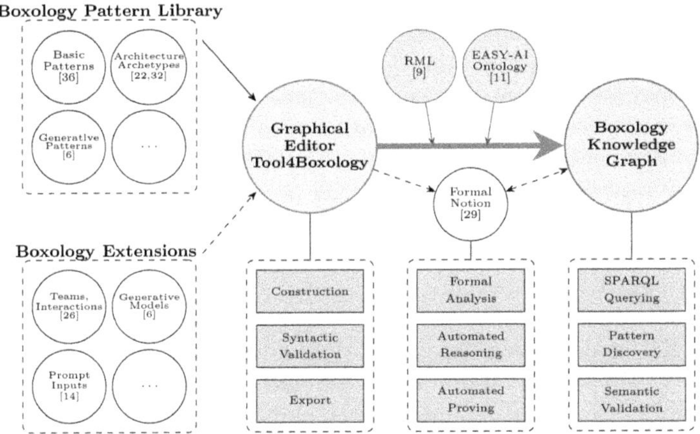

Fig. 3. Boxology Toolbox overview. Colours correspond to the layers in Fig. 1.

3.1   The Tool4Boxology Overview

The core of our toolbox is a graphical editor for boxologies, corresponding to the *Boxology Design Layer* shown in Fig. 1. The editor is implemented on two platforms: a plugin for the online diagramming tool draw.io and a standalone version based on GoJS. Both let users place and connect components such as models and processes. However, the standalone version provides additional functionality and is therefore the focus of this paper. Its primary purpose is to support the manual construction and syntactic validation of boxology diagrams. The editor enforces elementary pattern rules, such as determining which component types may serve as inputs or outputs for a given process. These validity checks correspond to the well-typed grammar of the Boxology language and are implemented directly in the editor using the rules listed in Appendix A. Beyond these syntactic constraints, semantic correctness is validated at the KG level using SHACL. For example, SHACL shapes ensure that a deduce process consumes an appropriate semantic model, e.g., an OWL ontology, along with a compatible input artefact, such as a KG or a predicate, respectively. This reflects domain-specific requirements that go beyond the purely structural grammar enforced by the editor.

On the top-left of the figure, the **Boxology Pattern Library** represents not only the currently existing patterns in the tool but also the general possibility of creating or extending such a library. Additional patterns, such as Kierner et al.'s *architecture archetypes* [22,31] (see Sect. 4), or patterns for generative AI systems as proposed by De Boer et al. [6] could be added to it.

The **Boxology Extensions** represent proposals for future additions to the language, such as a generative model subclass [6], prompts as a new type of input [14], or components and patterns for teams and interactions between actors [26]. These extensions have not yet been implemented in the current editor, and are therefore shown as dashed arrows to indicate their exploratory status.

The Tool4Boxology environment exports the content of the graphical editor in JSON format and transforms it into a structured, machine-readable knowledge graph. All visual elements—components, processes, and pattern instances—are mapped to classes and relations in the Tool4Boxology Ontology, forming the **Boxology Knowledge Graph**. This KG preserves the typed structure of the original diagram and enables downstream tasks such as querying, comparison, and semantic validation of neuro-symbolic architectures. The Knowledge Graph and downstream task correspond to the *Knowledge Representation Layer* and *Knowledge Graph Exploration Layer* in Fig. 1, respectively. A detailed description of the KG construction workflow is provided in Subsect. 3.3.

The *Formal Notion* node in Fig. 3 not only represents the aforementioned ontology but also attempts to formalise the boxology concept. These include the work of Till Mossakowski [29], who developed a notion of refining and combining boxology components and patterns using concepts from graph theory and category theory. Such formalisms could enable a more rigorous analysis of the boxology, as well as automated reasoning and proving of theorems.

3.2   The Tool4Boxology Ontology

The Tool4Boxology Ontology (T4B Ontology) provides the semantic layer that grounds the Boxology language in machine-interpretable terms. Although boxology diagrams specify architectures graphically, the ontology encodes their meaning via classes, object properties, and constraints, enabling automated validation, querying, and integration into the T4B Knowledge Graph. The ontology builds on and extends the EASY-AI Ontology [11], refining its conceptualisation to capture the core expressive range of Boxology patterns. Figure 4 shows an overview of the ontology, which is organised around three main concepts: *Boxology*, *Artefacts*, and *Design Patterns*. This structure mirrors the typed grammar of the Boxology and ensures that the meaning of each visual element is preserved in the T4B KG.

Main Classes and Hierarchies. At the core of the ontology is the class `t4b:Artefact`, which generalises all entities that appear as boxes in Boxology diagrams. Five key subclasses refine it: (i) `t4b:Data`, representing raw or processed information (e.g., text, images, embeddings); (ii) `t4b:Symbol`, representing structured knowledge items such as labels, RDF triples, or KGs; (iii)

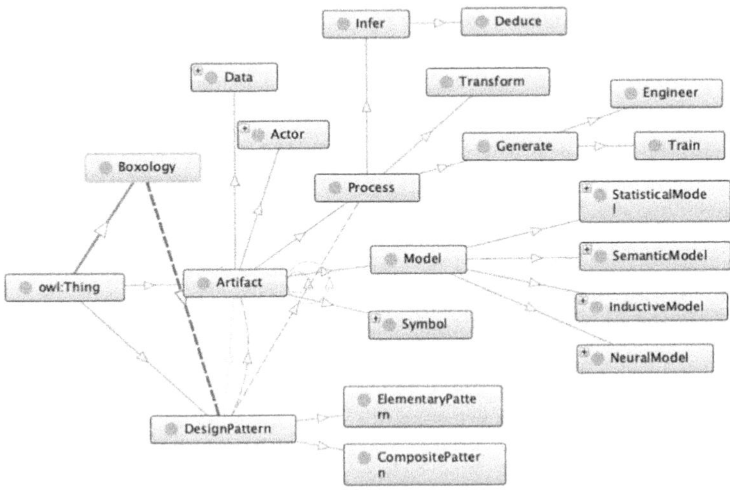

Fig. 4. The Tool4Boxology Ontology. Grey boxes denote ontology classes; arrows represent object properties. Arrow colours indicate: blue subclass relations; yellow `t4b:hasInput`; orange `t4b:hasOutput`; purple `t4b:hasPattern`; dark green `t4b:hasProcess`; green `t4b:outputRoleParticipatesInProcess`; and grey `t4b:inputRoleParticipatesInProcess`. Ontology created with Protégé [30].

`t4b:Actor`, representing humans or software agents; (iv) `t4b:Process`, modelling processes (e.g., generate, infer, transform); and (v) `t4b:Model`, representing the models (e.g. symbolic, statistical, semantic or neural) used for prediction or inference. All computational artefacts ultimately connect to `t4b:Boxology`, which aggregates the elements of a complete architecture. Models are represented under `t4b:Model`, further refined into `t4b:SemanticModel` (e.g., RDF, RDFS, OWL, SHACL, rule-based systems) and `t4b:NeuralModel` (e.g., NeuralNetwork, CNN, RNN, LSTM, Transformer, GNN). Following the inductive learning taxonomy of [18], the ontology also defines `t4b:InductiveModel` with two branches: `t4b:NumericModel` (e.g., unsupervised graph analytics, self-supervised embeddings, supervised GNNs) and `t4b:SymbolicLearningModel` (e.g., rule mining, axiom mining), representing both semantic and statistical learning methods. Processes appear under `t4b:Process` and include as subclasses the three fundamental Boxology's elementary patterns: `t4b:Generate`, `t4b:Transform`, and `t4b:Infer`. Training (`t4b:Train`) and engineering (`t4b:Engineer`) specialize `t4b:Generate`, while deduction (`t4b:Deduce`) specialises `t4b:Infer`.

Design Patterns. The graphical patterns of the Boxology language correspond to the class `t4b:DesignPattern`. Each pattern encodes the allowed input artefacts, the permitted process, and the required outputs. The ontology captures these constraints through a combination of subclasses and property restrictions, enabling the validation of whether a boxology instantiates a correct pattern. A `t4b:Boxology` instance links to its patterns via `t4b:hasPattern`.

Object Properties. The ontology specifies a set of object properties that formalise how Boxology components participate in design patterns. (i) `t4b:hasInput` (yellow arrows) and `t4b:hasOutput` (orange arrows) connect a `t4b:DesignPattern` to the artefacts it consumes, produces, and uses. (ii) `t4b:hasProcess` (dark green arrows) links a design pattern to the specific process it instantiates. (iii) how artefacts participate in processes and supporting type-correct modelprocess interactions are represented with `t4b:inputRoleParticipatesInProcess` (grey arrows) and `t4b:outputRoleParticipatesInProcess` (green arrows).

Semantic Role of the Ontology. Together, these classes and properties provide a complete semantic account of the Boxology grammar. The ontology defines which components may participate in patterns, how processes and models must interact, and what structural constraints must hold. When combined with RML mappings and SHACL constraints (Subsect. 3.3), the T4B Ontology enables automated validation of diagrams, creation of semantically consistent KGs, and pattern-based analysis of neuro-symbolic architectures.

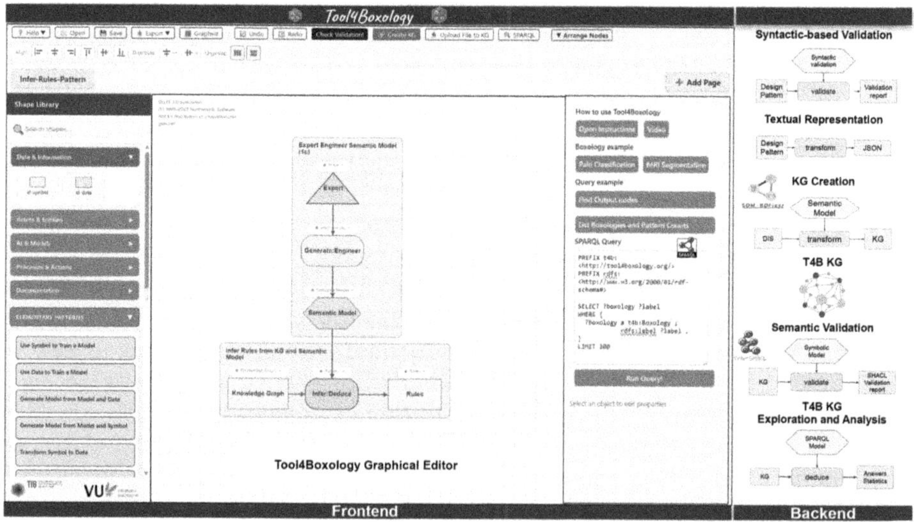

Fig. 5. Tool4Boxology Interface illustrating the Tool4Boxology graphical editor (Frontend) and T4B KG creation and management (Backend).

3.3   The Tool4Boxology Knowledge Graph

The Tool4Boxology Knowledge Graph (T4B KG) provides the machine-readable representation of Boxology designs created in the graphical editor. It is defined, populated, and validated through a declarative data-integration pipeline that ensures semantic consistency with the T4B Ontology and enables querying, analysis, and reuse of neuro-symbolic architectures.

KG as a Data Integration System. The T4B KG follows the data-integration specification $DIS_{\mathcal{G}} = \langle O, S, M, \Sigma \rangle$ from [19]. O is the T4B Ontology, which defines the semantics of the Boxology components, processes, and design patterns. The sources S are the JSON exports from the editor containing all nodes, edges, component types, and pattern annotations of a diagram. The mappings M specify how the JSON structures are transformed into RDF triples, while the constraint schema Σ ensures that the KG instances correspond to valid boxologies. The data integration system of T4B employs a structured transformation pipeline that converts the JSON specifications produced by the T4B Interface into a semantically enriched T4B KG. The integration process is grounded in a formal definition of the T4B ontology, which specifies the core concepts of boxology components, processes, relations, and their associated constraints. This establishes a controlled semantic vocabulary for the T4B KG. Although the system does not merge heterogeneous external sources, it adheres to the classical data-integration paradigm by explicitly distinguishing between (i) the source schema represented by the JSON export format, (ii) the target ontology defining the conceptual schema of the KG, and (iii) the declarative mappings that connect the source schema and the ontology. Specifically, RML mappings are employed to formally specify how boxology components in JSON are transformed into RDF triples, ensuring reproducibility, transparency, and semantic modeling. Therefore, the system complies with the methodological principles of data integration by offering an explicit schema, formal mappings, and constraint validation mechanisms.

Data Sources Generated by the Tool4Boxology Editor. Each JSON element exported by the editor—design pattern, process, model—is converted through RML rules into an RDF graph that is compliant with the T4B Ontology. For example, a visual `model:semantic` box becomes an instance of `t4b:SemanticModel`; a `deduce` process becomes an instance of the class `t4b:Deduce`; and each visual connection (an arrow in a boxology diagram) is represented by a relation such as `t4b:hasInput`, `t4b:hasProcess`, and `t4b:hasOutput`. Thus, the T4B KG contains a fine-grained and factual description of the design patterns encoded in the original diagram, fully preserving their semantics.

RML Rules. The mappings M define the correspondences between the JSON schema used by the editor and the T4B Ontology. A JSON node with `"type":` `"process"` and `"subtype": "train"` is mapped to `t4b:Train`, while `inputs` and `outputs` are mapped to `t4b:hasInput` and `t4b:hasOutput` relations. These mappings ensure a lossless transformation from visual structures (as represented in the JSON export) to semantic representations in a knowledge graph, preserving the typed grammar and compositional constraints of the Boxology language.

SHACL Shapes as Integrity Constraints. The schema Σ enforces the structure of valid Boxology design patterns. For instance, a `t4b:Train` process must consume at least one Data or Symbol artefact and produce exactly one Model; a `t4b:Deduce` process must take at least one Model as input; and a `t4b:Transform` process is prohibited from generating a Model. These integrity constraints ensure that every KG instance is both syntactically and semantically correct.

The T4B KG Creation. The pipeline that generates the T4B KG is shown in Fig. 5 and consists of four steps: (i) *export* the JSON representation of a Boxology diagram from the editor, including all components, processes, and pattern annotations; (ii) *map* these JSON elements to RDF triples using RML rules aligned with the T4B Ontology; (iii) *apply reasoning* to derive inferred types and relationships from the ontology's class hierarchy; and (iv) *validate* the resulting KG with SHACL shapes to ensure well-typed design. Whenever a diagram is modified in the editor, the pipeline automatically reruns, producing an updated KG instance that remains fully synchronised with the visual design.

Running Example: Figure 6 shows the different stages of this pipeline when applied to the diagram Fig. 2b: the JSON representation exported by the editor from the diagram Fig. 2b, the RML mapping rules that are executed by the SDM-RDFizer engine [20] to translate the JSON representation into RDF triples, and the resulting knowledge graph (including `rdf:type` links to classes defined in the ontology from Fig. 4. The KG correctly represents that Fig. 2b consists of two design patterns: an **engineer** pattern that produces a semantic model which is consumed by a **deduce** pattern.

Semantic grounding transforms Boxology diagrams into machine-interpretable KGs. This enables architecture-based search (e.g., retrieving all patterns combining neural and semantic models via deduction), structural comparison across systems, and constraint-based validation through SHACL. Additionally, ontology-based reasoning enables higher-level pattern detection. The T4B KG representation could facilitate downstream tasks, such as architecture recommendation and link prediction. These capabilities cannot be achieved with purely visual or informal diagramming tools.

Performance Evaluation: To evaluate the efficiency of our approach, we measured the execution time for the main operations. The runtimes, as summarized in Table 2, were measured on the aforementioned 61 boxology patterns. This performance scales linearly in the number of architectures since there is no combinatorial interaction. Each architecture can be represented by a few hundred triples, so the system easily scales with current semantic technologies.

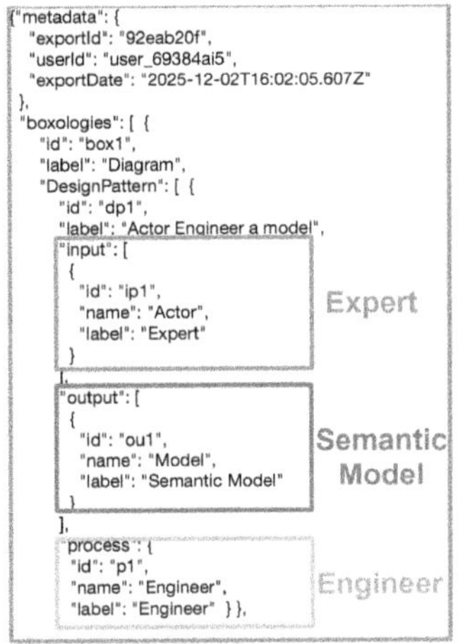

(a) JSON Representation

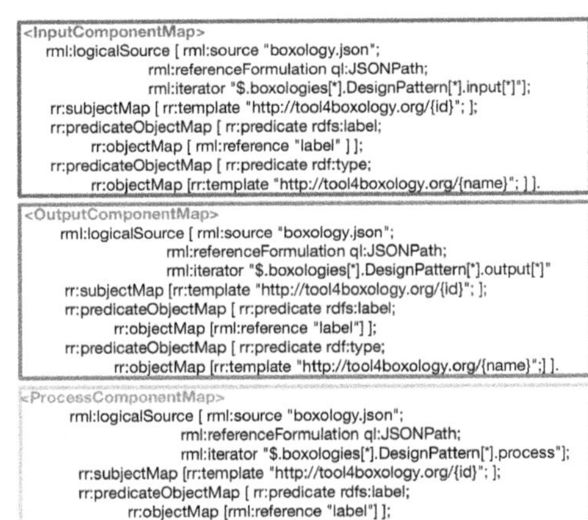

(b) RML Mapping Rules

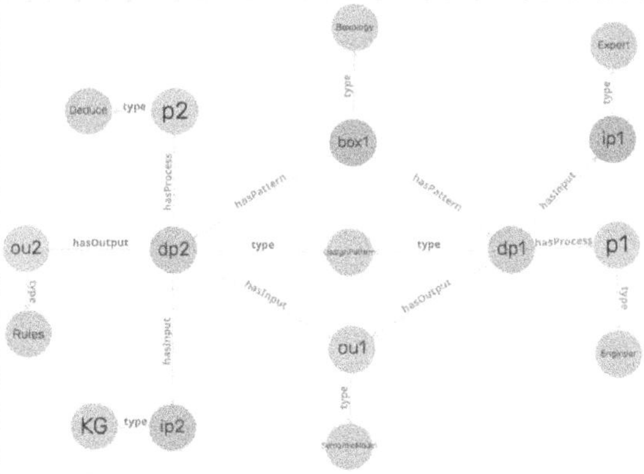

(c) T4B Knowledge Graph Powered by GraphDB

Fig. 6. Running Example: KG for design patterns from Fig. 2b.

Table 2. Runtime performance results for all 61 boxologies.

Task	Runtime
KG Generation	2.6 s
SHACL Validation	147.0 s
SPARQL Querying	0.2 s

4 Application to neuro-symbolic Clinical Decision Systems

Kierner et al. [22] present a corpus of neuro-symbolic clinical decision systems that combine medical knowledge (for example, in the form of clinical rules) with machine learning models. Their literature search (PubMed, IEEE Xplore, and Google Scholar, 19922022) identified 71 systems, which all fit into one of five so-called *architectural archetypes*: (i) rules embedded in machine learning models (the most common); (ii) machine learning methods used to preprocess data before rule-based inference; (iii) rule-based methods used to preprocess data before ML-based prediction; (iv) rules guiding or influencing the training of a machine learning model; and (v) a parallel ensemble of rule-based and machine learning methods. The authors of [31] analysed these archetypes and the underlying clinical decision systems reported by Kierner et al. They manually constructed the architectural patterns in the graphical boxology notation, both for the five archetypes and for 69 of the 71 medical neuro-symbolic decision support systems[1]. This showed that the archetypes and the individual systems could all be modelled using the graphical boxology notation, and the authors of [31] manually verified that all the systems reported were indeed a refinement of their corresponding archetype. This provided us with a substantial corpus of boxology diagrams. Subsequently, we applied Tool4Boxology to a manually constructed corpus of boxology diagrams from [31]. This confrontation of the manually constructed boxology diagrams with the syntactic and semantic constraints of the Tool4Boxology revealed that only 12 of the diagrams were syntactically correct, while 49 of the diagrams violated typing constraints. The most common errors were incorrect connections between artefacts and processes. We attribute this to the prior lack of clear connection/validation rules between boxes, thus highlighting the utility of the rules listed in Appendix A. We were able to detect and correct these using the toolbox. The remaining eight patterns all violated semantic constraints, and a return to the original paper would be needed to correct these semantic violations.

We subsequently translated the 61 validated models into a knowledge graph using the ontology from Subsect. 3.2, enabling us to query and analyse this collection of neuro-symbolic boxology diagrams of systems and their archetypes. For example, the query from Listing 1.1 counts the number of unique patterns that occur in each diagram, giving us insight into the frequency of use of the elementary patterns from Fig. 2a). Twelve further examples of SPARQL queries are available online in a Jupyter Notebook (see next section for the reference).

Application to Other Domains: In addition to neuro-symbolic system in the biomedical domain, the boxologies likewise finds usage in other domains such as systems for energy production and consumption [10,15], military applications [8], education [33], Large Language Models [23], and others [35,38]. We also modelled and verified these architectures using the Tool4Boxology.

```
1  PREFIX t4b: <http://tool4boxology.org/>
2  PREFIX rdfs: <http://www.w3.org/2000/01/rdf-schema#>
3
4  SELECT ?type COUNT(DISTINCT ?boxology) as ?num   WHERE {
5      ?boxology a t4b:Boxology.
6      ?boxology t4b:hasPattern ?pattern.
```

---

[1] Two papers were not accessible.

```
 7   ?pattern a t4b:DesignPattern .
 8   ?pattern t4b:hasProcess ?process.
 9   ?process a ?type.
10  } GROUP BY ?type ORDER by DESC(?num)
```

Listing 1.1. Example SPARQL query to list boxologies and pattern counts

5 Tool4Boxology as a Resource

Table 3. Resources Description

Resources	Description
Tool4Boxology Web Interface https://tool4boxology. service.tib.eu	**The landing page for the resource, a web service for the Tool4Boxology editor** **providing documentation to get started**
Extended Boxology Patterns and Syntactic Validation	Defines the extended Boxology modelling patterns and the syntactic rules used to validate the structure and consistency of Boxology patterns.
Boxology Ontology https://github.com/ SDM-TIB/ Tool4BoxologyOntology	provides the formal semantic representation of Boxology concepts, including classes, relations, and constraints, enabling reasoning and interoperability.
Boxology KG and RML Mapping Rules https://labs.tib.eu/ sdm/T4B-KG/sparql	Implements the Boxology Ontology as a machine-readable knowledge graph and uses RML mapping rules to transform source data into structured KG instances.
Tool4Boxology https://github.com/ SDM-TIB/ Tool4Boxology/	A software tool that supports the creation, editing, validation and visualization, of Boxology patterns using the underlying patterns, ontology, and KG. Elaborate on design patterns added to the KG and tool [a -g] in rebuttal
Tool4Boxology Jupyter Notebook https://edu.nl/khcf3	An example notebook demonstrating the query-answering and statistical analysis over the Tool4Boxology KG.
Video Tutorial https://www.youtube. com/watch? v=yr8KNgPh-Vw	A demo showcases an example in Tool4Boxology interface

Novelty: Tool4Boxology provides the first semantically grounded environment to design, analyse, and validate neuro-symbolic AI systems. In contrast to generic diagramming tools, it integrates semantic technologies—OWL ontologies, RML mappings, knowledge graphs, and SHACL constraints—to produce a machine-interpretable, type-consistent representation of Boxology design patterns. By enforcing ontology-backed design rules, Tool4Boxology eliminates structural ambiguity, and ensures that neuro-symbolic architectures are both syntactically well-formed and semantically interpretable.

Availability: Tool4Boxology is released under a permissive licence and is maintained by the Scientific Data Management Group at TIB and VU Amsterdam on a public GitHub repository. The codebase is built on an extensible JavaScript framework. Following the principles of FAIR data, each release is archived with a persistent identifier in Zenodo and the TIB-Leibniz Data Manager ensure reproducibility and long-term sustainability. The Tool4Boxology editor and a SPARQL endpoint for accessing the T4B KG are available through TIB.

Utility: Tool4Boxology is open-source software with documentation, a tutorial, and curated examples of 61 neuro-symbolic architectures. The public repository provides a step-by-step guide to modelling neuro-symbolic architectures, along with illustrative use cases. A video demonstrates an example of how to design, validate, and explore a boxology pattern in the Tool4Boxology. A Jupyter Notebook – available via the Tool4Boxology entry in the Leibniz Data Manager – provides runnable examples of KG exploration and analysis. See Table 3.

Predicted Impact: As argued in Sect. 1, the increasing architectural complexity in neuro-symbolic systems demands principled semantically enabled methods for documentation and validation. Our Tool4Boxology directly addresses this need by offering a visually coherent and semantically rigorous methodology to represent neuro-symbolic workflows. Beyond neuro-symbolic systems, the underlying semantic representation of typed components and design patterns generalises to other software architectures.

Adoption and Reusability: Tool4Boxology is already used in academia and industry. In research, it supports the specification of neuro-symbolic architectures in initiatives such as *TrustKG, CAIMed*, and the *Hybrid Intelligence* programme. Industrial partners apply the tool to smart building scenarios and manufacturing pipelines. The Boxology notation itself is seeing a rapidly increasing uptake in third-party publications, demonstrating a growing interest in principled and comparable neuro-symbolic representations. The integration of the T4B editor with the T4B KG is anticipated to further facilitate the adoption of T4B by offering a reproducible, extensible, and semantically interoperable framework for modeling complex AI architectures.

6   Related Work

Our work is situated in the subdiscipline of Semantic Software Engineering. Semantic Software Engineering aims to elevate software artefacts (in our case, architecture diagrams) from syntactic structures (the boxology diagrams) into semantic representations (an ontology and knowledge graph), to enable the application of inferential reasoning (RDFS and OWL) and constraint validation (SHACL). See [17] for an early survey, and [1,7] for proposals to formalise the classical Gang of Four design patterns [13] in OWL. Similar to the resource presented in this paper, recent work in Semantic Software Engineering has used SHACL to validate a system's architecture [25], applied ontologies to software engineering [28], and used semantic reasoning to automate design pattern selection [2]. Tool4Boxology is novel with respect to this state of the art in Semantic Software Engineering in two ways. First, we specifically address Semantic Software Engineering for AI systems, in particular, neuro-symbolic architectures. This is comparable to [24] for design patterns in Machine Learning, but the patterns described there remain entirely informal, in contrast to our formalisation of patterns as an ontology-based knowledge graph. Second, we provide an integrated resource to create, validate, and analyse these design patterns at both the syntactic and the semantic levels, both in a human-readable graphical form and a machine-processable formal representation. To the best of our knowledge, no other such integrated environment for SSE exists to date.

7   Conclusion and Future Work

We have presented Tool4Boxology, a resource that uses semantic technologies to construct, validate, and query architecture diagrams for neuro-symbolic systems. The resource consists of (i) a graphical editor to construct architectural diagrams, (ii) a typed grammar to check for syntactic correctness, (iii) an ontology to capture some of the semantics of the diagrams, (iv) an RML translation from diagrams into an ontology-compliant knowledge graph, (v) SHACL constraints that express semantic constraints on the architectural diagrams, and (vi) the ability to execute SPARQL queries for analysing the architectural patterns, (vii) a corpus of 61 neuro-symbolic architectures captured in validated diagrams and represented as a knowledge graph.

Some of the many possible extensions of the Tool4Boxology are already depicted in Fig. 3: a larger library of patterns [5]; an extended boxology notation to include generative AI components [6], formalisations of the boxology diagrams that capture more aspects than our own ontology [11,29], (semi-)automatically proving properties of an architecture and (semi-)automatically discovering archetypes across a corpus of boxology diagrams.

A broader vision is to generalise the Tool4Boxology to other software domains (e.g., microservices, embedded systems, real-time systems). We claim that our design is sufficiently generic for this. It would "simply" require a new set of primitive patterns, syntactic constraints to express their compositions, an ontology to capture the components and their connections, an RML script to map the patterns into a knowledge graph, and SHACL constraints to capture (in)valid patterns. The pipeline from Fig. 3 and the design of our toolbox would remain unchanged while reaching beyond the neuro-symbolic systems from this paper.

Acknowledgements. Work at TIB was supported by the "Leibniz Best Minds: Programme for Women Professors", through funding of the "TrustKG-Transforming Data in Trustable Insights" project (Grant P99/2020), and by the Lower Saxony Ministry of Science and Culture (MWK) with funds from the Volkswagen Foundation's zukunft.niedersachsen program (CAIMed - Lower Saxony Center for AI and Causal Methods in Medicine; GA No. ZN4257). Work at VUA was supported by the 'Hybrid Intelligence' (https://hybrid-intelligence-centre.nl) project, number 024.004.022 of the 'Gravitation' programme from the Dutch Research Council (NWO).

Appendix A: Validation Rules for Syntactical Correctness

- **{model[1..]} → transform → {model[1]}**
  "One or more *models* as input and exactly one *model* as output."
- **{(data|symbol)[1..]} → transform → {(data|symbol)[1]}**
  "One or more *data* or *symbols* as input and exactly one *data* or *symbol* as output."
- **{model[0..1], (data|symbol)[1..]} → transform:embed → {(data| model)[1]}**
  "Zero or one (embedding-) *models* and one or more *data* or *symbols* as input and exactly one *data* or *model* as output."
- **{model[1..], (data|symbol[1..]} → infer:deduce → {(data|symbol|model)[1] }**
  "One or more *models* and one or more *data* or *symbols* as input and exactly one *data*, *symbol*, or *model* as output."
- **{actor[1], (data|symbol|model)[0..]} → generate:engineer → { (data|symbol|model)[1]}**
  "Exactly one *actor* and zero or more *data*, *symbols* or *models* as input and exactly one *data*, *symbol* or *model* as output."
- **{model[0..1], (data|symbol)[1..]} → generate:train → {model[1]}**
  "Zero or one *models* and one or more *data* or *symbols* as input and exactly one *model* as output."
- **Sets of *data*, *symbols*, or *actors* are represented as a single *data*, *symbols*, or *actors* box respectively.**

References

1. Alnusair, A., Zhao, T.: Using ontology reasoning for reverse engineering design patterns. In: Ghosh, S. (ed.) MODELS 2009. LNCS, vol. 6002, pp. 344–358. Springer, Heidelberg (2010). https://doi.org/10.1007/978-3-642-12261-3_32
2. Attia, R.A., Ghoniemy, S., Hamdy, A.: Advanced design pattern selection methodology through weighted semantic ontologies for microservices. In: International Conference on Computer and Applications, ICCA 2024, pp. 1–10. IEEE (2024)

3. Beall, J.: The weaknesses of full-text searching. J. Acad. Librariansh. **34**(5), 438–444 (2008)

4. Bérubé, M., Giannelia, T., Vial, G.: Barriers to the implementation of AI in organizations: findings from a Delphi study. In: 54th Hawaii International Conference on System Sciences (HICSS) (2021)

5. Breit, A., et al.: Combining machine learning and semantic web: a systematic mapping study. ACM Comput. Surv. **55**(14s), 313:1–313:41 (2023)

6. de Boer, M.H.T., Smit, Q.T.S., van Bekkum, M., Meyer-Vitali, A., Schmid, T.: Modular design patterns for generative neuro-symbolic systems. In: Joint Proceedings of the ESWC 2024 Workshops and Tutorials co-located with 21th European Semantic Web Conference (ESWC 2024). CEUR Workshop Proceedings, vol. 3749. CEUR-WS.org (2024)

7. Dietrich, J., Elgar, C.: A formal description of design patterns using OWL. In: 16th Australian Software Engineering Conference (ASWEC 2005), pp. 243–250. IEEE Computer Society (2005)

8. Dijk, J., Schutte, K., Oggero, S.: A vision on hybrid AI for military applications. In: Dijk, J. (ed.) Artificial Intelligence and Machine Learning in Defense Applications, p. 30. SPIE (2019)

9. Dimou, A., Sande, M.V., Colpaert, P., Verborgh, R., Mannens, E., Van de Walle, R.: RML: a generic language for integrated RDF mappings of heterogeneous data. In: Workshop on Linked Data on the Web co-located with the 23rd International World Wide Web Conference (WWW 2014). CEUR Workshop Proceedings, vol. 1184. CEUR-WS.org (2014)

10. Eiraudo, S., et al.: Experimental application of a semi-parametric model for interpretable and accurate egression analysis of building energy consumption. Energy Build. **349**, 116495 (2025)

11. Ellis, A., Dave, B., Salehi, H., Ganapathy, S., Shimizu, C.: EASY-AI: sEmantic And compoSable glYphs for representing AI systems. In: Lorig, F., et al. (eds.) Frontiers in Artificial Intelligence and Applications. IOS Press (2024)

12. Hogan, A., et al.: Knowledge Graphs. Semantics, and Knowledge. Synthesis Lectures on Data. Springer Nature (2021)

13. Gamma, E., Helm, R., Johnson, R., Vlissides, J.: Design Patterns: Elements of Reusable Object-Oriented Software. Addison-Wesley Longman Publishing Co., Inc, USA (1995)

14. García-Barragán, Á., et al.: NSSC: a neuro-symbolic AI system for enhancing accuracy of named entity recognition and linking from oncologic clinical notes. Med. Biol. Eng. Comput. **63**(3), 749–772 (2025)

15. Gijón, A., et al.: Explainable hybrid semi-parametric model for prediction of power generated by wind turbines. In: Franco, L., et al. (eds.) ICCS 2024, pp. 299–306. Springer, Cham (2024)

16. Gusenbauer, M.: Google scholar to overshadow them all? Comparing the sizes of 12 academic search engines and bibliographic databases. Scientometrics **118**(1), 177–214 (2019)

17. Happel, H.-J., Maalej, W., Seedorf, S.: Applications of ontologies in collaborative software development. In: Mistrík, I., van der Hoek, A., Grundy, J., Whitehead, J. (eds.) Collaborative Software Engineering, pp. 109–129. Springer, Cham (2010)

18. Hogan, A., et al.: Knowledge graphs. CoRR, abs/2003.02320 (2020)

19. Iglesias, E., Jozashoori, S., Vidal, M.-E.: Scaling up knowledge graph creation to large and heterogeneous data sources. J. Web Semant. **75**, 100755 (2023)

20. Iglesias, E., Vidal, M.-E., Collarana, D., Chaves-Fraga, D.: Empowering the SDM-RDFizer tool for scaling up to complex knowledge graph creation pipelines1. Semant. Web 1–28 (2024)
21. Jones, N: This AI combo could unlock human-level intelligence. Nat. News Feature (2025)
22. Kierner, S., Kucharski, J., Kierner, Z.: Taxonomy of hybrid architectures involving rule-based reasoning and machine learning in clinical decision systems: a scoping review. J. Biomed. Inform. **144**, 104428 (2023)
23. Kubesch, J., Kiesling, E., Ekaputra, F.J., Serles, U., Toma, I., Havas, C.: Exploring the benefits of iterative retrieval-augmented generation for risk mitigation in LLM responses. In: Fischer, L., et al. (eds.) DEXA 2025, pp. 3–14. Springer, Cham (2026)
24. Lakshmanan, V., Robinson, S., Munn, M.: Machine Learning Design Patterns: Solutions to Common Challenges in Data Preparation, Model Building, and MLOps, 1st edn. O'Reilly, Sebastopol (2020)
25. Markaj, A., Gehlhoff, F., Fay, A.: Semantic-aware validation in model-driven requirements engineering using SHACL. In: Aveiro, D., Poggi, A., Bernardino, J. (eds.) 16th International Joint Conference on Knowledge Discovery, Knowledge Engineering and Knowledge Management, IC3K 2024, vol. 2, pp. 191–198. SCITEPRESS (2024)
26. Meyer-Vitali, A., Mulder, W., de Boer, M.H.T.: Modular design patterns for hybrid actors. CoRR, abs/2109.09331 (2021)
27. Montes, G.A., Goertzel, B.: Distributed, decentralized, and democratized artificial intelligence. Technol. Forecast. Soc. Chang. **141**, 354–358 (2019)
28. Moreira, J.L.R., Donkers, A., Pauwels, P., Bektas, E., van Ee, T.: Onto4reuse: towards an ontology reuse framework for knowledge-intensive software engineering. In: Joint Ontology Workshops (JOWO). CEUR Workshop Proceedings, vol. 3882. CEUR-WS.org (2024)
29. Mossakowski, T.: Modular design patterns for neural-symbolic integration: refinement and combination. In: d'Avila Garcez, A.S., Jiménez-Ruiz, E. (eds.) 16th International Workshop on Neural-Symbolic Learning and Reasoning as part of the 2nd International Joint Conference on Learning & Reasoning (IJCLR 2022). CEUR Workshop Proceedings, vol. 3212, pp. 192–201 (2022)
30. Musen, M.A.: The protégé project: a look back and a look forward. AI Matters **1**(4), 4–12 (2015)
31. Ng, C.H., ten Teije, A., van Harmelen, F.: A boxology-based analysis of design patterns for neuro-symbolic medical decision making systems. In: Bellazzi, R., Herrero, J.M.J., Sacchi, L., Zupan, B. (eds.) Artificial Intelligence in Medicine, AIME 2025. LNCS, vol. 15734. Springer, Cham (2025)
32. Ng, C.H., Teije, A.T., Van Harmelen, F.: A boxology-based analysis of design patterns for neuro-symbolic medical decision making systems. In: Bellazzi, R., Herrero, J.M.J., Sacchi, L., Zupan, B. (eds.) Artificial Intelligence in Medicine. LNCS, vol. 15734, pp. 333–343. Springer, Cham (2025)
33. Prater, R., Laurenzi, E.: A hybrid intelligent approach for the support of higher education students in literature discovery. In: AAAI 2022 Spring Symposium on Machine Learning and Knowledge Engineering for Hybrid Intelligence (AAAI-MAKE 2022), Palo Alto (2022)
34. Qudrat-Ullah, H.: Applications in various domains. In: Navigating Complexity: AI and Systems Thinking for Smarter Decisions, pp. 93–118. Springer, Cham (2025)
35. Rincon-Yanez, D., Senatore, S., O'Sullivan, D.: Scaling NeuroSymbolic AI integration for seismic event detection, pp. 108–113. Springer, Cham (2025)

36. Van Bekkum, M., De Boer, M., Van Harmelen, F., Meyer-Vitali, A., Ten Teije, A.: Modular design patterns for hybrid learning and reasoning systems: a taxonomy, patterns and use cases. Appl. Intell. **51**(9), 6528–6546 (2021)
37. van Harmelen, F., ten Teije, A.: A boxology of design patterns for hybrid learning and reasoning systems. J. Web Eng. **18**(1–3), 97–124 (2019)
38. Witschel, H.F., Pande, C., Martin, A., Laurenzi, E., Hinkelmann, K.: Visualization of patterns for hybrid learning and reasoning with human involvement. In: Dornberger, R. (ed.) New Trends in Business Information Systems and Technology. SSDC, vol. 294, pp. 193–204. Springer, Cham (2021). https://doi.org/10.1007/978-3-030-48332-6_13

Bench4KE: Benchmarking Automated Competency Question Generation

Paolo Ciancarini[1], Anna Sofia Lippolis[1,2]([✉]),
Andrea Giovanni Nuzzolese[2], Valentina Presutti[1,2],
and Minh Davide Ragagni[1,3]

[1] University of Bologna, Bologna, Italy
`minhdavide.ragagni3@unibo.it`
[2] CNR - Institute of Cognitive Sciences and Technologies, Rome and Bologna, Italy
`annasofia.lippolis2@unibo.it`
[3] University of Pisa, Pisa, Italy

Abstract. The availability of Large Language Models (LLMs) presents a unique opportunity to reinvigorate research on Knowledge Engineering (KE) automation. This trend is already evident in recent efforts developing LLM-based methods and tools for the automatic generation of Competency Questions (CQs), natural language questions used by ontology engineers to define the functional requirements of an ontology. However, the evaluation of these tools lacks standardisation. This undermines methodological rigor and hinders the replication and comparison of results. To address this gap, we introduce Bench4KE, an extensible API-based benchmarking system for KE automation. The presented release focuses on evaluating tools that generate CQs automatically. Bench4KE provides a curated gold standard consisting of CQ datasets from 17 real-world ontology engineering projects and uses a suite of similarity metrics to assess the quality of the CQs generated. We present a comparative analysis of 6 recent CQ generation systems, which are based on LLMs, establishing a baseline for future research. Bench4KE is also designed to accommodate additional KE automation tasks, such as SPARQL query generation, ontology testing and drafting. Code and datasets are publicly available under the Apache 2.0 license.

Resource type: Benchmarking system
License: Apache 2.0
DOI: 10.5281/zenodo.17817277
URL: https://github.com/fossr-project/ontogenia-cini.

Keywords: Benchmark · Knowledge Engineering · Competency Questions · Large Language Models

---

Author statement: P.C., A.G.N., V.P.: methodology and supervision; A.S.L. and M.D.R.: methodology, validation, investigation and data curation; all authors contributed to the conceptualization of the idea, wrote, reviewed, and approved the manuscript.

1 Introduction

Competency Questions (CQs) [19], formulated as natural language questions that outline and constrain the scope of an ontology, play a crucial role in Knowledge Engineering (KE). In fact, they are fundamental for defining functional requirements and unit tests [7] in ontology design processes, such as eXtreme Design [40] (XD). In addition, they guide the modeling of an ontology structure by informing the selection of relevant concepts and relationships, and support verification and validation processes of the encoded knowledge [3].

With the advent of Large Language Models (LLMs), several research efforts aim to reinvigorate research on KE automation, thanks to the models' ability to generate structured representations from natural language text. Many tasks have shown the potential of LLM-assisted KE, focusing, for example, on the generation of ontologies from requirements or on the automatic support to requirements elicitation [27,29]. Among these, three main approaches: LLM-based generation, reverse-engineering, and retrofitting of CQs, have been identified in the literature [3] and received significant attention [2,4,15,31,35,42,50]. These systems rely on LLMs to generate CQs from requirements such as scenario descriptions, or documents, or from existing ontologies that lack documentation.

As these tools evolve, the need for systematic and standardised evaluation methods becomes increasingly pressing. Although gold standards have been proposed to evaluate automated KE tasks involving CQs, the landscape of KE automation evaluation is scattered and unaligned: to the best of our knowledge, there is lack of a reference benchmarking platform to support the evaluation of CQ generation systems and a rigorous definition for this task is missing. They are both necessary to support advancement of the state of the art through systematic comparative analysis.

To address this gap, we introduce **Bench4KE**, a benchmarking system with the overarching aim to systematically evaluate KE automation tasks. BenchKE provides the community with a framework to evaluate the ability of their systems to act as an expert knowledge engineer, who can adapt their skills to any knowledge domain. In its first release, presented in this paper, Bench4KE supports the evaluation of tools that generate CQs. Our benchmark includes gold standard datasets derived from the related literature (see Sect. 2) and enhanced for the coverage of multiple generation scenarios, and combines multiple semantic and lexical similarity metrics to evaluate the semantic proximity and appropriateness of the generated CQs.

Bench4KE provides user stories, heterogeneous source datasets, ontologies, and PDF documents, as input. We test six existing tools through Bench4KE, establishing a baseline for future research on the CQ generation task; however, the objective of this work is primarily to launch a platform for future challenge and collaborative research within the Semantic Web community.

Our key contributions are:

1. **Bench4KE**: An extensible framework for benchmarking KE automation tasks, currently focused on the evaluation of CQ generation systems, along with usage and configuration instructions;

2. **Gold Standard Dataset**: A curated and extensible dataset of 843 manually crafted CQs from multiple real-world ontology projects;
3. A first **baseline** for the CQ generation task, based on the evaluation of six recent peer-reviewed systems.

The rest of the paper is organized as follows. Section 2 presents the related work; Sect. 3 describes Bench4KE, including its usage scenarios, benchmarking dataset, system architecture, evaluation metrics, and execution framework. Section 4 shows the evaluation setup and the experimental results. Section 5 describes the availability of the resource. Finally, Sect. 6 discusses the evaluation metrics, the impact and extensibility of the benchmark, while Sect. 7 concludes the paper.

2 Related Work

To the best of our knowledge this is the first proposal of a reference benchmarking framework for the systematic evaluation of tools that generate CQs. We identify four main categories of relevant related work: (i) systems for automated CQ generation, (ii) benchmarks for CQ generation with LLMs, (iii) existing approaches to CQ validation; and (iv) ontology engineering tools.

Automated CQ Generation. Automated and semi-automated CQ generation are long-standing tasks [17,32]. Recently, significant effort has been devoted to developing tools that automate CQ generation using LLMs. These systems differ in the type of input they handle, the prompting techniques adopted, and their intended use within the ontology engineering pipeline. They can fall into three main approach categories [3]: (i) Generating CQs; (ii) Reverse engineering of CQs; and (iii) Retrofitting CQs. Systems that follow the (i) approach are the most common. Among them, Ontochat [50,52], designed for assisting the whole KE workflow, supports CQ generation through conversational prompting and is designed to facilitate participatory ontology engineering. NeOn-GPT [15] generates CQs from requirements, by combining structured prompt templates with ontology metadata. AgOCQs [4,31] is an automatic corpus-based CQ generation tool that uses fine-tuned LLMs to generate CQs from domain-specific corpora. It applies lexical, syntactic, and semantic filters to ensure that generated CQs are well-formed, answerable, and reflect the intended semantics of the ontology. Pan *et al.* [35] introduce a retrieval-augmented generation (RAG) approach to CQ generation. This approach first retrieves relevant ontology content or background knowledge, which is then passed to an LLM to generate context-aware CQs. Rebboud *et al.* [42] investigate the use of LLMs to generate CQs given an existing ontology, comparing six different models and multiple prompt settings to evaluate how well LLMs can support knowledge engineers in this task. RevOnt [11] is a CQ reverse engineering system trained on specific domains that extracts CQs directly from knowledge graphs. It operates in three stages: (a) verbalisation abstraction (generalising instance-level triples to class-level text); (b) question generation using a T5-based model; and (c) question filtering via SBERT-based similarity and paraphrase detection, to derive a compact set of core CQs. For what concerns retrofitting, Alharbi *et al.* [2] introduce the RETROFIT-CQs approach to generate candidate CQs from ontology triples using GPT-3.5 and GPT-4. Di Nuzzo *et al.* [13] propose a pipeline that combines LLMs and Knowledge Graphs. We rely on these previous results to design the different input scenario of the CQ generation task supported by BEnch4KE.

Benchmarks for CQ Generation with LLMs. Recently, benchmarks have been proposed to evaluate LLM-based tools in KE and cover tasks such as ontology conceptualisation or CQ generation.

Rebboud *et al.* [41] propose a gold standard designed to evaluate how LLMs contribute to ontology conceptualisation tasks, including ontology generation, documentation, and CQ generation. Such a benchmark focuses on understanding the broader modeling capabilities of LLMs. CQsBEN [3] is a gold standard that supports the comparison of manual, semi-automatic, and automatic CQ construction approaches. The dataset is organized around tasks such as requirement scoping and CQ verification, providing guidance for method classification. It is oriented towards mapping and categorising CQ engineering approaches in the ontology life-cycle. We rely on these previous results to build the Bench4KE benchmarking datasets.

Approaches to CQ Validation. Early tools for CQ validation rely on manually crafted SPARQL queries to check CQs against an ontology. Notable examples are *OntologyTest* [18] or the *Ontology Testing* framework of Blomqvist *et al.* [7]. Other solutions, such as *CQChecker* [6] and the *CQ-Driven Authoring and Testing* plugin by Dennis *et al.* [12], integrate CQ validation directly into the ontology-development workflow. Another solution is *TESTaLOD* [10] that enables agile knowledge graph testing by providing an automated, adaptable framework based on a Web interface a REST API for assessing the quality and consistency of knowledge graphs via CQ validation, error provocation, and logic consistency checking. Instead, *CLaRO* [23] introduces a controlled natural language for authoring CQs, to assist ontology engineers in formulating requirements. Finally, Wiśniewski *et al.* [49] distilled a set of recurring CQ templates. Traditional systems are evaluated on benchmark ontologies by measuring (i) the accuracy of answering CQs or detecting violations and (ii) coverage. Some studies complement these metrics with user experiments to verify that automated checks align with human judgments of ontology completeness (e.g., Dennis *et al.* [12]).

LLMs remove the manual formalisation step by mapping a natural-language CQ directly to a query or answer. Tufek *et al.* [47] compare LLM-generated CQs with expert gold standards to assess precision, recall, and F_1. Alharbi *et al.* [2] rely on semantic similarity; Pan *et al.* [35] emphasise precision and answer consistency; and Di Nuzzo *et al.* [13] use cosine similarity between expected and produced answers along with the LLM-based rating of each CQ on a scale from 1–5 for what concerns relevance, clarity, and depth. Although Lippolis *et al.* [29] focus on ontology generation, they likewise treat CQs as a benchmark of completeness. Compared with earlier work, recent evaluations go beyond simple answer accuracy: researchers now analyze retrieval metrics for generated queries, semantic-similarity scores, and expert assessments in addition to the traditional coverage measure. Bench4K combines a suite of metrics for evaluating the generated CQs against the benchmarking dataset. We implement metrics that have been used in previous work and add novel ones, such as the hit rate, for allowing a more comprehensive and multi-perspective interpretation of the results. We also introduce and explore the use of LLMs as evaluators of CQs.

Ontology Engineering Tools. Bench4KE is broadly introduced as an ontology engineering tool in which the evaluation of CQ generation systems is only one step of the ontology engineering benchmarking pipeline. Recently, the need for automation with LLM has garnered interest in creating such tools. The already cited OntoChat [50] is a conversational ontology engineering framework that uses an LLM-based agent to elicit and manage requirements in various stages until ontology generation. NeOn-GPT [15] is a pipeline that integrates the NeOn ontology development methodology with GPT-based prompts to convert domain descriptions into OWL/Turtle ontologies.

In summary, existing studies have paved the way for reinvigorating research on knowledge engineering automation, but they are hardly comparable. A reusable end-to-end benchmark to automatically evaluate the semantic quality of generated CQs is missing and is arguably needed to systematically advance the state-of-the-art in this area. Bench4KE fills this gap by pairing a gold standard set of 843 CQs manually generated by expert ontology engineers from 17 real?world ontologies with a validation pipeline that combines classical similarity metrics and an LLM?based semantic judge, delivering the first systematic, extensible framework for comparing LLM?driven CQ generation tools.

3   Bench4KE

We present Bench4KE by detailing: (i) the requirements specification; (ii) the usage scenarios considered for evaluating the task of CQ generation; (iii) the benchmark dataset used as gold standard; (iv) the system architecture; (v) the execution setup for CQ validation, and (vi) the evaluation metrics supported (Fig. 1).

3.1   Requirements Specification

Requirements are a prerequisite for successful software development [48]. For Bench4KE, we derived requirements from related work (Sect. 2) to build on validated ideas, expose missing capabilities as requirements, and provide a focused direction for future development.

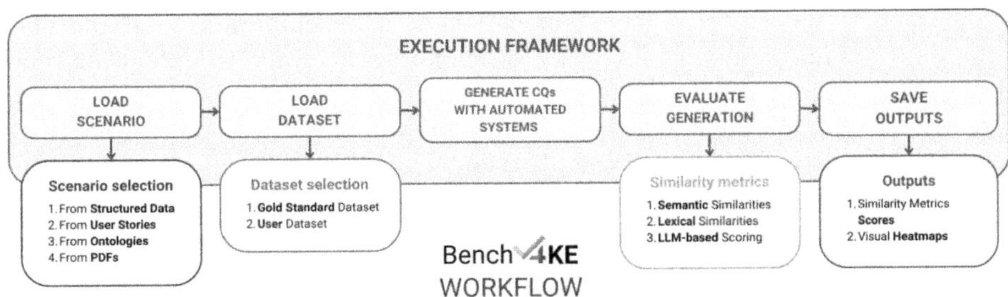

Fig. 1. Bench4KE's workflow for CQ validation.

Table 1. Functional (F) and non functional (N) requirements for Bench4KE. The Proven Feature column represents the successfully built and tested functionalities; the Current Deficit column shows areas that need future development.

	Requirement	Proven feature	Current deficit
F1	The system shall support the evaluation of CQ generation tools based on user story inputs.	✓	
F2	The system shall support the evaluation of CQ generation tools using raw structured data.	✓	
F3	The system shall support the evaluation of CQ extraction or retrofitting tools from ontologies.	✓	
F4	The system shall support the evaluation of CQ generation tools from academic PDF sources.	✓	
F5	The system shall allow external datasets to be uploaded as gold standards for benchmarking.	✓	
F6	The system shall expose a RESTful API for dataset access and benchmark submission.	✓	
F7	The system shall support multiple LLM models for validation.	✓	
F8	The system should provide configurable prompt templates for domain-specific CQ generation contexts.		✓
F9	The system should support incremental/streaming CQ generation with partial result visibility.	✓	
N1	The system shall ensure standardised formats and traceability of results.	✓	
N2	The system shall support the comparative benchmarking of multiple CQ generation systems.	✓	
N3	The system shall provide transparent evaluation metrics and benchmarking reports.	✓	
N4	The system will provide ontology coverage analytics comparing CQ sets against ontology terms.		✓

The elicited requirements capture the core features, functionalities, and quality characteristics expected from Bench4KE as a benchmarking system. They serve as a guide for us and for other developers who want to extend or reuse Bench4KE, supporting the systematic development of a robust and maintainable benchmark. We document both functional and non-functional requirements using the established MASTER template and The SOPHISTs [44]. This template distinguishes the degree of obligation: "shall" marks mandatory requirements, "should" marks desirable but optional improvements, and "will" marks planned extensions beyond the current scope.

Table 1 summarises the requirements for Bench4KE. Functional requirements (F) describe the essential capabilities of Bench4KE, while non-functional requirements (N) capture performance, usability, and other quality criteria. The Proven Feature column indicates which requirements are already implemented or empirically validated, and the Current Deficit column highlights open points and planned enhancements for future releases. Together, these dimensions support progress monitoring and gap analysis, ensuring that Bench4KE evolves in line with user needs and benchmarking goals. In line with this specification, Bench4KE already fulfills most of the defined functional and non-functional requirements, which demonstrates its suitability for a broad range of evaluation scenarios.

3.2   Usage Scenarios

Although our approach is independent on the type of tool to be evaluated, we currently focus on LLM-based tools, as they represent the most recent and prominent approach in the field. We consider two main scenarios distinguishing the type of input the CQ generation system processes: i) the existence, as input, of structured or semi-structured data that shall be conceptually and syntactically homogenised, which is common in data integration tasks, and ii) the elicitation of requirements through user stories, a common situation in most ontology development methodologies. These two scenarios are frequently combined. User stories are crucial to understanding the domain and identifying the boundaries of ontological requirements. Existing data are to be analysed (via reverse engineering) in order to uncover their underlying conceptual model, which is often implicit. These usage scenarios are designed to test the generalisability of CQ generation models across different input modalities.

3.3   Benchmarking Dataset

We build a gold standard dataset by collecting CQs from 17 real-world ontology-driven projects, combining various knowledge domains, as reported in Table 2 along with statistics. Each CQ is accompanied by a resource related to one or more usage scenarios, making the resulting dataset a carefully structured and semantically aligned collection, designed to support comparative analysis. As a result, we obtain 843 CQs.

Table 2. Bench4KE dataset statistics.

Project	Domain(s)	CQs	Stories	Datasets	Ontologies	PDFs
Polifonia [5]	Music	67	✓			
	Music	28			✓	
ArCo [9]	Cultural Heritage	10		✓		
WHOW [28]	Water	9		✓		
	Health	6		✓		
HACID [24,46]	Health	7	✓			
	Climate Services	2		✓		
SWO [33]	Life Sciences	88			✓	
Stuff [22]	Macroscopic Stuff	11			✓	
AWO [21]	African Wildlife	14			✓	
Dem@Care [16]	Health	107			✓	
OntoDT [36]	Computer Science	14			✓	
Wine Ontology [34]	Wine	7			✓	
DOREMUS [1]	Music	58			✓	
NORIA-O [45]	ICT	26			✓	
Odeuropa [30]	Cultural Heritage	74			✓	
VGO [16,38]	Videogames	68			✓	
VICINITY Core [16]	IoT	126			✓	
HCI [20]	Hum.-Comp. Interaction	15				✓
RE [14]	Requirement Engineering	106				✓
Total		**843**	74	27	621	121

3.4 Bench4KE Architecture and Execution Flow

Bench4KE is designed to systematise the evaluation of automatically generated CQs against a curated gold standard dataset. It follows an API-centric architecture and is implemented using the modular and extensible FastAPI[1] framework.

Figure 2 shows the system's architecture, structured to support comprehensive benchmarking capabilities with a focus on maintainability and scalability.

Validator. The implementation mirrors the modular layout shown in Fig. 2. The Bench4KE Validation API exposes a single REST endpoint, /validate, implemented through FastAPI. During an evaluation session, the Orchestrator acts as the central coordinator: it manages the evaluation lifecycle by loading the dataset, triggering the external CQ generation system, collecting its output, and forwarding the generated and gold standard CQs to the CQ-Validator for scoring. To integrate with the orchestrator, any external system only needs to expose

---

[1] https://fastapi.tiangolo.com/.

an HTTP endpoint that accepts a CSV file and returns a CSV with a generated column. That is the entire integration requirement, which is what makes the system extensible. Optionally, the caller may provide a custom input dataset, along with a corresponding set of gold standard CQs. This dual capability ensures flexibility, supporting both predefined and ad hoc evaluation scenarios tailored to experimental needs. The Validator then computes evaluation scores (Sect. 4). Currently, the Validator API can be run locally.

Evaluation Metrics. The platform enables the execution of multiple evaluation strategies that assess both syntactic and semantic correspondence between automatically generated and manually authored CQs. The evaluation results are presented both numerically and visually. The following evaluation metrics are currently implemented: Cosine similarity (computed over sBERT [43] sentence embeddings), BERTScore [51], Jaccard similarity, ROUGE-L [26], BLEU [37], Hit Rate and LLM-based semantic scoring. We also implement a metric called *Hit Rate*, where the validator counts how many gold CQs obtain at least one generated counterpart whose Cosine similarity exceeds the 0.60 threshold. Hit rate is therefore a coverage indicator: high values mean that the generator reliably produces at least one relevant CQ per requirement.

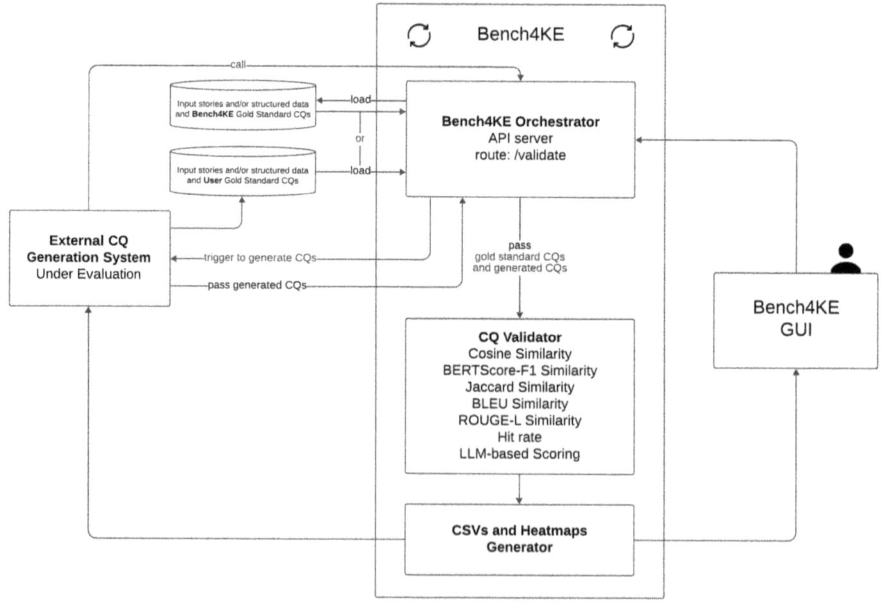

Fig. 2. Bench4KE's system architecture.

Hit Rate, along with other metrics, collectively help quantify both surface-level and structural overlaps between generated and gold-standard CQs. In fact,

BERTScore captures contextual and distributional semantics, Jaccard focuses on token-level overlap, while BLEU and ROUGE-L assess n-gram precision and recall. Nevertheless, these metrics often fail to reflect deep semantic equivalence, especially when paraphrasing is involved. To mitigate this issue, we enhance the analysis with an LLM-based semantic scoring that interprets the metric outputs and selects the most semantically aligned CQ pairs. The prompt asks the model to identify the closest matching CQs and to produce any essential questions missing from the gold standard, for human oversight[2]. In addition, we include an LLM-as-a-judge metric derived from [13], where the LLM is asked to judge the responses on a scale from 1 to 5 on the parameters of relevance, clarity and depth. This approach allows Bench4KE to benefit from the reproducibility of lexical metrics, while introducing semantic depth and human-like reasoning through LLM analysis. For what concerns LLMs, we support multiple models, including those provided by OpenAI[3], Anthropic[4], Meta[5], and Together AI[6].

Output Structure. The evaluation outputs are automatically stored in a user-defined output directory as JSON and visualised on the console.

Usage. There are two ways to use Bench4KE: via curl (API) or via the web front-end.

The following cURL command launches a complete validation call:

```
curl -X POST "http://127.0.0.1:8000/validate/" \
  -F "use_default_dataset=true" \
  -F "external_service_url=http://127.0.0.1:8001/newapi" \
  -F "api_key=your_key" \
  -F "validation_mode=all" \
  -F "model=gpt-4" \
  -F "save_results=true"
```

The parameters are defined as follows:

- `use_default_dataset`: Set to `true` to use the benchmark dataset.
- `external_service_url`: The URL of the external CQ generation API.
- `api_key`: If required by the external service.
- `validation_mode`: One of: `all`, `cosine_bertscore_judge`, `llm`, `cosine`, `jaccard`.
- `output_folder`: Directory where the outputs and heatmaps are stored.
- `model`: LLM to use.
- `save_results`: If set to `true`, stores the results as CSV.
- `save_every`: It is set to a number for incremental savings.

---

[2] https://github.com/fossr-project/ontogenia-cini/blob/main/restapi/app/services/cq_validator.py.

[3] https://openai.com/.

[4] https://www.anthropic.com/.

[5] https://ai.meta.com/meta-ai/.

[6] https://www.together.ai/.

To facilitate human interaction, Bench4KE includes two additional software components: (i) a CQ generation app[7], which simulates the behavior of a CQ generation system and its communication with Bench4KE, and (ii) a web front-end[8] that allows users to run the evaluation entirely on their local machine, returning the results in JSON format. The front-end accepts as input an external service URL. Once provided, the system computes the metrics and displays them immediately in the browser; the same results (together with heatmap images) are written to a dedicated output directory for later inspection.

3.5 Documentation

Comprehensive and detailed documentation can be accessed from the Github repository[9]. The documentation is organised into the following main parts: *Getting Started*, which provides all the necessary information to help users begin using Bench4KE; and *How to Use?*, which includes installation instructions and a quickstart guide for both creators of CQ generation systems and general users of the system, who might want to test or customize the tool with the information to make Bench4KE work.

4 Evaluation and Impact

To establish a baseline, we have selected the `cosine_bertscore_judge` configuration, i.e., cosine similarity over SBERT embeddings, BERTScore-F1, and an LLM-as-a-judge score. This choice is motivated by computational considerations as well as by the complementarity of these metrics and their suitability for the CQ generation task. Furthermore, through discussion and a Survey (cf. Sect. 4.3) we have assessed Bench4KE's potential impact, and how to improve it.

4.1 Analysed Systems

The current baseline is given from the existing systems interested in testing our tool. They have been evaluated against the entire Bench4KE dataset. This includes the datasets used originally for evaluating these tools as well as datasets from a set of European ontology-driven projects that provide manually crafted CQs associated with stories and source datasets, covering additional domains of knowledge.

To achieve integration of Bench4KE with existing systems, minor modifications were necessary to some of them. In particular, OntoChat required limiting the output to a maximum number of CQs, as the tool requires this setting, and allowing to inject also PDF, ontology, and dataset text instead of user stories; the

---

[7] https://github.com/fossr-project/ontogenia-cini/blob/main/restapi/
cq_generator_app.py.

[8] https://github.com/fossr-project/ontogenia-cini/blob/main/restapi/bench4ke-validate-ui.py.

[9] https://github.com/fossr-project/ontogenia-cini.

generation from user stories was preserved. NeOn-GPT originally was evaluated on the Wine Ontology, as its specific use case. Therefore, the authors proposed adapting the prompt to allow a more general evaluation of the generated CQs. They also slightly modified the output generation as it included additional text along the CQs, which would result in injecting noise in the metrics computation. In the case of the work by Pan *et al.*, we also expanded the system so that it could detect whether to use RAG (on PDFs) or chat (in the other cases) based on the input format.

Similarly, AgOCQs required some input reorganisation and adjustments to the requested output format. RETROFIT-CQs' few-shot templates were updated to reflect the details of each ontology analysed in the benchmarking system. Details of the changes for each system may be included in their respective projects' Github. We consider these modifications and the potential limitation they imply as part of the process of standardisation, as these systems have been developed with different scenarios and project goals in mind and evaluated in isolation with custom gold standards. They allow us to provide a reference baseline for the CQ generation task and identify the variations of this task that shall be considered in future developments of Bench4KE.

All evaluations have been executed on the same laptop with Intel(R) Core(TM) i5-8265U CPU @ 1.60GHz 1.80 GHz, with 8 Gb of RAM. For the LLM, when no explicit specification was set, we use GPT-4o as it is the most frequently used LLM in the literature at the moment, with temperature set to 0.

4.2 Results

Table 3. Performance comparison across usage scenarios on Bench4KE. Approaches are distinguished according to the classification proposed by [3] based on two classes, i.e. Generating and Retrofitting. Similarity metrics range 0 to 1, with Cosine similarity, computed on sBERT embeddings, and BERTScore-F1, balancing token-level precision and recall by rewarding both correct matches and adequate coverage of the reference answer. LLM-as-a-Judge scores represent averaged quality ratings on a 1–5 scale.

System	Approach	Similarity		LLM-as-a-Judge		
		Cosine	F1	Relev.	Clar.	Depth
OntoChat [50]	Generating	0.24	0.60	4.47	4.70	3.52
NeOn-GPT [15]	Generating	0.24	0.58	4.63	4.65	3.90
AgOCQs [31]	Generating	0.16	0.60	3.60	4.10	2.50
Pan et al. [35]	Generating	0.32	0.60	4.78	4.54	3.90
Rebboud et al. [42]	Generating	0.30	0.60	4.67	4.55	3.98
RETROFIT-CQs [25]	Retrofitting	0.21	0.57	4.49	4.48	3.97

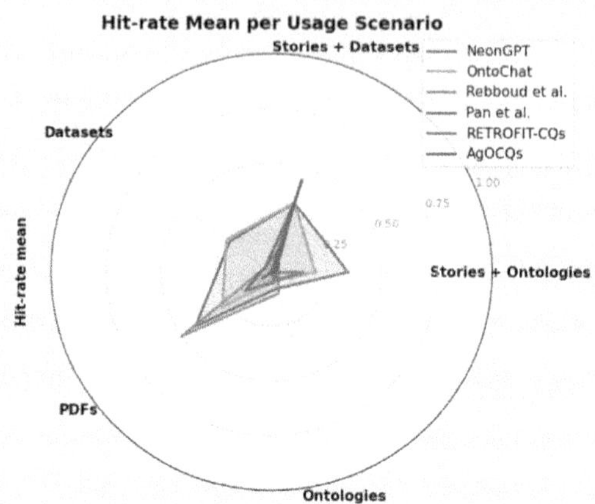

Fig. 3. Mean Hit Rate across systems according to the usage scenarios.

Table 3 reports the performance of the six evaluated systems using the Cosine similarity, BERTScore and LLM-as-a-judge metrics with a total average across projects. In Fig. 3, we also provide an overview of the systems with respect to the contemplated usage scenarios. These results provide a baseline, however we remark that community participation is paramount to add new datasets and domains to obtain a more robust and comprehensive baseline. The low absolute values observed may sometimes be due to either the fact that the systems are evaluated on a usage scenario or file formats that they do not support, or a possible limitation of similarity-based metrics in evaluating questions that might be fit but are not similar to the gold standard CQs. This topic is addressed in Sect. 6 of the paper.

4.3   Impact of Bench4KE

To test the beta version of Bench4KE, the authors of three of the systems involved in our experiments provided feedback through a survey, including an assessment of the usefulness and usability of Bench4KE and possible suggestions for its future improvement. The survey was adapted from the System Usability Scale [8]. It included a series of statements in which the raters express their agreement on a scale from 1 (strongly disagree) to 5 (strongly agree), covering the system's usefulness, extensibility, support for user-provided datasets, clarity of instructions, and usability of the interface. The Survey questionnaire's responses are available on GitHub and the questionnaire itself is online for future collection of users' feedback on Bench4KE. All users agreed that Bench4KE fills a critical gap by providing a community reference for evaluating CQ generation tools and is a vital contribution towards advancing research in this domain. Users stated that they plan to use Bench4KE to evaluate methods/tools for automatic CQ generation to evaluate other KE automation tasks that will be supported. They perceive that the system is easy to use, including its graphical user interface. The main suggestion for improvement is on the metrics supported by Bench4KE to assess the performance of the tool. These suggestions have been implemented in the first Bench4KE release, which we present in this paper.

5   Availability, Sustainability and Licensing

Benck4KE is available on a public GitHub repository, which aggregates all its material (software code, data, experiment configurations, results, etc.), and it is archived on Zenodo, which provides its DOI: 10.5281/zenodo.17817277. Bench4KE is released under the Apache 2.0 License. The documentation accompanying each release reduces the barrier for first?time users.

Sustainability is guaranteed by the University of Bologna and CNR?ISTC for at least five years; additionally, Bench4KE is intended to solicit community effort and has already received the interest and willingness to contribute from researchers external to the authors' group and institutions. Both maintenance plan and contribution guidelines are available on Github.

Standards Compliance. Bench4KE exposes a REST interface that returns JSON?LD and CSV. Dataset metadata in DCAT-AP is also available online.

6   Discussion

In this section, we discuss the envisioned current and future implications of the resource for the Semantic Web community.

6.1   Beyond Similarity. How Should CQs Be Evaluated?

CQ elicitation inherently requires human involvement. We consider automatic evaluation to be insufficient and non-exhaustive, yet a very useful supporting tool. The metrics we report follow conventions established in the related literature and should be understood as approximations rather than definitive judgements of quality. LLM-based scoring can contribute to this evaluation pipeline, especially for the evaluation of novelty in the generated CQs. For example, since in our system the LLM proposes possible additional questions to add, these can, upon scrutiny, be incorporated into an evolving set of gold standard CQs, effectively simulating an enrichment process. With the aim of enriching the gold standard dataset, generated CQs that do not match with the gold standard but are sufficiently similar to those proposed by the LLM can also be considered in the Hit Rate scoring and eventually included in the gold standard. Over time, this iterative refinement can help us move towards a more comprehensive coverage of the CQ space, while keeping human judgement at the center of the evaluation.

6.2   Registered Interest from the Semantic Web Community

The authors of existing CQ generation systems expressed interest in Bench4KE, including, for some of them, the willingness to participate in its future development. The systems evaluated with Bench4KE are now showing an "Evaluated with Bench4KE" text in their project's official README file (e.g. see https://github.com/King-s-Knowledge-Graph-Lab/OntoChat/tree/main). This is a first step towards a shared methodological approach to define and evaluate knowledge engineering automation tasks. Feedback from a first round of user testing supports the relevance and usability of Bench4KE. The respondents agreed that the system is a valuable reference for evaluating CQ generation tools and a promising foundation for future KE automation tasks. They highlighted the importance of extending the support to additional metrics and expressed a strong interest in using the system for future research.

6.3   Novelty of Bench4KE

To date, there is no dedicated benchmarking system for evaluating KE automation at large, making Bench4KE a novel contribution. Although gold standards for CQ validation exist, they are not tailored to the challenges and characteristics of LLM-generated outputs. Also, most of existing gold standards are use cases- and domain-specific. Bench4KE is designed to evaluate a system over multiple domains and use case scenarios. Moreover, coordinated, community-wide evaluation efforts in this space are lacking. Bench4KE aims to address this gap by providing a standardised and extensible evaluation framework, which can incorporate and evolve with datasets and gold standard resources from diverse projects and knowledge domains.

6.4   Extensibility

The framework is extensible in three main directions: (i) support for additional datasets, (ii) extension to additional metrics, and (iii) extension to other Knowledge Engineering tasks. Regarding datasets, users can either evaluate their own datasets locally or request integration into the official Bench4KE dataset following the contribution guidelines outlined in https://github.com/fossr-project/ontogenia-cini/blob/main/contributionguidelines.md. The latter requires meeting specific quality criteria to ensure consistency. Although the system is already comprehensive for what concerns the implemented metrics, additional ones are a planned improvement, which will be discussed within a community effort. As for other KE tasks, future work involves expanding Bench4KE into a suite of REST APIs that support the broader ontology development life cycle. According to user feedback on the first release (Sect. 4.3), Bench4KE could be extended beyond CQ validation: respondents expressed interest in new KE automation tasks such as CQ-to-SPARQL mapping, ontology alignment, ontology requirements extraction, and entity recognition from domain texts. These suggestions point to the potential for Bench4KE to evolve into a comprehensive evaluation suite for various stages of KE workflows.

6.5   Limitations and Future Work

The baseline we propose in this paper has some limitations. The benchmark default dataset corresponds to what has been proposed by the related works, which is imbalanced; however, the system can currently be configured to be more balanced and extensible with additional datasets. The variety of dataset formats also remains limited to XML, JSON, and CSV. There is also a risk of data leakage. This problem is relevant, and some works in the state of the art are addressing it, e.g. [25,39], for instance by running models without internet access or using generated benchmarks on the fly. In the future, we plan to add these solutions for data leakage.

Future work also includes extending the dataset with more data formats and usage scenarios, issuing a call for community collaboration to contribute on GitHub, and promoting wider adoption and a more robust baseline through a future challenge and through its integration into Knowledge Engineering courses at the University of Bologna. Additional developments will also focus on expanding coverage across more stages of the ontology engineering process.

7   Conclusion

We introduced Bench4KE, a benchmarking system that in this release evaluates tools that automatically generate Competency Questions. Bench4KE fills a critical gap, as it is intended to homogenise the evaluation of KE automation tasks and to ease comparative analysis: the existing effort has indeed been evaluated in isolation against project-specific gold standards. In general, Bench4KE is designed to evolve to support the evaluation of additional knowledge engineering automation tasks, such as SPARQL query generation, ontology generation, ontology testing, etc. With 17 gold standards from real-world ontology projects covering several domains, and its ability to load custom gold standards, Bench4KE can be used to evaluate CQ generation systems on their ability to act as expert knowledge engineers, while also recognising their strengths within specific domains. We establish a baseline for the LLM-based CQ generation task by performing a comparative analysis of six recently published systems. The results show that this task is challenging. Researchers from the community have provided useful feedback, including expressing willingness to contribute and use Bench4KE in their future research. They evaluated the system as a necessary resource, finding it easy to use.

Acknowledgments and Disclosure of Funding. We gratefully acknowledge the authors of NeOn-GPT, RETROFIT-CQs, OntoChat, AgOCQs, the work by Pan *et al.* and Rebboud *et al.* for their interest in our work, contribution, and collaborative exchanges. This CINI project has received funding from NextGenerationEU under NRRP Grant agreement n. MUR IR0000008 - FOSSR (CUP B83C22003950001). This work was also supported by the PhD scholarship "Discovery, Formalisation and Re-use of Knowledge Patterns and Graphs for the Science of Science", funded by CNR-ISTC through the WHOW project (EU CEF programme - grant agreement no. INEA/CEF/ICT/ A2019/2063229). Finally, we thank OpenAI's Researcher Access Program Grant for the API credits. This work was supported by the INFINITY project, funded by the European Unions Horizon Europe research and innovation programme under grant agreement No. 101233051

Use of Generative AI. ChatGPT was used to enhance the readability of some of the text and improve the language of this paper, after the content was first added manually. After using this service, the author(s) reviewed and edited the content as needed and take(s) full responsibility for the content of the published article.

References

1. Achichi, M., Lisena, P., Todorov, K., Troncy, R., Delahousse, J.: DOREMUS: a graph of linked musical works. In: Vrandečić, D., et al. (eds.) ISWC 2018. LNCS, vol. 11137, pp. 3–19. Springer, Cham (2018). https://doi.org/10.1007/978-3-030-00668-6_1

2. Alharbi, R., Tamma, V., Grasso, F., Payne, T.: An experiment in retrofitting competency questions for existing ontologies. In: Proceedings of the 39th ACM/SIGAPP Symposium on Applied Computing, pp. 1650–1658 (2024)

3. Alharbi, R., Tamma, V., Grasso, F., Payne, T.R.: A review and comparison of competency question engineering approaches. In: International Conference on Knowledge Engineering and Knowledge Management, pp. 271–290. Springer, Cham (2024)

4. Antia, M.J., Keet, C.M.: Automating the generation of competency questions for ontologies with agocqs. In: Iberoamerican Knowledge Graphs and Semantic Web Conference, pp. 213–227. Springer, Cham (2023)

5. de Berardinis, J., et al.: The Polifonia ontology network: building a semantic backbone for musical heritage. In: International Semantic Web Conference, pp. 302–322. Springer, Cham (2023)

6. Bezerra, C., Freitas, F., Santana, F.: Cqchecker: a tool to check the satisfaction of description logic competency questions on ontologies. In: 2013 X National Conference on Artificial and Computational Intelligence (ENIAC) (2013)

7. Blomqvist, E., Seil Sepour, A., Presutti, V.: Ontology testing - methodology and tool. In: ten Teije, A., et al. (eds.) EKAW 2012. LNCS (LNAI), vol. 7603, pp. 216–226. Springer, Heidelberg (2012). https://doi.org/10.1007/978-3-642-33876-2_20

8. Brooke, J., et al.: Sus-a quick and dirty usability scale. Usability Eval. Ind. **189**(194), 4–7 (1996)

9. Carriero, V.A., et al.: ArCo: the Italian cultural heritage knowledge graph. In: Ghidini, C., et al. (eds.) ISWC 2019. LNCS, vol. 11779, pp. 36–52. Springer, Cham (2019). https://doi.org/10.1007/978-3-030-30796-7_3

10. Carriero, V.A., Mariani, F., Nuzzolese, A.G., Pasqual, V., Presutti, V.: Agile knowledge graph testing with testalod. In: ISWC2019-Satellites: proceedings of the ISWC 2019 Satellite Tracks (Posters & Demonstrations, Industry, and Outrageous Ideas) co-located with 18th International Semantic Web Conference (ISWC 2019), Auckland, New Zealand, 26–30 October 2019, pp. 221–224. CEUR-WS (2019)

11. Ciroku, F., de Berardinis, J., Kim, J., Meroño-Peñuela, A., Presutti, V., Simperl, E.: Revont: reverse engineering of competency questions from knowledge graphs via language models. J. Web Semant. **82**, 100822 (2024)

12. Dennis, M., van Deemter, K., Dell'Aglio, D., Pan, J.Z.: Computing authoring tests from competency questions: experimental validation. In: d'Amato, C., et al. (eds.) ISWC 2017. LNCS, vol. 10587, pp. 243–259. Springer, Cham (2017). https://doi.org/10.1007/978-3-319-68288-4_15
13. Di Nuzzo, D., Vakaj, E., Saadany, H., Grishti, E., Mihindukulasooriya, N.: Automated generation of competency questions using large language models and knowledge graphs. In: SEMANTiCS Conference (2024)
14. Dornelas Costa, S., Barcellos, M., Falbo, R., Conte, T., Oliveira, K.: A core ontology on the human-computer interaction phenomenon. Data Knowl. Eng. **138**, 101977 (2022). https://doi.org/10.1016/j.datak.2021.101977
15. Fathallah, N., Das, A., Giorgis, S.D., Poltronieri, A., Haase, P., Kovriguina, L.: Neon-GPT: a large language model-powered pipeline for ontology learning. In: European Semantic Web Conference, pp. 36–50. Springer, Cham (2024)
16. Fernández-Izquierdo, A., Poveda-Villalón, M., García-Castro, R.: CORAL: a corpus of ontological requirements annotated with lexico-syntactic patterns. In: Hitzler, P., et al. (eds.) ESWC 2019. LNCS, vol. 11503, pp. 443–458. Springer, Cham (2019). https://doi.org/10.1007/978-3-030-21348-0_29
17. Gangemi, A., Lippolis, A.S., Lodi, G., Nuzzolese, A.G.: Automatically drafting ontologies from competency questions with frodo. In: Towards a Knowledge-Aware AI, pp. 107–121. IOS Press (2022)
18. García-Ramos, S., Otero, A., Fernández-López, M.: OntologyTest: a tool to evaluate ontologies through tests defined by the user. In: Omatu, S., et al. (eds.) IWANN 2009. LNCS, vol. 5518, pp. 91–98. Springer, Heidelberg (2009). https://doi.org/10.1007/978-3-642-02481-8_13
19. Grüninger, M., Fox, M.S.: The role of competency questions in enterprise engineering. In: Rolstadås, A. (ed.) Benchmarking — Theory and Practice. IAICT, pp. 22–31. Springer, Boston (1995). https://doi.org/10.1007/978-0-387-34847-6_3
20. Karras, O., Wernlein, F., Klünder, J., Auer, S.: Divide and conquer the empire: a community-maintainable knowledge graph of empirical research in requirements engineering (2023). https://arxiv.org/abs/2306.16791
21. Keet, C.: The African wildlife ontology tutorial ontologies. J. Biomed. Semant. **11** (2020). https://doi.org/10.1186/s13326-020-00224-y
22. Keet, C.M.: A core ontology of macroscopic stuff. In: Janowicz, K., Schlobach, S., Lambrix, P., Hyvönen, E. (eds.) EKAW 2014. LNCS (LNAI), vol. 8876, pp. 209–224. Springer, Cham (2014). https://doi.org/10.1007/978-3-319-13704-9_17
23. Keet, C.M., Mahlaza, Z., Antia, M.-J.: CLaRO: a controlled language for authoring competency questions. In: Garoufallou, E., Fallucchi, F., William De Luca, E. (eds.) MTSR 2019. CCIS, vol. 1057, pp. 3–15. Springer, Cham (2019). https://doi.org/10.1007/978-3-030-36599-8_1
24. Kurvers, R.H., Nuzzolese, A.G., Russo, A., Barabucci, G., Herzog, S.M., Trianni, V.: Automating hybrid collective intelligence in open-ended medical diagnostics. Proc. Natl. Acad. Sci. U.S.A. **120**(34) (2023). https://doi.org/10.1073/pnas.2221473120
25. Li, M., Alharbi, R., de Berardinis, J., Tamma, V., Payne, T.R.: Retrofit-CQ revisited: leveraging ontology triples and context in few-shot LLM prompting. In: ISWC 2025, 24rd International Semantic Web Conference (2025)
26. Lin, C.Y.: Rouge: a package for automatic evaluation of summaries. In: Text Summarization Branches Out, pp. 74–81 (2004)
27. Lippolis, A.S., Ceriani, M., Zuppiroli, S., Nuzzolese, A.G.: Ontogenia: ontology generation with metacognitive prompting in large language models. In: European Semantic Web Conference, pp. 259–265. Springer, Cham (2024)

28. Lippolis, A.S., Lodi, G., Nuzzolese, A.G.: The water health open knowledge graph. Sci. Data **12**(1), 274 (2025)
29. Lippolis, A.S., Saeedizade, M.J., Keskisarkka, R., Gangemi, A., Blomqvist, E., Nuzzolese, A.G.: Assessing the capability of large language models for domain-specific ontology generation. In: European Semantic Web Conference 2025 (2025)
30. Lisena, P., et al.: Capturing the semantics of smell: the odeuropa data model for olfactory heritage information. In: Groth, P., et al. (eds.) The Semantic Web, pp. 387–405. Springer, Cham (2022)
31. Mahlaza, Z., Keet, C.M., Chahinian, N., Haydar, B.: On the feasibility of LLM-based automated generation and filtering of competency questions for ontologies. In: Alam, M., et al. (eds.) Proceedings of the 5th Conference on Language, Data and Knowledge, pp. 136–146. Unior Press, Naples, Italy (2025). https://aclanthology.org/2025.ldk-1.15/
32. Malheiros, Y., Freitas, F.: Unification in EL for competency question generation. In: Description Logics (2017)
33. Malone, J., et al.: The software ontology (SWO): a resource for reproducibility in biomedical data analysis, curation and digital preservation. J. Biomed. Semant. **5**, 25 (2014). https://doi.org/10.1186/2041-1480-5-25
34. Noy, N., Mcguinness, D.: Ontology development 101: a guide to creating your first ontology. Knowl. Syst. Laboratory **32** (2001)
35. Pan, X., Ossenbruggen, J., de Boer, V., Huang, Z.: A rag approach for generating competency questions in ontology engineering. In: Research Conference on Metadata and Semantics Research, pp. 70–81. Springer, Cham (2024)
36. Panov, P., Soldatova, L., Džeroski, S.: Generic ontology of datatypes. Inf. Sci. **329** (2015). https://doi.org/10.1016/j.ins.2015.08.006
37. Papineni, K., Roukos, S., Ward, T., Zhu, W.J.: Bleu: a method for automatic evaluation of machine translation. In: Proceedings of the 40th Annual Meeting of the Association for Computational Linguistics, pp. 311–318 (2002)
38. Parkkila, J., et al.: An ontology for videogame interoperability. Multimed. Tools Appl. **76** (2017). https://doi.org/10.1007/s11042-016-3552-6
39. Paulheim, H.: Ontologies, knowledge graphs, and LLMs: how do we get evaluations done right? In: ISWC 2025, 24rd International Semantic Web Conference (2025)
40. Presutti, V., Daga, E., Gangemi, A., Blomqvist, E.: Extreme design with content ontology design patterns. In: Proceedings of the Workshop on Ontology Patterns, pp. 83–97. CEUR-WS (2009)
41. Rebboud, Y., Lisena, P., Tailhardat, L., Troncy, R.: Benchmarking LLM-based ontology conceptualization: a proposal. In: ISWC 2024, 23rd International Semantic Web Conference (2024)
42. Rebboud, Y., Tailhardat, L., Lisena, P., Troncy, R.: Can LLMs generate competency questions? In: European Semantic Web Conference, pp. 71–80. Springer, Cham (2024)
43. Reimers, N., Gurevych, I.: Sentence-BERT: sentence embeddings using Siamese BERT-networks. In: Proceedings of the 2019 Conference on Empirical Methods in Natural Language Processing. Association for Computational Linguistics (2019). https://doi.org/10.18653/v1/D19-1410
44. Rupp, C., et al.: Requirements-engineering und-management: Das Handbuch für Anforderungen in jeder Situation. Carl Hanser Verlag GmbH Co KG (2020)
45. Tailhardat, L., Chabot, Y., Troncy, R.: Noria-o: an ontology for anomaly detection and incident management in ICT systems. In: Meroño Peñuela, A., et al. (eds.) The Semantic Web, pp. 21–39. Springer, Cham (2024)

46. Trianni, V., et al.: Hybrid collective intelligence for decision support in complex open-ended domains. In: HHAI 2023: Augmenting Human Intellect, pp. 124–137. IOS Press (2023)
47. Tufek, N., et al.: Validating semantic artifacts with large language models. In: European Semantic Web Conference, pp. 92–101. Springer, Cham (2024)
48. Wiegers, K., Beatty, J.: Software Requirements. Pearson Education (2013)
49. Wiśniewski, D., Potoniec, J., Ławrynowicz, A., Keet, C.M.: Analysis of ontology competency questions and their formalizations in sparql-owl. J. Web Semant. **59**, 100534 (2019)
50. Zhang, B., et al.: Ontochat: a framework for conversational ontology engineering using language models. In: European Semantic Web Conference, pp. 102–121. Springer, Cham (2024)
51. Zhang, T., Kishore, V., Wu, F., Weinberger, K.Q., Artzi, Y.: Bertscore: evaluating text generation with BERT. arXiv abs/1904.09675 (2019). https://api.semanticscholar.org/CorpusID:127986044
52. Zhao, Y., et al.: Improving ontology requirements engineering with ontochat and participatory prompting. In: Proceedings of the AAAI Symposium Series, vol. 4, pp. 253–257 (2024)

Traqula: Providing a Foundation for the Evolving SPARQL Ecosystem Through Modular Query Parsing, Transformation, and Generation

Jitse De Smet[(✉)] and Ruben Taelman

IDLab, Department of Electronics and Information Systems, Ghent University imec, Ghent, Belgium
`{jitse.deSmet,ruben.taelman}@ugent.be`

Abstract. SPARQL engines continue to diverge in the versions and language extensions they support. With SPARQL 1.2 nearing completion and the working group's plans toward future maintenance, this divergence is likely to grow. As a result, language tools that aim to support multiple SPARQL versions and dialects face increasing maintenance costs. Furthermore, areas such as SPARQL federation become harder to manage. To address these challenges, we introduce Traqula, a modular TypeScript-based framework for parsing, transforming, and generating SPARQL queries. This article presents Traqula's architecture and its builder-based dependency-injection design, and we show how these components provide modular support for SPARQL 1.1 and 1.2 parsing and generation. Traqula enables round-tripping, supports generic algebraic and AST transformations, and delivers significantly faster parsing compared to current JavaScript-based SPARQL parsing. Traqula's modular architecture enables flexible experimentation for researchers, and it is production-ready thanks to extensive testing and modern development techniques. Looking forward, Traqula lays the groundwork for supporting additional query languages, developing generic rewriting modules for SPARQL federation engines, and building robust tooling for an increasingly diverse SPARQL ecosystem.

Keywords: SPARQL · query · algebra · modularity · parser · generator

Canonical version: https://traqula-resource.jitsedesmet.be/. **Resource type**: software package · **Licence**: MIT · **DOI**: 10.5281/zenodo.17793334 · **URL**: https://github.com/comunica/traqula.

1 Introduction

The SPARQL query language [30] is the standard way to query RDF [17] data. While SPARQL endpoints [21] are commonly used to expose Knowledge

Graphs through a highly expressive query API, alternative APIs have been proposed [7,8,32,44,55,68] that offer different trade-offs between client and server effort for query execution. This heterogeneity in server APIs introduces challenges when executing federated SPARQL queries [6,27,29,58,59] over multiple of these APIs. While this API-based heterogeneity has been an active field of research [14,37,46,64] in recent years, the heterogeneity of the SPARQL language itself lacks understanding. In practice, many SPARQL dialects exist [9,11,31,52] each introducing their own extensions or limitations. Virtuoso, for example, extends SPARQL with full-text search capabilities [3], Apache Jena adds support for constructing quads [1], and Oxigraph provides an additional built-in function, ADJUST [2,70]. Furthermore, some SPARQL endpoints might limit the SPARQL language by removing expensive operators such as OPTIONAL [54]. In the near future, this heterogeneity is expected to increase even further as the RDF and SPARQL W3C Working Group enters its maintenance mode [28,62], which could deliver more rapid successions of SPARQL once SPARQL 1.2 is finalized. As a result, the growing diversity of SPARQL versions and dialects introduces several challenges:

1. **Query evaluation**: A user query written in one version might not be executable by a SPARQL engine supporting another version.
2. **Tooling**: Linters, formatters, editors, and language servers often assume a specific SPARQL version.
3. **Maintainability**: Tools that support multiple SPARQL versions typically do so by maintaining multiple software versions, or one version with many conditions, making maintainability highly challenging.

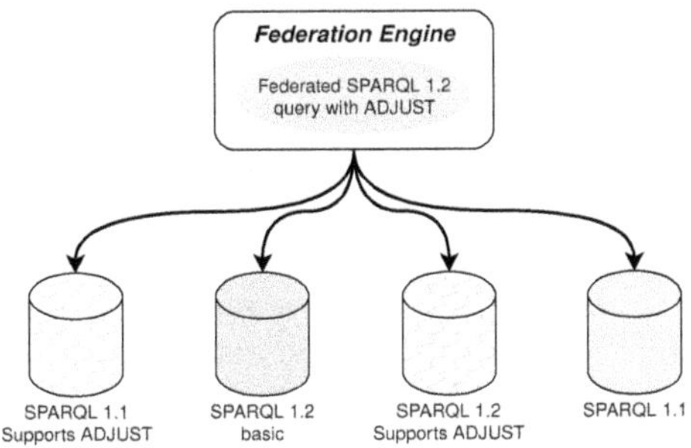

Fig. 1. The federated SPARQL query (blue) uses SPARQL 1.2 features such as triple terms and uses the non-standard builtin function ADJUST [70]. The query targets four SPARQL endpoints, all supporting different SPARQL version, RDF profiles [33], and language extensions. To allow query engines to integrate this heterogeneity, frameworks such as Traqula are necessary to bridge between these SPARQL dialects.

The query evaluation problem within a heterogeneous SPARQL ecosystem becomes especially visible in federated SPARQL query execution. While language dialects are common in many technologies, SQL [40] being a well-known example, the RDF data model [17] is explicitly designed for seamless integration of distributed datasets. SPARQL reflects this distributed nature through support for federated queries, where a single query may involve multiple endpoints, each with its own capabilities, limitations, and language features.

In such a setting, a SPARQL query written in one SPARQL version or dialect may not be executable on all federation members. For instance, a query formulated in SPARQL 1.2 might rely on features, functions, or syntactic constructs that an older endpoint does not support, or that are only available as vendor-specific extensions. Figure 1 illustrates this scenario: a query is written in SPARQL 1.2 and uses the non-standard ADJUST function [70], yet the endpoints differ in their supported SPARQL versions, RDF profiles [33], and feature sets. Although the SPARQL Service Description specification [69] allows endpoints to declare their supported features, it does not offer mitigation paths to resolve these language mismatches. To solve this problem, there is a need for a parsing, transformation, and generation framework such as Traqula that can handle various SPARQL dialects. Such a framework provides a foundation for future SPARQL federation research towards new techniques and algorithms to manage these dialects.

In prior work [62], we introduced our vision of a modular SPARQL parser to address the growing heterogeneity of SPARQL dialects using builder-based dependency injection [23]. We envisioned a design with a prototype implementation demonstrating parser composability, showing how grammar modules could be combined to modularly define multiple SPARQL parsers covering different versions and dialects. In this article, we fully realize that vision with a complete implementation as well as applying the modular architecture to query generation and transformation. These features allow queries to be parsed, transformed, and regenerated reliably, while preserving their structure and semantics across different dialects.

This article presents Traqula, a modular SPARQL toolkit implementing these ideas. Traqula has already been integrated into the widely used Comunica SPARQL querying framework [64], demonstrating its maturity and practical applicability. Traqula's modular and composable architecture enables researchers and practitioners to: 1. Experiment with grammar changes, including adding, modifying, or removing rules; 2. tackle the complexities of federated SPARQL querying in a heterogeneous SPARQL ecosystem; and 3. lay the foundation for future query formatting and rewriting tools. Traqula is implemented in TypeScript and is available as open-source software under the MIT license on GitHub and npm, providing the community with a robust and flexible resource for building tools and engines for a heterogeneous SPARQL environment.

The remainder of this paper is structured as follows. Section 2 details the requirements of Traqula, Sect. 3 reviews related work, Sect. 4 presents the high-level architecture of Traqula, and Sect. 5 discusses its implementation. Section 6 evaluates its performance, and Sect. 7 concludes the paper.

2 Requirements

Traqula was designed with a set of concrete requirements in mind. In this section, we outline the most important ones:

1. **Flexibility**: allowing composition of parser, transformer, or generator out of small components that can be easily added, removed, or replaced.
2. **Round-tripping**: enable parsing into an AST and generating back to exactly the same string. When manipulations are made, only the manipulated parts change.
3. **Web-based**: harnessing web technologies and enabling execution within and outside browsers, to ensure wide usability.
4. **Language-agnostic**: by providing a generic core, Traqula facilitates the support of many query languages.

2.1 Flexibility

Given the increasing heterogeneity of SPARQL versions and dialects, Traqula must support a highly flexible and composable architecture. As outlined in the introduction and Fig. 1, differences in supported features, functions, and syntactic constructs already complicate query evaluation and tooling, and these discrepancies are expected to grow as SPARQL enters a phase of more rapid evolution through the SPARQL working group's maintenance mode [62]. Addressing this landscape requires a toolkit in which parsers, generators, and transformers can be assembled from small, reusable modules rather than fixed monolithic grammars.

Flexibility is essential for two main reasons. First, researchers and practitioners need the ability to experiment with the language itself, by adding, modifying, or removing grammar rules, and evaluating the impact of those changes. Traqula's modular design aims to make such experimentation routine rather than burdensome, supporting rapid prototyping of new language features or alternative syntactic constructs. This capability has already been demonstrated in practice: Traqula's adoption within the modular Comunica query engine allowed us to contribute effectively to the standardization work for SPARQL 1.2.

Second, SPARQL federation engines require a way to mediate between heterogeneous dialects of the SPARQL endpoints they federate over. A single query may need to be adapted to the capabilities of multiple endpoints, each exposing different SPARQL versions or dialects. Traqula's modularity is therefore not merely a convenience but a prerequisite for constructing reliable transformations between dialects, since each dialect will require a parser and generator. By decoupling parsing, generation, and algebraic transformation into flexible and composable components, Traqula positions itself as a key tool for SPARQL rewriting across versions and dialects.

2.2 Round-Tripping

Because SPARQL is a structured language that is both written and inspected by humans, it benefits from the same ecosystem of tooling found around programming languages, which includes editors, code highlighters, linters, refactoring tools, and formatters. Many of these tools rely on *round-tripping*, the ability to parse a query into an internal representation, apply a transformation, and regenerate a query string that is identical except for the intentional change. Achieving this property requires the parser and generator to preserve all user-visible details, such as comments, spacing, punctuation, or keyword capitalization, even when these details have no semantic impact on query evaluation.

Round-tripping is essential for practical tooling. A linter that renames a variable, for instance, should not unexpectedly rewrite unrelated parts of the query or normalize stylistic elements that the user intended to keep. A simple example of SPARQL reformatting is rewriting the variable *'s'* to *'subject'* in `SelECT * { ?s ?p ?o }`, which should result in `SelECT * { ?subject ?p ?o }`, without altering the spacing or keyword casing. Figure 2 illustrates the different representations involved: the query string; its abstract syntax tree (AST) as received from the parser, and used by linters and formatters; and its algebraic form, constructed from the AST [30], and used by query engines. Transformations may occur at any of these levels, but we only require the round-tripping property for AST-level transformations.

Supporting round-tripping therefore shapes Traqula's design. The parser must retain sufficient information to reproduce the original query string, while the generator must reconstruct that structure without introducing unintended changes. This requirement ensures that Traqula can serve as a foundation for reliable SPARQL editing, formatting, and rewriting tools.

2.3 Web-Based

Web-based technologies offer a natural foundation for widely adoptable SPARQL tooling. A Web-first JavaScript (or TypeScript after transpilation) implementation can run natively in the browser, where many SPARQL editors [19,57], and even query engines [64] operate, while also executing on backend environments such as Node.js, Bun, or Deno, and inside application frameworks like Electron or Tauri. Moreover, the Web ecosystem lowers the barrier to implementing a SPARQL language server, similar to Qlue-ls [48], via the Language Server Protocol (LSP) using Microsoft's *vscode-languageserver-node*. Such an LSP or more broadly any Traqula derived (system) service could directly leverage the modular architecture, enabling rule-level extensibility and round-tripping capabilities within interactive editors, especially since Traqula's APIs comunicate using JSON Data Transfor Objects (DTOs), which simplifying crossing language boundries. In principle, WebAssembly (WASM) provides the same portability, enabling code written in languages such as Rust (wasm-pack), C++ (Emscripten), or even JavaScript (Javy) to run across the Web platform.

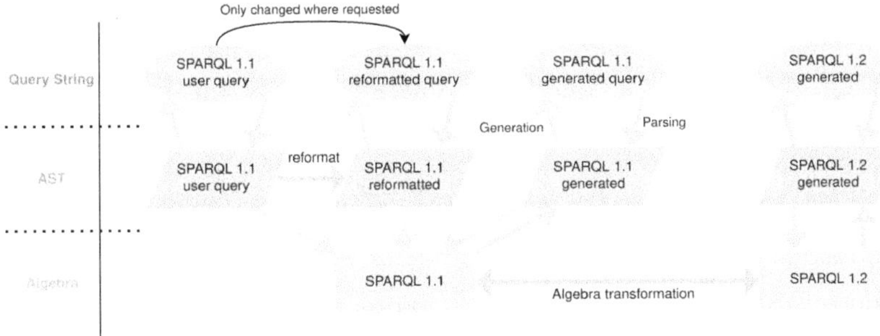

Fig. 2. Various representations of an algebraically equivalent SPARQL 1.1 query and its SPARQL 1.2 counterpart. The figure illustrates three layers: the concrete query string (blue), the Abstract Syntax Tree (AST, green), and the algebraic representation (orange). Round-tripping requires that parsing a query string into an AST and regenerating it produces the identical string. Transforming the leftmost query at the AST level preserves all syntactic information, so regenerating the transformed AST changes only those parts intentionally modified. By contrast, converting the AST to algebra discards syntactic detail: algebraically equivalent queries map to the same algebra and thus to the same canonicalized query string. Finally, the figure shows how transformations between SPARQL 1.1 and SPARQL 1.2 can be performed at the algebra level, where engines operate and where syntactic information is no longer preserved.

However, for Traqula we prioritise not only portability, but also accessibility, quick prototyping, and ease of contribution. TypeScript provides the same core advantages as WASM, execution across browser and server environments, while remaining far more familiar to the broad developer community, and without requiring ahead-of-time compilation. As the most widely used language on GitHub in 2025 [24], TypeScript maximises the likelihood that researchers and practitioners can both use and extend Traqula without specialised compilation toolchains or language-specific runtimes.

By choosing a Web-based, TypeScript implementation, Traqula satisfies the requirement of broad, frictionless adoption while integrating naturally into both front-end applications and backend query engines.

2.4 Language-Agnostic

To address query language heterogeneity, Traqula provides a generic core that can be reused to implement modular parsers, generators, and transformers for a variety of query languages. The SPARQL 1.1 and 1.2 parser, generator, and algebra transformations all build on this single core library, which enables the creation of modular systems beyond SPARQL. For example, the SHACL Compact Syntax developed by W3C's Data Shapes Working Group [5] shares syntactic constructs with SPARQL, allowing it to reuse some components of Traqula's SPARQL grammar. Similarly, mapping languages such as RML [18] for Knowledge Graph constructions allow for representation in SPARQL algebra [43]. Another example is the LDQL query language [35], which enables directed web navigation for Link Traversal Query Processing [34,65] over Linked Data documents. In contrast, the newly standardized ISO Graph Query Language (GQL) [39] defines a completely different syntax, but could still leverage Traqula's language-agnostic core to implement modular parsers, generators, and transformations.

In summary, Traqula's design is guided by four core requirements: flexibility, round-tripping, Web-based execution, and language-agnosticism. Together, these requirements ensure that Traqula can support modular, maintainable tooling for heterogeneous SPARQL dialects while remaining extensible to other query languages and deployment environments. With these principles in place, we now turn to existing approaches in parser design and query language tooling to position Traqula within the broader landscape.

3   Related Work

This section first covers the related work on parsers and compilers from a theoretical perspective; afterward, we list the most relevant existing parsers for SPARQL and other query languages. After covering the parsing itself, we take a look at common abstract syntax tree (AST) structures, specifically focussing on approaches that support round-tripping.

3.1   Parsers

Parsing a structured language involves transforming it from a string into a desired data structure. Conceptually, parsing can be divided into three steps [4]:

1. **Lexical analysis** (or *scanning*) performed by a *lexer*: the source character stream is read and broken into non-overlapping sequences called lexemes. Each lexeme is classified by a token type, often defined using regular expressions. Lexemes may be represented as strings, ranges in the source, or both. The lexer outputs a stream or list of tokens, which are the lexeme representation together with its token type.
2. **Syntax analysis** (or *parsing*) performed by a *parser*: after having generated a 'flat' stream or list of tokens, the next part is to create a data structure, typically a tree, often called the Abstract Syntax Tree (AST), representing the syntactic structure of the language. Some parsers can automatically parse into a Concrete Syntax Tree (CST) which traces all grammar information without generalization or abstraction.
3. **Semantic analysis**: the AST is checked for non-structural constraints. In SPARQL, for example, the variable assigned in a BIND clause cannot be used in the immediately preceding tripleBlock within the same group. In programming languages, type checking is a common example of semantic analysis.

One example of joined execution is the case of a streaming parser where the output data structure of the parser is itself a stream. Streaming parsers are specifically interesting when parsing large amounts of data, where semantic analysis is either not required or required in limited form. Example data formats that tailor themselves to streaming, specifically RDF data, are Jelly [63] and n-triples [13]. Another way of joined execution is where syntax analysis and semantic analysis are joined, allowing the parser to fail fast in case a semantic constraint would be broken, an approach partially taken by Traqula.

The **parser construction technique**, i.e., the actual programmatic definition of the different parsing steps, can happen in various ways; we identify the following three:

1. **Generated**: Code generation tools come with their Domain Specific Language (DSL) that will typically share similarities with Extended Backus–Naur Form (EBNF) and some Regular Expression dialect. The build process of the software that uses the custom parser should then compile the parser definition file (in the custom DSL) to the desired target language, which is typically the same language as the software being built. Example parser generators are Bison [25] and ANTLR [51].
2. **Hand-built**: Code generation can come with optimization limitations, when the grammar allows for optimizations not taken by the code generator. Writing a handwritten parser is powerful but very challenging to get right, because compilers often know powerful, very specific, non-trivial optimizations. On top of programming language-specific optimizations, other powerful optimizations exist for specific sets of grammars, such as LL(1) and LL(k) [4,51].
3. **Toolkits/libraries**: A compromise between hand-built parsers and generators exists in the form of toolkits/libraries. Parser building toolkits, e.g. Chevrotain [15], are software libraries that provide an API that facilitates the construction of parsers within a specific programming language. The benefit of constructing a parser within the programming language the parser would be called from is that the project's code can be more coherent, allowing better integration, while also providing abstraction for powerful optimizations that can be made for specific grammars such as LL(1) and LL(k). Additionally, it allows the usage of language-specific features, e.g. a type system, which are often not present in DSLs used by parser generators, since DSLs are often limited in complexity. However, toolkits miss out on compiler-based optimizations since they cannot fully optimize the user's code, and similarly miss out on certain optimizations that could be possible in hand-built parsers.

In the next section, we argue that parser-building toolkits offer a practical middle ground between generated and hand-built parsers, achieving good performance while providing flexibility for modular, language-specific systems like Traqula.

3.2   Existing Query Parsers

Table 1 provides an overview of the parsing frameworks used by popular open-source query language parsing software and query engines. The table extends upon our previous work [62], extending our analysis to open-source implementations of the SQL [40], GraphQL [20], GQL [39], and Neo4J's cypher [22] query languages. We can clearly see that parser generators are the dominant approach, with 13 of the 16 systems listed using generated parsers. Only one implementation uses a handwritten parser, and two employ a parser toolkit, namely Chevrotain.

Some parser generators support limited flexibility through grammar composition. ANTLR [51], for example, supports grammar imports, enabling composition at the parser-level, but not at the rule-level. Parsers can be extended in a manner similar to object-oriented class inheritance, but individual rules cannot be deleted or patched while preserving references to the original implementation. Furthermore, mainstream parser generators, like ANTLR require a compilation step and typically produce only a Concrete Syntax Tree (CST), which reflects the full grammar without abstracting to a higher-level AST. Producing an AST therefore requires an additional traversal and transformation pass, effectively splitting parsing into two stages and increasing complexity for tools that operate on the AST. While ANTLR theoretically supports targeting multiple host languages, including JavaScript and thus the Web, this generally produces only a CST without an AST. Crucially, none of these frameworks provides the combination of fine-grained extensibility, AST-level round-tripping, Web based execution, and query language agnostic design demanded by Traqula, motivating the need for a new modular approach.

To address these requirements, we selected Chevrotain [15] as the foundation for Traqula. Chevrotain can uniquely satisfy our requirements for flexibility, round-tripping, Web-based execution, and language-agnostic extensibility, while maintaining a sufficiently high-level abstraction. As a parsing toolkit, Chevrotain supports rule-level modularity: parser construction is expressed directly in TypeScript rather than a separate grammar language, enabling dynamic manipulation of individual rules, runtime extensions, and distribution of language-agnostic helpers as ordinary TypeScript modules. Unlike most parser generators, Chevrotain can parse directly into AST rather than CSTs, allowing round-trippable ASTs without an intermediate transformation. Finally, Chevrotain is highly optimised for Web-based execution environments, producing fast parsers that run efficiently in both browsers and server-side JavaScript. Together, these properties make Chevrotain a particularly strong match for Traqula's design goals.

3.3 AST Structures for Round-Tripping

To support Traqula's requirement for AST-level round-tripping, we examine two popular tools designed for reformatting and rewriting: Babel [61] and ESLint [49].

Babel [61] is a compiler that enables developers to write next-generation JavaScript and transpile it to earlier versions, ensuring compatibility with older environments. To support transformations while preserving the original source structure, Babel annotates its AST nodes with source location information, specifying the range of offsets each node represents in the original string. When a

Table 1. List of parsing software packages for various query languages, including the underlying parsing framework and construction technique. The table also indicates how each package satisfies the requirements established in Sect. 2: *flexibility (F)*, *round-tripping (R)*, *web-based (W)*, and *language agnosticism (L)*. Each requirement is marked as fully met ($\checkmark$), partially met ($\sim$), or not met ($\times$); question marks indicate uncertainty.

Parsing Software	Query Language	Parsing Framework	Construction Technique	Req:F R W L
Apache Jena - arq(v5.6.0) [60]	SPARQL	JavaCC	Generated	$\times$ $\times$ $\times$ $\times$
Blazegraph (commit 829ce82) [66]	SPARQL	JavaCC	Generated	$\times$ $\times$ $\times$ $\times$
Oxigraph (v0.5.2) [53]	SPARQL	rust-peg	Generated	$\times$ $\times$ $\times$ $\times$
QLever (commit 6ec0a5e) [41]	SPARQL	ANTLR	Generated	$\sim$ $\times$ $\sim$ $\times$
RDF4J (v5) [50]	SPARQL	JavaCC	Generated	$\times$ $\times$ $\times$ $\times$
SPARQL.js (v3.7.1) [67]	SPARQL	Jison	Generated	$\times$ $\times$ $\checkmark$ $\times$
Stardog - Millan (commit 6109984) [10]	SPARQL	Chevrotain	Toolkit	$\checkmark$ $\times$ $\checkmark$ $\times$
Virtuoso opensource (commit 23cff67) [42]	SPARQL	Bison	Generated	$\times$ $\times$ $\times$ $\times$
Yasgui (v4.0.113) [57]	SPARQL	SWI Prolog	Generated	? $\times$ $\checkmark$ $\times$
Traqula (v1.0.0)	SPARQL	Chevrotain	Toolkit	$\checkmark$ $\checkmark$ $\checkmark$ $\checkmark$
DuckDB (v1.4.2) [56]	SQL	Bison	Generated	$\times$ $\times$ $\times$ $\times$
PostgreSQL (v18) [45]	SQL	Bison	Generated	$\times$ $\times$ $\times$ $\times$
SQLite (v3.51.0) [16]	SQL	Lemon	Generated	$\times$ $\times$ $\times$ $\times$
GraphQL-js (v16.12.0) [26]	GraphQL		Hand written	$\checkmark$ $\times$ $\checkmark$ $\times$
opengql grammar (v1.9.0) [12]	GQL	ANTLR	Generated	$\sim$ $\times$ $\sim$ $\times$
Neo4J (v25) [22]	Cypher	ANTLR	Generated	$\sim$ $\times$ $\sim$ $\times$

node is replaced, it can either be substituted with a raw source string, preserving user-visible syntax exactly, or with a new AST node, whose string representation is automatically generated. Auto-generation produces a valid string for the node; for example, a SPARQL string literal 'a' could be generated as `'a'`, `"a"`, `'''a'''`, `"""a"""`, `'a'^^xsd:string`, etc. Similarly, SPARQL keywords are case-insensitive, so an auto-generated keyword may have any casing.

ESLint [49] is a widely used linter that also supports automatic fixes. Like Babel, ESLint requires AST nodes to carry source location information. Unlike Babel, fixes are applied externally through a helper that patches specific source ranges with new strings, rather than modifying the AST nodes directly. Despite this difference, the key principle of associating each node with its position in the source text remains central to enabling precise, reliable transformations.

4   Architecture

In this section, we go into more detail on the architecture implemented in Traqula, and how we extended the architectural vision towards query generation and transformation. We explain how Traqula exists of many small, interlinkable packages and provide more detail on the core package, which contains the language-agnostic builders and transformers.

4.1   Builder-Based Dependency Injection

At the core of Traqula's flexibility is its builder-based dependency injection architecture. Dependency injection [23] is a software design that promotes flexible software by creating objects or functions that receive the components they rely on, rather than instantiating them directly, allowing them to treat their dependencies more conceptually. In Traqula each functional element, such as a parser rule, generator rule, or transformation step, is declared under a symbolic name, and refers to other rules by name rather than by concrete implementation. This indirection enables users to replace or patch functionality simply by adjusting the mapping from names to functions. Figure 3 (left) illustrates this idea, while Fig. 3 (right) shows Traqula's TypeScript definition of the `iri` parser and generator rules, each depending on named subrules.

To manage and share these rule dictionaries, Traqula adopts the builder design pattern. Builders construct complex, modular objects step by step, and expose a uniform API for composing, extending, and patching rules. By implementing the builder pattern directly in the host language, Traqula avoids the need for additional compilation stages and allows builders to be referenced, extended, and combined programmatically. This choice also makes it possible to integrate language-specific features, most notably static type checking, into the dependency-injection mechanism itself. Traqula provides three task-specific builders: a parser builder, a generator builder, and a general indirection builder. All share the same underlying rule dictionary, differing only in the `build` artifact they produce, and the supplementary dependencies they inject.

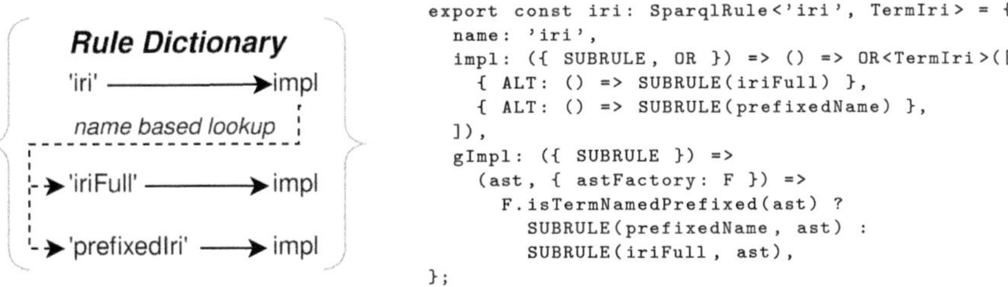


```
export const iri: SparqlRule<'iri', TermIri> = {
  name: 'iri',
  impl: ({ SUBRULE, OR }) => () => OR<TermIri>([
    { ALT: () => SUBRULE(iriFull) },
    { ALT: () => SUBRULE(prefixedName) },
  ]),
  gImpl: ({ SUBRULE }) =>
    (ast, { astFactory: F }) =>
      F.isTermNamedPrefixed(ast) ?
        SUBRULE(prefixedName, ast) :
        SUBRULE(iriFull, ast),
};
```

Fig. 3. Left: A mapping from rule names to their implementations, as registered by the user. In this example, the `iri` rule depends on the named rules *iriFull* and *prefixedIri*. At execution time, the implementation of `iri` consults the user-provided map to resolve these names to their concrete implementations. **Right:** Traqula's Type-Script declaration of the same `iri` rule. The rule is defined under the name *iri*, and provides both a parser implementation (`impl`) and a generator implementation (`gImpl`). The use of *'SUBRULE'* reflects the same *name-based lookup* shown in the left figure, delegating execution to the implementations registered under the names `iriFull` and `prefixedIri`.

Each builder exposes the same operations to manage the rule dictionary. Figure 4 illustrates the core management functions, creating a builder, adding a rule, patching a rules, and building the final artifact .

```
const iriGeneratorBuilder = GeneratorBuilder
  .create([iriFull, prefixedIri])
  .addRule(iri);
const specialIriGenerator = GeneratorBuilder
  .create(iriGeneratorBuilder)
  .patchRule(alternativePrefixedRule)
  .build();
```

Fig. 4. Construction of a 'special IRI generator' based on an existing builder for building a IRI generator, patching it with an alternative rule implementation for the *'prefixedIri'* rule.

4.2   Modular Packages

To maximize independent evolution and minimise unnecessary dependencies, Traqula is composed of many small, interlinked packages rather than a single big monolithic package. A user only need to depend on what packages they actually use. For example, if a user only requires SPARQL 1.1 parsing, they can depend solely on *'@traqula/parser-sparql-1-1'*, without needing to pull in, and consider the SPARQL 1.2 parser, the generators, or the transformers Traqula maintains. Beyond flexibility and extensibility, this modular architecture is crucial for controlling bundle size, a key requirement for modern Web applications. By allowing developers to import only the necessary components, Traqula avoids the overhead of shipping unused functionality to the browser, improving load times and performance.

Modularity also reinforces well-defined and stable package interfaces: because Traqula itself is built from these packages, its APIs must remain open for extension. For maintainability and version control, all packages are managed as a monorepo in a single GitHub repository under the Comunica organisation: https://github.com/comunica/traqula.

4.3   Language-Agnostic Core Package

Following the modular package design, Traqula provides a central language-agnostic core package containing the generic builders and transformers. These components are designed to be reusable across different query languages, whether for dialects of SPARQL or entirely different languages such as SHACL-compact or GQL. The core package underpins Traqula's flexibility, enabling parsers, generators, and transformations to be composed in a language-independent manner.

A key feature of the core package is the generic transformers and visitors, which facilitate AST- and algebra-level transformations/visiting. This transformer is type-safe, ensuring correctness, leveraging TypeScript's generics to adapt dynamically to the AST or algebra representation being manipulated. Beyond enabling rewrites and translations between query languages, the transformers also support AST-level reformatting, since reformatting can be expressed as a transformation on the AST level in the case of Traqula.

To use the transformer, one instantiates a new transformer object, specifying the types of AST nodes or algebra operations, and optionally providing a default transformation context. Transformations can be guided through preVisitors, allowing to 1. stop traversal completely; 2. continue traversal into descen-

```
const transformed = new TransformerTyped<Sparql11Nodes>()
  .transformNode({ // A query as AST
    type: Algebra.Types.SLICE,
    input: {
      type: Algebra.Types.PROJECT,
      input: {
        type: Algebra.Types.JOIN,
        input: [{ type: Algebra.Types.PROJECT }, { type: Algebra.Types.BGP }],
      },
    },
  }, {
    [Algebra.Types.PROJECT]: { // Transformation of projections
      preVisitor: () => ({ continue: false }),
      transform: projection => algebraFactory.createDistinct(projection),
    },
  });
```

Fig. 5. Example usage of the transformer to wrap a *'distinct'* node around the first *'project'* node in an algebraic expression. Providing input query **SELECT * {** **{ Select * { ?s ex:a ?o } } . ?s ex:b ?o . }** LIMIT 20 would result in rewritten query **SELECT DISTINCT * { { Select * { ?s ex:a ?o } } . ?s ex:b ?o . }** LIMIT 20.

dants; 3. skip specified properties of the current node; 4. copy certain keys without traversal; and 5. shallowly copy the current node. The transformer maintains a stack of nodes to be transformed, ensuring that descendant nodes are processed before their parents. Figure 5 illustrates a simple algebraic transformation that wraps a *'distinct'* operation around the first *'project'* node it encounters. In addition, we provide a catalogue of known existing transformations powered by Traqula, thereby also demonstrating how Traqula can be used for traditional query optimizing or rewriting.

5   Implementation

This section outlines the technological foundations of Traqula, the SPARQL languages and dialects currently supported, and the maintenance plan that ensures the system remains reliable and extensible as query languages continue to evolve.

5.1   TypeScript

Traqula is implemented in TypeScript, following the rationale outlined in Sect. 2: TypeScript offers broad accessibility, quick prototyping, browser-server portability, and a familiar development environment for a large set of developers. Within Traqula, TypeScript empowers the dynamic, and generic APIs used by the builders and transformers, through its expressive type system, allowing IDEs to surface context-specific function signatures, prevent common integration mistakes, and support safe extensibility.

Traqula is developed openly on GitHub under the MIT licence, making the project straightforward to adopt, audit, modify, and extend. Documentation is provided through: 1. documentation pages maintained by the development team as part of exposed package READMEs (e.g. @traqula/parser-sparql-1-1), 2. dedicated documentation guides, and 3. JSDoc annotations that are directly visible within IDEs while also being exposed through an automatically generated documentation website. Together, these choices reduce the barrier for integrating Traqula into research prototypes, production systems, and community-driven extensions.

5.2   Current Language Support

Traqula currently supports SPARQL 1.1 [30], SPARQL 1.2 [36], and SPARQL 1.1 + ADJUST [70], providing the ability to transform queries to an AST and to their algebraic representation. Given an algebra, a corresponding AST can be created that produces the same algebra, and given an AST, a matching query string can be generated. Both the AST and algebra for all supported query languages can be transformed using type-safe functions from the core library. When an AST includes source location information, the generator produces a query string that preserves user-visible details, ensuring that semantically irrelevant information, such as spacing or keyword capitalization, is maintained.

To ensure correctness, Traqula includes extensive integration tests. The project currently incorporates 6269 tests, achieving 94% line coverage, combining tests authored specifically for Traqula with those inherited from SPARQL.js [67] and SPARQLAlgebra.js [38], two widely used projects that have been deprecated in favor of Traqula. In addition, Traqula leverages the RDF Tests Community Group's test suites, which provide a broad set of positive and negative tests for implementers of RDF technologies. To further increase maturity, reliability, and facilitate future contributions, the authors are committed to further expanding Traqula's test coverage to 100% through additional unit and integration tests.

5.3 Maintenance and Sustainability Plan

A clear maintenance plan is essential for both the authors of a software project and for users who rely on it. Traqula's source code is openly available as free software under the MIT license, which allows anyone to contribute, fork, or maintain their own version. This ensures that, even if the original maintainers step away, the project can continue to evolve.

However, unmaintained software remains a concern for dependent projects regardless of licensing. To address this, Traqula is maintained under the Comunica association, which distributes responsibility across multiple contributors rather than relying on a single individual. Furthermore, the association provides a mechanism for financial incentives, such as bounties, allowing organizations or users to fund the implementation of specific features or changes. This structured support increases the trustworthiness and long-term sustainability of the project, providing confidence to both researchers and practitioners who depend on Traqula.

6 Performance Analysis

While the primary focus of Traqula is flexibility and expressivity, sufficient performance is still important in broadly used software tools. As such, we evaluate the execution-time performance of Traqula, comparing its different parsing modes, contrasting it with the SPARQL.js [67] parser plus the algebra transformation software SPARQLAlgebra.js [38], and highlighting the impact of parser reuse. All measurements were conducted on Fedora Linux 43 (Workstation Edition), equipped with an Intel® Core^{TM} Ultra 7 165U $\times$ 14 processor, having 16 GiB of RAM. We measured the execution times of parsing 50 real-world SPARQL 1.1 queries from the DBpedia SPARQL Benchmark [47] in each of the tools and Fig 6 shows the execution times of Traqula across different targets and parser configurations. Patching a modular SPARQL 1.1 parser to support SPARQL 1.2 results in a small but statistically significant increase in execution time ($p < 0.05$), with a mean increase of 10.8% . Tracking source information to support round-tripping introduces additional overhead due to the more complex lexing process, with mean increases of 18.8% and 16.6% for SPARQL 1.1 and 1.2 parsing, respectively (both $p < 0.05$). Parsing directly into algebra, effectively

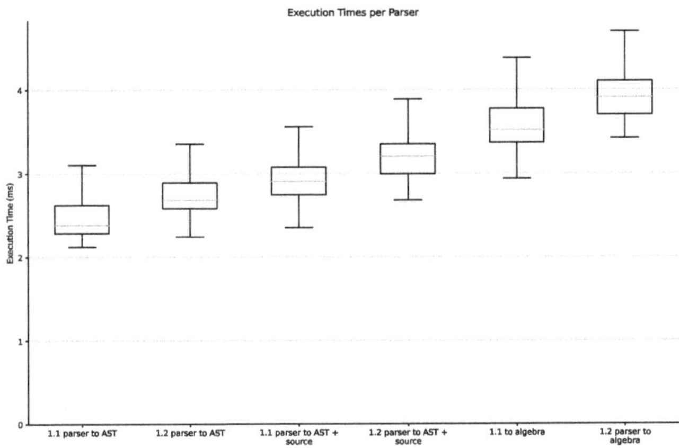

Fig. 6. Variance of execution times (ms) for Traqula's SPARQL 1.1 and 1.2 parsers, when parsing 50 real world SPARQL 1.1 queries [47] using different configurations. Parsing SPARQL 1.2 is always slightly slower than parsing SPARQL 1.1 due to increased grammar complexity, and source tracking or algebra transformation further increase effort.

performing an AST transformation after parsing, further increases execution time, with mean increases of 45.6% for SPARQL 1.1 and 43.0% for SPARQL 1.2 ($p < 0.05$). We also carried out these same experiments for the 199 manually curated SPARQL 1.1 queries within Traqula's unit tests and reach the same conclusions. For space reasons, we omit these results from this paper.

Despite these relative execution time increases, Fig. 7 shows Traqula remains substantially faster than existing solutions. The mean execution time for parsing into an AST using Traqula is 2.5 ms, compared to 23.93 ms for SPARQL.js, a 957.2% difference, even though Traqula generates a more complete and correct AST. Parsing into algebra shows a similar trend: Traqula takes 3.64 ms, while SPARQLAlgebra.js (using SPARQL.js) requires 24.04 ms, a 660.4% difference. Much of this advantage can be attributed to the performance of Chevrotain, the parser toolkit underlying Traqula.

Finally, it is important to recognise the cost associated with using a parser toolkit in a non-pre-compiled language like TypeScript. Traqula's underlying parser toolkit, Chevrotain, performs its parser optimizations at runtime, making parser construction computationally expensive. To illustrate, parsing the 50 queries with a prebuilt and reused parser has a mean execution time of 2.5 ms, whereas creating a new parser for each query results in a mean execution time of 689.31 ms; a slowdown of 275.7 times. Moreover, because Chevrotain is optimised for JavaScript's V8 engine, many performance benefits rely on JIT compilation. Recreating the parser repeatedly may prevent or invalidate JIT optimizations, further degrading performance. This makes parser reuse essential in practice, as already adopted in systems such as the Comunica query engine.

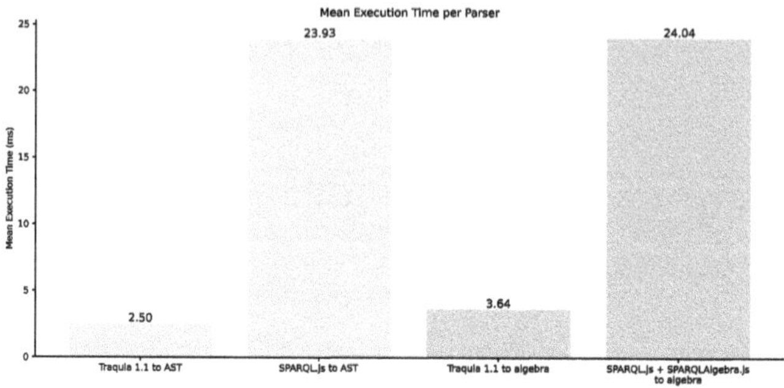

Fig. 7. Execution time (ms) comparison between Traqula and the SPARQL.js [67] parser plus the algebra transformation software SPARQLAlgebra.js [38] when parsing 50 real-world SPARQL 1.1 queries [47]. Parsing to both an AST (green) and to algebra (orange) is significantly faster using Traqula.

7   Conclusion

In this paper, we introduced Traqula, a framework for modular query parsing, generation, and transformation. We argue that a modular parser is essential for the future of SPARQL querying, particularly in increasingly heterogeneous federated environments. By decoupling the various steps in query manipulation, Traqula enables rapid experimentation with new or modified language features. Furthermore, it facilitates query rewriting between different languages and supports the creation of modular query tooling, such as linters and reformatters, through easy AST and algebra manipulation.

Traqula is designed for widespread adoption. It has already been integrated into the modular querying framework Comunica. In addition, we demonstrate its usability for extending language features as part of a Comunica tutorial. The framework is openly available on GitHub under the MIT license and is maintained by the Comunica association, ensuring long-term sustainability.

Looking forward, Traqula's modular architecture and developer-friendly APIs pave the way for supporting additional query languages, such as SHACL Compact Syntax and GQL.

Acknowledgments. Jitse De Smet is a predoctoral fellow of the Research Foundation – Flanders (FWO) (1SB8525N). Ruben Taelman is a postdoctoral fellow of the Research Foundation – Flanders (FWO) (1202124N).

References

1. Apache jena: Arq - construct quad (2024). https://jena.apache.org/documentation/query/construct-quad.html#Grammar
2. Oxigraph: Sep 0002: calendar and duration operations (2024). https://github.com/oxigraph/oxigraph/wiki/SPARQL#sep-0002-calendar-and-duration-operations

3. Virtuoso: Using full text search in sparql (2024). https://docs.openlinksw.com/virtuoso/sparqlextensions/#rdfsparqlrulefulltext

4. Aho, A.V., Sethi, R., Ullman, J.D.: Compilers: Principles, Techniques, and Tools. Addison-Wesley series in computer science/World student series edition, Addison-Wesley (1986). https://www.worldcat.org/oclc/12285707

5. Alexiev, V., Wright, J.: Shacl compact syntax. Tech. rep. (2025). https://w3c.github.io/data-shapes/shacl12-cs/

6. Aranda, C.B., Prud'hommeaux, E.: SPARQL 1.1 federated query. W3C recommendation, W3C (2013). https://www.w3.org/TR/2013/REC-sparql11-federated-query-20130321/

7. Azzam, A., Aebeloe, C., Montoya, G., Keles, I., Polleres, A., Hose, K.: Wisekg: Balanced access to web knowledge graphs. In: Proceedings of the Web Conference 2021, pp. 1422–1434 (2021)

8. Azzam, A., Fernández, J.D., Acosta, M., Beno, M., Polleres, A.: Smart-kg: hybrid shipping for sparql querying on the web. In: Proceedings of The Web Conference 2020, pp. 984–994 (2020)

9. Barbieri, D.F., Braga, D., Ceri, S., Della Valle, E., Grossniklaus, M.: C-sparql: Sparql for continuous querying. In: Proceedings of the 18th international conference on World wide web, pp. 1061–1062 (2009)

10. Barrett, A., Rogers, J.: Stardog: Millan. https://github.com/stardog-union/millan

11. Battle, R., Kolas, D.: Geosparql: enabling a geospatial semantic web. Seman. Web J. 3(4), 355–370 (2011)

12. Burbidge, M., Wileński, D.: Opengql: Gql antlr. https://github.com/opengql/grammar

13. Carothers, G., Seaborne, A.: RDF 1.1 n-triples. W3C recommendation, W3C (2014). https://www.w3.org/TR/2014/REC-n-triples-20140225/, https://www.w3.org/TR/2014/REC-n-triples-20140225/

14. Cheng, S., Hartig, O.: Fedqpl: A language for logical query plans over heterogeneous federations of RDF data sources. In: Indrawan-Santiago, M., Pardede, E., Salvadori, I.L., Steinbauer, M., Khalil, I., Kotsis, G. (eds.) iiWAS '20: The 22nd International Conference on Information Integration and Web-based Applications & Services, Virtual Event / Chiang Mai, Thailand, November 30–December 2, 2020, pp. 436–445. ACM (2020). https://doi.org/10.1145/3428757.3429120, https://doi.org/10.1145/3428757.3429120

15. Chevrotain, Soel, S.: Chevrotain - parser building toolkit for javascript (2025). https://github.com/Chevrotain/chevrotain/

16. community, S.: Sqlite. https://github.com/sqlite/sqlite/

17. Cyganiak, R., Wood, D., Lanthaler, M.: Rdf 1.1: Concepts and abstract syntax. Recommendation, W3C (2014). https://www.w3.org/TR/2014/REC-rdf11-concepts-20140225/

18. Dimou, A., Vander Sande, M., Colpaert, P., Verborgh, R., Mannens, E., Van de Walle, R.: Rml: a generic language for integrated RDF mappings of heterogeneous data. Ldow 1184 (2014)

19. Emonet, V., Sima, A.C., Mendes de Farias, T.: A user-friendly sparql query editor powered by lightweight metadata. In: The Semantic Web: ESWC 2025 Satellite Events: Portoroz, Slovenia, June 1–5, 2025, Proceedings, pp. 3–7. Springer-Verlag, Berlin, Heidelberg (2025). https://doi.org/10.1007/978-3-031-99554-5_1

20. Facebook: Graphql. Rec. (2025). https://spec.graphql.org/September2025/#

21. Feigenbaum, L., Todd Williams, G., Grant Clark, K., Torres, E.: SPARQL 1.1 protocol. Rec., W3C (2013). https://www.w3.org/TR/2013/REC-sparql11-protocol-20130321/

22. Finné, M., Demianenko, M., Melke, P., Nawroth, A.: Neo4j. https://github.com/neo4j/neo4j
23. Fowler, M.: Inversion of control containers and the dependency injection pattern. Tech. rep. (2004). https://martinfowler.com/articles/injection.html
24. GitHub: Octoverse: A new developer joins github every second as ai leads typescript to #1. Tech. rep. (2025). https://github.blog/news-insights/octoverse/octoverse-a-new-developer-joins-github-every-second-as-ai-leads-typescript-to-1/
25. GNU: Gnu bison. Tech. rep. (2025). https://www.gnu.org/software/bison/
26. Goncharov, I., Byron, L.: Graphql.js. https://github.com/graphql/graphql-js
27. Görlitz, O., Staab, S.: Splendid: Sparql endpoint federation exploiting void descriptions. COLD (2011)
28. Gschwend, A., Lassila, O.: Rdf & sparql working group charter (2025). https://www.w3.org/2025/04/rdf-star-wg-charter.html
29. Hanski, J., Crum, E., Taelman, R.: Observations on automated client-side query federation over wikidata sparql endpoints. In: Proceedings of the Wikidata Workshop 2025 co-located with 24th International Semantic Web Conference (ISWC 2025) (2025). https://www.rubensworks.net/raw/publications/2025/hanski_wikidata_federation_2025.pdf
30. Harris, S., Seaborne, A., Prud'hommeaux, E.: SPARQL 1.1 query language. Recommendation, W3C (2013). https://www.w3.org/TR/2013/REC-sparql11-query-20130321/
31. Hartig, O.: Querying trust in RDF data with tsparql. In: European Semantic Web Conference, pp. 5–20. Springer (2009)
32. Hartig, O., Buil-Aranda, C.: Bindings-restricted triple pattern fragments. In: Debruyne, C., et al. (eds.) OTM Confederated International Conferences On the Move to Meaningful Internet Systems, pp. 762–779. Springer, Cham (2016). https://doi.org/10.1007/978-3-319-48472-3_48
33. Hartig, O., Champin, P.A., Kellogg, G., Seaborne, A.: SPARQL 1.2 query language. Tech. rep. (2025). https://www.w3.org/TR/rdf12-concepts/
34. Hartig, O., Freytag, J.C.: Foundations of traversal based query execution over Linked Data. In: Proceedings of the 23rd ACM conference on Hypertext and social media, pp. 43–52. ACM (2012)
35. Hartig, O., Pérez, J.: Ldql: a query language for the web of linked data. J. Web Semant. **41**, 9–29 (2016)
36. Hartig, O., Seaborne, A., Taelman, R., Williams, G., Tanon, T.P.: SPARQL 1.2 query language. Tech. rep. (2025). https://www.w3.org/TR/sparql12-query/
37. Heling, L., Acosta, M.: Federated SPARQL query processing over heterogeneous linked data fragments. In: Laforest, F., et al. (eds.) WWW '22: The ACM Web Conference 2022, Virtual Event, Lyon, France, April 25–29, 2022, pp. 1047–1057. ACM (2022). https://doi.org/10.1145/3485447.3511947
38. Herwegen, J.V.: Sparqlalgebra.js. https://github.com/joachimvh/SPARQLAlgebra.js
39. Information technology — database languages — gql. Standard, International Organization for Standardization, Geneva, CH (2024). https://www.iso.org/standard/76120.html
40. Information technology - database languages - sql. Standard, International Organization for Standardization, Geneva, CH (2023). https://www.iso.org/standard/76583.html
41. Kalmbach, J., Kalmbach, J.: Qlever. https://github.com/ad-freiburg/qlever/
42. van Kleef, P., Iliev, M.: Virtuoso open-source edition. https://github.com/openlink/virtuoso-opensource

43. Min Oo, S., De Meester, B., Taelman, R., Colpaert, P.: Algebraic mapping operators for knowledge graph generation. Semant. Web J. (2025). https://www.semantic-web-journal.net/system/files/swj3806.pdf
44. Minier, T., Skaf-Molli, H., Molli, P.: Sage: web preemption for public sparql query services. In: The World Wide Web Conference, pp. 1268–1278 (2019)
45. Momjian, B., Eisentraut, P., Lane, T.: Postgresql database management system. https://github.com/postgres/postgres
46. Montoya, G., Aebeloe, C., Hose, K.: Towards efficient query processing over heterogeneous RDF interfaces. In: Verborgh, R., Kuhn, T., Berners-Lee, T. (eds.) Proceedings of the 2nd Workshop on Decentralizing the Semantic Web co-located with the 17th International Semantic Web Conference, DeSemWeb@ISWC 2018, Monterey, California, USA, October 8, 2018. CEUR Workshop Proceedings, vol. 2165. CEUR-WS.org (2018). https://ceur-ws.org/Vol-2165/paper4.pdf
47. Morsey, M., Lehmann, J., Auer, S., Ngomo, A.N.: Dbpedia SPARQL benchmark - performance assessment with real queries on real data. In: Aroyo, L., et al. (eds.) The Semantic Web - ISWC 2011 - 10th International Semantic Web Conference, Bonn, Germany, October 23–27, 2011, Proceedings, Part I. Lecture Notes in Computer Science, vol. 7031, pp. 454–469. Springer, Cham (2011). https://doi.org/10.1007/978-3-642-25073-6_29
48. Nezis, I.: Qlue-ls, a powerful sparql language server (2025). https://ad-publications.cs.uni-freiburg.de/theses/Bachelor_Ioannis_Nezis_2025.pdf
49. Nicholas, C.Z.: Eslint (2025). https://github.com/eslint/eslint/
50. Ottestad, H.M., Broekstra, J.: Rdf4j. https://github.com/eclipse-rdf4j/rdf4j
51. Parr, T.J., Quong, R.W.: ANTLR: A predicated - ll(k) parser generator. Softw. Pract. Exp. **25**(7), 789–810 (1995). https://doi.org/10.1002/SPE.4380250705
52. Pelgrin, O., Taelman, R., Galárraga, L., Hose, K.: The need for better RDf archiving benchmarks. In: Proceedings of the 9th Workshop on Managing the Evolution and Preservation of the Data Web (2023). https://www.rubensworks.net/raw/publications/2023/TheNeedForBetterRdfArchivingBenchmarks.pdf
53. Pellissier Tanon, T.: Oxigraph (2025). https://doi.org/10.5281/zenodo.7408022
54. Pérez, J., Arenas, M., Gutierrez, C.: Semantics and complexity of SPARQL. ACM Trans. Database Syst. **34**(3), 16:1–16:45 (2009). https://doi.org/10.1145/1567274.1567278
55. Pham, T.H.T., Montoya, G., Nédelec, B., Skaf-Molli, H., Molli, P.: Passage: ensuring completeness and responsiveness of public sparql endpoints with sparql continuation queries. In: Proceedings of the ACM on Web Conference 2025, pp. 47–58 (2025)
56. Raasveldt, M., Muehleisen, H.: DuckDB. https://github.com/duckdb/duckdb
57. Rietveld, L., Hoekstra, R.: The YASGUI family of SPARQL clients. Semant. Web **8**(3), 373–383 (2017). https://doi.org/10.3233/SW-150197
58. Saleem, M., Ngomo, A.N.: Hibiscus: Hypergraph-based source selection for SPARQL endpoint federation. In: Presutti, V., d'Amato, C., Gandon, F., d'Aquin, M., Staab, S., Tordai, A. (eds.) The Semantic Web: Trends and Challenges - 11th International Conference, ESWC 2014, Anissaras, Crete, Greece, May 25–29, 2014. Proceedings. Lecture Notes in Computer Science, vol. 8465, pp. 176–191. Springer, Cham (2014). https://doi.org/10.1007/978-3-319-07443-6_13
59. Schwarte, A., Haase, P., Hose, K., Schenkel, R., Schmidt, M.: Fedx: Optimization techniques for federated query processing on linked data. In: Aroyo, L., et al.(eds.) The Semantic Web - ISWC 2011 - 10th International Semantic Web Conference,

Bonn, Germany, October 23–27, 2011, Proceedings, Part I. Lecture Notes in Computer Science, vol. 7031, pp. 601–616. Springer, Cham (2011). https://doi.org/10.1007/978-3-642-25073-6_38
60. Seaborne, A.: Apache: Jena. https://github.com/apache/jena
61. sebmck, nicolo ribaudo, hzoo: Babel (2025). https://github.com/babel/babel
62. Smet, J.D., Taelman, R.: Towards tackling SPARQL heterogeneity through modular parsing. In: Chaves-Fraga, D., Heibi, I., Garijo, D., Collarana, D., Salatino, A.A., Vahdati, S. (eds.) Joint Proceedings of Posters, Demos, Workshops, and Tutorials of the 21st International Conference on Semantic Systems co-located with 21st International Conference on Semantic Systems (SEMANTiCS 2025), Vienna, Austria, September 3–5, 2025. CEUR Workshop Proceedings, vol. 4064. CEUR-WS.org (2025). https://ceur-ws.org/Vol-4064/SEMDEV-paper4.pdf
63. Sowinski, P., Bogacka, K., Danilenka, A., Kozlov, N.: Jelly: a fast and convenient RDF serialization format. In: Chaves-Fraga, D., Heibi, I., Garijo, D., Collarana, D., Salatino, A.A., Vahdati, S. (eds.) Joint Proceedings of Posters, Demos, Workshops, and Tutorials of the 21st International Conference on Semantic Systems co-located with 21st International Conference on Semantic Systems (SEMANTiCS 2025), Vienna, Austria, September 3–5, 2025. CEUR Workshop Proceedings, vol. 4064. CEUR-WS.org (2025). https://ceur-ws.org/Vol-4064/SEMDEV-paper3.pdf
64. Taelman, R., Herwegen, J.V., Sande, M.V., Verborgh, R.: Comunica: a modular SPARQL query engine for the web. In: Vrandecic, D., et al. (eds.) The Semantic Web - ISWC 2018 - 17th International Semantic Web Conference, Monterey, CA, USA, October 8–12, 2018, Proceedings, Part II. Lecture Notes in Computer Science, vol. 11137, pp. 239–255. Springer, Cham (2018). https://doi.org/10.1007/978-3-030-00668-6_15
65. Taelman, R., Verborgh, R.: Evaluation of link traversal query execution over decentralized environments with structural assumptions. CoRR abs/2302.06933 (2023). https://doi.org/10.48550/ARXIV.2302.06933
66. Thompson, B.: Blazegraph. https://github.com/blazegraph/database/
67. Verborgh, R., et al.: Rubenverborgh/sparql.js (2025). https://doi.org/10.5281/zenodo.597487
68. B.D., Haesendonck, G., Colpaert, P.: Triple pattern fragments: a low-cost knowledge graph interface for the web. J. Web Semant. 37-38, 184–206 (2016). https://doi.org/10.1016/J.WEBSEM.2016.03.003
69. Williams, G.: SPARQL 1.1 service description. W3C recommendation, W3C (2013). https://www.w3.org/TR/2013/REC-sparql11-service-description-20130321/
70. Williams, G.T.: Sparql dev: Add support durations, dates, and times. Tech. rep. (2025). https://github.com/w3c/sparql-dev/blob/7350808a4f0b7ca7c2a8d823b4e7d8958dfb073b/SEP/SEP-0002/sep-0002.md

SIDEKICK: A Semantically Integrated Resource for Drug Effects, Indications, and Contraindications

Mohammad Ashhad[1], Olga Mashkova[2], Ricardo Henao[3],
and Robert Hoehndorf[2(✉)]

[1] BESE, King Abdullah University of Science and Technology,
Thuwal, Kingdom of Saudi Arabia
[2] CEMSE, King Abdullah University of Science and Technology,
Thuwal, Kingdom of Saudi Arabia
`robert.hoehndorf@kaust.edu.sa`
[3] Department of Bioinformatics and Biostatistics, Duke University, Durham, USA

Abstract. Pharmacovigilance and clinical decision support systems utilize structured drug safety data to guide medical practice. However, existing datasets frequently depend on terminologies such as MedDRA, which limits their semantic reasoning capabilities and their interoperability with Semantic Web ontologies and knowledge graphs. To address this gap, we developed SIDEKICK, a knowledge graph that standardizes drug indications, contraindications, and adverse reactions from FDA Structured Product Labels. We developed and used a workflow based on Large Language Model (LLM) extraction and Graph-Retrieval Augmented Generation (Graph RAG) for ontology mapping. We processed over 50,000 drug labels and mapped terms to the Human Phenotype Ontology (HPO), the MONDO Disease Ontology, and RxNorm. Our semantically integrated resource outperforms the SIDER and ONSIDES databases when applied to the task of drug repurposing by side effect similarity. We serialized the dataset as a Resource Description Framework (RDF) graph and employed the Semanticscience Integrated Ontology (SIO) as upper level ontology to further improve interoperability. Consequently, SIDEKICK enables automated safety surveillance and phenotype-based similarity analysis for drug repurposing.

Keywords: Drug Safety · Knowledge Graph · Drug Information Dataset

1 Introduction

Drug safety information is fundamental to clinical decision-making, yet accessing this knowledge and integrating it with other clinical information remains a

Resource type: Knowledge Graph
License: CC BY 4.0 International
DOI: https://doi.org/10.5281/zenodo.17779317
URL: http://sidekick.bio2vec.net.

challenge. Adverse drug reactions (ADRs) represent a leading cause of morbidity and mortality, with medication errors estimated to cost \$42 billion annually in the United States [7]. Despite their clinical importance, the systematic organization and computational representation of drug indications, contraindications, and side effects has lagged behind other areas of biomedical informatics.

Several existing drug safety datasets have contributed to pharmacovigilance research. SIDER [20] was the first open database to systematically extract adverse drug reactions from drug labels, mapping both side effects and indications to the Medical Dictionary for Regulatory Activities (MedDRA) [4]. However, SIDER's indication data were extracted primarily as a byproduct to filter false positives during adverse event extraction. OnSIDES [34] employed fine-tuned PubMedBERT models to extract over 3.6 million drug-adverse event pairs from FDA labels but focused exclusively on adverse events without extracting therapeutic indications or contraindications; OnSIDES also uses MedDRA for standardization. OFFSIDES and TWOSIDES [35], while important for pharmacovigilance, employ a different method by mining post-marketing surveillance data from the FDA adverse event reporting system (FAERS) to identify off-label effects and drug–drug interactions using statistical techniques. Notably, no existing resource provides comprehensive coverage of contraindications that are critical safety information that indicates when drugs should not be used.

Although these resources are widely used in biomedical informatics research, they share a common limitation: reliance on MedDRA for the standardization of drug effects and indications. Although MedDRA is widely adopted in a regulatory context, MedDRA imposes constraints on computational interoperability due to its architectural design. Fundamentally, MedDRA functions as a terminology rather than an ontology; terminologies primarily facilitate data access and indexing, while ontologies exist to structure the underlying domain through explicit semantic relationships and axioms [12].

Drug repurposing, which seeks to identify new therapeutic uses for existing medications, increasingly relies on automated reasoning about drug–disease relations [29], and phenotype-based approaches to drug repurposing use the observation that drugs with similar side effects likely share mechanisms of action [6]. Such analyzes benefit from detailed phenotypic descriptors and the ability to compute semantic similarity between side effects, and semantic similarity measures improve with the ability to deductively infer relations between terms [21,27].

We developed SIDEKICK to address the limitations of existing resources focused on drug safety. Our contributions are: (1) We extended the clinical scope beyond side effects and integrated indications, contraindications, and adverse reactions from over 50,000 FDA Structured Product Labels; (2) We used an extraction method based on Large Language Models combined with Graph-RAG to improve both granularity and coverage of extracted information; (3) We normalized clinical entities to standardized ontologies, mapping side effects and phenotypic contraindications to the Human Phenotype Ontology (HPO) [17], disease terms to the MONDO Disease Ontology [36], and drug entities to RxNorm; (4) We serialized SIDEKICK as a Resource Description Framework (RDF) [26] graph using the Semanticscience Integrated Ontology (SIO) [9] as an upper-level framework, further improving interoperability and adherence to FAIR data principles [37]; and (5) We evaluated SIDEKICK against the OnSIDES [34] baseline to demonstrate that SIDEKICK improves the prediction of shared drug targets by side effect similarity.

2   Related Work

2.1   Drug Safety Databases and Their Applications

SIDER [20] established the precedent for extracting adverse drug reactions (ADRs) from package inserts, providing frequency data and mappings to the Medical Dictionary for Regulatory Activities (MedDRA) [5]. The resource also includes a dataset of drug indications, though these should not be considered authoritative therapeutic use data. They were extracted primarily to reduce false positives during ADR extraction by helping to identify medical terms not corresponding to adverse events. Most recently, OnSIDES [34] employed fine-tuned PubMedBERT models to extract over 3.6 million drug-adverse event pairs from FDA labels, achieving high extraction performance. OnSIDES uses MedDRA for standardization and focuses exclusively on adverse drug events; it does not extract therapeutic indications or contraindications.

Despite these advancements, these resources universally employ MedDRA for standardization. MedDRA functions as a terminology designed to facilitate regulatory data access rather than a formal ontology designed to structure the domain. Consequently, MedDRA aggregates terms based on retrieval expectations rather than shared biological properties or logical definitions. Consequently, MedDRA's hierarchical structure may aggregate ontologically distinct terms. Furthermore, this terminology lacks explicit logical linkage to other types of biological entities, such as anatomical structures defined in Uberon [23] or physiological processes defined in the Gene Ontology (GO) [8]. In contrast, phenotype ontologies such as the Human Phenotype Ontology (HPO) utilize logical definitions based on the Entity-Quality (EQ) design pattern [11,22]. This axiomatic structure of phenotype ontologies therefore enables the semantic interoperability required to link clinical phenotypes with their underlying biological mechanisms, a capability that the access-oriented architecture of MedDRA does not support.

2.2   Phenotype and Disease Ontologies

To overcome the limitations of using terminologies, modern biomedical informatics employs formal ontologies that support deductive reasoning [15]. The Human Phenotype Ontology (HPO) [17] provides a standardized vocabulary of phenotypic abnormalities organized as an axiomatic theory (a formal ontology) using the Web Ontology Language (OWL) [2]. HPO enables the calculation of semantic similarity between terms by exploiting not only a graph structure but also its axioms [33]. The axioms are important for identifying drugs with similar side effect profiles, where similarity may be based on shared mechanisms of action or shared anatomical structures. Similarly, the MONDO Disease Ontology [36] harmonizes disease definitions across different databases using equivalence axioms. By mapping clinical entities to HPO and MONDO, we enable the application of algorithms that require deductive reasoning and axiomatic theories, a capability absent in current MedDRA-based resources.

2.3 Semantic Interoperability and Upper-Level Ontology

Semantic interoperability across heterogeneous biomedical datasets requires more than standardized vocabularies; it also requires a structural framework to define relationships between entities. Upper-level ontologies provide such a framework by defining general classes (*e.g.*, processes, objects, attributes) that remain consistent across domains and providing relationships that can be used across resources. The Semanticscience Integrated Ontology (SIO) [9] serves as a lightweight upper-level ontology widely adopted in scientific applications of the Semantic Web.

Upper-level ontologies also provide Ontology Design Patterns (ODPs) [10]— reusable modeling solutions for recurrent ontology design problems. For example, SIO employs ODPs to model associations as reified entities (*i.e.*, classes). This practice allows for the attachment of metadata (such as provenance or evidence strength) directly to the association. Consequently, by utilizing classes and relations from SIO, we improve interoperability because the use of an upper level ontology allows users and agents to query the knowledge graph using standard SIO-based query patterns, and without requiring explicit knowledge of how each interaction is specifically represented. This approach contrasts with ad-hoc schema designs that may be found in standalone databases, which limit data federation and integration.

3 Methodology

3.1 Overview

We developed a workflow to extract, structure, and map drug-related clinical information from FDA Structured Product Labels (SPLs) to standardized biomedical ontologies. The workflow consists of seven main stages: (1) identification and filtering of human prescription drugs, (2) deduplication of redundant SPL documents, (3) automated extraction of clinical entities using large language models (LLMs), (4) mapping of side effects to the Human Phenotype Ontology (HPO), (5) classification of contraindications and indications, (6) mapping of disease and phenotype terms to MONDO and HPO ontologies, respectively, and (7) mapping of drug–drug interaction terms to RxNorm. Finally, all extracted and mapped information is integrated into an RDF knowledge graph. Figure 1 illustrates the full workflow.

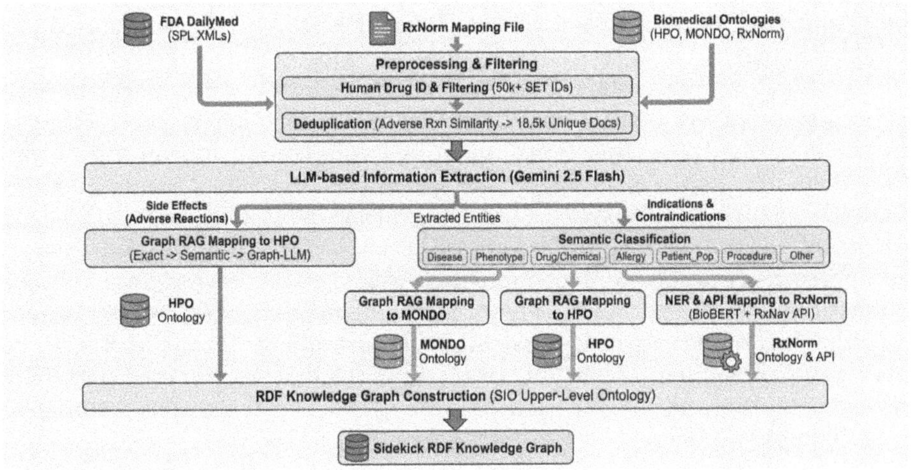

Fig. 1. Workflow for creating the SIDEKICK knowledge graph.

3.2   Data Collection and Preprocessing

We acquired the latest version of drug labeling data from the FDA DailyMed database[1], which serves as the official repository for FDA-approved package inserts. These documents adhere to the Health Level Seven (HL7) [30] Structured Product Labeling (SPL) standard[2], an XML-based format used for the exchange of product information. To isolate relevant data, we processed the DailyMed human prescription label archives to extract the SPL Set Identifier (SET ID), a globally unique identifier that persistently represents a specific labeling document version. This process yielded 50,559 unique SET IDs corresponding to human prescription drugs. Subsequently, we standardized these entries by linking the SPL documents to RxNorm, the normalized naming system for clinical drugs produced by the National Library of Medicine (NLM). Using the NLM mapping files, we associated each document with its specific RxNorm Unique Identifier (RxCUI). Finally, to map branded products back to their fundamental bioactive components, we queried the RxNav REST API; for each product RxCUI, we retrieved the related terms and filtered strictly for active ingredients (term type 'IN').

For standardized term mapping, we utilized three major biomedical ontologies: the Human Phenotype Ontology (HPO) [17], which provides standardized vocabulary of phenotypic abnormalities with over 16,000 terms; the MONDO Disease Ontology [36], which integrates disease terminologies from multiple

---

[1] https://dailymed.nlm.nih.gov/.

[2] https://www.fda.gov/industry/fda-data-standards-advisory-board/structured-product-labeling-resources.

sources; and RxNorm, accessed via the RxNav REST API[3] for normalized clinical drug names.

To mitigate computational overhead, we implemented a deduplication strategy centered on the textual similarity of adverse reaction profiles. First, we extracted the "Adverse Reactions" sections from the SPL documents, identifying them via their specific Logical Observation Identifiers Names and Codes (LOINC) identifiers. Within each grouping defined by a Product RxCUI, we performed pairwise comparisons of these sections. We used a two-tiered matching approach: (1) exact content matching utilizing MD5 cryptographic hashes to identify identical text; and (2) fuzzy matching utilizing the SequenceMatcher algorithm [31] to identify near-duplicates with a similarity threshold of 95%. For every group of similar documents, we retained the first encountered SET ID as the representative entry. This filtering process reduced the corpus from 50,559 to 18,500 unique documents (*i.e.*, 63.4% reduction).

3.3 Information Extraction from SPL Documents

Text Preprocessing. Prior to LLM-based extraction, we preprocessed SPL XML documents to create structured text suitable for language model processing. We excluded sections not relevant to clinical decision-making using a blacklist of LOINC section codes, including package labeling (51945-4), product data elements (48780-1), *etc.* For retained sections, we preserved the hierarchical structure with section titles, body text, and tables. We converted tables to a text representation with rows separated by newlines and cells separated by pipe characters.

Large Language Model Extraction. We used the Google Gemini 2.5 Flash model via the OpenRouter API for automated extraction of clinical entities from processed SPL text. The model was prompted to extract three categories of information: (1) Indications: medical conditions, diseases, or symptoms for which the drug is indicated; (2) Contraindications: conditions, situations, or patient populations in which the drug should not be used; (3) Adverse reactions (side effects): undesirable effects associated with drug use. The extraction prompt instructed the model to search over all sections (not just those with dedicated section headers), extract items in concise form while maintaining accuracy. The model returned structured XML output with nested tags for each entity category. We processed the deduplicated SPL documents in batches of 500 with 10-second sleep intervals to adhere to the API rate limits.

3.4 Ontology Mapping

After extracting clinical entities from SPL documents, we mapped the terms to standardized biomedical ontologies to enable interoperability and integration with existing knowledge bases. We then mapped the extracted adverse reactions

---

[3] https://rxnav.nlm.nih.gov/.

to the Human Phenotype Ontology (HPO) [17] using a Graph-Retrieval Augmented Generation (Graph RAG) approach. We loaded the HPO ontology into the OBO format[4] and constructed a NetworkX [13] directed graph with a bidirectional dictionary that maps the term names and synonyms to HPO IDs. We computed dense vector embeddings for all HPO terms and synonyms using the `all-MiniLM-L6-v2` sentence transformer model, processing in batches of 1,000 terms and caching to disk. The embedding matrix contained representations for over 45,000 terms and synonyms.

For each side effect term, we used a three-stage matching strategy. First, we used exact string matching against primary term names and synonyms (case-insensitive). Second, we used a semantic similarity search based on the cosine similarity against all embeddings, retrieving the top-10 matches and deduplicated by HPO ID. In the third step, we used graph-enhanced LLM mapping where the top semantic matches served as seed nodes for graph traversal to identify related classes (parents, children, siblings), limited to 15 nodes. We then constructed new prompts for Gemini 2.5 Flash that contained the top 5 semantic matches with similarity scores, definitions, related terms with relationship annotations, and instructions to return HPO ID and term name in JSON format. The responses were validated to ensure that the returned IDs exist and that the term names match canonical names, with automatic retries (up to 3 attempts). Unmappable terms were assigned to `HP:0000001` ("All", the top-level class of the HPO).

We implemented an ontology-based mapping workflow to handle the heterogeneous nature of contraindication and indication terms. First, we classified terms that occur in the indication and contraindication sections of the SPL into seven categories: *Disease* (medical conditions, disorders, syndromes), *Phenotype* (observable clinical signs, symptoms), *Drug or Chemical* (drug interactions, concomitant medications), *Allergy or Hypersensitivity* (allergic reactions), *Patient Population* (demographics, life stages), *Procedure* (medical/surgical procedures), and *Other* (which corresponds to any other type for which we could not identify a specific category). We implemented a two-stage classification: (1) keyword-based auto-classification for allergy-related terms containing the strings "hypersensitivity", "allergic", or "anaphylaxis"; (2) LLM-based classification using Gemini 2.5 Flash Lite with terms batched in groups of 15, receiving detailed category definitions and returning JSON-formatted classifications.

We mapped the terms classified as *Disease* to the MONDO Disease Ontology using the same Graph RAG approach as we used for HPO mapping. We constructed a NetworkX graph from the MONDO OBO file and generated embeddings using `all-MiniLM-L6-v2`. The MONDO ontology contains over 25,000 disease classes. We applied the three-stage matching strategy (exact match, semantic similarity with top-10 retrieval, graph-enhanced LLM mapping with Gemini 2.5 Flash), with validation and retries. Unmappable terms defaulted to `MONDO:0000001` (*Disease or disorder*).

---

[4] https://owlcollab.github.io/oboformat/doc/obo-syntax.html.

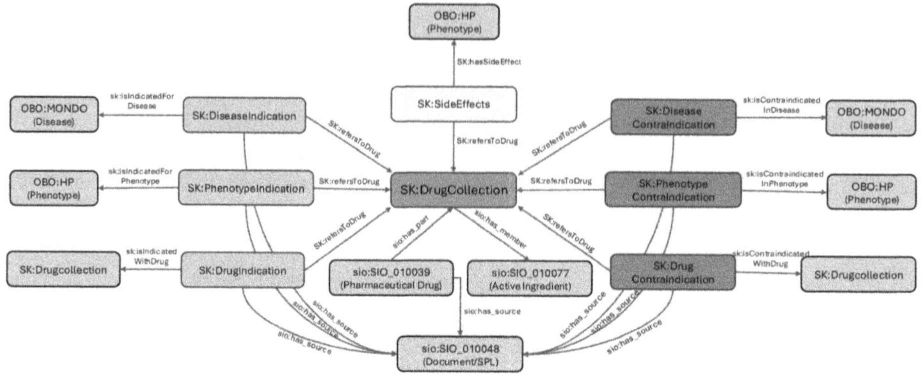

Fig. 2. RDF schema for the SIDEKICK knowledge graph.

Terms classified as *Phenotype* were mapped to HPO using the identical Graph RAG pipeline described above for the mapping of side effects. We employed BioBERT-based chemical NER[5] (`alvaroalon2/biobert_chemical_ner`) [1] to identify drug entities within text, with fallback heuristic parsing to remove phrases like "coadministration with" and splitting on delimiters. For each entity, we queried the RxNav REST API using: (1) exact match via `/rxcui.json`, and (2) approximate match via `/approximateTerm.json` (top 10 candidates) if exact match failed. Terms without recognized entities or ingredient-level RxCUIs were assigned a group labeled "other".

4 Knowledge Graph Construction and Validation

4.1 Data Model and Schema

We constructed the SIDEKICK knowledge graph (KG) using the Resource Description Framework (RDF) [26]. We selected the Semanticscience Integrated Ontology (SIO) [9] as the upper-level ontology due to its wide use in similar resources in the biomedical domain. Figure 2 illustrates the schema that we developed for the SIDEKICK knowledge graph.

We distinguish between the pharmaceutical product and the collection of active ingredients it contains. A *Pharmaceutical Drug* is a specific branded or generic product (e.g., "Naproxen 500 MG Oral Tablet"), which are typed as *PharmaceuticalDrug* (`SIO:010039`) and identified by their Product RxCUI using the Internationalized Resource Identifiers (IRIs) for the RxNorm version provided by the NCBO BioPortal [24].

We introduced a new class, *DrugCollection* (`sk:DrugCollection`), which is a subclass of *Collection* (`SIO:000616`), to represent the set of active ingredients that

---

are independent of the manufacturer or specific formulation. For instance, "Amoxicillin/Clavulanate" as a drug collection (`sk:ingredient_set_723_48203`) represents the combination of amoxicillin and clavulanate as active ingredients, regardless of the brand name. This abstraction is necessary for three reasons: (1) pharmaceutical products may contain multiple active ingredients, (2) clinical relationships such as side effects and therapeutic indications apply to the ingredient combination rather than specific branded products, and (3) it prevents redundant assertions across multiple branded or generic versions of the same drug formulation. In practice, most drug collections are singleton sets containing a single active ingredient, but the uniform representation allows consistent handling of both single-ingredient and combination products. Finally, individual active ingredients are instances of *Active ingredient* (`SIO:010077`) and use the RxNorm IRIs from BioPortal.

The relationships are formalized such that a product *has part* (`SIO:000028`) a *Drug Collection*, which in turn *has member* (`SIO:000059`) specific ingredients. We also assert a property chain, *'has part'* ∘ *'has member'* ⊆ *'has part'*.

We modeled clinical relationships (indications, contraindications, and side effects) as reified associations (where relationships are represented as explicit entities) rather than simple edges. This approach, consistent with a corresponding SIO design pattern, treats the relationship itself as a node (an instance of *Association*, `SIO:000897`). We use this reified class for adding provenance, specifically the source SPL document, directly to the assertion via *has source* relation (`SIO:000253`).

To distinguish the two main participants in each clinical relationship, we use two distinct properties: *refers to drug* (`sk:refersToDrug`), a subproperty of *refers to* (`SIO:000628`), always points to the drug collection being described (the subject of the clinical finding), while properties such as `sk:hasSideEffect`, `sk:isIndicatedForDisease`, etc., point to the clinical entity (phenotype, disease or another drug collection) that is the target of the relationship.

We defined seven specific subclasses of *Association* (`SIO:000897`) to capture drug associations with their side effects, their indications and contraindications (either diseases or phenotypes), other drugs to be used in combination or to be avoided (a drug contraindication). Provenance is captured using the *has source* property (`SIO:000253`), which links each association node directly to the specific SPL document node from which the information was extracted.

4.2 Triple Generation Workflow

The graph generation workflow was implemented in Python using rdflib [18]. We established Internationalized Resource Identifiers (IRIs) in the SIDEKICK namespace for drug collections, associations, and documents. For phenotypes and diseases, we use identifiers in the OBO namespace, for drugs we use the namespace used in the BioPortal version of RxNorm, and for upper-level classes and relations we use the SIO IRIs.

We use the Dublin Core [3] vocabulary for additional metadata, in particular `dcterms:description` for textual descriptions of classes and properties, `dcterms:created` for a creation timestamp, `dcterms:creator` to specify author information, and `dcterms:license` to provide licensing information. Additionally, we created a VOID file that provides basic metadata of the entire dataset.

Table 1. SIDEKICK knowledge graph statistics.

Component	Count
Graph Structure	
Total RDF Triples	3,155,094
Drug Collections (Hubs)	3,036
Active Ingredients	2,370
Pharmaceutical Products	14,689
SPL Source Documents	48,458
Clinical Associations	
Total Associations	205,683
Side Effects (Drug→Phenotype)	184,512
Disease Indications (Drug→Disease)	6,478
Phenotype Indications (Drug→Phenotype)	1,748
Drug Indications (Drug→Drug)	71
Disease Contraindications	5,891
Phenotype Contraindications	4,047
Drug Contraindications	2,936
Ontology Coverage	
Unique HPO Terms	6,243
Unique MONDO Terms	2,168
Unique RxNorm Ingredients	2,370
Ontology Mapping Coverage[*]	
Total Unique Terms	134,515
Mapped	129,848 (96.5%)
Unmapped	4,667 (3.5%)

[*]Across all unique extracted side effects, disease/phenotype indications and contraindications, and drug/chemical indications and contraindications.

The final SIDEKICK knowledge graph is serialized in Turtle format and can be converted to any RDF syntax. Table 1 summarizes basic statistics of the SIDE-KICK knowledge graph. The resulting graph structure enables SPARQL queries across SIDEKICK and the referenced ontologies. Beyond retrieval of directly asserted relationships (such as identifying all drugs contraindicated for a specific phenotype), the RDF structure enables reasoning over the HPO and MONDO ontology hierarchies. For example, a SPARQL query can retrieve all drugs with cardiac-related side effects by querying for phenotypes that are subclasses of abnormality of the cardiovascular system (HP:0001626) or anatomically associated with heart structures via HPO's axioms that use the UBERON anatomy ontology [23].

4.3   Schema and Semantic Validation

To validate whether the SIDEKICK knowledge graph complies with the schema in Fig. 2, we formalized the schema, and additional constraints, using Shape Expressions (ShEx) [28]. We validated all 205,683 clinical associations with respect to rules that ensure that SIDEKICK (1) maintains structural integrity with proper entity relationships, (2) include complete provenance linking to FDA SPL source documents, (3) enforce ontology type safety with pattern-matched IRIs ensuring correct use of MONDO and HPO terminologies, and (4) satisfy all cardinality constraints. We further validated that our data is semantically consistent using the ELK reasoner [16] (v0.5.0). Although SIO contains axioms that exceed the OWL 2 EL profile [19], we were only able to validate the EL subset due to the computational complexity of OWL 2 DL reasoning [19]. At release time, SIDEKICK is both consistent with respect to the schema, and semantically consistent based on ELK.

5   Experiments

We conducted three evaluations to validate the quality and utility of SIDEKICK: (1) human expert evaluation of mapping accuracy, (2) prediction of shared drug targets using side effect similarity, and (3) a competency question-based evaluation to demonstrate the reasoning capabilities of the ontology-driven schema.

5.1   Mapping Quality Evaluation

To assess the accuracy of our Graph RAG mapping methodology, we performed a human expert evaluation study. We randomly selected 50 unique SPL SET IDs from our dataset and extracted all unique side effects (identified by HPO ID) from these documents, producing 980 distinct side effect mappings. Two domain experts with backgrounds in pharmacology and biomedical informatics independently reviewed each mapping. For each side effect, experts evaluated whether the assigned HPO term accurately represented the original text from the drug label. Mappings where the side effect name and HPO term matched exactly were automatically validated without manual review. In cases of disagreement between experts, they collaboratively discussed the mapping until reaching consensus. Experts marked mappings as correct if the HPO term was either an exact match or, when the original term did not exist in HPO, represented the closest semantically appropriate class available in the ontology.

Of the 980 mappings being evaluated, 948 were judged correct (overall accuracy of **96.7%**, 32 incorrect mappings). This accuracy demonstrates the effectiveness of our Graph RAG methodology. 13 of the 32 incorrect mappings involved terms with multiple possible interpretations, requiring disambiguation. 10 errors occurred with newly coined terms or drug-specific terminology not represented in HPO.

5.2   Drug Target Prediction via Side Effect Similarity

Drugs with similar side effect profiles often share molecular targets [6,25]. We hypothesized that SIDEKICK's ontology-based representation would better capture mechanistic similarity compared to existing vocabularies, enabling more accurate prediction of shared drug targets.

We compared SIDEKICK against OnSIDES [34], the current state-of-the-art drug side effect database. OnSIDES has been previously shown to outperform SIDER [20] on the same task; therefore, comparison to OnSIDES represents comparison against the best available resource.

We used DrugBank 5.1.12 [38] as the ground truth for drug–target associations. For each drug, we extracted all protein targets and created drug pairs. Positive pairs were defined as drugs that share at least one protein target; negative pairs were drugs with no shared targets.

To ensure a fair comparison, we identified drugs present in all three datasets: SIDEKICK, OnSIDES (US labels only), and DrugBank. This yielded 1,105 matched drugs, resulting in 13,983 positive pairs (shared targets) and 595,977 negative pairs (no shared targets). We applied identical semantic similarity measures to both datasets. Both SIDEKICK (HPO-based) and OnSIDES (MedDRA-based) were evaluated using Best Match Average (BMA) [14] with Resnik semantic similarity [32].

Resnik similarity uses ontological hierarchical structure and information content:

$$\text{Resnik}(t_1, t_2) = \max_{t \in \text{CA}(t_1, t_2)} \text{IC}(t) \tag{1}$$

where $\text{CA}(t_1, t_2)$ represents the common ancestors of terms t_1 and t_2, and $\text{IC}(t)$ is the information content based on term frequency in the dataset. BMA combines Resnik scores bidirectionally:

$$\text{BMA}(D_1, D_2) = \frac{1}{2} \left(\frac{\sum_{t_1 \in D_1} \max_{t_2 \in D_2} \text{Res}(t_1, t_2)}{|D_1|} + \frac{\sum_{t_2 \in D_2} \max_{t_1 \in D_1} \text{Res}(t_1, t_2)}{|D_2|} \right) \tag{2}$$

For OnSIDES, we constructed the MedDRA hierarchy from *Preferred Terms, High Level Terms, High Level Group Terms*, and *System Organ Class* levels. OnSIDES MedDRA codes were normalized to Preferred Terms, and Resnik similarity was computed over the MedDRA hierarchy using the same information content calculation approach as SIDEKICK.

We computed similarity scores for all drug pairs and evaluated classification performance using the Area Under the ROC (AUC-ROC), with higher AUC indicating better discrimination between positive and negative pairs. Table 2 present the comparative results. SIDEKICK achieved an AUC of 0.7174, representing 8.5% relative improvement over OnSIDES (AUC 0.6612). The superior performance of SIDEKICK is evident in the discrimination between positive and negative pairs. SIDEKICK achieves a mean similarity difference (Δ) of 0.1711, indicating 14% better separation of true drug pairs. This demonstrates that HPO's ontology structure captures mechanistically relevant patterns more effectively than MedDRA's hierarchy.

Table 2. Drug target prediction performance comparing SIDEKICK (HPO) and OnSIDES (MedDRA) using identical BMA+Resnik similarity. Mean(+) and Mean(−) represent average similarity scores for positive and negative pairs respectively. Δ shows the discrimination between classes.

Method	AUC-ROC	Mean (+)	Mean (−)	Δ
OnSIDES (MedDRA+BMA+Resnik)	0.6612	0.4412	0.2700	0.1711
SIDEKICK (HPO+BMA+Resnik)	**0.7174**	**0.5935**	**0.3978**	**0.1957**

Discussion. The performance difference between SIDEKICK and OnSIDES, when both use identical similarity methodology, demonstrates that ontology design fundamentally impacts pharmacological applications. Although MedDRA possesses a hierarchical structure, it was designed for regulatory adverse event reporting with broad categorical groupings optimized for safety surveillance rather than mechanistic reasoning. In contrast, HPO was designed to ontologically structure the domain of abnormal phenotypes in humans [17], with axioms that explicitly relate to potential mechanisms and anatomical structures affected in the abnormality.

Our results validate both the quality of our HPO mappings and the utility of HPO as a representational framework for adverse drug reactions. SIDEKICK's improved performance for finding drugs with shared mechanisms of action also shows it is useful for generating hypotheses for new therapeutic indications, and can provide a foundation for computational pharmacovigilance systems that reason over drug safety profiles.

5.3   Competency Question Evaluation

To further validate SIDEKICK, we executed a series of Competency Questions (CQs) that require traversing ontological hierarchies, and answered them using SPARQL queries. We evaluated two forms of reasoning: (1) hierarchical traversal over the transitive closure of `rdfs:subClassOf` relations in HPO and MONDO, and (2) federated reasoning over ontology axioms that queries the logical definitions of phenotypes and anatomical parthood relations stored in ontology repositories such as Ubergraph[6]. We formulated questions that represent real-world pharmacovigilance scenarios where a clinician searches for a general class of adverse events (e.g., "cardiovascular abnormalities") rather than enumerating hundreds of specific symptoms (*e.g.*, "atrial fibrillation", "myocardial infarction", "bradycardia").

Table 3 summarizes the results. The evaluation demonstrates several reasoning capabilities. For anatomical subsumption, the query for "Abnormality of the cardiovascular system" (HP:0001626) successfully retrieved 1,903 unique drugs. These drugs were rarely annotated with this general term; rather, they were linked to specific descendants such as "Hypertension" or "Myocardial infarction" via the inference $Drug \rightarrow SpecificPhenotype \xrightarrow{is\_a} ParentPhenotype$. Regarding physiological grouping, the query for "Arrhythmia" (HP:0011675) retrieves drugs based on physiological mechanisms rather than gross anatomy, aggregating disjoint subtypes (e.g., Tachycardia, Bradycardia) into a single axiomatic group without requiring manual enumeration. We also demonstrated contextual safety reasoning by constraining the relation type to *contraindication*, successfully identifying drugs unsafe for patients with renal conditions. This highlights the graph's ability to distinguish between drugs that *cause* a phenotype and those that are *contraindicated* by it using the same vocabulary. We further utilized cross-ontology inference to query drugs indicated for "Infectious Disease" (MONDO:0005550), leveraging the MONDO hierarchy to aggregate drugs for specific infections.

Table 3. Results of Competency Questions (CQs) demonstrating ontology-based reasoning. "Unique Drugs" indicates the number of distinct drug formulations identified.

Competency Question	Reasoning Type	Unique Drugs
Find all drugs with any cardiovascular side effect	Anatomical Hierarchy	1,903
Find all drugs with any nervous system side effect	Anatomical Hierarchy	1,907
Find drugs affecting cardiac rhythm (Arrhythmia)	Physiological Grouping	1,216
Find drugs with metabolic side effects	Multi-level Hierarchy	1,915
Find drugs contraindicated in kidney conditions	Cross-System Safety	184
Find drugs indicated for any infectious disease	Disease Taxonomy	434
Find drugs affecting specific anatomical parts of heart	Federated Axiomatic	504

Finally, we performed a federated query across SIDEKICK and Ubergraph[7] to demonstrate federated axiomatic reasoning, identifying drugs affecting any anatomical part of the heart. This query utilizes the logical axioms of HPO, which define phenotypes in terms of the entities they affect (via UPHENO:0000001), and the transitive parthood relations (BFO:0000050) in the UBERON anatomy ontology. This allows SIDEKICK to retrieve drugs causing

---

abnormalities in specific heart structures (e.g., heart valves, ventricles) based on implicit biological knowledge not present in the simple class hierarchy. The federated query identified 504 drugs associated with abnormalities in 40 distinct cardiac anatomical structures, including myocardium, cardiac valves, and coronary vessels, yielding 2,676 drug-structure associations.

6   Conclusion

We developed SIDEKICK, a knowledge graph of drug safety information that addresses limitations in existing pharmacovigilance resources through ontology-based standardization. By mapping drug side effects to HPO, disease terms to MONDO, and drug interactions to RxNorm, SIDEKICK provides the semantic richness essential for computational reasoning in modern biomedical applications. Comparative evaluation against the state of the art OnSIDES database demonstrates the practical impact of ontology-based representation; SIDEKICK can improve the performance in predicting shared drug targets based on side effects (ROC AUC increase of 8.5%). The knowledge graph, containing information extracted from over 50,000 drug labels, is serialized as RDF using the Semanticscience Integrated Ontology as upper level ontology. It will interact with the Semantic Web ecosystem and can be used to answer complex and federated SPARQL queries.

6.1   Limitations and Future Work

Our mapping methodology achieved high accuracy; however, few ambiguous terms and newly coined drug-specific terminology necessitated manual disambiguation. Furthermore, we prioritized the extraction of binary clinical associations; consequently, the current release lacks the quantitative frequency data for adverse reactions available in databases such as SIDER [20]. We also observed challenges in normalizing terms related to patient populations within contraindications to standard ontologies, which limits the ability to reason over complex demographic exclusions. Regarding semantic validation, the computational complexity of OWL 2 DL hindered full validation of SIO axioms, necessitating a restriction to the EL subset using the ELK reasoner. Finally, the resource currently relies exclusively on FDA Structured Product Labels; therefore, it lacks complementary signals for rare adverse events often derived from post-marketing surveillance data, electronic health records, or biomedical literature.

Future work prioritizes integrating genomic data via HPO's disease–gene annotations to further improve pharmacogenomic applications. Additionally, we aim to incorporate clinical trial databases to validate drug repurposing hypotheses generated through phenotype-based similarity.

6.2   Availability

SIDEKICK is freely available under the CC BY 4.0 International license. We deposited the complete dataset at Zenodo (DOI: 10.5281/zenodo.17779317), and intend to update the dataset yearly. Users can access the knowledge graph via a web interface at https://sidekick.bio2vec.net/ or query it programmatically through the SPARQL endpoint at https://sidekick.bio2vec.net/sparql, which includes example queries. We provide source code and documentation at https://github.com/bio-ontology-research-group/sidekick/.

A   LLM Extraction Prompt

The following prompt was used with Google Gemini 2.5 Flash to extract clinical entities from FDA Structured Product Labels:

```
You are an expert in extracting information from FDA drug
labels. I have provided the text from a drug package label.
Please extract the following information in a structured
format:
IMPORTANT: Only respond with the extracted XML. Do not repeat
any part of these instructions or the input text in your
response.
1. Indications (what the drug is used for)
2. Contraindications (when the drug should not be used)
3. Side effects (with frequencies if available)
For each indication, contraindication, side effect, provide
only one item per tag and include the exact line from the
text that contains this information. Extract any and all
indications, contraindications, side effects you find.
It is important to note that these side effects, indications
and contraindications can be found in sections other than the
ones
specifically dedicated for them so search carefully across
the entire text and find all of them.
Try to keep the indication, contraindication and side-effect
names that you extract as short and straightforward as
possible but accuracy is important.
Provide your response in the following XML format:

<drug_information>
  <indications>
    <indication>
      <indication_name>INDICATION NAME</indication_name>
    </indication>
  </indications>
  <contraindications>
    <contraindication>
      <contraindication_name>CONTRAINDICATION NAME
      </contraindication_name>
    </contraindication>
```

```
    </contraindications>
    <side_effects>
      <side_effect>
        <side_effect_name>SIDE EFFECT NAME</side_effect_name>
      </side_effect>
    </side_effects>
  </drug_information>
```

B   Excluded LOINC Section Codes

During SPL text preprocessing, the following LOINC section codes were excluded
as they do not contain clinically relevant information for extraction (Table 4):

Table 4. LOINC section codes excluded from SPL text preprocessing

LOINC Code	Section Description
51945-4	Package Label - Principal Display Panel
48780-1	SPL Product Data Elements Section
44425-7	Storage and Handling Section
34069-5	How Supplied Section
34093-5	References Section
59845-8	Instructions for Use Section
42230-3	SPL Patient Package Insert Section
42231-1	SPL MedGuide Section
53413-1	OTC - Questions Section
50565-1	OTC - Keep Out of Reach of Children Section
34076-0	Information for Patients Section
55106-9	OTC - Active Ingredient Section
50569-3	OTC - Ask Doctor Section
50568-5	OTC - Ask Doctor/Pharmacist Section
34068-7	Dosage & Administration Section
51727-6	Inactive Ingredient Section
55105-1	OTC - Purpose Section
50570-1	OTC - Do Not Use Section
50567-7	OTC - When Using Section
50566-9	OTC - Stop Use Section
53414-9	OTC - Pregnancy or Breast Feeding Section
34090-1	Clinical Pharmacology
43682-4	Pharmacokinetics
43679-0	Mechanism of Action
43681-6	Pharmacodynamics

C   Entity Classification Categories

Terms extracted from indication and contraindication sections were classified into seven categories using LLM-based classification:

1. **Disease**: Medical conditions, disorders, syndromes (e.g., "diabetes mellitus", "hypertension", "myocardial infarction")
2. **Phenotype**: Observable clinical signs, symptoms, or abnormalities (e.g., "seizures", "hypotension", "bradycardia")
3. **Drug or Chemical**: Drug interactions, concomitant medications, or chemical substances (e.g., "monoamine oxidase inhibitors", "warfarin", "alcohol")
4. **Allergy or Hypersensitivity**: Allergic reactions or hypersensitivity conditions. Terms automatically classified if containing keywords: "hypersensitivity", "allergic", or "anaphylaxis"
5. **Patient Population**: Demographics, life stages, or patient groups (e.g., "pregnancy", "pediatric patients", "elderly")
6. **Procedure**: Medical or surgical procedures (e.g., "surgery", "hemodialysis", "cardiac catheterization")
7. **Other**: Any terms that do not fit into the above categories

Each category was mapped to appropriate standardized vocabularies: *Disease* terms to MONDO Disease Ontology, *Phenotype* terms to Human Phenotype Ontology (HPO), and *Drug or Chemical* terms to RxNorm The unclassified terms were excluded in the final Knowledge Graph.

D   API Configuration Parameters

Table 5 summarizes the hyperparameters used for the Information extraction and Graph RAG ontology mapping process.

Table 5. API configuration parameters for information extraction and ontology mapping

Parameter	Value
Information Extraction	
API Provider	OpenRouter
Model	google/gemini-2.5-flash
Temperature	0.1
Max Tokens	50,000
Batch Size	500 documents
Sleep Between Batches	10 s
Max Retries	3
Retry Delay	5 s
Ontology Mapping (Graph-RAG)	
API Provider	OpenRouter
Model	google/gemini-2.5-flash
Temperature	0.1
Max Tokens	500 (mapping), 1,000 (side effects)
Batch Size	1 term
Max Retries	3
Retry Delay	2 s
Embedding Model	all-MiniLM-L6-v2
Top-K Semantic Matches	10
Top-K Graph Nodes	15

E   Example SPARQL Queries

This section provides representative SPARQL queries demonstrating SIDE-KICK's ontology-based reasoning capabilities.

E.1   Query 1: Cardiovascular Side Effects

This query demonstrates hierarchical reasoning by finding all drugs with any cardiovascular-related side effects through transitive traversal of the HPO hierarchy:

```
PREFIX sio: <http://semanticscience.org/resource/>
PREFIX sk: <http://sidekick.bio2vec.net/>
PREFIX obo: <http://purl.obolibrary.org/obo/>
PREFIX rdfs: <http://www.w3.org/2000/01/rdf-schema#>

SELECT DISTINCT ?drug_name ?phenotype_label ?phenotype_id
WHERE {
```

```
    # Root term: Abnormality of the cardiovascular system
    BIND(obo:HP_0001626 AS ?cardiac_root)

    # Find all phenotypes that are subclasses (transitive)
    ?phenotype_id rdfs:subClassOf* ?cardiac_root .
    ?phenotype_id rdfs:label ?phenotype_label .

    # Find drugs with these side effects
    ?assoc a sk:SideEffect ;
           sk:refersToDrug ?collection ;
           sk:hasSideEffect ?phenotype_id .

    ?collection rdfs:label ?drug_name .
}
ORDER BY ?drug_name ?phenotype_label
```

Result: Identified 1,903 unique drugs with cardiovascular-related side effects, automatically aggregating specific conditions (e.g., hypertension, myocardial infarction, arrhythmia) without manual enumeration.

E.2   Query 2: Renal Contraindications

This query demonstrates cross-system reasoning by identifying drugs contraindicated in any kidney-related condition (combining phenotypes and diseases):

```
PREFIX sio: <http://semanticscience.org/resource/>
PREFIX sk: <http://sidekick.bio2vec.net/>
PREFIX obo: <http://purl.obolibrary.org/obo/>
PREFIX rdfs: <http://www.w3.org/2000/01/rdf-schema#>

SELECT DISTINCT ?drug_name ?contraindication_type ?
    phenotype_label
WHERE {
    # Root term: Abnormality of the kidney
    BIND(obo:HP_0000077 AS ?kidney_root)

    # Find kidney-related phenotypes
    ?phenotype_id rdfs:subClassOf* ?kidney_root .
    ?phenotype_id rdfs:label ?phenotype_label .

    # Phenotype contraindications
    {
        ?assoc a sk:PhenotypeContraindication ;
               sk:refersToDrug ?collection ;
               sk:isContraindicatedInPhenotype ?phenotype_id
               .
        BIND("Phenotype" AS ?contraindication_type)
    }
```

```
    UNION
    {
        # Disease contraindications with "renal" or "kidney"
        ?assoc a sk:DiseaseContraindication ;
               sk:refersToDrug ?collection ;
               sk:isContraindicatedInDisease ?disease_id .
        BIND("Disease" AS ?contraindication_type)
        ?disease_id rdfs:label ?phenotype_label .
        FILTER(CONTAINS(LCASE(?phenotype_label), "renal") ||
               CONTAINS(LCASE(?phenotype_label), "kidney"))
    }

    ?collection rdfs:label ?drug_name .
}
ORDER BY ?drug_name
```

Result: Identified 184 drugs contraindicated in kidney-related conditions, demonstrating the ability to distinguish drugs that *cause* kidney problems from those that are *contraindicated* in existing kidney disease.

E.3   Query 3: Infectious Disease Indications

This query demonstrates disease taxonomy reasoning using the MONDO Disease Ontology:

```
PREFIX sio: <http://semanticscience.org/resource/>
PREFIX sk: <http://sidekick.bio2vec.net/>
PREFIX obo: <http://purl.obolibrary.org/obo/>
PREFIX rdfs: <http://www.w3.org/2000/01/rdf-schema#>

SELECT DISTINCT ?drug_name ?disease_label ?disease_id
WHERE {
    # Root term: Infectious disease (MONDO)
    BIND(obo:MONDO_0005550 AS ?infectious_root)

    # Find all infectious diseases via hierarchy
    ?disease_id rdfs:subClassOf* ?infectious_root .
    ?disease_id rdfs:label ?disease_label .

    # Find drugs indicated for these diseases
    ?assoc a sk:DiseaseIndication ;
           sk:refersToDrug ?collection ;
           sk:isIndicatedForDisease ?disease_id .

    ?collection rdfs:label ?drug_name .
}
ORDER BY ?drug_name
```

Result: Identified 434 drugs indicated for infectious diseases, automatically aggregating antimicrobial, antiviral, and antifungal agents through disease taxonomy reasoning.

E.4   Query 4: Federated Axiomatic Reasoning

This query demonstrates federated reasoning across SIDEKICK and the Ubergraph knowledge graph, utilizing HPO's logical axioms that link phenotypes to anatomical structures via UBERON:

```
PREFIX sio: <http://semanticscience.org/resource/>
PREFIX sk: <http://sidekick.bio2vec.net/>
PREFIX obo: <http://purl.obolibrary.org/obo/>
PREFIX rdfs: <http://www.w3.org/2000/01/rdf-schema#>

SELECT DISTINCT ?drug_name ?phenotype_label ?heart_part_label
WHERE {
    # Local query: Find drugs and side effects
    ?assoc a sk:SideEffect ;
           sk:refersToDrug ?collection ;
           sk:hasSideEffect ?phenotype_id .

    ?collection rdfs:label ?drug_name .
    ?phenotype_id rdfs:label ?phenotype_label .

    # Federated query: Find anatomical parts
    SERVICE <https://ubergraph.apps.renci.org/sparql> {
        # What entity does this phenotype affect?
        ?phenotype_id obo:UPHENO_0000001 ?affected_entity .

        # Is it part of the heart (transitively)?
        ?affected_entity obo:BFO_0000050* obo:UBERON_0000948
        .

        # Get the anatomical part label
        ?affected_entity rdfs:label ?heart_part_label .

        FILTER(?affected_entity != obo:UBERON_0000948)
    }
}
ORDER BY ?drug_name ?heart_part_label
```

Result: Identified 504 drugs associated with abnormalities in 40 distinct cardiac anatomical structures (e.g., myocardium, cardiac valves, coronary vessels), yielding 2,676 drug-structure associations through axiomatic reasoning not present in the simple class hierarchy.

References

1. Alonso Casero, Á.: Named entity recognition and normalization in biomedical literature: a practical case in SARS-CoV-2 literature. Ph.D. thesis, ETSI_Informatica (2021)
2. Antoniou, G., Harmelen, F.: Web ontology language: OWL. In: Staab, S., Studer, R. (eds.) Handbook on Ontologies. IHIS, pp. 91–110. Springer, Heidelberg (2009). https://doi.org/10.1007/978-3-540-92673-3_4
3. Baker, T.: Maintaining dublin core as a semantic web vocabulary. In: Hemmje, M., Niederée, C., Risse, T. (eds.) From Integrated Publication and Information Systems to Information and Knowledge Environments. LNCS, vol. 3379, pp. 61–68. Springer, Heidelberg (2005). https://doi.org/10.1007/978-3-540-31842-2_7
4. Brown, E.G.: Using MedDRA: implications for risk management. Drug Saf. **27**(8), 591–602 (2004). https://doi.org/10.2165/00002018-200427080-00010
5. Brown, E.G., Wood, L., Wood, S.: The medical dictionary for regulatory activities (meddra). Drug Saf. **20**(2), 109–117 (1999)
6. Campillos, M., Kuhn, M., Gavin, A.C., Jensen, L.J., Bork, P.: Drug target identification using side-effect similarity. Science **321**(5886), 263–266 (2008). https://doi.org/10.1126/science.1158140
7. Classen, D.C., Pestotnik, S.L., Evans, R.S., Lloyd, J.F., Burke, J.P.: Adverse drug events in hospitalized patients: excess length of stay, extra costs, and attributable mortality. JAMA **277**(4), 301–306 (1997). https://doi.org/10.1001/jama.1997.03540280039031
8. Consortium, G.O.: The gene ontology resource: 20 years and still going strong. Nucleic Acids Res. **47**(D1), D330–D338 (2019)
9. Dumontier, M., et al.: The Semanticscience Integrated Ontology (SIO) for biomedical research and knowledge discovery. J. Biomed. Semant. **5**(1), 1–11 (2014). https://doi.org/10.1186/2041-1480-5-14
10. Gangemi, A., Presutti, V.: Ontology design patterns. In: Handbook on Ontologies, pp. 221–243. Springer, Heidelberg (2009)
11. Gkoutos, G.V., Schofield, P.N., Hoehndorf, R.: The anatomy of phenotype ontologies: principles, properties and applications. Brief. Bioinf. **19**(5), 1008–1021 (2018)
12. Guarino, N.: Formal ontology in information systems. In: Proceedings of the First International Conference (FOIS'98), Trento, Italy, 6–8 June 1998, vol. 46. IOS press (1998)
13. Hagberg, A., Swart, P.J., Schult, D.A.: Exploring network structure, dynamics, and function using networkx. Technical report, Los Alamos National Laboratory (LANL), Los Alamos, NM (United States) (2008)
14. Harispe, S., Ranwez, S., Montmain, J., et al.: Semantic Similarity from Natural Language and Ontology Analysis. Springer, Cham (2022)
15. Hoehndorf, R., Schofield, P.N., Gkoutos, G.V.: The role of ontologies in biological and biomedical research: a functional perspective. Brief. Bioinf. **16**(6), 1069–1080 (2015)
16. Kazakov, Y., Krötzsch, M., Simančík, F.: The incredible elk: from polynomial procedures to efficient reasoning with $\mathcal{EL}$ ontologies. J. Autom. Reason. **53**(1), 1–61 (2014)
17. Köhler, S., et al.: The human phenotype ontology in 2021. Nucleic Acids Res. **49**(D1), D1207–D1217 (2021). https://doi.org/10.1093/nar/gkaa1043
18. Krech, D., et al.: Rdflib. Zenodo (2002)

19. Krötzsch, M.: OWL 2 profiles: an introduction to lightweight ontology languages. In: Eiter, T., Krennwallner, T. (eds.) Reasoning Web 2012. LNCS, vol. 7487, pp. 112–183. Springer, Heidelberg (2012). https://doi.org/10.1007/978-3-642-33158-9_4

20. Kuhn, M., Letunic, I., Jensen, L.J., Bork, P.: The SIDER database of drugs and side effects. Nucleic Acids Res. **44**(D1), D1075–D1079 (2016). https://doi.org/10.1093/nar/gkv1075

21. Kulmanov, M., Smaili, F.Z., Gao, X., Hoehndorf, R.: Semantic similarity and machine learning with ontologies. Brief. Bioinf. **22**(4), bbaa199 (2021)

22. Mungall, C.J., Gkoutos, G.V., Smith, C.L., Haendel, M.A., Lewis, S.E., Ashburner, M.: Integrating phenotype ontologies across multiple species. Genome Biol. **11**(1), R2 (2010)

23. Mungall, C.J., Torniai, C., Gkoutos, G.V., Lewis, S.E., Haendel, M.A.: Uberon, an integrative multi-species anatomy ontology. Genome Biol. **13**(1), R5 (2012)

24. Noy, N.F., et al.: Bioportal: ontologies and integrated data resources at the click of a mouse. Nucleic Acids Res. **37**(suppl_2), W170–W173 (2009)

25. Nugent, T., Plachouras, V., Leidner, J.L.: Computational drug repositioning based on side-effects mined from social media. PeerJ Comput. Sci. **2**, e46 (2016). https://doi.org/10.7717/peerj-cs.46

26. Pan, J.Z.: Resource description framework. In: Staab, S., Studer, R. (eds.) Handbook on Ontologies. IHIS, pp. 71–90. Springer, Heidelberg (2009). https://doi.org/10.1007/978-3-540-92673-3_3

27. Pesquita, C., Faria, D., Falcao, A.O., Lord, P., Couto, F.M.: Semantic similarity in biomedical ontologies. PLoS Comput. Biol. **5**(7), e1000443 (2009)

28. Prud'hommeaux, E., Labra Gayo, J.E., Solbrig, H.: Shape expressions: an RDF validation and transformation language. In: Proceedings of the 10th International Conference on Semantic Systems, pp. 32–40 (2014)

29. Pushpakom, S., et al.: Drug repurposing: progress, challenges and recommendations. Nat. Rev. Drug Disc. **18**(1), 41–58 (2019). https://doi.org/10.1038/nrd.2018.168

30. Quinn, J.: An HL7 (health level seven) overview. J. AHIMA **70**(7), 32–4 (1999)

31. Ratcliff, J.W., Metzener, D.E., et al.: Pattern matching: the gestalt approach. Dr. Dobb's J. **13**(7), 46 (1988)

32. Resnik, P.: Using information content to evaluate semantic similarity in a taxonomy. arXiv preprint cmp-lg/9511007 (1995)

33. Smaili, F.Z., Gao, X., Hoehndorf, R.: Formal axioms in biomedical ontologies improve analysis and interpretation of associated data. Bioinformatics **36**(7), 2229–2236 (2020)

34. Tanaka, Y., et al.: OnSIDES database: extracting adverse drug events from drug labels using natural language processing models. Cell Rep. Methods **5**(4), 100698 (2025). https://doi.org/10.1016/j.medj.2025.100642

35. Tatonetti, N.P., Ye, P.P., Daneshjou, R., Altman, R.B.: Data-driven prediction of drug effects and interactions. Sci. Transl. Med. **4**(125), 125ra31–125ra31 (2012). https://doi.org/10.1126/scitranslmed.3003377

36. Vasilevsky, N.A., et al.: Mondo: unifying diseases for the world, by the world. medRxiv, pp. 2022–04 (2022). https://doi.org/10.1101/2022.04.13.22273750

37. Wilkinson, M.D., et al.: The fair guiding principles for scientific data management and stewardship. Sci. Data **3**(1), 1–9 (2016)

38. Wishart, D.S., et al.: Drugbank 5.0: a major update to the drugbank database for 2018. Nucleic Acids Res. **46**(D1), D1074–D1082 (2018)

HERTy-Wiki: A Benchmark for Hierarchical Entity Reasoning and Typing in Wikidata

Nicole Obretincheva[1]([✉]) [ID], Helen Yannakoudakis[1] [ID], and Elena Simperl[1,2] [ID]

[1] King's College London, Strand, London WC2R 2LS, UK
nicole.obretincheva@kcl.ac.uk
[2] Institute for Advanced Study, Technical University Munich, Lichtenbergstrasse 2a, D-85748 Garching, Germany

Abstract. Entity typing is central to knowledge engineering, and as large language models (LLMs) are increasingly used to support knowledge graph construction, it becomes essential to assess how reliably they can perform this task. We present HERTy-Wiki (Hierarchical Entity Reasoning and Typing in Wikidata) - a human-verified benchmark for evaluating hierarchical reasoning in LLMs for knowledge graph entity typing (KGET). Unlike existing probing benchmarks that test factual recall, HERTy-Wiki evaluates whether models can select the most specific valid type from limited contextual evidence, reflecting realistic Wikidata scenarios, where editors often work with incomplete or noisy information. HERTy-Wiki comprises 8,767 multiple-choice questions derived from 3,776 Wikidata entities across five domains, with an optional multimodal extension. Each question requires models to discriminate between semantically related types within the same subclass hierarchy. We evaluate several state-of-the-art, reasoning-oriented LLMs under zero-shot, few-shot, and chain-of-thought prompting settings and compare them against a no-context baseline to quantify model reliance on memorised priors. Across models, macro-F1 scores remain around 0.50–0.60, with only modest improvements over the no-context baseline. Our analysis indicates that the emerging hierarchical reasoning abilities of models are overshadowed by strong dependence on memorised priors. These findings highlight fundamental challenges in using LLMs for reasoning in knowledge engineering pipelines, particularly in rapidly evolving or specialised domains with limited pre-training coverage. HERTy-Wiki provides a human-verified benchmark for developing and evaluating genuine reasoning-driven approaches to KGET.

Keywords: Knowledge Graph Entity Typing · Hierarchical Reasoning · Large Language Models · Wikidata

Resource type: Evaluation benchmark
License: CC-BY 4.0
DOI: https://doi.org/10.5281/zenodo.17828539
URL: https://doi.org/10.5281/zenodo.17828539
Code: https://github.com/nobretincheva/HERTy-Wiki.

1   Introduction

With the increasing use of Large Language Models (LLMs) in knowledge graph construction [35,57,61], assessing how reliably they perform core construction tasks becomes essential. One such key task is **knowledge graph entity typing** (KGET), the task of correctly classifying entities based on a set of closely related types from a pre-defined ontology. Existing KGET approaches primarily rely on embeddings that capture structural relationships in the graph [21–23,50], and the resources supporting them are largely triple-based, with limited textual context to support LLM-based inference.

Hierarchical Reasoning, i.e. the ability to navigate hierarchical structures that are latent within language, is central to fine-grained KGET. Existing work has shown, however, that language models may struggle in understanding hierarchical relationships [11,25,44]. While prior work in knowledge engineering has evaluated LLMs on data modelling tasks, e.g. taxonomy discovery and term typing, these studies are predominantly framed as knowledge probing, treating LLMs as static knowledge bases rather than assessing their ability to reason over structured context [4,46]. Consequently, existing benchmarks do not support evaluation of the hierarchical inference required in realistic KGET settings, where information is often incomplete or noisy.

Mirroring how human editors approach entity typing, an LLM performing this task should be able to recognise valid types, navigate type hierarchies, and reason about what the available context entails. To jointly evaluate these capabilities, we introduce **HERTy-Wiki, a crowdsourced KGET benchmark** explicitly designed to test whether LLMs can navigate and reason over hierarchical type structures, as summarised in Fig. 1. We construct the benchmark from Wikidata [49], whose deep type hierarchy and broad domain coverage make it a demanding and realistic setting for assessing fine-grained KGET.

HERTy-Wiki contains 3,776 human-verified Wikidata entities spanning five domains and yields 8,767 multiple-choice questions. Our benchmark provides:

- **Controlled hierarchical reasoning tasks**, requiring models to identify the most specific valid type or generalise appropriately.
- **Self-contained minimal domain-specific context**, designed to suppress memorisation.
- **Multimodal extension**, enabling analysis of whether visual information supports hierarchical reasoning.
- **No-context baseline**, that isolates the contribution of contextual reasoning from memorised priors.

To assess the relevance of our benchmark, we evaluate several state-of-the-art reasoning-oriented LLMs under zero-shot, chain-of-thought, and in-context learning settings and compare them against the no-context baseline to quantify model reliance on memorised priors. Our analysis reveals **systematic limitations in hierarchical reasoning** across model families and underscores the need for text-based and multimodal resources to advance LLM-driven knowledge graph construction and entity typing.

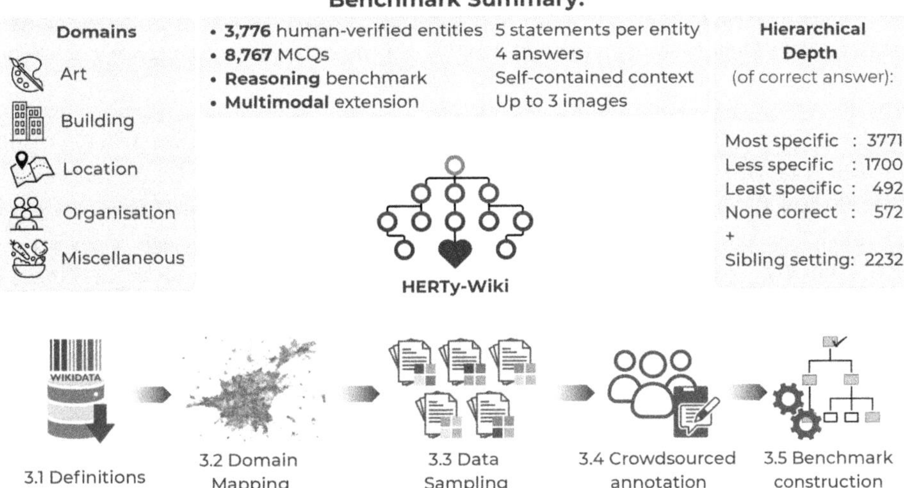

Fig. 1. Overview of the HERTy-Wiki benchmark and our construction methodology.

2   Background

2.1   Knowledge Graph Entity Typing

Knowledge graph entity typing (KGET) is a knowledge integration task with the goal of predicting the type t for an entity in an existing knowledge graph, where t is defined in the knowledge graph's ontology T. Most KGET approaches focus on utilising embeddings that capture the structural relationships within the graph [21–23,50], with recent work leveraging additional semantic information (such as the textual descriptions) [7,24].

KG entity typing works are evaluated on one of two sets of baselines: FIGER + DBpedia360k and FB15K-ET + YAGO43K-ET. FIGER and DBpedia360k are frequently chosen for fine-grained KGET tasks [7,8,23,42], however offer limited hierarchical depth and restricted domain coverage. FB15K-ET and YAGO43K-ET [24,36,50] cover a wide variety of possible types, however, they are overly specific and lack a structured hierarchy. Table 1 summarises the hierarchical depth characteristics of these datasets, highlighting the shallow hierarchies present in commonly used KGET benchmarks. For FIGER and DBpedia360k, depth statistics are computed over the reconstructed versions of the datasets, published by Biswas et al. [8]. For HERTy-Wiki, depth statistics are computed over the seed entity types used to construct the benchmark, traced through the Wikidata hierarchy from predefined domain roots.

While these resources can be verbalised for use with LLMs, they were not designed for reasoning over semantically close or hierarchically related types. Moreover, no existing KGET benchmark is based on Wikidata, despite its breadth, rich domain coverage, and large user base.

Table 1. Depth-wise comparison of commonly used KGET datasets. Columns "Max" and "Median" denote the maximum and median hierarchical depth of types in each dataset. Note that greater depths in FB15K-ET often reflect namespace or schema paths rather than meaningful subclass hierarchies.

Dataset	# Types	Max	Median	Example
FIGER [8,23,54]	102	2	2	organization → airline
DBpedia360k [8,23,59]	14 coarse / 37 fine	5	3	Athlete → Person → Agent
FB15K-ET [9,36]	3,584	8	3	/base/magic/magician
YAGO43K-ET [30,36]	45,182	-	-	Football_clubs_in_Ghana
HERTy-Wiki	**3,776**	**17**	**7**	Felidae → cat → working cat

2.2   LLMs in the Context of Data Modelling

Recent work has explored the role of LLMs in ontology-related tasks, often through a knowledge probing lens. Giglou et al. [4] introduce LLMs4OL, a framework evaluating the emergent ontology learning capabilities of LLMs using prompt-based language modelling. Sun et al. [46] continue this line of work with TaxoGlimpse, a benchmark on hierarchical structure discovery that compares LLM performance at different taxonomic depths. Notably, both works rely solely on label-level probing with no additional context.

Mai et al. [29] investigate whether LLMs can perform ontology learning tasks over perturbed data by removing surface lexical cues while preserving structural relationships. Their findings suggest that off-the-shelf models rely heavily on lexical familiarity, and only show signs of structural reasoning after fine-tuning.

In contrast, our work addresses a different challenge: reasoning under real-world ambiguity and sparse but realistic context. Rather than testing knowledge recall or removing lexical cues, we assess whether LLMs can disambiguate fine-grained types in noisy, hierarchical settings like Wikidata.

2.3   Wikidata and Data Modelling Challenges

Wikidata [49] is a collaborative, open-source knowledge graph containing over 110 million entities [51] across a wide range of domains. Its taxonomy is structured via two key properties: "instance of" (P31), which links entities to their types (e.g. Paris "instance of" city), and "subclass of" (P279), which expresses a taxonomic relationship between classes (e.g. city "subclass of" human settlement).

Unlike traditional knowledge graphs, Wikidata does not adhere to a strict, predefined ontology, which can introduce challenges in maintaining consistency and accuracy [3,10,12,39,43]. These problems can lead to classification errors, issues with automated inferencing and anti-patterns that undermine the taxonomy's structure.

Despite the importance of these challenges, few works have explored the use of LLMs for data modelling in the case of Wikidata. Notably, Peng et al. [37]

explore LLM-based taxonomy refinement, whilst Erenrich [14] introduces Psychiq, a fine-tuned model for entity typing that focuses on the most frequent entity types in Wikidata. To date, a controlled evaluation setting for assessing whether LLMs can reason over Wikidata's heterogeneous, noisy, and often irregular type structures is yet to be curated.

3   The HERTy-Wiki Benchmark

HERTy-Wiki contains 3,776 human-verified Wikidata entities across five balanced domains, resulting in 8,767 multiple-choice questions (MCQs). Each question is self-sufficient, providing all necessary context for its solution, and offers hierarchically related and thus semantically close distractors. By restricting the original Wikidata context prior to human annotation, the assigned type may become less specific or invalid, creating settings where models must genuinely reason rather than rely on memorised associations.

Figure 1 provides a general summary of HERTy-Wiki as well as an overview of our construction methodology.

3.1   Definitions

We define an entity type as any Wikidata entity that appears as the value of an "instance of" statement (P31), and is part of Wikidata's taxonomy via at least one "subclass of" (P279) relation. Applying this definition to a January 2024 dump yields 85,384 candidate types. Preliminary analysis showed that the distribution is long-tailed with roughly two-thirds of these entity types having been used fewer than 10 times.

3.2   Domain Selection

We acknowledge that certain domains in Wikidata may be easier to type or may attract more experienced contributors. To capture this complexity while avoiding a potentially biased manual domain selection, we cluster candidate types via topic modelling.

As topic modelling requires textual input, we first verbalise each candidate type by concatenating all of its Wikidata statements into a short textual representation. These representations are embedded using all-mpnet-base-v2[1] and clustered using BERTopic [18], which groups semantically similar text via sentence embeddings and DBSCAN, a density-based clustering algorithm.

To ground our dataset design in established entity typing research, we map the identified topics to the FIGER domains - an entity recognition dataset and label schema [27], that is widely adopted in both fine-grained entity typing [31, 34,41] and KGET research [7,55,56].

---

[1] all-mpnet-base-v2. Sentence Transformers model. Accessed via huggingface.

We only consider FIGER domains which are well-represented in our topic clusters and that exhibit sufficiently deep type hierarchies. This yields four primary domains: **Art, Building, Location, Organisation**, plus a **Miscellaneous** domain comprising large topic clusters that did not map neatly to a single FIGER class: Vehicles, Food, Clothing and Furniture, and Computers.

We map topics to domains in two passes: we perform (i) **keyword matching** against high-level FIGER domain names and their subtypes, and (ii) **semantic neighbourhood expansion** by examining a visualisation of the embedding space and adding adjacent topics that lack explicit keywords.

We exclude two FIGER domains from this release, namely **Person** and **Event**. We omit **Person** as all individuals in Wikidata are modelled as instances of "human" (Q5) - including this category could therefore risk violating established modelling conventions and reduce the alignment of our benchmark with Wikidata's schema. **Event** was not prominently represented in our generated topics and showed comparatively shallow hierarchical structure in preliminary analysis.

Ontology-Based Filtering. Topic modelling alone cannot fully resolve domain membership. For example, "music band" may be semantically clustered in the Art domain, but is structurally an Organisation, as it represents a collective of individuals.

To align our topic-based domains with Wikidata's ontology, we filter each domain by identifying a set of domain-specific root types and retaining only types that descend from them. We identify these root types by examining the top 10 most frequently used types per domain and following their superclass chains to appropriate high-level ancestors. When an unusually large portion of types is filtered out, we manually inspect the domain to ensure that no important root was missed.

Notably, two of our 30 domain roots correspond to meta-classes in Wikidata (e.g. genre). Following Wikidata practice, we treat these as valid nodes in the subclass hierarchy, rather than enforcing a strict separation between classes and meta-classes.

This step only removes structurally inconsistent assignments (e.g. filtering "music band" out of Art) and does not automatically place such types in their correct domain. Thus, we additionally expand each domain with manually selected topics from other domains that we expect may contain relevant types.

3.3   Data Collection

To ensure the high quality of data present in our benchmark, we implement a number of heuristic checks in the sampling process itself. We begin by retaining only domain-specific types that have both an English label and description.

Sampling Representative Entity per Type. We adopt an initial oversampling strategy, retrieving up to 100 entities per type (or however many are available). Entities that lack either an English label or description are discarded.

Wikidata descriptions frequently restate an entity's type, in line with common modelling guidelines. This can risk trivialising our task by making the correct answer recoverable through lexical overlap alone. To avoid such trivial cases, we prioritise entities with high-quality descriptions defined as descriptions that satisfy the following:

- **Minimum description length**:

$$N = \max(\mu_t - \ell_t,\ 1.5 \times \ell_t)$$

  where: μ_t is the mean description length over all candidate entities for this type and ℓ_t is type label length.
- **No type leakage**: using a sliding-window Ratcliff-Obershelp string similarity check matching the type label length, we exclude descriptions that surpass a 0.9 similarity threshold. We adopt this conservative lexical similarity measure as semantic similarity methods risk filtering legitimate samples by conflating leakage with valid semantic cues.

To preserve coverage over rare types, we relax these constraints progressively if needed (first allowing shorter descriptions without leakage, then allowing leakage with sufficient length). If no valid entity remains, the type is discarded. Where an entity appears under multiple types, we keep it only for the rarest type to better represent the long-tail distribution of Wikidata.

All heuristics and thresholds were tuned empirically to balance coverage and quality.

Selecting Domain-Specific Statements. For each sampled entity, we extract all Wikidata statements whose values fall into modalities we can process - images, text, geographical coordinates, links, numerical values (including quantities and dates), and other Wikidata entities (represented via their English labels and descriptions). Any entity with fewer than five usable statements is removed.

We additionally pre-process coordinates and images. Coordinate statements are reverse-geocoded into readable locations. For the Location domain, where types may depend on geographic form, we instead generate static map visualisations via OpenStreetMap's API [33]. Images are sourced from Wikimedia Commons, filtered for permissive licences compatible with our CC-BY-4.0 release, and passed through a content-safety check to remove sensitive or restricted material.

Although images carry limited signal for fine-grained typing [28], they still aid annotators by making the task less demanding and support our multimodal extension. We prioritise including up to two image-based statements, although additional images may still appear through random sampling.

We sample five statements per entity (domain-relevant where possible) to form a consistent reasoning context for both annotation and model evaluation. When fewer than five domain-relevant statements are available, the remaining statements are sampled randomly. Domain relevance is ensured by manually selecting a set of properties per domain, based on existing Wikidata entity schemas [52] and a small manual inspection of representative entities per topic.

By standardising the number of statements, we avoid information imbalance and reduce cognitive load for annotators. Additionally, this can suppress trivial memorisation by making the original Wikidata type unrecoverable from the reduced context, forcing models to infer a valid type rather than rely on memorised associations.

Final Entity Sample. Following these data filtering efforts, we are left with a minimum of 5,000 entities per domain. We sample **756 entities per domain** to ensure sufficient coverage per domain whilst also fitting annotation budget constraints and enabling even annotator batch sizes. The sampling process is two-fold. We select one representative entity per type based on a tuple scoring function: (number of domain-specific properties, description length). Entities are sorted lexicographically by this tuple, prioritising higher property coverage. We balance representation across topics through proportional random sampling. If the target of 756 samples is not met after this pass, the remaining entities are pooled and sampled at random.

3.4   Crowdsourced Annotation

To verify the correctness of Wikidata types and to account for potential shifts in type granularity caused by our statement selection process, we conduct a crowdsourced validation study on Prolific [40]. We recruit fluent English speakers with ≥ 750 completed tasks, a $\geq 90\%$ approval rate, and a self-reported preference for logic-based tasks.

Annotation Protocol and Interface. An annotation task contains 35 entities, and has an expected completion time of 30 min. Annotators are given the entity's label, description, five statements, and the assigned type with its description. They are asked (i) Is the proposed type correct given provided context and common sense, and (ii) If not, which is the most specific fallback type, if any, that fits. The fallback types are direct parents and grandparents of the proposed type and are shown in a tree-style drop-down to reduce cognitive load in hierarchical decisions, following literature on crowdsourcing interface design [45].

In total, we collect annotations for 3,780 entities and pay annotators the UK minimum wage (£12.21/hr). The total annotation cost was £1,979 paid to workers (£2,771.5 including platform fees and VAT). To mitigate cognitive overload and ensure task clarity, we assign entities from a single domain per task when possible. We include three attention checks and an onboarding example to screen for inattentive responders and ensure task is communicated clearly. Annotations are collected via Argilla [13], for which we implement a custom UI supporting hierarchical fallback selection.

Label Aggregation. Each entity is annotated independently by three crowdworkers. To validate our aggregation strategy, we provide 1,080 expert annotations (10 per annotation group) as gold-standard data.

We aggregate responses using behaviour-weighted majority voting. Because most proposed types are correct, No labels are rarer and more prone to noise.

Table 2. Aggregation outcomes and quality metrics. Strict routing criteria ensure high precision on the minority (*No*) class. Gold evaluation uses 1,080 expert-labelled items.

Metric Category	Metric	Value
Decision distribution		
Auto-accepted items	Count	2,165 (1,694 Yes / 471 No)
Expert-routed items	Count	1,615
Gold evaluation (n=1,080)		
Yes class	Precision/Recall	P=0.88, R=0.58
No class	Precision/Recall	P=0.50, R=0.84
MCC	Excl. routed items	0.62

As they determine the quality of our negative cases, we apply stricter reliability checks. For each batch, we compute Positive and Negative Agreement (PA/NA) and Gwet's AC1 [19,20], and use these metrics to scale annotator weights, identify low-effort behaviour, and detect Yes/No spamming.

An item is labelled Yes when the weighted Yes-score exceeds a reliability-adjusted threshold with no substantial disagreement. It is labelled No when the weighted No-score dominates and is supported by either at least two No votes or a single high-confidence No. If No voters propose fallback types, we accept a fallback type if at least two annotators agree, or when a single high-confidence annotator proposes it without any opposing Yes votes. All remaining cases, including low-confidence batches and disagreements about fallback types, are routed for expert adjudication. We provide an overview of our aggregation outcomes in Table 2.

3.5   Benchmark Construction

Following aggregation, each entity is assigned the most specific type that can be reliably inferred from its restricted textual context. This yields 2,358 Yes annotations, 1,188 No annotations with a fallback type, and 234 No annotations without a fallback. Four annotations were found to be logically inconsistent with annotators disagreeing with the provided type, but also selecting it as a fallback. These were therefore removed, resulting in a final set of 3,776 entities.

Answer Set Construction. Each benchmark question requires a multiple-choice answer set consisting of the verified type, two hierarchically related distractors, and a "None of the types listed" option. To maintain consistency with the annotation setup, we draw distractors exclusively from the parent and grandparent types presented to annotators as fallback options.

Depending on the granularity of the verified type, we generate answer sets using one of three strategies:

1. **Verified type is the most specific**: When the original Wikidata type remains valid, we construct one answer set per parent type. This captures

possible alternative parent-level interpretations and prevents this predominant case from dominating the benchmark.

2. **Verified type is less specific than the original**: When annotators select a fallback type, we generate all hierarchical paths that include the verified type. Each path results in a separate answer set.
3. **No type is correct**: When neither the original nor any fallback type applies, we generate all possible answer sets and specify "None of the types listed" as correct. These cases provide a strong adversarial test against memorisation, as models may recognise the entity but must still conclude that none of the listed types fit the restricted context.

We additionally construct sibling sets to assess difficulty when all options lie at the same granularity. These are created only for types verified as most specific by choosing three random siblings sharing the same parent. This serves as a baseline for comparison with the hierarchical setting.

Our construction strategy resulted in 3,771 questions where the most specific type is correct, 1700 where the less specific type is correct, 492 where the least specific type is correct, and 572 where none of the listed types is correct. Our sibling setting resulted in 2232 questions.

Problem Formulation. Each benchmark item is a multiple-choice question (MCQ) containing a brief task introduction, the entity's label, description, and five curated statements, as well as a pre-shuffled answer set, with definitions for all referenced entities and candidate answer types. An example full prompt is provided in Fig. 2.

Explicitly asking for the most specific valid type frames the task as a form of natural language inference: the model must determine which types are supported by the context and identify the deepest applicable one. The inclusion of definitions for all entities present, both in the statements and among the candidate answer types, makes every question logically self-contained, ensuring the task can be solved through logic alone, rather than through knowledge probing or memorised associations. Semantically close distractors enforce this further, requiring the model to reason over the given evidence rather than to rely on surface similarity.

4   Evaluation Framework

4.1   Metrics

Given the high class imbalance, we report macro-F1 as our primary evaluation metric.

Despite our curation strategy, which focuses on rarely used types and the reduction of surface cues, there is still a risk that models may exploit memorised associations, as many Wikidata entities appear in Wikipedia, a common pre-training source.

To estimate the degree of such contamination we include a no-context baseline, following the design of existing reasoning-oriented benchmarks [5, 15, 48].

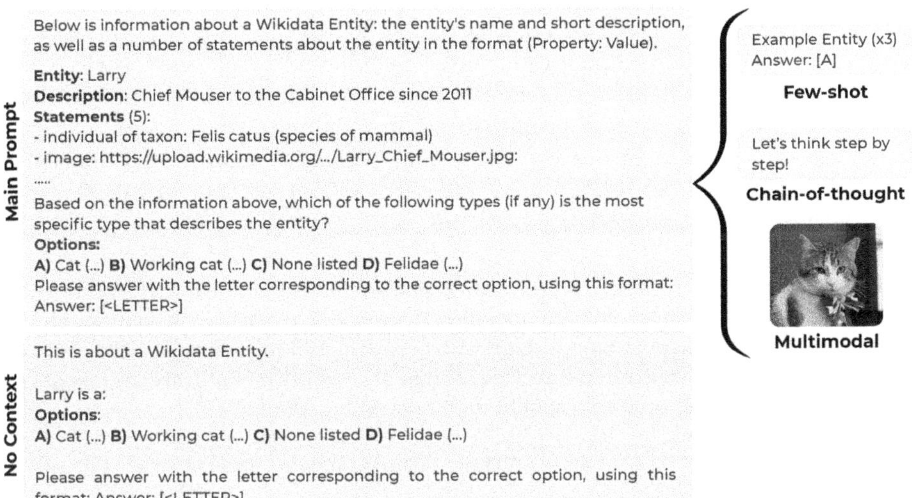

Fig. 2. Example of the prompt settings and the no-context prompt used in HERTy-Wiki. (...) indicates that entities and entity types are accompanied by definitions in brackets, omitted here for space. Entity descriptions and statements are drawn from Wikidata (Photo "Larry, Chief Mouser to the Cabinet Office", His Majesty's Government, reproduced under the Open Government Licence v1.0. Available at https://commons.wikimedia.org/wiki/File:Larry_Chief_Mouser.jpg. Last accessed 10 March 2026.).

In this setting, the model is provided with only the entity label, an indication that this entity comes from Wikidata, and the answer set, with no additional supporting context. The model is then asked to choose one of the four answers in a classification-style setup (see Fig. 2). The minimal amount of information forces the model to rely on its pre-training knowledge or to choose an answer at random.

Similarly to Bean et al. [5], we report the difference (Δ) between the full-context and no-context settings. A higher Δ indicates a better capability of the model to reason over the provided context rather than relying solely on memorised knowledge.

4.2   Models

We evaluate five open-weight, reasoning-oriented models: Llama3.1 8B and 70B [17], Gemma-3 12B [47], Phi-4 mini [1] and GPT-OSS-20B [2], as well as a closed-source model, GPT5-mini [32]. We run multimodal experiments on Gemma-3 12B and GPT5-mini as well as on Phi-4-multimodal [1], a multimodal version of Phi-4-mini with added vision adapters. We utilise four prompting settings: the aforementioned no-context setting, zero-shot, few-shot, and chain-of-thought (CoT). The few-shot setting uses three expert-written demonstrations with each correct answer covering a different granularity.

Our experiments were run on a single A100 with 80 GB VRAM, and no sampling to retain deterministic output. The Llama3.1 70B model was quantised for inference to fit within memory constraints. All batching configurations, quantisation settings, system prompts, and approximate runtime statistics per model are documented in our accompanying GitHub repository. Evaluation of the closed-source model (GPT5-mini) incurred $43.78 in total API costs across all experimental and preliminary test runs.

Table 3. Macro-F1 scores across prompting settings for hierarchical and sibling tasks. Δ denotes improvement over the no-context baseline. For each task and prompting setting, the best-performing model (highest macro-F1) and highest non-negative Δ are shown in **bold**. In the multimodal section, scores marked with † indicate improvements over the corresponding text-only model under the same setting.

Model	Task	0-shot	Δ	Few-shot	Δ	CoT	Δ
Text-only input							
Gemma-3 12B	hierarchical	0.371	0.123	0.525	0.277	0.413	0.165
	sibling	0.848	0.327	0.938	0.417	0.879	0.357
GPT-OSS-20B	hierarchical	0.558	0.135	0.570	0.148	0.580	0.157
	sibling	0.948	0.057	0.943	0.052	0.952	0.060
Phi-4 mini	hierarchical	0.559	0.166	0.503	0.110	0.509	0.115
	sibling	0.938	0.107	0.933	0.102	0.921	0.090
Llama3.1 8B	hierarchical	0.533	0.123	0.515	0.147	0.501	0.134
	sibling	0.948	0.054	0.909	0.037	0.940	0.068
Llama3.1 70B	hierarchical	0.594	0.185	0.589	0.179	0.590	0.180
	sibling	0.960	0.067	0.947	0.053	0.958	0.065
GPT5-mini	hierarchical	**0.608**	0.137	**0.624**	0.153	**0.599**	0.128
	sibling	0.967	0.024	0.967	0.024	0.962	0.020
Multimodal input							
Gemma-3 12B	hierarchical	0.552†	**0.304**	0.546†	**0.297**	0.518†	**0.270**
	sibling	0.948†	**0.427**	0.947†	**0.426**	0.936†	**0.414**
Phi-4 Multimodal	hierarchical	0.459	0.065	0.486	0.093	0.441	0.048
	sibling	0.861	0.030	0.906	0.074	0.777	-0.055
GPT5-mini	hierarchical	0.588	0.117	**0.624**	0.153	0.598	0.127
	sibling	**0.972**†	0.030	**0.973**†	0.030	**0.970**†	0.028

5   Results

5.1   Overall Performance

Overall, models achieve moderate performance on HERTy-Wiki, with macro-F1 scores for the hierarchical task typically between 0.50 and 0.60 across evaluation settings. Sibling classification is substantially easier for all models, with macro-F1 being between 0.80 and 0.90 across evaluation settings. In both tasks, performance varies little across zero-shot, few-shot and chain-of-thought settings, with macro-F1 even falling in the latter two cases for some models. This suggests that performance is driven mainly by memorised knowledge rather than by a model's reasoning capabilities over the provided context.

Our no-context baseline supports this interpretation, with most models scoring between 0.25 and 0.42 for the hierarchical setting and over 0.8 for the sibling setting. The resulting Δ improvements are small, on average around 0.1, with Gemma-3 being an outlier in this trend. Detailed scores for all models can be found in Table 3.

The impact of images on hierarchical reasoning varies drastically across models. Gemma-3 shows a large improvement, suggesting that images act as grounding cues even when they contribute little explicit hierarchical information. GPT5-mini shows a slight degradation in the hierarchical case, whilst Phi-4 multimodal performs worse across all prompting settings.

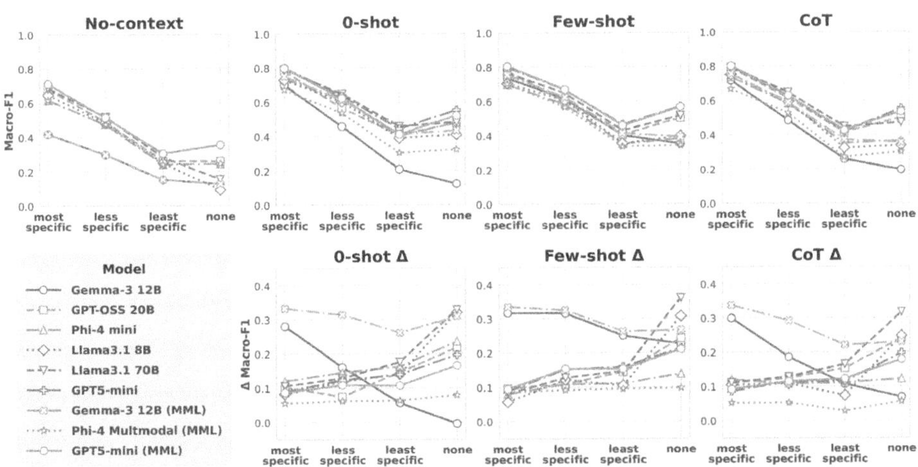

Fig. 3. Macro-F1 across hierarchy levels (top) and corresponding improvements over the no-context baseline (bottom) for different prompting settings.

5.2   Performance Across Hierarchy

Model performance varies substantially across hierarchy levels, as can be seen in Fig. 3. All models achieve their highest scores on the most specific types, with performance steadily decreasing as types become more general. This pattern indicates that fine-grained cases align more closely with memorised priors, while broader types require more abstract reasoning that models struggle to perform reliably. The same pattern can be observed in our no-context baseline, with most models scoring a macro-F1 of $\approx$0.25 in the least-specific and none classes, close to random guessing, indicating little memorised knowledge for those cases.

Model-specific behaviour emerges from the Δ values we compute. Gemma-3 12B's performance is highest on the most-specific cases and falls as the types become less granular, suggesting it mainly uses the provided context to reinforce fine-grained types it already knows. In contrast, GPT-OSS-20B and Llama3.1 8B show the largest improvement in the "none" class, suggesting models utilise context to improve abstention but not hierarchical knowledge.

These patterns suggest that hierarchical reasoning is present within models yet is weak and highly model-dependent - different models exploit the contextual information in systematically different and often shallow ways.

Table 4. Standard deviation of macro-F1 across domains for each model and prompting setting. Lower values indicate more uniform domain performance. Best values per column are in bold.

Model	No context	0-shot	Few-shot	CoT
Text-only input				
Gemma-3 12B	**0.025**	0.042	0.067	0.051
GPT-OSS-20B	0.054	0.039	0.058	0.040
Phi-4 mini	0.062	0.051	0.038	0.051
Llama3.1 8B	0.056	0.048	0.064	0.052
Llama3.1 70B	0.043	0.065	0.054	0.043
GPT5-mini	0.055	0.049	0.034	0.040
Multimodal input				
Gemma-3 12B	**0.025**	0.069	0.069	0.074
Phi-4 Multimodal	0.062	0.044	**0.034**	**0.034**
GPT5-mini	0.055	**0.040**	0.031	0.041

5.3   Performance Across Domains

Performance varies across domains but not dramatically, with standard deviation typically falling between 0.04 and 0.07 macro-F1 (see Table 4). Domain performance follows broadly similar patterns across models - with some domains (e.g. Building, Location) being consistently easier. However, when examining the per-domain Δ values, clear model-specific patterns emerge. Models appear to perform better in domains where their no-context performance is lower, suggesting that models may be compensating for missing prior knowledge, rather than performing genuine domain-specific hierarchical reasoning. These trends are visually summarised in the radar charts in Fig. 4.

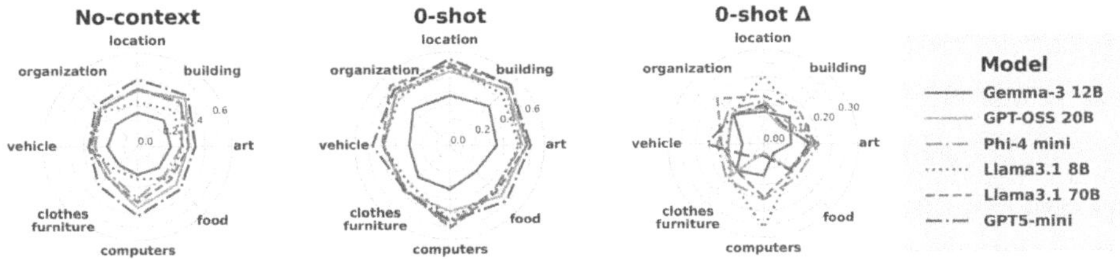

Fig. 4. Domain-wise macro-F1 for no-context, zero-shot, and zero-shot Δ. Please note that the Δ panel uses a smaller radial scale (0.0–0.3). For readability we omit multimodal settings.

6   Discussion

Key Findings. Our evaluation reveals that LLMs show limited hierarchical reasoning capabilities. Despite the logically straightforward nature of our task, few-shot or CoT prompting yield minimal improvements over zero-shot. This suggests that whilst LLMs may possess emerging reasoning capabilities, they are dominated by memorised priors. While models can often identify the most specific valid type, they frequently fail to abstain or generalise when the context is insufficient, pointing to brittle reasoning capabilities. Analysis over our no-context baseline further suggests that memorised priors vary across domains, model families, and hierarchical depth.

Performance on the sibling task is significantly higher, reinforcing that our hierarchical task cannot be solved through semantic similarity alone. Finally, the heterogeneous effects of multimodal information across models highlight that visual grounding does not enhance hierarchical inference reliably, underscoring the need for more robust reasoning approaches across modalities.

Implications for Knowledge Engineering. LLMs are increasingly used in knowledge engineering pipelines, from probing-based ontology learning to agentic and retrieval-augmented KG construction. Our findings underscore limitations in both paradigms - namely, memorised knowledge is uneven and model-specific, and reasoning over structured context remains unreliable. These limitations are especially important in fast-evolving or proprietary domains, where pre-training coverage is limited. In such cases, brittle reasoning can lead to overconfident misclassification, overgeneralisation, or semantic inconsistency within the knowledge graph.

HERTy-Wiki is designed to expose and quantify these failure modes. By using hierarchically close distractors and self-contained inputs designed to minimise knowledge contamination from pre-training, it enables researchers to disentangle recall from reasoning. Our benchmark offers the knowledge engineering community a means to evaluate structure-aware performance and to develop more robust approaches for the automation of knowledge graph construction.

Limitations. HERTy-Wiki is an MCQ-based benchmark to allow for precise control over distractor types. While widely adopted, MCQs make LLMs susceptible to "selection bias" [26,38,60], and answer shuffling alone cannot fully eliminate this bias [60]. Our no-context baseline provides an estimate of training-data contamination, but it cannot perfectly isolate memorisation, with factors such as positional bias potentially influencing outputs. Due to page constraints, we do not include prompt-ablation studies. Experiments showed that prompt wording shifts the distribution of predicted hierarchy levels (e.g. towards less specific types), but overall macro-F1 remains stable and our main conclusions persist.

Reusability. HERTy-Wiki is released under a CC-BY-4.0 licence, with all code and data openly available. Built on Wikidata, it is easily extendable to new domains or languages, and our evaluation suite supports diverse prompting and model configurations. We additionally release the aggregated annotation data, enabling further research on alternative answer-set constructions. To preserve the benchmark as an evaluation-only resource and guard against pre-training contamination [58], we include a canary string in the dataset following established best practices [6,16].

7   Future Work

Multilingual Support. At present, HERTy-Wiki is based on the English language, with a number of our heuristic filters relying explicitly on the availability and quality of English-language labels and descriptions within Wikidata. As such, the benchmark does not yet address the increasing demand for multilingual resources or evaluations in a non-English context [53]. However, Wikidata's language-agnostic data model makes it well-suited for cross-linguistic extension, with a possible extension of HERTy-Wiki to additional languages only requiring additional translation effort. To support this, we intend to release both a language-agnostic version of the dataset and tools for automatic population of the benchmark in other languages.

Hierarchical Modelling Challenges. While HERTy-Wiki focuses on fine-grained KGET, there are a number of challenges Wikidata contributors encounter [3] that require structured reasoning, such as navigating transitive subclass chains, identifying modelling antipatterns, or distinguishing between the Wikidata properties "instance of" (P31) and "subclass of" (P279). Extending HERTy-Wiki to cover such tasks would better reflect real-world curation needs and enable more comprehensive evaluation of LLMs in ontology construction workflows. To that end, we intend to release an extension of HERTy-Wiki catered to a number of these challenges.

8   Conclusion

As knowledge engineering workflows move toward increasingly autonomous and LLM-driven KG construction, reliable evaluation resources become essential. This work introduces the first human-verified, text-based benchmark for fine-grained entity typing in Wikidata, as well as a novel framing of the KGET task as one of hierarchical reasoning. By shifting the focus from memorisation to reasoning over structured context, HERTy-Wiki evaluates LLMs under conditions that mirror real-world modelling scenarios.

Our experiments provide the first systematic test of LLMs on hierarchical inference over KGET, revealing consistent limitations and highlighting the gap between memorised knowledge and genuine contextual reasoning. We hope HERTy-Wiki will provide a solid foundation for the evaluation of reasoning-driven entity typing methods and a key resource for advancing trustworthy, context-informed KG construction within the Semantic Web community.

Supplementary Materials. The HERTy-Wiki benchmark dataset is publicly available at https://doi.org/10.5281/zenodo.17828539. The code to run the benchmark is available at https://github.com/nobretincheva/HERTy-Wiki.

Acknowledgments. This study was funded by King's College London. Co-funded by the SIEMENS AG and the Technical University of Munich, Institute for Advanced Study, Germany.

Disclosure of Interests. The authors have no competing interests to declare that are relevant to the content of this article.

References

1. Abouelenin, A., et al.: Phi-4-mini technical report: Compact yet powerful multimodal language models via mixture-of-loras. arXiv preprint arXiv:2503.01743 (2025)
2. Agarwal, S., et al.: gpt-oss-120b & gpt-oss-20b model card. arXiv preprint arXiv:2508.10925 (2025)
3. Ammalainen, D.: Wikidata ontology issues — suggestions for prioritisation. Presentation, Wikimedia Commons (2023). https://commons.wikimedia.org/wiki/File:Wikidata_ontology_issues_%E2%80%94_suggestions_for_prioritisation_2023.pdf. Accessed 18 Oct 2024
4. Babaei Giglou, H., D'Souza, J., Auer, S.: Llms4ol: large language models for ontology learning. In: Payne, T.R., et al. (eds.) The Semantic Web - ISWC 2023, pp. 408–427. Springer Nature Switzerland, Cham (2023)
5. Bean, A., et al.: Lingoly: A benchmark of olympiad-level linguistic reasoning puzzles in low resource and extinct languages. In: Globerson, A., et al., (eds.) Advances in Neural Information Processing Systems. vol. 37, pp. 26224–26237. Curran Associates, Inc. (2024)
6. Bean, A.M., et al.: Measuring what matters: Construct validity in large language model benchmarks. arXiv preprint arXiv:2511.04703 (2025)
7. Biswas, R., Portisch, J., Paulheim, H., Sack, H., Alam, M.: Entity type prediction leveraging graph walks and entity descriptions. In: The Semantic Web – ISWC 2022: 21st International Semantic Web Conference, Virtual Event, October 23–27, 2022, Proceedings, pp. 392–410. Springer-Verlag, Berlin, Heidelberg (2022). https://doi.org/10.1007/978-3-031-19433-7_23

8. Biswas, R., Sofronova, R., Sack, H., Alam, M.: Cat2type: Wikipedia category embeddings for entity typing in knowledge graphs. In: Proceedings of the 11th Knowledge Capture Conference, pp. 81–88. K-CAP '21, Association for Computing Machinery, New York, NY, USA (2021). https://doi.org/10.1145/3460210.3493575

9. Bordes, A., Usunier, N., Garcia-Duran, A., Weston, J., Yakhnenko, O.: Translating embeddings for modeling multi-relational data. Adv. Neural Inform. Process. Syst. **26** (2013)

10. Brasileiro, F., Almeida, J.a.P.A., Carvalho, V.A., Guizzardi, G.: Applying a multi-level modeling theory to assess taxonomic hierarchies in wikidata. In: Proceedings of the 25th International Conference Companion on World Wide Web, pp. 975–980. WWW '16 Companion, International World Wide Web Conferences Steering Committee, Republic and Canton of Geneva, CHE (2016). https://doi.org/10.1145/2872518.2891117

11. Cai, G., Gong, J., Du, J., Liu, H., Kai, A.: Investigating hierarchical term relationships in large language models. J. Comput. Sci. Softw. Appl. **5**(4) (2025)

12. Dadalto, A.A., Almeida, J.P.A., Fonseca, C.M., Guizzardi, G.: Evidence of large-scale conceptual disarray in multi-level taxonomies in wikidata. Semantic Web **15**(6), 2253–2270 (2024)

13. Daniel, V.S., Francisco, A.: Argilla - Open-source framework for data-centric NLP (Jan 2023). https://github.com/argilla-io/argilla

14. Erenrich, D.: Psychiq and wwwyzzerdd: Wikidata completion using wikipedia. Semantic Web **15**(6), 2145–2158 (2024)

15. Geva, M., Khashabi, D., Segal, E., Khot, T., Roth, D., Berant, J.: Did aristotle use a laptop? a question answering benchmark with implicit reasoning strategies. Trans. Assoc. Comput. Linguist. **9**, 346–361 (04 2021). https://doi.org/10.1162/tacl_a_00370

16. Ghazal, A., et al.: Bigbench: Towards an industry standard benchmark for big data analytics. In: Proceedings of the 2013 ACM SIGMOD International Conference on Management of Data, pp. 1197–1208 (2013)

17. Grattafiori, A., et al.: The llama 3 herd of models. arXiv preprint arXiv:2407.21783 (2024)

18. Grootendorst, M.: Bertopic: Neural topic modeling with a class-based TF-IDF procedure. arXiv preprint arXiv:2203.05794 (2022)

19. Gwet, K.: Handbook of inter-rater reliability. Gaithersburg, MD: STATAXIS Publishing Company pp. 223–246 (2001)

20. Gwet, K., et al.: Inter-rater reliability: dependency on trait prevalence and marginal homogeneity. Stat. Methods Inter-rater Reliab. Assess. Series **2**(1), 9 (2002)

21. Hu, Z., Gutierrez-Basulto, V., Xiang, Z., Li, R., Pan, J.: Transformer-based entity typing in knowledge graphs. In: Goldberg, Y., Kozareva, Z., Zhang, Y. (eds.) Proceedings of the 2022 Conference on Empirical Methods in Natural Language Processing, pp. 5988–6001. Association for Computational Linguistics, Abu Dhabi, United Arab Emirates (Dec 2022). https://doi.org/10.18653/v1/2022.emnlp-main.402, https://aclanthology.org/2022.emnlp-main.402

22. Hu, Z., Gutiérrez-Basulto, V., Xiang, Z., Li, R., Pan, J.Z.: Cotet: Cross-view optimal transport for knowledge graph entity typing (2024). https://arxiv.org/abs/2405.13602

23. Jin, H., Hou, L., Li, J., Dong, T.: Fine-grained entity typing via hierarchical multi graph convolutional networks. In: Inui, K., Jiang, J., Ng, V., Wan, X. (eds.) Proceedings of the 2019 Conference on Empirical Methods in Natural Language Processing and the 9th International Joint Conference on Natural Language Processing (EMNLP-IJCNLP), pp. 4969–4978. Association for Computational Linguistics,

Hong Kong, China (Nov 2019). https://doi.org/10.18653/v1/D19-1502, https://aclanthology.org/D19-1502

24. Li, M., Hu, M., King, I., Leung, H.f.: The integration of semantic and structural knowledge in knowledge graph entity typing. In: Duh, K., Gomez, H., Bethard, S. (eds.) Proceedings of the 2024 Conference of the North American Chapter of the Association for Computational Linguistics: Human Language Technologies (Volume 1: Long Papers), pp. 6625–6638. Association for Computational Linguistics, Mexico City, Mexico (Jun 2024). https://doi.org/10.18653/v1/2024.naacl-long.369, https://aclanthology.org/2024.naacl-long.369

25. Li, N., et al.: CR-LLM: A dataset and optimization for concept reasoning of large language models. In: Ku, L.W., Martins, A., Srikumar, V. (eds.) Findings of the Association for Computational Linguistics: ACL 2024, pp. 13737–13747. Association for Computational Linguistics, Bangkok, Thailand (Aug 2024). https://doi.org/10.18653/v1/2024.findings-acl.815, https://aclanthology.org/2024.findings-acl.815/

26. Li, W., Li, L., Xiang, T., Liu, X., Deng, W., Garcia, N.: Can multiple-choice questions really be useful in detecting the abilities of LLMs? In: Calzolari, N., Kan, M.Y., Hoste, V., Lenci, A., Sakti, S., Xue, N. (eds.) Proceedings of the 2024 Joint International Conference on Computational Linguistics, Language Resources and Evaluation (LREC-COLING 2024), pp. 2819–2834. ELRA and ICCL, Torino, Italia (May 2024). https://aclanthology.org/2024.lrec-main.251/

27. Ling, X., Weld, D.: Fine-grained entity recognition. Proc. AAAI Conf. Artif. Intell. **26**(1), 94–100 (2012). https://doi.org/10.1609/aaai.v26i1.8122

28. Ma, C., Shen, A., Yoshikawa, H., Iwakura, T., Beck, D., Baldwin, T.: On the (in)effectiveness of images for text classification. In: Merlo, P., Tiedemann, J., Tsarfaty, R. (eds.) Proceedings of the 16th Conference of the European Chapter of the Association for Computational Linguistics: Main Volume, pp. 42–48. Association for Computational Linguistics, Online (Apr 2021). https://doi.org/10.18653/v1/2021.eacl-main.4, https://aclanthology.org/2021.eacl-main.4/

29. Mai, H.T., Chu, C.X., Paulheim, H.: Do LLMs really adapt to domains? an ontology learning perspective. In: Demartini, G., et al. (eds.) The Semantic Web - ISWC 2024, pp. 126–143. Springer Nature Switzerland, Cham (2025)

30. Moon, C., Jones, P., Samatova, N.F.: Learning entity type embeddings for knowledge graph completion. In: Proceedings of the 2017 ACM on Conference on Information and Knowledge Management, pp. 2215–2218. CIKM '17, Association for Computing Machinery, New York, NY, USA (2017). https://doi.org/10.1145/3132847.3133095

31. Mtumbuka, F., Schockaert, S.: EnCore: Fine-grained entity typing by pre-training entity encoders on coreference chains. In: Graham, Y., Purver, M. (eds.) Proceedings of the 18th Conference of the European Chapter of the Association for Computational Linguistics (Volume 1: Long Papers), pp. 1768–1781. Association for Computational Linguistics, St. Julian's, Malta (Mar 2024). https://aclanthology.org/2024.eacl-long.106/

32. OpenAI: Gpt-5-mini. https://openai.com (2025), large language model. Accessed via OpenAI API

33. OpenStreetMap Foundation: Nominatim: Search and reverse geocoding service (2025). https://nominatim.org. Accessed 25 May 2025

34. Ouyang, S., Huang, J., Pillai, P., Zhang, Y., Zhang, Y., Han, J.: Ontology enrichment for effective fine-grained entity typing. In: Proceedings of the 30th ACM SIGKDD Conference on Knowledge Discovery and Data Mining, pp. 2318–2327 (2024)

35. Pan, S., Luo, L., Wang, Y., Chen, C., Wang, J., Wu, X.: Unifying large language models and knowledge graphs: a roadmap. IEEE Trans. Knowl. Data Eng. **36**(7), 3580–3599 (2024). https://doi.org/10.1109/TKDE.2024.3352100

36. Pan, W., Wei, W., Mao, X.L.: Context-aware entity typing in knowledge graphs. In: Moens, M.F., Huang, X., Specia, L., Yih, S.W.t. (eds.) Findings of the Association for Computational Linguistics: EMNLP 2021, pp. 2240–2250. Association for Computational Linguistics, Punta Cana, Dominican Republic (Nov 2021). https://doi.org/10.18653/v1/2021.findings-emnlp.193, https://aclanthology.org/2021.findings-emnlp.193

37. Peng, Y., Bonald, T., Alam, M.: Refining wikidata taxonomy using large language models. In: ACM International Conference on Information and Knowledge Management. Boise, Idaho, United States (Oct 2024). https://doi.org/10.1145/3627673.3679156, https://hal.science/hal-04687258

38. Pezeshkpour, P., Hruschka, E.: Large language models sensitivity to the order of options in multiple-choice questions. In: Duh, K., Gomez, H., Bethard, S. (eds.) Findings of the Association for Computational Linguistics: NAACL 2024, pp. 2006–2017. Association for Computational Linguistics, Mexico City, Mexico (Jun 2024). https://doi.org/10.18653/v1/2024.findings-naacl.130, https://aclanthology.org/2024.findings-naacl.130/

39. Piscopo, A., Simperl, E.: Who models the world? collaborative ontology creation and user roles in wikidata. Proc. ACM Hum.-Comput. Interact. **2**(CSCW) (Nov 2018). https://doi.org/10.1145/3274410

40. Prolific: Prolific. https://www.prolific.com (2014), version: <August 2025>. Location: London, UK. Copyright © 2024

41. Qian, J., et al.: Fine-grained entity typing without knowledge base. In: Moens, M.F., Huang, X., Specia, L., Yih, S.W.t. (eds.) Proceedings of the 2021 Conference on Empirical Methods in Natural Language Processing, pp. 5309–5319. Association for Computational Linguistics, Online and Punta Cana, Dominican Republic (Nov 2021). https://doi.org/10.18653/v1/2021.emnlp-main.431, https://aclanthology.org/2021.emnlp-main.431/

42. Riaz, A., Abdollahi, S., Gottschalk, S.: Entity typing with triples using language models. In: European Semantic Web Conference, pp. 169–173. Springer (2023)

43. Shenoy, K., Ilievski, F., Garijo, D., Schwabe, D., Szekely, P.: A study of the quality of wikidata. J. Web Semantics **72**, 100679 (2022)

44. Sosa, R.U., Ramamurthy, K.N., Chang, M., Singh, M.: Reasoning about concepts with LLMs: Inconsistencies abound. In: First Conference on Language Modeling (2024)

45. Stureborg, R., Dhingra, B., Yang, J.: Interface design for crowdsourcing hierarchical multi-label text annotations. In: Proceedings of the 2023 CHI Conference on Human Factors in Computing Systems. CHI '23, Association for Computing Machinery, New York, NY, USA (2023). https://doi.org/10.1145/3544548.3581431

46. Sun, Y., et al.: Are large language models a good replacement of taxonomies? Proc. VLDB Endow. **17**(11), 2919–2932 (Jul 2024). https://doi.org/10.14778/3681954.3681973

47. Team, G., et al.: Gemma 3 technical report. arXiv preprint arXiv:2503.19786 (2025)

48. Uddin, M.N., et al.: UnSeenTimeQA: Time-sensitive question-answering beyond LLMs' memorization. In: Che, W., Nabende, J., Shutova, E., Pilehvar, M.T. (eds.) Proceedings of the 63rd Annual Meeting of the Association for Computational Linguistics (Volume 1: Long Papers), pp. 1873–1913. Association for Computational Linguistics, Vienna, Austria (Jul 2025). https://doi.org/10.18653/v1/2025.acl-long.94, https://aclanthology.org/2025.acl-long.94/

49. Vrandečić, D., Krötzsch, M.: Wikidata: a free collaborative knowledgebase. Commun. ACM **57**(10), 78–85 (2014). https://doi.org/10.1145/2629489, https://doi.org/10.1145/2629489

50. Wang, Y.C., Ge, X., Wang, B., Kuo, C.C.J.: Asyncet: asynchronous representation learning for knowledge graph entity typing. In: Proceedings of the 30th ACM SIGKDD Conference on Knowledge Discovery and Data Mining, pp. 3267–3276. KDD '24, Association for Computing Machinery, New York, NY, USA (2024). https://doi.org/10.1145/3637528.3671832, https://doi.org/10.1145/3637528.3671832

51. Wikidata: Statistics (2025). https://www.wikidata.org/wiki/Wikidata:Statistics, Accessed 05 Dec 2025

52. Wikidata contributors: Wikidata: Entity schema directory (May 2025). https://www.wikidata.org/wiki/Wikidata:Schemas

53. Wu, M., et al.: The bitter lesson learned from 2,000+ multilingual benchmarks (2025). https://arxiv.org/abs/2504.15521

54. Yaghoobzadeh, Y., Adel, H., Schütze, H.: Corpus-level fine-grained entity typing. J. Artif. Int. Res. **61**(1), 835–862 (2018)

55. Yaghoobzadeh, Y., Schütze, H.: Corpus-level fine-grained entity typing using contextual information. In: Màrquez, L., Callison-Burch, C., Su, J. (eds.) Proceedings of the 2015 Conference on Empirical Methods in Natural Language Processing, pp. 715–725. Association for Computational Linguistics, Lisbon, Portugal (Sep 2015). https://doi.org/10.18653/v1/D15-1083, https://aclanthology.org/D15-1083/

56. Yaghoobzadeh, Y., Schütze, H.: Multi-level representations for fine-grained typing of knowledge base entities. In: Lapata, M., Blunsom, P., Koller, A. (eds.) Proceedings of the 15th Conference of the European Chapter of the Association for Computational Linguistics: Volume 1, Long Papers, pp. 578–589. Association for Computational Linguistics, Valencia, Spain (Apr 2017). https://aclanthology.org/E17-1055/

57. Ye, H., Gui, H., Zhang, A., Liu, T., Hua, W., Jia, W.: Beyond isolation: Multi-agent synergy for improving knowledge graph construction (2023). https://arxiv.org/abs/2312.03022

58. Zhang, H., et al.: A careful examination of large language model performance on grade school arithmetic. Adv. Neural. Inf. Process. Syst. **37**, 46819–46836 (2024)

59. Zhang, X., Zhao, J., LeCun, Y.: Character-level convolutional networks for text classification. Adv. Neural Inform. Process. Syst. **28** (2015)

60. Zheng, C., Zhou, H., Meng, F., Zhou, J., Huang, M.: Large language models are not robust multiple choice selectors (2024). https://arxiv.org/abs/2309.03882

61. Zhu, Y., et al.: LLMs for knowledge graph construction and reasoning: recent capabilities and future opportunities. World Wide Web **27**(5), 58 (2024)

AgentO: An Ontology for Modeling Agentic AI Systems

Andreas Ekelhart[1,2], Kabul Kurniawan[3(✉)], Fajar J. Ekaputra[4],
and Elmar Kiesling[4]

[1] University of Vienna, Vienna, Austria
[2] SBA Research, Favoritenstraße 16, Vienna, Austria
[3] Universitas Gadjah Mada, Yogyakarta, Indonesia
`kabul.kurniawan@ugm.ac.id`
[4] Vienna University of Economics and Business, Vienna, Austria

Abstract. Agentic AI systems are rapidly being deployed as autonomous, goal-directed entities to manage the orchestration of complex, multi-step workflows across diverse domains. Despite their growing adoption, current frameworks often lack a formalized model and architecture. Hence, many implementations remain ad-hoc, relying on simplistic data structures and monolithic designs that hinder scalability, reusability, and interoperability. This paper addresses these limitations by introducing AgentO, an OWL/RDF-based ontology and accompanying knowledge graph that formally represent the core concepts, components, and interactions that underpin agentic AI workflows. Our ontology provides a standardized vocabulary for modeling agentic patterns including agents, tasks, workflows, and resource dependencies. To build and evaluate AgentO, we developed an automated LLM-driven process and translated 66 agentic workflows from four different agentic AI frameworks. We further evaluated our approach through three real-world use cases: declarative reconstruction of agentic patterns, cross-context reuse of tasks and agents, and agentic AI workflow auditing. Our results demonstrate the potential of semantic technologies to bring structure, reusability, and transparency to agentic AI systems.

Keywords: Ontology · Agentic AI · LLMs · Knowledge Graph

Resource type: Ontology

License: CC BY 4.0 International

DOI: 10.5281/zenodo.18342624

URL: https://w3id.org/agentic-ai/onto

1 Introduction

Agentic AI systems, characterized by their autonomy, goal-directed behavior, and capacity for adaptive decision-making, are increasingly deployed in domains that demand the orchestration of complex, multi-step workflows, ranging from cybersecurity [16], finance [22], industrial automation [3], and healthcare [15] to scientific research [11]. These systems promise enhanced flexibility, coordination, and scalability by delegating tasks to intelligent agents capable of interacting, reasoning, and responding to dynamic environments [30].

Agentic AI has seen widespread adoption and implementation across various sectors[1] and is expected to drive development of AI-based systems in the next years[2]. However, achieving impact requires clearly defined use cases and demonstrable business value. While early setbacks are likely[3],[4], both agent frameworks and the underlying models are evolving at a rapid pace.

Despite growing adoption, current implementations of such systems typically rely on frameworks that are characterized by tight coupling of ad-hoc data structures and configurations. This includes modern agentic AI frameworks such as Microsoft's AutoGen [31] CrewAI[5], LangGraph [29] and Mastra AI[6], which typically encode logic and workflows directly into code, resulting in hard-coded and brittle monolithic software architectures. This lack of standardization and formalization not only limits system scalability and maintainability, but it also hampers reusability, traceability, and interoperability across tools, platforms, and domains. Furthermore, the absence of a shared vocabulary and well-defined design patterns prevents practitioners from reusing existing agentic solutions and systematically managing workflows, goals, and resource dependencies.

To address these challenges, we introduce AgentO, an ontology for modeling agentic AI systems[7]. The ontology provide concepts of agentic AI systems and formally defines the key components of agentic patterns such as agents, goals, tasks, tools, resources as well as their interrelations and workflows. The ontology serves as a standardized vocabulary (built based on Resource Description Framework (RDF)/Web Ontology Language (OWL)) that enables modular design, reusability, facilitates reasoning over agentic configurations, and supports system auditing and traceability. Accompanying the ontology is a knowledge graph that instantiates existing agentic patterns and allows integration with heterogeneous agentic AI frameworks and application contexts. This resource enables the declarative construction, reuse, and analysis of agentic workflows, empowering developers and researchers to design more organized, transparent, and interoperable agentic AI systems.

We validate the feasibility of the proposed ontology and Knowledge Graph (KG) through three real-world use cases: *(i)* declarative reconstruction of agentic

---

[1] https://www.gartner.com/en/articles/intelligent-agent-in-ai.

[2] https://www.deloitte.com/global/en/about/press-room/deloitte-globals-2025-predictions-report.html.

[3] https://www.gartner.com/en/newsroom/press-releases/2025-06-25-gartner-predicts-over-40-percent-of-agentic-ai-projects-will-be-canceled-by-end-of-2027.

[4] https://www.forbes.com/sites/jasonsnyder/2025/08/26/mit-finds-95-of-genai-pilots-fail-because-companies-avoid-friction.

[5] https://crewai.com.

[6] https://mastra.ai.

[7] https://w3id.org/agentic-ai/onto.

AI patterns for visual inspection and reasoning, *(ii)* reusability of agent-task configurations across problem contexts, and *(iii)* auditing of execution traces for transparency and explainability.

To summarize, our **main contributions** of this paper are as follows: *(i)* we propose the first standardized conceptual model for agentic AI as a reusable resource, formalized as an ontology that captures core concepts, relationships, workflow structures, and agentic interaction patterns;[8] *(ii)* we transform and integrate 66 existing agentic AI patterns from heterogeneous frameworks across various application domains into a unified KG and share the Large Language Model (LLM)-based pipeline[9], and *(iii)* we demonstrate the practical utility of our semantic model through three real-world use cases.

The remainder of this paper is organized as follows: Sect. 2 presents related work in the area of agentic AI systems and Semantic Web; Sect. 3 discusses the conceptual modeling and the development of the ontology; Sect. 4 describes the overall KG construction and refinement; Sect. 5 presents three use-case applications; and Sect. 6 discusses limitations, concludes, and gives an outlook on future work.

2   Related Work

2.1   Semantic Web

The original Semantic Web (SW) proposal already envisioned "intelligent agents" performing tasks such as searching for information, completing transactions, and making decisions on behalf of humans [6]. To enable software agents to interact with the web on behalf of their users, explicit semantic representations were supposed to create an environment to (eventually) enable large-scale, agent-based mediation [26]. However, the envisioned SW environment for agents never fully materialized. A related line of work focused on adding a semantic layer to traditional web services in order to enable automatic discovery, execution, and composition of web services [7,18] based on technologies such as WSMO/WSML (Web Service Modeling Ontology) [25], SWSO/SWSL [4,5], and WSDL-S [2]. This was supposed to enable Agent-based Semantic Web Services [10]. Due to the high modeling complexity, the wide range of proposed standards, limited tool support, as well as lacking incentives for adoption, this vision did not materialize either. A survey of critiques of the Semantic Web in 2020 [14] suggested that there is more recently a growing recognition that these visions may be made redundant by advances in Machine Learning.

The arrival of LLMs provides strong evidence for this hypothesis, given their impressive ability to process natural language without explicit "machine-readable" semantics. The role of semantic descriptions and agents in this context can now be considered inverted w.r.t. the original SW vision – reasoning agents leveraging knowledge graphs may participate in agentic workflows (e.g.,

---

[8] Available at: https://w3id.org/agentic-ai/onto.
[9] Available at: https://agentic-patterns.github.io/.

in hybrid artificial intelligence (AI) scenarios), but more commonly, the issue of "machine-interpretability" and interfaces is now largely solved with natural language as a lingua franca. In this context, semantic descriptions are becoming increasingly critical for describing, composing, and coordinating agentic workflows, surpassing their traditional role in representing inputs and outputs or facilitating interoperability, as was the vision for Semantic Web Services initiatives [10].

2.2 Agentic AI Systems

Agentic AI Systems consist of autonomous, goal-oriented software agents capable of perceiving their environment, making decisions, and executing actions to achieve specific objectives [19]. Unlike traditional AI pipelines that operate in a single-pass or stateless manner, agentic AI incorporates several elements such as planning, tool usage, reflection, and multi-agent collaboration to enable iterative and adaptive behavior. This paradigm enables agents to move beyond reactive tasks into more strategic and emergent problem solving.

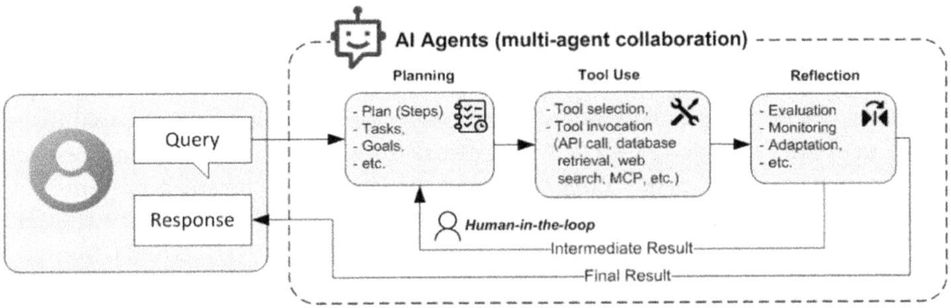

Fig. 1. Agentic AI System Architecture.

Figure 1 shows an agentic AI architecture, which comprises four prevalent ideas for constructing such intelligent systems [20]: *(i) Planning*: the agent decomposes high-level goals into actionable sub-tasks, sequences them, and executes them iteratively; *(ii) Tool Use*: the agent leverages external tools or APIs (e.g., web search, databases, MCP, etc.) to extend its capabilities beyond its native reasoning abilities; *(iii) Reflection*: the agent periodically evaluates its progress toward a goal, updates its internal memory, and adjusts its plan accordingly; *(iv) Multi-Agent Collaboration*: multiple agents with specialized roles coordinate, communicate, and collaborate to solve complex problems. Incorporating human oversight at critical junctures within multi-agent systems (human-in-the-loop) enables the provision of feedback and guidance.

A wealth of recent work has explored agentic AI systems based on LLMs with the idea to allow LLMs not only to generate text, but also to plan actions and use tools. The ReAct (Reason+Act) paradigm [32], for instance, combines chain-of-thought reasoning with task-specific actions. Building on this idea, the possibility of coordinating agents with heterogeneous knowledge and expertise has further been explored and formalized [12,13,27]. Frameworks such as Auto-Gen [31] allow multiple LLM agents to converse and coordinate to solve tasks. This follows the principle of division of labor, mirroring the concept of teamwork in human collaboration. Other approaches structure agents into specialized roles with defined interaction protocols. ChatDev [24], for example, simulates a software team by assigning different roles (e.g., developer, tester, manager) to LLM agents and orchestrating their communication through a predefined workflow.

Another line of work focuses on extending LLM agents with short-term and long-term memory concepts and planning subsystems [1,23,24]. One example is [23], where the agent records each experience in a natural language memory stream, synthesizes higher-level reflections, and retrieves relevant memories to inform its plans. Similarly, Voyager [28] demonstrated an LLM-driven agent that iteratively learns in an environment (Minecraft) by generating code, executing it, and storing new skills in a growing library.

Researchers have also begun to formalise agent interactions: e.g., Google's Agent-to-Agent (A2A)[10] communication protocol and Anthropic's Model-Context Protocol (MCP)[11] aim to define standard interfaces for LLM agents to communicate or use tools. Despite this progress, current academic frameworks remain limited in their semantic expressiveness and reusability. Many systems rely on natural language prompts or hard-coded pipelines to orchestrate agent behavior. There is no widely adopted ontology for describing agent capabilities, goals, or world knowledge – each framework uses its own task definitions and role descriptors, making it difficult to transfer knowledge between systems.

Several frameworks have emerged from both research and industry that put agentic concepts into practice. Among them, AutoGen [31] stands out as a prominent open-source multi-agent framework. It allows agents to communicate in natural language to coordinate on tasks, supporting customizable conversation patterns such as sequential, group and nested chats. CrewAI[12] is an open source Python framework that orchestrates role-playing autonomous AI agents to complete tasks. Mastra[13] is a cloud-native, open-source TypeScript framework designed for building AI agents with persistent memory and structured workflows, and LangGraph [29], a framework based on the LangChain ecosystem designed for stateful orchestration of LLM agents using graph structures. We will focus on these four popular frameworks in this work.

3   Conceptualization

In this section, we present a conceptual foundation for developing a semantic model for agentic AI systems. Based on Sect. 2, we identify and define essential

---

[10] https://developers.googleblog.com/en/a2a-a-new-era-of-agent-interoperability/.
[11] https://www.anthropic.com/news/model-context-protocol.
[12] https://www.crewai.com.
[13] https://mastra.ai.

components, interactions and architectural patterns that characterize agentic AI systems. The goal of this section is to provide a standardized ontology for representing agentic patterns, constructing an agentic pattern KG, and support building agentic AI systems.

3.1   Modelling Agentic AI Systems

We follow a bottom-up conceptual modeling approach [21] guided by a combination of domain analysis, literature review, design patterns and implementation techniques observed in widely used agentic AI frameworks. The conceptualization is carried out through four systematic stages:

1. *Component Identification.* In this initial stage, we extract core components that appear consistently across existing implementations of agentic AI systems. For example, in the *CrewAI* framework, agents are instantiated with a defined `Role` (e.g., *Writer, Reviewer*, etc.), specific `Tasks` (e.g., *Writing literature, Evaluate written content*, etc.), and associated `Tools` (e.g., *SearchTool*, etc.). The identified concepts are then systematically organized and characterized with associated attributes and properties relevant to their function within a system. For example, the `Agent` concept is described using metadata such as `title, description, role`, etc. Figure 2 shows examples of identified concepts and their attributes.

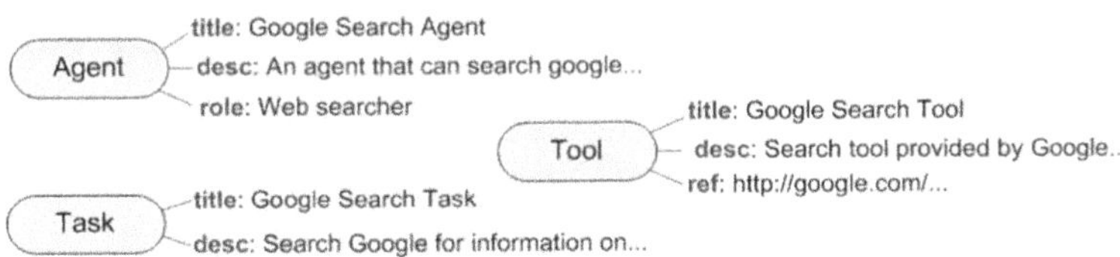

Fig. 2. Agentic AI concepts and their attributes (excerpt).

2. *Relationship Modeling.* Once the core components are identified, we define the key relationships and dependencies between these concepts, forming a coherent semantic network. These relationships are critical for modeling how each concept interacts with others and representing the behavior of an agentic AI System. For example, in agentic frameworks such as *AutoGen or CrewAI*, an `Agent` is typically performs one or more `Tasks`, where each task represents a concrete action or objective. Agents may also use specific `Tools` (e.g., databases, APIs, web search) to achieve a specific `Goal` set by an agent. To formally capture this interaction, we defined semantic relationships

such as: *(i)* `responsibleAgent` - links a `Task` with a responsible `LLMAgent`; *(ii)* `performedBy` - captures the tools or external resources that are used by an `LLMAgent` to accomplish `Tasks`. Figure 3 shows an example of the agent relationship model.

Fig. 3. Agentic relationship model (excerpt).

3. Pattern Identification. In this stage, we capture reusable behavioral patterns that represent recurring sequences of actions, decisions and interactions commonly found in agentic AI systems. These patterns represent abstract workflow or coordination structures that can be reused across tasks, domains, and frameworks.

For example, a `literature review` task may involve multiple agents collaborating in a sequential workflow: *(i)* a `Google Search` agent is tasked with searching articles on the web; *(ii)* an `Arxiv Search` agent then refines the search by querying an academic database; finally *(iii)* a `Report` agent compiles and synthesizes the results into a structured report. This sequence *Search → Refine → Report* forms a *Sequential Pattern*, where the output of one agent serves as the input to the next. To this end we identified three main workflow patterns i.e., *(i)* Sequential Pattern, a workflow where tasks are executed one after another, in a linear order. *(ii)* Parallel Pattern, involves multiple agents working independently at the same time on different sub-tasks. The results are later combined or aggregated. *(iii)* Nested Pattern, refers to workflows where one or more agents invoke internal sub-workflows that can themselves be sequential, parallel, or nested. Figure 4 shows agentic patterns composed of the identified concepts and relationships. (See Apendix for other patterns, i.e., Fig. 8 and Fig. 7)

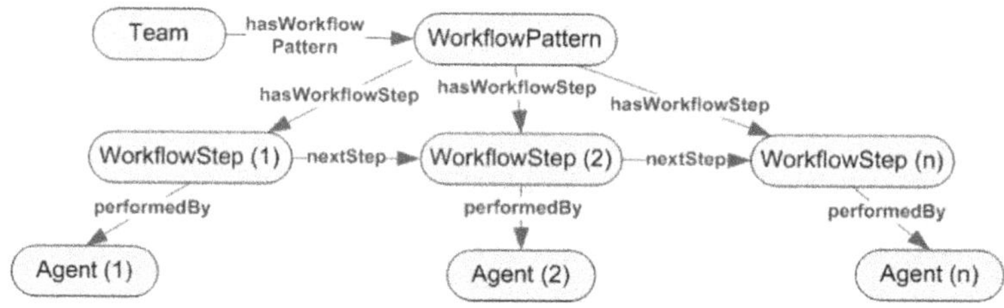

Fig. 4. Sequential pattern workflow model example.

4. Ontology Mapping. This final stage of the conceptualization process involves translating the previously defined model into a formal ontology using Semantic Web Standards based on RDF/OWL. Each core component identified in the *Component Identification* phase—such as **Agent, Task, Tool**, etc.—is formally defined as a **Class** in the ontology. These classes are enriched with **Data Properties** (e.g., agent `title, description, role`, etc.) that capture attributes and metadata. Meanwhile, the relationships identified during the *Relationship Modelling* phase—such as `hasAssociatedTask, performedBy, hasAgentGoal`—are formalized as **Object Properties**, which link instances of one class to another.

3.2   Agentic AI Ontology

Building upon the conceptual model established in the previous section, we propose a formal ontology that defines the structure and semantics of agentic AI systems based on RDF/OWL. This ontology referred to as `AgentO` (Agentic AI Ontology) is designed to enable modular integration and interoperability across agentic frameworks. To this end, we built upon concepts from existing ontologies, specifically PROV-O [17](i.e., `prov:Agent`), P-Plan [9] (`pplan:Plan` and `pplan:Step`), and BEAM [8] (`beam:System`,`beam:Context` and `beam:Resource`).

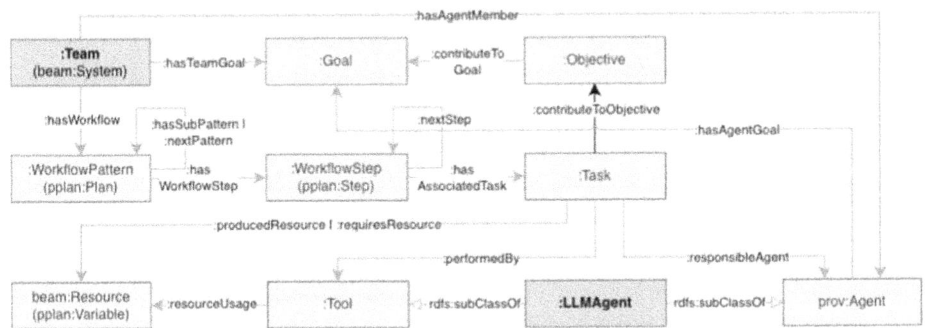

Fig. 5. Overview of the Agentic AI Ontology (AgentO).

Figure 5 shows an overview of the core classes *AgentO* and their relations, capturing the essential components of agentic systems including `Team`, `Agent`, `Task`, `Goal`, `Objective`, `WorkflowPattern`, `WorkflowStep`, `Tool`, and `Resource`. To support various types of agentic patterns (i.e., sequential, parallel, and nested), we introduce a combination of classes and relationships in the ontology. Specifically, the class `:WorkflowPattern` is used to model composite structures of workflows; individual actions are captured as instances of `:WorkflowStep`. The relationships `:hasWorkflowStep`, `:hasNextStep`, and `:hasSubPattern` allow us to represent linear sequences, branching parallel flows, and recursive nesting of sub-patterns. Due to space limitations, full documentation and the complete ontology specification are provided online[14].

4   Knowledge Graph Construction and Refinement

To operationalize the ontology and support the retrieval over agentic AI design patterns, we construct a KG that instantiates the ontology with representative patterns extracted from existing agentic AI frameworks. We designed an automated process including LLMs to populate the ontology and discover current

---

[14] Agentic AI Ontology (AgentO): http://w3id.org/agentic-ai/onto.

issues and potential extensions. The results were then manually evaluated and used to refine the ontology. The KG construction and refinement process consisted of several stages:

1. Source Pattern Collection. We collected 66 existing agentic AI templates and design patterns from several framework repositories (AutoGen[15], CrewAI[16], LangGraph[17], and Mastra AI[18]). An overview can be found in Table 6. These templates are typically hardcoded configurations within *Python* or *JavaScript* files. Listing 1.1 shows an illustrative example of an AI agent configuration (written in *Python*) for a `literature review` task in the CrewAI framework. The agent is defined with a role, goal, tools and is tasked with a specific objective with expected outputs. These are then assembled into a "crew" which coordinates the execution.

```
from crewai import Agent, Task, Crew
from tools.search_tools import SearchTool
# Define an AI agent to act as a research assistant
literature_reviewer = Agent(
    role="AI Research Assistant",
    goal="Review and summarize recent academic papers on Agentic AI...",
    backstory="An expert in AI research who analyzes and synthesizes...",
    tools=[SearchTool()], memory=True  )
# Define a task for the agent to perform
review_task = Task(
    description="Identify, retrieve, and summarize at least five...",
    expected_output="A concise literature review summarizing
    objective, methodologies, and unique features of each framework.",
    agent=literature_reviewer)
# Initialize the Crew with the agent and the task
crew = Crew(agents=[literature_reviewer], tasks=[review_task])
crew.kickoff() # Start the process
```

Listing 1.1. Agentic AI pattern configuration in the CrewAI framework

2. Pattern Extraction and Ontology Mapping. Next, we developed an automated workflow[19] to parse the raw agentic pattern files and related source code to extract relevant concepts. We use prompt-guided LLM generation to semantically map and transform the framework patterns into RDF statements based on the AgentO schema. Since the frameworks use different structures, all source code files from each pattern were passed as context to the LLM. All extractions were done with `gpt-5-mini` via API access. The extraction time per pattern varied between 47.72 and 166.60 s. Approximately 2.8 million input and 570,000 output tokens were processed for parsing the 66 patterns[20].

A consistent output structure was generated for each pattern translation, including execution time and the used model, issues and assumptions from the LLM, followed by the instance definition in .ttl format. In Listing 1.2 we show an excerpt of the CrewAI trip planner output as an example.

---

[15] https://github.com/ksm26/AI-Agentic-Design-Patterns-with-AutoGen 6 patterns.

[16] https://github.com/crewAIInc/crewAI-examples 16 patterns.

[17] https://github.com/langchain-ai/langgraphjs-gen-ui-examples 9 patterns.

[18] https://github.com/mastra-ai/mastra 35 patterns.

[19] The source code is available here: https://agentic-patterns.github.io/.

[20] The API costs for transforming the 66 patterns to AgentO totaled $2.72.

```
1  # Execution time: 107.83 seconds
2  # Model used: gpt-5-mini
3
4  # Issues / Assumptions:
5  # - The ontology lacks explicit properties for "backstory", "verbose"
       flags, "tool endpoints", and "agent prompt/backstory" fields. These
       are represented using dcterms:description literals on Agent and Tool
       individuals.
6  # - There is no dedicated property to link an Agent to the API keys or
       environment variables;
7  ...
8
9  @prefix : <http://www.w3id.org/agentic-ai/onto#> .
10 ...
11
12 :TripPlannerCrew rdf:type :Team ;
13     dcterms:title "Trip Planner Crew" ;
14     dcterms:description "A team composed of multiple agent roles (city
        selector, local expert, travel concierge) collaborating to identify a
        city, gather local knowledge, and produce a 7-day travel itinerary." ;
15     :hasAgentMember :Agent_CitySelection ,
16                    :Agent_LocalExpert ,
17                    :Agent_TravelConcierge ;
18     :hasWorkflowPattern :Workflow_TripPlanning .
19 ...
```

Listing 1.2. Translation output for the Crew AI trip planner (abbreviated)

3. Human-in-the-Loop Verification and Refinement. Next, we manually selected and reviewed 6 translated patterns from each framework (24 in total). This included a review of the results, and a discussion of the identified issues. For changes we considered helpful, we extended the ontology. In Table 1 we show an excerpt of the 27 common issues we discovered and how we addressed them.

Table 1. Identified issues and resolutions during ontology refinement (excerpt)

#	Source	Issue	Resolution
1	All	Missing axioms: Prompt & Config Classes and properties; temporal properties	Add Prompt class and relevant properties (e.g., prompt, role); introduce a Config class (runtime/design-time).
2	CrewAI, AutoGen, Mastra	Missing axioms: datatype properties for Agent runtime (e.g., allow_delegation, verbose).	We will not model these implementation details.
3	All	Confusion between Agent and Tool.	Tool is a superclass of LLMAgent.
4	All	No structured representation for function calls, loops, or invocation semantics.	We will not model these implementation details.

308   A. Ekelhart et al.

```
PREFIX :     <http://www.w3id.org/agentic-ai/onto#>
PREFIX dct: <http://purl.org/dc/terms/>

CONSTRUCT {  ?team  :hasAgentMember     ?agent ;
              # similar to triple pattern after WHERE clause..
} WHERE {?team a :Team ; :hasAgentMember     ?agent ;
                :hasWorkflowPattern ?workflowPattern ;
                dct:title    ?teamTitle .
        ?agent :hasAgentGoal ?goal .
  OPTIONAL { ?agent :agentToolUsage ?tool . }
  OPTIONAL { ?workflowPattern :hasWorkflowStep ?workflowStep .
            ?workflowStep :hasAssociatedTask ?task ;
                              :relatedStep   ?nextWorkflowStep .
            ?task :performedByAgent ?agent .
  OPTIONAL { ?task :requiresResource ?resource . }
  } FILTER (regex(str(?teamTitle), "Trip Planner", "i"))
}
```

Listing 1: SPARQL Query for Constructing Agentic AI Patterns

After the extension of the ontology, we re-ran the automated extraction process with the updated ontology and compared them to the initial round. The new concepts and changes were picked up and integrated.

4. KG Generation and Publishing. The result triples from the translations were compiled into a coherent RDF graph. The resulting KG is in a triple store and publicly available through a SPARQL endpoint[21]. This provides access to the agentic AI patterns for downstream applications such as agent composition, design inspection, and explainability reasoning.

5   Use Case Scenarios

In this section, we present three practical use-cases that demonstrate the utility and expressiveness of AgentO in real-world scenarios. The KG provides access to agentic AI patterns for downstream applications such as agent composition, design inspection, and explainability reasoning. For example, developers can query the KG to retrieve all agent configurations involved in a data analysis pipeline or inspect common tools used in report generation tasks[22].

Use-Case 1: Agentic AI Pattern Reconstruction. In this use case, we demonstrate how the AgentO ontology and its associated KG can be used to reconstruct agentic patterns in a fully declarative manner. This is particularly useful for users who want to understand the characteristics and behaviour of

---

[21] http://w3id.org/agentic-ai/sparql.

[22] An excerpt of the generated KG in RDF turtle format can be found in Listing 1.3 in the Appendix.

particular agentic workflows in solving specific problems. Users can more easily identify the interconnected elements of agentic AI systems (e.g., goals, the tasks to perform, tool use) rather than manually inspecting hard-coded agentic templates and source code. By formulating a SPARQL CONSTRUCT query, as shown in Listing 1, users can retrieve and generate a subgraph pattern from existing agentic workflows.

The query retrieves and reconstructs a subgraph of an agentic workflow related to a specific objective, in this case, "Trip Planner"—including the agents, their tasks, goals, tools, and resources involved. Figure 6 shows a graph visualization of the resulting query in Listing 1. It highlights how each agent contributes to a shared team objective through assigned tasks and workflow steps. The generated subgraph shows the connected elements of agentic patterns necessary for achieving the "Travel Planning" objective.

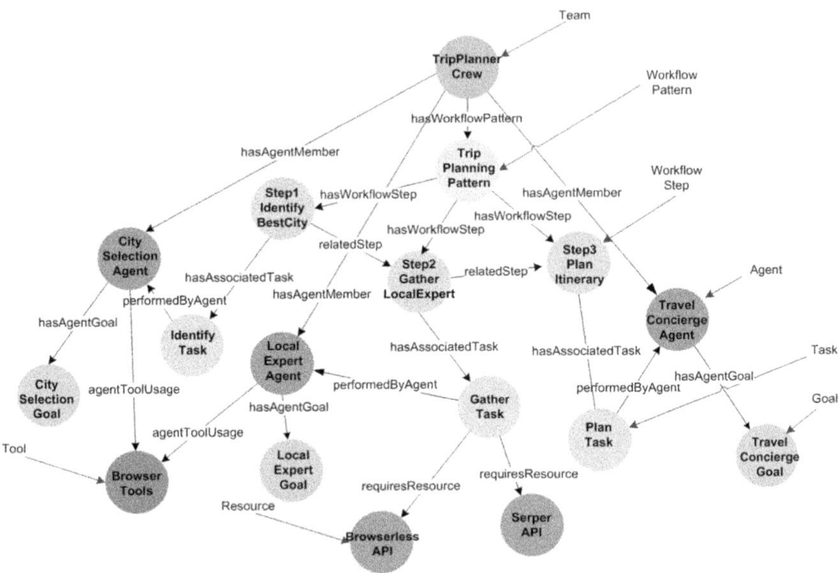

Fig. 6. Visualization of the constructed agentic pattern for Travel Planning.

Use-Case 2: Reusability of Agentic AI components. This use case showcases how the AgentO ontology and its KG facilitate the discovery and reuse of agentic components such as tasks, agents, tools, and resources for addressing complex problem scenarios. For example, a travel application may require a system that selects the best destination for a traveler, gathers local insights, and composes a detailed multi-day itinerary. Instead of designing the agents and their configurations from scratch, we can query the KG and reuse the existing concepts and agentic patterns.

```
PREFIX : <http://www.w3id.org/agentic-ai/onto#>
PREFIX dct: <http://purl.org/dc/terms/>

SELECT DISTINCT ?agent ?dtask ?dlm ?dtool ?dresource
WHERE { ?team a :Team ; dct:description ?Objdesc ;
                  :hasAgentMember ?agent .
        ?task :performedByAgent ?agent ; dct:description ?dtask .
        OPTIONAL {  ?agent :useLanguageModel ?lm .
                    ?lm dct:description ?dlm . }
        OPTIONAL {  ?agent :agentToolUsage ?tool.
                    ?tool dct:description ?dtool .}
        OPTIONAL {   { ?tool :resourceUsage ?resource } UNION
                     { ?task :requiresResource ?resource }
                 ?resource dct:description ?dresource .}
        FILTER(regex(?Objdesc, "Trip", "i"))
}
```

Listing 2: SPARQL Query for existing agentic patterns

As shown in Listing 2, the formulated SPARQL query enables users to retrieve previously instantiated Agentic-AI pattern compositions and configurations aligned with a specific objective. The ability to semantically query reusable components based on high-level objectives provides modularity, reduces duplication, and accelerates system development in agentic AI workflows. Table 2 shows the result with three agent definitions along with their associated tasks, language models, tools, and resources required to perform trip planning.

Table 2. Retrieved Agentic-AI Pattern Components for Travel Agents

Agent	Task	LLM	Tool	Resource
:CitySelection Agent	Analyze and choose best city for trip based on weather, events, and travel cost	OpenAI LLM client	Internet search (Serper)	Organic search results
:LocalExpert Agent	Create in-depth city guide with attractions, culture, hidden gems, and tips	OpenAI LLM client	Internet search (Serper)	Organic travel insights
:TravelConcierge Agent	Expand guide into 7-day itinerary: daily plans, hotels, restaurants, budget	OpenAI LLM client	Internet search + itinerary planning tools	Real-world travel info

Use-Case 3: Agentic AI Implementation Auditing. In this use case, we demonstrate how the Agentic AI Ontology and its KG enable the auditing of agent-driven workflow patterns, i.e., to support explainability, transparency, and traceability in multi-agent AI systems. The SPARQL query in Listing 3 targets a specific workflow pattern e.g., `:RecruitmentWorkflow` and retrieves the

```
PREFIX :<http://www.w3id.org/agentic-ai/onto#>
PREFIX dcterms:<http://purl.org/dc/terms/>

SELECT DISTINCT ?step ?stepOrder ?taskDescription ?agent ?tool
WHERE { :RecruitmentWorkflow :hasWorkflowStep ?step .
        ?step :stepOrder ?stepOrder ; :hasAssociatedTask ?task .
        ?task  :performedByAgent ?agent .
    ?task dcterms:description ?taskDescription .
    OPTIONAL {?agent :agentToolUsage ?tool .}
} ORDER BY ?stepOrder
```

Listing 3: SPARQL Query to investigate a specific workflow pattern

sequence of workflow steps `:hasWorkflowStep` along with their order, associated task descriptions, responsible agents, and optionally the tools used.

The results, illustrated in Sect. 3, present a traceable view of the AI-driven recruitment process, detailing a multi-step workflow including researching job candidates, evaluating suitability, engaging selected candidates, and summarizing hiring recommendations (Table 3).

Table 3. Retrieved Agentic-AI Workflow Components: Recruitment Use Case

Step	Order	Task	Agent	Tool
`ws_research`	1	Research job candidates using online sources; ensure requirement match	Researcher Agent	SerperDev Search API
`ws_match_and_score`	2	Evaluate and score candidate-job suitability	Matcher Agent	SerperDev Search API
`ws_outreach`	3	Engage and reach selected candidates	Communicator Agent	SerperDev Search API
`ws_report`	4	Summarize findings and recommend best candidates	Reporter Agent	—

6   Conclusions and Future Work

In this paper, we introduced the AgentO ontology and the accompanying KG to address the limitations in designing current agentic AI systems, which are often implemented in an ad-hoc, monolithic manner. By capturing the fundamental concepts such as agents, tasks, goals, and interactions in a standardized RDF/OWL based representation, our approach facilitates modularity, interoperability, and reusability across diverse agentic AI frameworks. Through three use cases, we demonstrated how our ontology enables the declarative construction of agentic AI workflows, the reuse of existing agent-task configurations for new problems, and the auditing of agentic AI implementations for explainability and traceability.

Future work will extend the ontology to cover additional aspects of agentic AI systems. In particular, we will explore dynamic execution traces and enable the semantic representation of runtime behaviors and decision flows. We also plan to enrich the modeling of interactions and behavior patterns, as well as to explore alignment and deeper integration with emerging tool interface protocols such as MCP and communication protocols such as A2A. Supporting discoverability and reuse of agents and tools is another area of focus. AgentO already reuses standards including PROV-O, P-Plan, and BEAM. We will further investigate mappings to complementary models to strengthen interoperability. The ontology will continue to be developed and maintained as part of the BILAI (Bilateral AI Clusters of Excellence) project, which runs until autumn 2029. In addition, as part of another research project, we aim to develop and maintain an open-source software that can (semi-automatically) translate AgentO workflows to specific target frameworks. Users can then model agentic workflows independently using AgentO and select a target framework for execution. We further plan to use AgentO as basis for the automated construction of agentic workflows from natural-language specifications.

Declaration of use of Generative AI

During the preparation of this work, the authors used Grammarly and ChatGPT to assist with grammar and spelling checks. After using these tools, the authors reviewed and edited the content and take full responsibility for the publication's content.

Acknowledgments. This research is funded by the Indonesian Endowment Fund for Education (LPDP) on behalf of the Indonesian Ministry of Higher Education, Science and Technology and managed under the EQUITY Program (Contract Number: 4301/B3/DT.03.08/2025 and 10107/UN1.P/Dit-Keu/HK.08.00/2025). This work was supported by the Austrian Science Fund (FWF) Bilateral AI project (Grant Nr. 10.55776/COE12) and the Austrian Research Promotion Agency (FFG) FAIR-AI project (Grant Nr. FO999904624). This work is related to the AIAGENT4CYBER project (Project 101249698), funded by the European Union. Views and opinions expressed are however those of the author(s) only and do not necessarily reflect those of the European Union or the granting authority. Neither the European Union nor the granting authority can be held responsible for them. SBA Research (SBA-K1 NGC) is a COMET Center within the COMET Competence Centers for Excellent Technologies Programme and funded by BMIMI, BMWET, and the federal state of Vienna. The COMET Programme is managed by FFG.

Appendix

See Tables 4 and 5.

Prompt Instructions

You are an expert in agent systems and ontology population.
You will be given: 1) An existing ontology file in Turtle format (http://www.w3id.org/agentic-ai/onto) ontology 2) The source code and configuration of an agent-based solution (with agents, tasks, tools, workflows, prompts, etc.) {{source_code}}
Your task:
1. Study the ontology file. - Treat it as a fixed schema with all classes and properties already defined. - Do NOT add, modify, or remove any classes or properties in the schema!
2. Study the source code and configuration. - Extract all instance-level information needed to fully describe the solution (agents, tasks, tools, workflows, prompts, parameters, data/artifacts, etc.).
3. Populate the ontology with individuals based on the extracted information. The goal is that another LLM can reconstruct the agent solution from the ontology instances you create. Do not add source code because we do not know the target framework to recreate the agent solution, so keep the semantic meaning and logic instead. - Use ONLY the existing classes and properties from the ontology! - Create individuals for all relevant entities and connect them with the appropriate properties. - Preserve all information from the source code (including prompts, parameters, and important logic) as literals or links between individuals. Fidelity is important, do not change or condense information.
4. If some aspects of the solution cannot be modeled with the current ontology: - Do NOT invent new classes or properties! - Do NOT model programming information (e.g., specific code structures, language-specific constructs, or implementation details), we are interested in the semantic meaning. Do NOT model framework specific functions and SDKs! Do NOT model UI components! - Model them as closely as possible with the existing schema but do not abuse the schema (e.g., do not press different classes and info in literals and descriptions). - If concepts are missing: list any missing concepts, limitations, or necessary extensions in an "Issues / Assumptions" comment block at the top of the Turtle output. Do not put it into the ontology in descriptions, etc.
Output format:
- Respond with the instance information in .ttl format. - At the very top of the Turtle content, write a comment block:
Issues / Assumptions: - <issue 1 or "No issues detected"> - <issue 2> ...

Do not ask for confirmation or clarifications, just produce the output as specified.

```
@prefix : <http://www.w3id.org/agentic-ai/onto#> .
@prefix pp: <http://purl.org/net/p-plan#> .
@prefix owl: <http://www.w3.org/2002/07/owl#> .
@prefix rdf: <http://www.w3.org/1999/02/22-rdf-syntax-ns#> .
@prefix xml: <http://www.w3.org/XML/1998/namespace> .
@prefix xsd: <http://www.w3.org/2001/XMLSchema#> .
@prefix beam: <http://w3id.org/beam/core#> .
@prefix rdfs: <http://www.w3.org/2000/01/rdf-schema#> .
@base <http://www.w3id.org/agentic-ai/onto#> .

:TripPlannerCrew a :Team ;
    <http://purl.org/dc/terms/title> "Trip Planner Crew" ;
    <http://purl.org/dc/terms/description> "A crew composed of multiple
    LLM agents and tools to plan trips: city selection, local expertise
    gathering, and travel concierge planning." ;
    :hasAgentMember :CitySelectionAgent , :LocalExpertAgent , :
    TravelConciergeAgent ;
    :hasWorkflowPattern :TripPlanningPattern ;
    :hasSystemConfig :EnvConfig .

:CitySelectionAgent a :LLMAgent ;
    :agentID "city_selection_agent" ;
    :agentRole "City Selection Expert" ;
    <http://purl.org/dc/terms/description> "An expert in analyzing travel
    data to pick ideal destinations" ;
    :agentPrompt :CitySelectionAgent_BackstoryPrompt ;
    :hasAgentGoal :CitySelectionGoal ;
    :hasAgentCapability :CitySelectionCapability ;
    :agentToolUsage :SearchTools , :BrowserTools ;
    :useLanguageModel :OpenAI_LanguageModel ;
    :hasAgentConfig :OpenAIConfig ;
    :interactsWith :LocalExpertAgent , :TravelConciergeAgent .
...

:CitySelectionAgent_BackstoryPrompt a :Prompt ;
    :promptInstruction "An expert in analyzing travel data to pick ideal
    destinations" ;
    :promptContext "Role: City Selection Expert; purpose: select the best
    city based on weather, season, and prices." ;
    :promptOutputIndicator "Use tools to find weather patterns, seasonal
    events, and travel costs. Produce a detailed report." .
...

:CitySelectionGoal a :Goal ;
    <http://purl.org/dc/terms/title> "Select the best city" ;
    <http://purl.org/dc/terms/description> "Select the best city based on
    weather patterns, seasonal events, and travel costs" .
...
:CitySelectionCapability a :Capability ;
    <http://purl.org/dc/terms/description> "Analyze and compare cities by
    weather conditions, events, and travel costs; deliver a detailed city
    selection report." .
```

Listing 1.3. Agentic AI pattern for "Trip Planner" in RDF Turtle. (excerpt)

Table 4. Agentic AI Core Concepts. (Ag = AutoGen, Cr = CrewAI, Lg = LangGraph, Ma = Mastra AI); ✓ = supported; × = not explicitly supported by the respective framework.

Class	Description	Ag	Cr	Lg	Ma
Agent	An entity capable of perceiving its environment, processing information, and acting to achieve goals.	✓	✓	✓	✓
LLMAgent	An AI agent based on a language model capable of autonomous task execution.	✓	✓	✓	✓
HumanAgent	A human participating or collaborating with agentic workflows.	✓	✓	×	✓
Capability	An ability that can be performed by an agent.	✓	✓	✓	✓
Tool	An instrument used by an agent to enhance its capabilities.	✓	✓	✓	✓
Resource	Any asset that can be utilized by an agent or tool to perform tasks.	✓	✓	✓	✓
Task	A specific activity that contributes to achieving objectives or goals.	✓	✓	✓	✓
WorkflowStep	An individual action or phase within a workflow pattern.	✓	✓	✓	✓
StartStep	The initial step in a workflow pattern.	✓	✓	✓	✓
EndStep	The final step in a workflow pattern.	✓	✓	✓	✓
WorkflowPattern	A reusable structured workflow template.	✓	✓	✓	✓
Goal	A desired state that an agent or team aims to achieve.	✓	✓	✓	✓
Objective	A collective objective assigned to a team.	✓	✓	×	×
Team	A coordinated group of agents.	✓	✓	×	×
Environment	The surroundings or context in which an agent operates.	✓	✓	✓	✓
KnowledgeBase	A structured collection of information that an agent can reference.	✓	✓	✓	✓
Memory	Stores and retrieves past information to support reasoning and decision-making.	✓	✓	✓	✓
Prompt	A structured instruction or input given to an agent.	✓	✓	✓	✓
Constraint	A rule or condition that restricts or guides agent decisions or actions.	✓	✓	✓	✓
Config	Configuration settings for agents, tools, or environments.	✓	✓	✓	✓
Context	A conceptual frame containing relevant situational data for reasoning and action.	✓	✓	✓	✓
Instance	A data object or content referenced/used within workflows.	✓	✓	✓	✓
LanguageModel	The model underlying LLM Agents.	✓	✓	✓	✓

Table 5. Identified Relationships Model for Agentic AI Systems

Property	Domain	Range	Description
participatedIn	Agent (prov)	Task	Agent participated in task execution.
humanParticipatedIn	HumanAgent	Task	Human agent participated in a task.
agentPrompt	LLMAgent	Prompt	Prompt used by an agent independently of tasks.
taskPrompt	Task	Prompt	Prompt used within a task.
hasPrompt	Agent or Task	Prompt	Associates prompts with agents or tasks.
agentResourceUsage	LLMAgent	Resource	Agent directly consumes a resource.
resourceUsage	Tool	Resource	Tool accesses a resource on behalf of the agent.
containsResource	Environment	Resource	Environment includes resources.
producedResource	Task	Resource	Resource produced as task output.
requiresResource	Task	Resource	Task requires a resource.
agentToolUsage	LLMAgent	Tool	Agent directly uses a tool.
toolUsage	Tool	Tool	A tool uses another tool during execution.
hasAgentCapability	LLMAgent	Capability	Capability an agent can perform.
requiresCapability	Task	Capability	Capability needed by a task.
hasCapability	Tool	Capability	Capability provided by a tool.
hasAgentConfig	LLMAgent	Config	Configuration settings for an LLM agent.
hasEnvironmentConfig	Environment	Config	Environment configuration settings.
hasSystemConfig	Team	Config	Configuration settings specific to team system.
hasToolConfig	Tool	Config	Configuration for tool.
hasConfig	Any	Config	Configuration associated with any component.
hasAgentGoal	LLMAgent	Goal	Goal assigned to an agent.
hasTeamGoal	Team	Goal	Goal assigned to a team.
hasGoal	Agent or Team	Goal	Relates entity to its goals.
contributesToGoal	Objective	Goal	Objective contributes to goal.
contributesToObjective	Task	Objective	Task contributes to objective.
hasObjective	Agent or Team	Objective	Assigned objective of a team/agent.
hasKnowledge	LLMAgent	KnowledgeBase	Knowledge possessed by an agent.
hasWorkflowPattern	Team	WorkflowPattern	Assigned workflow pattern of a team.
hasWorkflowStep	WorkflowPattern	WorkflowStep	Step that is part of workflow pattern.
hasAssociatedTask	WorkflowStep	Task	Workflow step corresponds to a task.
relatedStep	WorkflowStep	WorkflowStep	Relationship between workflow steps.
nextStep	WorkflowStep	WorkflowStep	Next step in workflow sequence.
hasSubPattern	WorkflowPattern	WorkflowPattern	Workflow sub-pattern relationship.
nextPattern	WorkflowPattern	WorkflowPattern	Next workflow pattern in sequence.
hasRelatedPattern	WorkflowPattern	WorkflowPattern	Pattern relationship.
hasAgentMember	Team	LLMAgent	Team includes agent members.
interactsWith	LLMAgent	LLMAgent	Agents interacting with each other.
operatesIn	LLMAgent	Environment	Agent's operational environment.
performedBy	Task	Tool	Tool that performs the task.
performedByAgent	Task	LLMAgent	Agent that performs the task.
useLanguageModel	LLMAgent	LanguageModel	Associates agent with language model.

Table 6. Agentic Pattern Collected from AutoGen, CrewAI, LangGraph and Mastra AI Repository (✓ = provided; × = not explicitly provided).

#	Title	Framework	Tool Use	Multi-Agent	Workflow Pattern
	AutoGen				
1	Multi-Agent Conversation and Stand-up Comedy	AutoGen	×	✓	Sequential
2	Sequential Chats and Customer Onboarding	AutoGen	×	✓	Sequential
3	Reflection and Blogpost Writing	AutoGen	×	✓	Nested-sequential
4	Tool Use and Conversational Chess	AutoGen	✓	✓	Hybrid
5	Coding and Financial Analysis	AutoGen	×	✓	Hybrid
6	Planning and Stock Report Generation	AutoGen	×	✓	Hybrid
	CrewAI				
7	Game Builder Crew	CrewAI	×	✓	Sequential
8	Industry Agents	CrewAI	✓	✓	Sequential
9	Instagram Post	CrewAI	✓	✓	Sequential
10	Job Posting	CrewAI	✓	✓	Sequential
11	Landing Page Generator	CrewAI	✓	✓	Sequential
12	Markdown Validator	CrewAI	✓	×	Single-step
13	Match Profile to Positions	CrewAI	✓	✓	Sequential
14	Marketing Strategy	CrewAI	✓	✓	Sequential
15	Meta Quest Knowledge	CrewAI	×	×	Sequential
16	Prep for a Meeting	CrewAI	✓	✓	Sequential
17	Recruitment	CrewAI	✓	✓	Sequential
18	Screenplay Writer	CrewAI	✓	×	Single-step
19	Starter Template	CrewAI	✓	×	Single-step
20	Stock Analysis	CrewAI	✓	✓	Sequential
21	Surprise Trip	CrewAI	✓	✓	Sequential
22	Trip Planner	CrewAI	✓	✓	Sequential
	LangGraph				
23	Chat Agent	LangGraph	×	×	Single-step
24	Email Agent	LangGraph	×	×	Hybrid
25	Open Code	LangGraph	✓	✓	Sequential
26	Pizza Orderer	LangGraph	✓	×	Sequential
27	Stockbroker	LangGraph	✓	×	Sequential
28	Supervisor	LangGraph	✓	✓	Hybrid
	Mastra AI				
29	A2A	Mastra AI	×	✓	Sequential
34	Bird Checker with Express	Mastra AI	✓	×	Single-step
35	Bird Checker with NextJS	Mastra AI	✓	×	Sequential
36	Bird Checker with NextJS and Eval	Mastra AI	✓	×	Sequential
37	Client Side Tools	Mastra AI	✓	×	Single-step
38	Crypto Chatbot	Mastra AI	✓	✓	Single-step
39	DANE	Mastra AI	✓	✓	Nested-sequential
41	Fireworks R1	Mastra AI	×	×	Single-step
42	Heads Up Game	Mastra AI	✓	✓	Sequential
43	Memory per Resource Example	Mastra AI	✓	×	Sequential
44	Memory Todo Agent Memory	Mastra AI	✓	×	Sequential
45	Memory with Context	Mastra AI	×	×	Single-step
46	Memory with LibSQL	Mastra AI	×	✓	Sequential
47	Memory with MongoDB	Mastra AI	✓	✓	Sequential
48	Memory with Postgres	Mastra AI	✓	✓	Sequential
49	Memory with Processors	Mastra AI	✓	✓	Sequential
50	Memory with Upstash	Mastra AI	✓	✓	Sequential
51	MCP Configuration	Mastra AI	✓	×	Single-step
52	MCP Registry	Mastra AI	✓	×	Sequential
53	OpenAPI Spec Writer	Mastra AI	✓	×	Sequential
54	Quick Start	Mastra AI	✓	×	Single-step
55	Stock Price Tool	Mastra AI	✓	×	Single-step
56	Travel App	Mastra AI	✓	✓	Sequential
57	Weather Agent	Mastra AI	✓	×	Sequential
58	Workflow AI Recruiter	Mastra AI	✓	×	Sequential
59	Workflow with Inline Steps	Mastra AI	×	✓	Sequential
60	Workflow with Memory	Mastra AI	✓	×	Hybrid
62	Workflow with Separate Steps	Mastra AI	×	×	Sequential
63	Workflow with Suspend Resume	Mastra AI	×	×	Hybrid
64	YC Directory	Mastra AI	✓	×	Sequential
65	AI SDK v5	Mastra AI	✓	×	Single-step
66	AI SDK UseChat	Mastra AI	✓	✓	Sequential

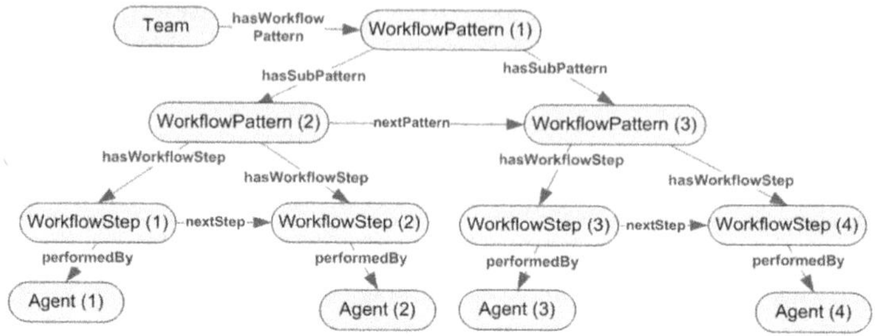

Fig. 7. An example of "nested pattern" workflow model

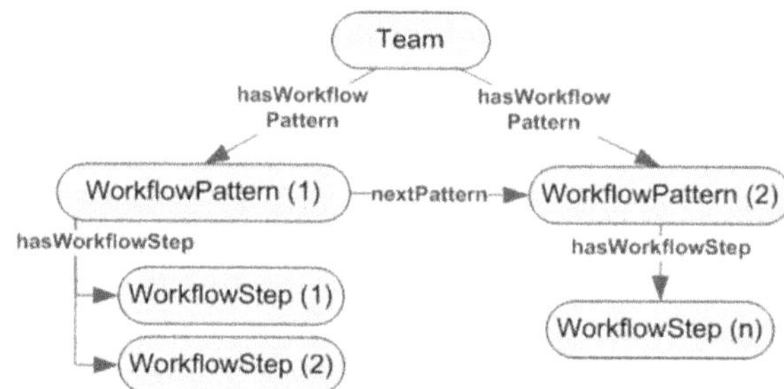

Fig. 8. An example of "parallel pattern" workflow model

References

1. Aadhithya, A.A., Kumar, S.S., Soman, K.P.: Enhancing long-term memory using hierarchical aggregate tree for retrieval augmented generation (2024). https://arxiv.org/abs/2406.06124
2. Akkiraju, R., Farrell, J., Miller, J.A., Nagarajan, M., Sheth, A.P., Verma, K.: Web service semantics-wsdl-s (2005). https://corescholar.libraries.wright.edu/knoesis/69
3. Aylak, B.L.: Sustai-scm: Intelligent supply chain process automation with agentic ai for sustainability and cost efficiency. Sustainability **17**(6) (2025). https://doi.org/10.3390/su17062453, https://www.mdpi.com/2071-1050/17/6/2453
4. Battle, S., et al.: Semantic web services language (swsl). W3C Member Submission **9**, 1–61 (2005)
5. Battle, S., et al.: Semantic web services ontology (swso). Member submission, W3C (2005)
6. Berners-Lee, T., Hendler, J., Lassila, O.: The semantic web. Sci. Am. **284**(5), 34–43 (2001)
7. Cabral, L., Domingue, J., Motta, E., Payne, T., Hakimpour, F.: Approaches to semantic web services: an overview and comparisons. In: Bussler, C.J., Davies, J., Fensel, D., Studer, R. (eds.) The Semantic Web: Research and Applications, pp. 225–239. Springer, Berlin Heidelberg, Berlin, Heidelberg (2004)

8. Ekaputra, F.J., Prock, A., Kiesling, E.: Towards supporting ai system engineering with an extended boxology notation. In: The 2nd International Workshop on Knowledge Graphs for Responsible AI (KG-STAR) Co-located with the Extended Semantic Web Conference (ESWC 2025), CEUR-WS (2025)

9. Garijo Verdejo, D., Gil, Y.: Augmenting prov with plans in p-plan: scientific processes as linked data. CEUR Workshop Proceedings (2012)

10. Gibbins, N., Harris, S., Shadbolt, N.: Agent-based semantic web services. In: Proceedings of the 12th International Conference on World Wide Web., pp. 710–717 (2003)

11. Gridach, M., Nanavati, J., Abidine, K.Z.E., Mendes, L., Mack, C.: Agentic ai for scientific discovery: A survey of progress, challenges, and future directions. arXiv preprint arXiv:2503.08979 (2025)

12. Han, S., Zhang, Q., Yao, Y., Jin, W., Xu, Z., He, C.: LLM multi-agent systems: challenges and open problems (2024). https://arxiv.org/abs/2402.03578

13. Hao, R., Hu, L., Qi, W., Wu, Q., Zhang, Y., Nie, L.: Chatllm network: More brains, more intelligence (2023). https://arxiv.org/abs/2304.12998

14. Hogan, A.: The semantic web: two decades on. Semant. Web **11**(1), 169–185 (2020)

15. Karunanayake, N.: Next-generation agentic AI for transforming healthcare. Inf. Health **2**(2), 73–83 (2025). https://doi.org/10.1016/j.infoh.2025.03.001, https://www.sciencedirect.com/science/article/pii/S2949953425000141

16. Kshetri, N.: Transforming cybersecurity with agentic ai to combat emerging cyber threats. Telecommun. Policy 102976 (2025). https://doi.org/10.1016/j.telpol.2025.102976, https://www.sciencedirect.com/science/article/pii/S0308596125000734

17. Lebo, T., et al.: Prov-o: The prov ontology (2013)

18. McIlraith, S.A., Martin, D.L.: Bringing semantics to web services. IEEE Intell. Syst. **18**(1), 90–93 (2005)

19. Miehling, E., et al.: Agentic ai needs a systems theory. arXiv preprint arXiv:2503.00237 (2025)

20. Ng, A.: Agentic design patterns part 1 (2024)

21. Noy, N.F., McGuinness, D.L., et al.: Ontology development 101: a guide to creating your first ontology (2001)

22. Okpala, I., Golgoon, A., Kannan, A.R.: Agentic ai systems applied to tasks in financial services: modeling and model risk management crews. arXiv preprint arXiv:2502.05439 (2025)

23. Park, J.S., O'Brien, J.C., Cai, C.J., Morris, M.R., Liang, P., Bernstein, M.S.: Generative agents: Interactive simulacra of human behavior (2023). https://arxiv.org/abs/2304.03442

24. Qian, C., et al.: ChatDev: Communicative agents for software development. In: Ku, L.W., Martins, A., Srikumar, V. (eds.) Proceedings of the 62nd Annual Meeting of the Association for Computational Linguistics (Volume 1: Long Papers). pp. 15174–15186. Association for Computational Linguistics, Bangkok, Thailand (2024). https://doi.org/10.18653/v1/2024.acl-long.810, https://aclanthology.org/2024.acl-long.810

25. Roman, D., et al.: Web service modeling ontology. Appl. Ontol. **1**(1), 77–106 (2005)

26. Shadbolt, N., Berners-Lee, T., Hall, W.: The semantic web revisited. IEEE Intell. Syst. **21**(3), 96–101 (2006)

27. Talebirad, Y., Nadiri, A.: Multi-agent collaboration: harnessing the power of intelligent LLm agents (2023). https://arxiv.org/abs/2306.03314

28. Wang, G., et al.: Voyager: an open-ended embodied agent with large language models (2023). https://arxiv.org/abs/2305.16291

29. Wang, J., Duan, Z.: Agent ai with langgraph: a modular framework for enhancing machine translation using large language models (2024). https://arxiv.org/abs/2412.03801
30. Wang, L., et al.: A survey on large language model based autonomous agents. Front. Comp. Sci. **18**(6), 186345 (2024)
31. Wu, Q., et al.: Autogen: Enabling next-gen llm applications via multi-agent conversation. arXiv preprint arXiv:2308.08155 (2023)
32. Yao, S., et al.: React: Synergizing reasoning and acting in language models (2023). https://arxiv.org/abs/2210.03629

BOLD: A Simulation Framework
for Dynamic Linked Data Environments
and a Benchmark for Linked Data User
Agents

Tobias Käfer[1]([✉])(iD), Victor Charpenay[2](iD), and Andreas Harth[3](iD)

[1] Karlsruhe Institute of Technology (KIT), Karlsruhe, Germany
`tobias.kaefer@kit.edu`
[2] Mines Saint-Etienne and UMR 6158 LIMOS, Saint-Étienne, France
`victor.charpenay@emse.fr`
[3] Friedrich-Alexander-Universität Erlangen-Nürnberg, Nuremberg, Germany
`andreas.harth@fau.de`

Abstract. The paper presents the BOLD (Buildings on Linked Data) framework to simulate dynamic Linked Data environments, next to an instantiation, a benchmark for Linked Data agents built using the framework. The BOLD benchmark provides a read-write Linked Data interface to a smart building with simulated time, occupancy movement and sensors and actuators around lighting. On the Linked Data representation of this environment, agents carry out specified tasks, such as controlling illumination. The BOLD framework runs the environment and provides means to check for the correct execution of tasks and to measure agent performance. We conduct measurements on rule-based Linked Data agents.

1 Introduction

Technologies from the Semantic Web stack are nowadays the technologies of choice to provide interoperable interfaces to information systems, including data ecosystems, and cyber-physical systems (CPS). Yet, programming applications for data ecosystems and controlling web-based cyber-physical systems requires novel approaches, as the Linked Data environment [2] has peculiar characteristics usually not found in other environments:

1. Hypermedia, i.e. the possibility of following links, allowing agents to discover new system components and possible actions (cf. HATEOAS in REST [19]),
2. Semantic alignments, allowing agents to bridge between terminologies of different system components using reasoning, and
3. RESTful interaction via resource representations, allowing agents to read/write information from/to various systems.

URL https://github.com/bold-benchmark/ **License** GPLv3.

To address those peculiarities, Solid developers [34], who write apps for data ecosystems, try novel approaches to program [35,49], and researchers investigate the combination of results from Agent-Oriented Programming and Semantic Web [6,7]. This paper aims to address the lack of testing and benchmarking frameworks for this environment.

Several initiatives illustrate the adoption of Semantic Web technologies in data ecosystems and for CPS. In government services, the adoption of Solid is documented[1], and other domains such as financial services and data ecosystems are investigating Solid [9]. The Linked Web Storage working group at the World Wide Web Consortium (W3C) has started in 2024. In Industry 4.0, the International Data Space uses semantic technologies [40]. In building information management, data can be represented using vocabularies such as Brick [1] or the work of the Linked Building Data community group at the W3C. Orthogonal works include the Semantic Sensor Network ontology (SSN) [23], a widely adopted W3C Recommendation to describe sensor and actuator data, and the Web of Things (WoT) Architecture and Thing Description [28,33] W3C Recommendations, which allow to access devices' sensor and actuator data. On top, we believe that Agentic Artificial Intelligence, which is often built on remote procedure calls (RPC) in the MCP protocol[2], will –similarly as RPC-style Web Services vanished– switch to REST, and shall require corresponding testing and gym environments.

Most attention in Linked Data research has been given to perfecting the environment, that is, storing and serving data at scale by servers [25,31,45]. Comparatively little work has been done on methodically programming Solid Apps or Linked Data agents, the flip side of the same coin. This gap could be filled in collaboration with the software engineering and the agent research community [14], which has been developing Multi-Agent System (MAS) architectures for manufacturing and building automation for decades [12,22,26].

However, MAS architectures feature agents that are physically situated in their environment. Thus, existing frameworks and benchmarks that already exist for MAS and that assume situated agents (e. g. [50]) are not applicable. For situatedness, agents are physically constrained in perceiving and acting [17]. In contrast, Linked Data agents are never strictly situated:

1. Linked Data servers make sensor data visible to all agents without distinction. Linked Data agents must restrict the scope of their perception to data relevant to their individual goal, e.g. by following certain links only.
2. Servers expose and consume symbolic representations of the physical environment. Different servers use different vocabularies that can be semantically aligned using mappings. Linked Data agents must interpret such alignments.
3. All potential actions may be exposed to Linked Data agents simultaneously. To take full advantage of parallel actions and avoid conflicts, Linked Data agents must restrict the scope of their actions.

---

[1] https://athumi.be/technologieen/solid.
[2] https://modelcontextprotocol.io/specification/2025-11-25.

Those points influence all three phases of an agent's cognitive loop: sensing, reasoning, and action; and echo the characteristics of Linked Data (link following, reasoning, RESTful interaction) that a Solid app should support.

In this paper, we thus introduce a framework to test apps and agents, and a benchmark to evaluate and compare agents, their architectures and runtimes. We call both BOLD (Buildings on Linked Data[3]). Our contributions are:

- A formal method for modelling dynamic Linked Data environments, and evaluating agent performances over such environments (Sect. 3);
- A corresponding execution framework (Sect. 5);
- An instantiation of our method and framework:
  - A dynamic Linked Data model of a building (building 3 of IBM Research Dublin) with simulated occupancy and lighting (Sect. 4); and
  - A set of agent evaluation tasks with increasing complexity (Sect. 6)

We start our exposition of BOLD with an example (Sect. 2), and close with a showcase evaluation (Sect. 7), related work (Sect. 8), and closing remarks (Sect. 9).

2   Example

We illustrate the main aspects of BOLD with the help of an example[4]. We first characterize the resources the server makes available as part of the benchmark, in the example the Coffee Dock room of IBM Research Dublin's building 3. Next, we give an overview of how an agent carries out a simple task, in the example switching on the light in the Coffee Dock room. Then, we say how the server tracks the agent's progress over time. Last, we draw an analogy to Solid Apps.

The resources relevant for the example are: The Coffee Dock Room, `:Room_CoffeeDesk`, the room's lighting system, `:Lighting_System_42GFLCoffeeDock`, and the lighting system's on/off property, `:property-Lighting_System_42GFL-CoffeeDock#it`. The description of the building, including rooms, lighting systems, and properties, uses the Brick ontology [1], plus terms from SOSA/SSN [23] to describe sensors and actuators, in particular the properties of sensing and actuation systems that agents can read and write. Thus, to understand the data, all an agent needs to know are those well-known ontologies, next to RDF(S).

For an agent to switch on the light in the Coffee Dock Room, the agent has to first discover the IRI of the light. Thus, the agent performs a HTTP `GET` request on `:Room_CoffeeDesk`, which leads to a response in RDF that contains a link to the lighting system. The agent next performs a traversal of this link and dereferences the IRI of the lighting system. The thus obtained graph contains a

---

[3] As we had started this work with an agent and Linked Data about a building.

[4] We assume the reader has a basic understanding of HTTP, RDF and SPARQL. The interested reader can find more details on these technologies e.g. in [24]. We follow the prefix practices recorded in http://prefix.cc/, except : as short for `http://localhost:8080/gsp/`, `brick:` as short for http://buildsys.org/ontologies/ Brick#, and `bf:` as short for http://buildsys.org/ontologies/BrickFrame#.

link to the lighting system's on/off property, where the agent traverses, finding out in the response's payload that the property's `rdf:value` is `"off"` and that the property is both observable and actuable. Finally, to achieve the task, the agent must set the property's value to `"on"`, which is done by an HTTP `PUT` request with a payload derived from the previous response, in which the value's triple is changed to `<#it> rdf:value "on"` ..

To test whether an agent succeeds in a task, each task is defined by 'faults' in the environment that the agent must fix, in MAS terms, a norm violation. In the simple task of turning on the light, there is one fault to fix, defined as the Coffee Dock lighting system having the property value `off`, expressed as a SPARQL `ASK` query. An agent has succeeded in the task if the query evaluates to false.

The simple task of turning on the light illustrates the basic case which just needs a single agent loop and for which a simple success metric suffices. The benchmark also includes tasks in which the agent has to loop and act continuously, which requires a more elaborate server setup (e.g. to provide dynamic data), together with a fault rate metric to capture the success of agents appropriately.

Similarly to finding the coffee dock's lighting system and turning it on, a Solid App often starts out with a WebID profile, to which it makes an HTTP `GET` request, and then discovers via link traversal a relevant storage or inbox, where the App can update data by writing RDF in an HTTP `PUT` request.

3   Formal Definitions

We now introduce a series of concepts to formally define the datasets, updates, tasks, and metrics used in the BOLD framework and benchmark. We build on [13] for the definitions. A task is defined by a set of faults that agents have to fix. Such tasks correspond to the MAS concept of norms, i. e. constraints to be satisfied. The evaluation of a task depends on a simulation environment. We define the simulation environment and metrics for faults and performance.

3.1   Dataset and Simulation Run

We use the usual abstract notation for RDF. I, B, L are disjoint sets of IRIs, blank nodes and literals. An RDF quad is an instance of $I \cup B \times I \times I \cup B \cup L \times I$ and an RDF dataset is a set of RDF quads. D denotes the set of RDF datasets.

Definition 1 (simulation run, simulation environment). *A simulation run is a finite sequence of datasets* $\langle d_0, d_1, \ldots, d_k \rangle$, $k \in \mathbb{N}$. *A simulation environment can be defined as the pair* $\langle d, u \rangle$, *where* $d \in D$ *and* u *is some update function* $u : D \mapsto D^n$, *such that for all possible simulation runs* $\langle d_0, d_1, \ldots, d_k \rangle$ *of the environment the following holds:* $d_0 = d$ *and* $\forall t < k : d_{t+1} \in u(d_t)$.

The update function u corresponds to a (finite set of) SPARQL UPDATE(s). A simulation environment can thus be seen as a transition system that generates a set of simulation runs (of arbitrary size). SPARQL updates may include non-deterministic function calls. The output of u is therefore a set of RDF datasets.

The above definition only defines 'dry runs', during which the environment is updated automatically without agent intervention. Between each environmental update, though, agents can act on the environment by updating a dataset d_i. Linked Data interfaces only allow a limited set of operations on Web resources (PUT, POST, DELETE), as formalized below (GET corresponds to the identity function id, as GET is considered safe and idempotent [18]). We note that our formalisation corresponds to the SPARQL graph store protocol [39] and the relationship between RDF datasets and Linked Data as outlined in Sec. 3.5 of [51], where the graph name (cf. 4th entry of a quad) denotes the IRI where the RDF graph can be dereferenced (or, in our case, also written), see also Sect. 5.2.

Definition 2 (agent operation). *Let $\delta : D \times D \mapsto D$ be the function calculating the symmetric difference between two datasets: $\delta(d, d') = (d \cup d') \setminus (d \cap d')$. Moreover, let π be a projection function such that $\pi(d)$ is the set of resource IRIs (graph names) appearing in d. A function $op : D \mapsto D$ is called an agent operation if $\pi(\delta(d, op(d)))$ is a singleton. The identity function $id(d) = d$ is an agent operation. On top, we use the following terminology:*

- *if $\pi(d) \subset \pi(op(d))$, op is a 'create' operation*
- *if $\pi(op(d)) \subset \pi(d)$, op is a 'delete' operation*
- *otherwise, $\pi(d) = \pi(op(d))$ and op is a 'replace' operation*

A full benchmark run is thus the interleaving of agent operations and environmental updates, obtained through composition, see $\circ$ in the following definition.

Definition 3 (benchmark run). *Let $\langle d, u \rangle$ be a simulation environment. A benchmark run ρ is a finite sequence $\langle d_0, d_1, \ldots, d_k \rangle$, such that: (1) $d_0 = d$, and (2) $\forall t < k$, there is a finite sequence of agent operations $\langle op_1, op_2, \ldots op_l \rangle$, such that $d_{t+1} \in op_1 \circ op_2 \circ \ldots op_l \circ u(d_t)$.*

3.2 Faults and Performance Metrics

Definition 4 (fault sequence). *A fault sequence γ is a finite sequence of datasets $\langle d_0, d_1, \ldots, d_l \rangle$, $l \in \mathbb{N}$. A run $\rho = \langle d_0, d_1, \ldots, d_k \rangle$ matches a fault sequence $\gamma = \langle d'_0, d'_1, \ldots, d'_l \rangle$ at time t if a) $l \leq t \leq k$ and b) $\forall i \mid 0 \leq i \leq l : d'_{l-i} \subseteq d_{t-i}$.*

Fault sequences can be defined in terms of sequences of SPARQL queries. Several fault sequences may occur in the environment at a given time. In the following, we denote Γ the (possibly infinite) set of fault sequences defined for a task. We also denote $\Gamma_{\rho,t}$ the set of all $\gamma \in \Gamma$ such that ρ matches γ at time t.

Every task is associated with potential fault sequences in the environment that agents must fix as quickly as possible. In the Coffee Dock room example, the SPARQL ASK query would be associated to one task (as a fault sequence of size one). Given a run ρ and a a set of fault sequences Γ (defining a task), one can specify performance metrics to evaluate ρ against Γ.

Definition 5 (success rate). *Let ρ be a benchmark run of size k and Γ a set of fault sequences of equal size l. The fault rate is the number of times at which ρ matches some $\gamma \in \Gamma$ divided by the length of ρ:*

$$\frac{|\{t \mid \Gamma_{\rho,t} \neq \emptyset\}|}{k - l} \tag{1}$$

When the task requires a single action loop (see Sect. 6), the fault rate gives an indication of 'how fast' an agent is at solving the task. When the task requires continuous action, the metric gives an indication of 'how efficient' an agent is over the whole duration of the task. Counting faults gives an indication of 'how close' an agent has been to solving the task on average.

Definition 6 (average fault count). *Let ρ be a benchmark run of size k and Γ be a set of fault sequences of equal size l. The average fault count is the sum of $\Gamma_{\rho,t}$ over ρ:*

$$\frac{\sum_{l \leq t \leq k} |\Gamma_{\rho,t}|}{|\{t \mid \Gamma_{\rho,t} \neq \emptyset\}|} \tag{2}$$

We do not only want to compare different agent architectures, but also the same architecture *across* tasks. To this end, we further define a task-independent fault count for a run that is normalized against a certain dry run.

Definition 7 (normalized fault count). *Let $\langle u, d \rangle$ be a simulation environment. Let ρ and $\bar{\rho}$, both of size k and both starting with d, be respectively a benchmark run and a simulation run in the environment. For all $\overline{d_t}$ of $\bar{\rho}$ ($t < k$), it holds that $\overline{d_{t+1}} \in u(\overline{d_t})$. The ratio*

$$\frac{\sum_{l \leq t \leq k} |\Gamma_{\rho,t}|}{\sum_{l \leq t \leq k} |\Gamma_{\bar{\rho},t}|} \tag{3}$$

is a normalized fault count if for all d_t of ρ ($t < k$): $d_{t+1} \cap \overline{d_{t+1}} = u(d_t) \cap u(\overline{d_t})$.

The above definition ensures that ρ and $\bar{\rho}$ are comparable in how non-deterministic functions behave during their execution, regardless of agent operations. In practice, it is enough to use a pseudo-random number generator with the same seed for those runs that are to be compared. The resulting normalized fault count then measures closeness to success for a task regardless of how many potential faults the task defines.

While the metrics defined so far relate to minimizing faults, they do not penalize agents that would perform unnecessary operations to reach their goal. To take this aspect into account, we finally introduce read/write ratio as a metric.

Definition 8 (read/write ratio). *Let ρ be a benchmark run of size k. Let there be a sequence $\langle op_{t,1}, op_{t,2}, \ldots op_{t,l_t} \rangle \ \forall t < k$, such that $d_{t+1} \in op_{t,1} \circ op_{t,2} \circ \ldots op_{t,l_t} \circ u(d_t)$. Each operation sequence at time t may include multiple occurrences of id, capturing read operations performed by agents. We denote i_t the number of occurences of id at time t. We thus define the read/write ratio for ρ as follows:*

$$\frac{\sum_{t<k} i_t}{\sum_{t<k} l_t - i_t} \tag{4}$$

4   Dataset

Our benchmark builds on Balaji et al.'s datasets that describe real buildings, which they published using their Brick ontology [1]. Of the datasets in [1], we chose building 3 of IBM Research Dublin, as it has diverse building systems with discrete and continuous state, including lighting, and built a Linked Data version.

4.1   The Brick Ontology

The Brick ontology is a simple model to represent buildings and their associated automation systems [1]. We base our considerations on version 1.0.0 of the Brick ontology, as this is the last version under which building 3 has been published.

The main classes we use from the Brick ontology are [10]:

- `brick:Equipment` refers to any technical equipment as part of some building automation system, including fire safety systems, HVAC (heating, ventilation and air conditioning), and lighting systems
- `brick:Point` refers to the data points (sensors, commands, set points) to monitor and control building automation systems
- `brick:Location` refers to any location inside a building (such as floors and rooms) that is of relevance for automation

The Brick ontology also defines properties:

- `bf:hasPoint` can relate locations and equipment to a `brick:Point`.
- `bf:hasPart` and its inverse `bf:isPartOf`, are transitive properties to define equipment and locations in a hierarchical fashion
- `bf:isLocatedIn`, a transitive property to specify the location of some equipment or a data point
- `bf:feeds`, a property to specify what locations are covered (fed) by some automation system, either directly or indirectly via other systems.

4.2   Description of Building 3 in Brick

Building 3 of IBM Research Dublin is a two-storey office building. The building has a rectangular ground plot with the long side oriented from north to south, i. e. most rooms are oriented eastwards or westwards. In the building description[5], we see a subdivision into floors and wings, to which the rooms are assigned using `bf:hasPart` relationships. The rooms and wings have `brick:Lighting_Systems` assigned. Such a light system comes in different forms: Some come with a `brick:Occupancy_Sensor` that determines whether there are people in its vicinity, others with a `brick:Luminance_Command`, in other words, a switch to control the lights, and others with a `brick:Luminance_Sensor` that we consider to be triggered by daylight. We provide basic statistics in Table 1.

Table 1. Basic counts for Building 3 and the benchmark.

Rooms	281	Lighting systems	278
with occupancy sensors	66	with occupancy sensors	156
with luminance commands	38	with luminance commands	105
with luminance sensors	20	with luminance sensors	48
Floors	2	Triples in IBM_B3.ttl	24 947
Wings	3	Resources IRIs	3 281
		Dynamic resources	551

4.3   Building 3 in Read-Write Linked Data

The description of building 3 as found online is one single static RDF graph with no dereferenceable URIs and no entity with a temporal extent (such as sensor measurements). To make it suitable for a Read-Write Linked Data benchmark, we extended this monolothic RDF graph in two ways. First, we scoped every triple in the graph with a dereferenceable resource IRI, such that the result is a proper RDF dataset, as defined in Sect. 3. Second, we extended the definition of the building's data points, in order to provide resources that change over time.

Resource Partitioning. As our aim is to provide a benchmark for Web agents, resource IRIs must be dereferenceable. Morever, we assume that cyber-physical systems with read-write Linked Data interfaces will be composed of large amounts of resources of small size (individually exposed by connected devices). To match this assumption, we partitioned the original RDF graph into smaller RDF graphs, each providing information about one room, one floor, one data point, etc. Specifically, we created an RDF dataset such that from every triple

---

[5] https://github.com/BuildSysUniformMetadata/GroundTruth/blob/2e48662/
building_instances/IBM_B3.ttl.

$\langle s, p, o \rangle$ from the graph, we derived the quads $\langle s, p, o, s \rangle$ and $\langle s, p, o, o \rangle$. We thus derived about 3k dereferenceable IRIs from 25k triples, with graphs of 5–50 triples (see Table 1).

Adding Dynamics to the Dataset. Although the original RDF graph includes resources like sensors and actuators, it does not provide means to expose actual sensor measurements (such as occupancy or an illuminance value) nor does it provide means to expose the state of actuators (such as the status of a light switch). To complete the dataset on which BOLD is defined, we included time-varying and writable resources for such sensor and actuator data. To that end, we aligned the Brick ontology with SOSA/SSN as illustrated on Fig. 1. We then extended our dataset with resources defined as instances of `ssn:Property`. For all data points of the building 3 lighting systems, we created a property resource, cf. the example payload of Sect. 2. We only consider lighting systems since all tasks of BOLD are defined according to a lighting scenario (see Sect. 6). Overall, we added 551 dereferenceable IRIs with dynamic representations, see Tab. 1.

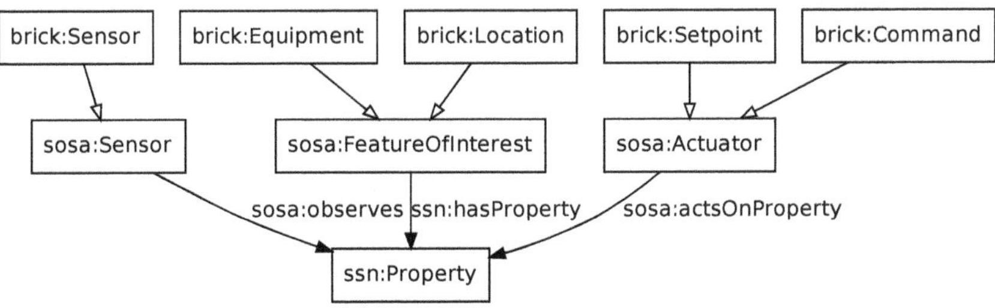

Fig. 1. Mapping from the Brick ontology to the SSN ontology. In the UML Class Diagram, UML classes depict RDFS classes, UML inheritance `rdfs:subClassOf` relationships, and UML associations `rdfs:domain` and `rdfs:range` of properties.

Simulating Sunlight and Occupancy. Some BOLD tasks rely on sensor measurements that result from two physical processes to simulate: the illumination of the building by the sun and the movements of occupants in the building. Although they take different parameters into account, the simulation of both processes have the following characteristics:

1. Non-determinism of simulation transitions so that agents may not optimize using out-of-band information. Note that agent writes are deterministic.
2. Predictable evolution (e. g. linear) to allow for model-based agent optimization

An example is provided in Fig. 2. In the case of sunlight, outside illuminance (measured in lux) evolves quadratically over time, as the effect of the sun rising (6 am), reaching a zenith and then setting (9 pm). Illuminance at zenith (i. e. the maximum value) equals 40,000 lux. A cloud coverage factor is applied to that

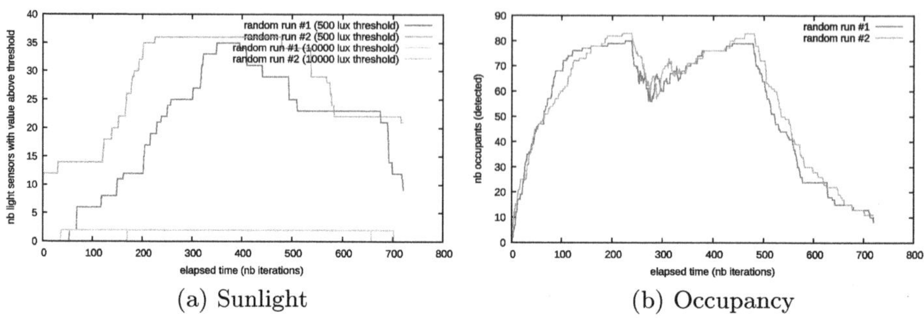

(a) Sunlight   (b) Occupancy

Fig. 2. Example of (plain) random runs for sunlight and occupancy.

baseline, as follows: a random coverage in $[0, 1]$ is generated at sunrise and at sunset, then coverage evolves linearly between these two values. Coverage either increases (cf. run 1 in Fig. 2a) or decreases (cf. run 2 in Fig. 2a) during the day.

Inside a building, the intensity of light that is incident on walls and floors is significantly lower than sunlight as measured outside. In BOLD, we randomly assign to each room an 'occlusion factor' in $[0.5, 1]/10$ to account for this phenomenon. As a result, the maximum illuminance inside the building is 4,000 lux.

To simulate the movements of occupants, time is divided in 4 periods, see Fig. 2b: At night, the building has no occupant. From 8 am, occupants start arriving at random times and occupy one room each. After 12 pm, occupants leave the building for lunch, about 1 h, after which they may come back for another 4 h of work. Starting from 4 pm, occupants start leaving the building.

5   Simulation Environment and Execution Framework

The simulation is an evolving RDF dataset that is provided via a Linked Data interface, where agents or apps can access the latest state by dereferencing resources, i.e. for a dataset d_i, agents can send HTTP requests to any resource IRI $r \in \pi(d_i)$. A well-known resource `</sim>` holds information about the simulation, including simulated time. Technically, the simulation environment is defined by an initial dataset, an update function that updates the dataset, and a definition of faults. We first describe how time evolves in the simulation, and then provide configuration and implementation details of the execution framework.

5.1   Time

The simulation environment increments time by triggering a timer at fixed intervals. The timer increments a number and progresses an OWL Time [15] description of the simulated time, all of which is made available to agents under a `</sim>` resource. A HTTP POST request to `</sim>` starts a simulation run.

Time, as exposed to agents and apps, corresponds to simulated time. Real execution time can be parameterized on the simulation server as the period of a timeslot between two successive environmental updates. However, care must be taken that all updates can be computed in less than the duration of a timeslot, to prevent time drift. Our configuration of BOLD for the benchmark can compute all environmental updates in $< 100\,ms$ (avg. $67 \pm 28\,ms$ for tasks w/sunlight and occupancy, see Sect. 6). We obtained all results (Sect. 7) with this configuration.

5.2   Execution Framework for Dynamic Linked Data Simulations

Depending on the use-case, a generic framework for dynamic Linked Data simulations needs to fulfil different purposes: a) provide a Linked Data interface to the latest state of an RDF dataset for agents or Apps to read and write, b) simulate an evolution of the RDF dataset and control the simulation (termination, speed), and c) export measurements to later calculate metrics. While benchmarking with BOLD requires all purposes, other use-cases require less and also benefit from the framework: e.g. developing Solid apps or link-traversal-based queries, on static (purpose a suffices), or dynamic (purposes a + b suffice) data – dynamics introduced by, e.g., updates to specific data, or new resources in an inbox.

For all three purposes, the BOLD server can be configured: Using `.properties` files, people can register datasets, and, if required, initialisation queries, simulation update queries, and/or fault check queries. RDF datasets are expected in TriG format, the initialisation/update or fault check queries are SPARQL `UPDATE` or `ASK` queries, resp. The TriG and the queries must make use of relative URIs in the graph names and data (e.g. graph name `<Room_CoffeeDesk>`), because during startup of the BOLD server, those URIs get resolved to the base URI of the server (e.g. `<http://localhost:8080/gsp/>`), to yield dereferencable URIs (e.g. `<http://localhost:8080/gsp/Room_CoffeeDesk>`). Links between relative URIs in the TriG are thus preserved. This way, people can build a repeatable standard-based dynamic Linked Data environment.

Our execution framework is based on Eclipse RDF4J (formerly Sesame [11]), and Jakarta-REST (formerly JAX-RS) running on Jetty, and fulfils the above purposes as follows: a): RDF4J maintains the registered RDF dataset, to which we built a REST interface using Jakarta-REST to implement the SPARQL Graph Store Protocol (with direct addressing) [39]. b): Other code initialises and periodically updates the RDF dataset (according to registered SPARQL `UPDATE` queries) for the simulation. On top, our user-defined functions allow for using random variables during simulation. The conditions for termination (formulated as registered SPARQL `ASK` queries) are also periodically evaluated. c): Last, we record the changes to the dataset, to calculate metrics according to faults (formulated as registered SPARQL `ASK` queries) after a simulation run.

Our implementation is multi-threaded: one thread triggers environmental updates at the end of each timeslot while a thread pool concurrently accesses the dataset to carry out agent operations.

6   BOLD Benchmark Tasks and Fault Definitions

Buildings account for 36 % of the global energy use [27] and the top two sinks during operation are HVAC (heating, ventilation, air conditioning) and lighting [48]. The tasks currently defined in BOLD focus on lighting. Agents must find and operate the building's lighting system in a fully automated manner in order to minimze electric consumption, thus, in MAS terms, fulfilling simple norms.

We consider different task types with increasing complexity: single-loop tasks and continuous-loop tasks. All tasks are designed for 24-hour runs. Tasks differ in the number of retrievals and updates they require, as well as in the number of perception-action loops agents must perform. These dimensions directly echo our discussion on the non-situatedness of Linked Data agents (Sect. 1, items 1 and 3). Moreover, certain tasks require that agents perform reasoning to make informed decisions (item 2). Table 2 summarizes the 10 tasks of BOLD.

6.1   Single-Loop Tasks

Single-loop tasks require to carry out operations only once in the environment. Formally, a single-loop task defined by some faults Γ on a simulation environment $\langle d, u \rangle$ has the following property: if $\Gamma_{\rho,t} = \emptyset$ for some t and $d_{t+1} \in u(d_t)$ (i. e. in the absence of agent operations), then it holds that $\Gamma_{\rho,t+1} = \emptyset$.

Table 2. BOLD tasks with lower bounds for read/write operations and loops (non-deterministic start/end ($*$) or loop count ($\dagger$)) to correctly achieve the task.

Task	Reads	Writes	Loops	Reasoning	Task	Reads	Writes	Loops	Reasoning
TS1	0	146	1		TC3	147	146	2*	
TS2	146	146	1		TC4	128	64	2*	
TS3	0	6	1	✓	TC5	128	64	$\sim$4*†	
TC1	146	146	2		TC6	192	64	$\sim$4*†	
TC2	146	146	2	✓	TC7	256	64	$\sim$4*†	

As scenario, we consider the simple control of lights, e. g. that a janitor would trigger when the whole building closes (TS1) or when they test functionalities of the system (TS2 and TS3).

TS1 A light that is on is considered a fault. This task does not require any data point to be read, the agent must merely find all lights and turn them off.
TS2 In this task, a light that has not been toggled since the beginning of the run is considered a fault. In contrast to TS1, this task involves perception. The agent must first read the state of a luminance command and then toggle.
TS3 Similar to TS1, but only a subset of the lights has to be switched off, namely those in rooms dedicated to 'personal hygiene' (toilet and shower).

Agents must properly classify rooms, which requires that they first read subclass axioms defined in the environment in a custom RDF vocabulary. These axioms specify, e.g. that 'disabled toilets' are a kind of 'toilets', themselves a kind of 'personal hygiene' rooms. As in TS1, a light in a toilet or shower that is on is considered a fault. To ensure agents correctly classify rooms, any light that has been toggled in other rooms is also considered a fault.

6.2   Continuous-Loop Tasks

In single-loop tasks, changes in the environment have no effect on success. This is not true of continuous-loop tasks. In continuous-loop tasks, changes in the environment may cause faults to appear, agents must thus continuously monitor the environment. The initial value of all lights is randomized so that fault rate always equals 0 in the absence of agent operation, regardless of the task.

Continuous tasks TC1 and TC2 involve time only. TC3 and TC4 involve illuminance sensors. TC5-TC7 introduce occupancy sensors.

TC1 In this scenario, a weather report provides sunrise and sunset times. A light that is off during the day is considered a fault, under the assumption that the building is likely to be occupied. Conversely, lights should be off during the night as no one is expected to be in the building. An agent needs to retrieve sunrise and sunset timestamps, and turn lights on or off accordingly.

TC2 In this scenario, in addition to sunrise and sunset times, parts of the building (floors) expose different opening and closing times beyond which no light should be on. The ground floor is assumed to close later (11 pm) than other floors (7 pm). All floors open at 8 am. When a floor is open, any light on the floor that is off is considered a fault. In this task, the agent must perform automated reasoning to infer what room belongs to each floor in order to decide whether lights in the room should be on or off at a given point in time.

TC3 In task TC3, a light that is off is a fault if outside illuminance is below a certain threshold (10,000 lux). In this scenario, we assume the building is equipped with a weather station mounted on its rooftop that includes a light sensor. By applying a threshold, we want to determine whether the lights in the entire building should be on or off. In contrast to TC1 and TC2, the times at which light should be on or off is randomly generated to simulate cloud coverage. Yet, agents could anticipate when to perform an action, as illuminance on the surface of the building varies from sunrise until sunset.

TC4 In task TC4, a fault is defined as in TC3 but at the level of a single room. We only consider the rooms in the building that are equipped with luminance sensors so as to decide for each room whether lights should be on or off by applying a global threshold (500 lux).

TC5 In TC5, a fault is any light that is off while an occupant is detected in the room. In this scenario, we assume that the rooms in the building are equipped with occupancy sensors. Using those sensors only, we want to determine whether the lights should be on or off. The challenge for agents in this task is to continuously monitor occupants coming in and leaving the building in a non-deterministic fashion. Although the simulated environment displays a clear occupancy pattern throughout the day, an agent cannot know it upfront, and can only build a model of occupancy by repeated observation.

TC6 This task combines TC5 with the constraints of TC4: a fault is any light that is off while an occupant is detected in the room *and* illuminance is below 500 lux. In this scenario, we want to raise energy efficiency and only turn on light in rooms with occupants and low illuminance. An agent faces less potential faults but overcoming them requires a more advanced model of the environment (or faster perception-action loops).

TC7 TC7, the most challenging task of the benchmark adds one more constraint to TC6: instead of a global illuminance threshold, agents have to make decision based on the preferences of occupants, which they can update at any time during the simulation via `brick:Setpoint` resources. In this scenario, we want to raise the individual comfort of occupants. The environment includes a set points for each illuminance sensor in the building. Thus, the agent must read the value that was last set by the room's occupant before deciding.

Table 3. Performance on single-loop and continuous-loop tasks of a baseline agent.

Task	ldfu/ldfu-prefetch				
	FR	AFC	NFC	RWR	
TS1	0.04/0.03	140/98	0.96/0.67	19/0	(0)
TS2	0.09/0.05	137/121	0.94/0.83	19/1	(1)
TS3	0.08/0.02	6/6	1/1	167/0	(0)
TC1	0.12/0.03	100/98	0.18/0.04	458/102	(1)
TC2	0.13/0.05	66/59	0.11/0.04	540/85	(1)
TC3	0.08/0.03	61/58	0.08 / 0.03	1831/282	(1)
TC4	0.23/0.09	17/16	0.11 / 0.05	983/211	(2)
TC5	0.42/0.29	16/15	0.15/0.10	395/68	(2)
TC6	0.26/0.31	11/10	0.09/0.11	800/105	(2)
TC7	0.40/0.31	16/12	0.20/0.11	628/127	(4)

FR: fault rate; AFC: average fault count; NFC: normalized fault count;

RWR: read/write ratio (ideal ratio in parenthesis, calculated from Table 2).

Future variations on TC3-TC7 could be based on temperature instead of illumination, to consider more elaborate control with feedback loops. Further versions of the benchmark could also consider incomplete environmental descriptions and defective devices. Agents could e.g. leverage additional topological information to infer probable values for the missing data points. We believe, however, that the nine BOLD tasks provided here represent a significant challenge to Linked Data agents, which must combine fast retrieval with non-trivial decision making.

7   BOLD Benchmark Showcase

We now provide an evaluation of the tasks introduced in the previous section, and discuss as the relevance of the metrics we defined for the benchmark. This showcase evaluation relies on the Linked-Data-Fu engine [46], which can perform HTTP operations, do reasoning over Linked Data, and execute rule-based programs [29]. Linked-Data-Fu is a multi-threaded engine optimized towards inference and communication. However, it does not directly provide support for agent-oriented programming features such as planning or memorization. In the following, we compare the performances of two Linked-Data-Fu agents and show that both features (planning, memorization) are desired for performance improvements.

The two Linked-Data-Fu agents we compare are configured as follows: a first agent (`ldfu`) is seeded with the building's URI and needs to discover all data points by following links. The second agent (`ldfu-prefetch`) behaves as if it had prefetched the model of the building, which allows immediate access to data points. Both agents execute a program that encodes condition-action rules, where the condition is a fault, as indicated in Tab. 2, and the action is a HTTP PUT request sent to the environment as a fix. Both agents repeat program execution as fast as possible until a run is over.

Table 3 summarizes the performance of our two baseline agents against our BOLD server. We ran experiments on a machine with an 8-core Intel Core i5 CPU shared between agent and server processes. The table features all metrics presented in Sect. 3: fault rate (FR), average fault count (AFC), normalized fault count (NFC) and read/write ratio (RWR).

As expected, `ldfu-prefetch`, which is directly provided with the building information, consistently outperforms `ldfu`.

Both agents show low fault rates for TS1, TS2 and TS3. As Linked-Data-Fu is optimised for such workloads, we regard those rates as lower bound. Rather, TS1-TS3 can be regarded as test cases to assert the correctness of an implementation.

Results for TC1 and TC2 are also showing satisfactory results regarding FR and NFC. However, the number of reads could be significantly reduced by optimising the agents for the task at hand. In those two tasks, agents essentially only synchronize with the environment. Similarly, in TC3, outside illuminance evolves linearly with time. It should thus be straightforward to calculate when a fault will occur and thus reduce the number of reads (see Sect. 6).

Regarding TC4-TC7, it is worth comparing the two agents. Although `ldfu-prefetch` sends $> 4\times$ fewer requests to the server, `ldfu-prefetch` only gains $< 14\%$ performance on fault rate, implying that fetching static data represents a rather low overhead. In contrast, the overhead related to exchanging dynamic readings is much more significant. Both agents perform readings on all data points, regardless of past values. For task TC4, for instance, they perform 128 reads which they complete only after the server has executed 10 iterations, on average. The agent's reaction time is therefore limited because of a high number of reads. Yet, the task only requires a total of 128×2 reads during the whole run (one for each light after sunrise and before sunset). To reduce agent-server interactions, an agent can remember past readings for statistical inference (e. g. to predict future illuminance levels). Such optimization is even more crucial for TC5-7, as indicated by significantly higher fault rates for `ldfu-prefetch` despite low fault counts.

8   Related Work

Our work is at the intersection of (Read-Write) Linked Data, MAS and building simulation. While we went to lengths to make realistic assumptions for the simulation, we do not claim to compete with commercial building simulation tools. We see building automation mostly a vehicle for evaluating efficient MAS architectures and not as an end in itself, similarly as the recent ProcTHOR [16].

There is considerable body of work on benchmarking triple stores with a read-only SPARQL interface, e. g. SP^{2}BENCH [45], BSBM [4], LUBM [21], and projects such as LDBC [8] and Hobbit [37]. Yet, BSBM contains a case that writes via SPARQL. A benchmark that sends read-only SPARQL requests to multiple sources is FedBench [44]. Closer to BOLD are approaches that consider the Linked Data interface, i. e. dereferencing of URIs in multiple requests: Hartig et al. [25] turned the dateset of BSBM into dereferencable URIs, and DLUBM [31] distributes the dataset from LUBM. SolidBench[6] serves a static social network via HTTP. None of these works consider state-changing operations that send RDF via HTTP. Thus, to our knowledge, BOLD is the first benchmark to provide a dynamic Read-Write Linked Data environment for benchmarking user agents.

The intersection of agent technologies and (Semantic) Web technologies includes: JASDL [32] combines Jason with the facility to process semantic data given in description logics (DL) ontologies. Using BOLD, we want to address agent technologies that also consider the other aspects of Semantic Web Technologies: interaction in HTTP and hypermedia, and tone down expressive DL reasoning. REST-A [20] is an agent framework providing an abstraction for perception and actions based on REST operations. Conversely, Ricci et al. built a framework to build environments and agent organizations based on JaCaMo (a mature MAS framework [5]), REST and RDF [41]. Both approaches focus on lowering the engineering effort to design MAS architectures. MAS engineering tools are complementary to BOLD: REST-A or JaCaMo could serve as the basis for MAS implementations evaluated with BOLD. BOLD does not evaluate inter-agent communication, which is an orthogonal topic with specialised benchmarks [36].

9   Conclusion

We have presented BOLD, a simulation framework for dynamic read-write Linked Data, next to a benchmark for measuring the performance of Linked Data agents.

---

[6] https://github.com/SolidBench/SolidBench.js

We hope that BOLD fosters research and development in the area of Solid apps, (Read-Write) Linked Data and Linked Data agents. Agent runtimes next to Linked Data-Fu are scarce (or slow [3], see [46]), but can be built based on existing link-traversal-based engines, e. g. [25,47] or generic Linked Data clients, e. g. N3.js+fetch, LDO [35], and LDflex [49], as used by the Solid community. MAS-based runtimes include [38], `bold-jacamo`[7]. As such, BOLD originated in the evaluation of [29], has served as an environment for ISWC 2021's All-the-Agents Challenge [30], sparked discussions at a Dagstuhl seminar [7], served as basis for a transportation showcase [43], and a research paper on planning with Semantic Web technologies [42].

Find the BOLD server source code, the building data (and scripts for its generation), building update and fault queries for the BOLD tasks online[8]. Rule-based agents that perform the BOLD tasks can also be found online[9].

Acknowledgements. This work was partially supported by the German Federal Ministry of Research, Technology, and Space (BMFTR) through the MOSAIK project (FKZ 01IS18070A), the NeSyPlan project (FKZ 16IS23052B), and by the French-German University (DFH-UFA) in the SeReCo project (CDFA-02-22).

References

1. Balaji, B., et al.: Brick: towards a unified metadata schema for buildings. In: Proceedings of the 3rd International Conference on Systems for Energy-Efficient Built Environments (BuildSys) (2016)
2. Berners-Lee, T.: Read-Write Linked Data. Design Issues (2009). http://www.w3.org/DesignIssues/ReadWriteLinkedData.html
3. Berners-Lee, T., Connolly, D., Prud'hommeaux, E., Scharf, Y.: Experience with N3 rules. In: W3C Workshop on Rule Languages for Interoperability, 27–28 April 2005, Washington, DC, USA. W3C (2005)
4. Bizer, C., Schultz, A.: The Berlin SPARQL benchmark. Seman. Web Inf. Syst. **5**(2) (2009)
5. Boissier, O., Bordini, R.H., Hübner, J.F., Ricci, A.: Multi-agent Oriented Programming: Programming Multi-agent Systems Using JaCaMo. MIT Press, Cambridge (2020)
6. Boissier, O., Ciortea, A., Harth, A., Ricci, A.: Autonomous agents on the web (Dagstuhl Seminar 21072). Dagstuhl Rep. **11**(1) (2021)
7. Boissier, O., Ciortea, A., Harth, A., Ricci, A., Vachtsevanou, D.: Agents on the web (Dagstuhl Seminar 23081). Dagstuhl Rep. **13**(2) (2023)
8. Boncz, P.A., Fundulaki, I., Gubichev, A., Larriba-Pey, J.L., Neumann, T.: The linked data benchmark council project. Datenbank-Spektrum **13**(2) (2013)
9. Both, A., et al.: Towards solid-based B2B data value chains. In: Meroño Peñuela, A., et al. (eds.) ESWC 2024. LNCS, vol. 15344, pp. 138–142. Springer, Heidelberg (2024). https://doi.org/10.1007/978-3-031-78952-6_14

---

[7] https://github.com/bold-benchmark/bold-jacamo – built on Linked Data-Fu.
[8] https://github.com/bold-benchmark/bold-server.
[9] https://github.com/bold-benchmark/bold-agents.

10. Brick Community: Metadata Schema for Buildings. https://brickschema.org/docs/Brick-Leaflet.pdf. Accessed 18 Dec 2022

11. Broekstra, J., Kampman, A., van Harmelen, F.: Sesame: a generic architecture for storing and querying RDF and RDF schema. In: Horrocks, I., Hendler, J. (eds.) ISWC 2002. LNCS, vol. 2342, pp. 54–68. Springer, Heidelberg (2002). https://doi.org/10.1007/3-540-48005-6_7

12. van Brussel, H., Wyns, J., Valckenaers, P., Bongaerts, L., Peeters, P.: Reference architecture for holonic manufacturing systems: PROSA. Comput. Ind. **37**(3) (1998)

13. Charpenay, V., Käfer, T., Harth, A.: A unifying framework for agency in hypermedia environments. In: Proceedings of the 9th International Workshop on Engineering Multi-Agent Systems (EMAS) at the 21st International Conference on Autonomous Agents and Multi-Agent Systems (AAMAS) (2021)

14. Ciortea, A., Mayer, S., Gandon, F.L., Boissier, O., Ricci, A., Zimmermann, A.: A decade in hindsight: the missing bridge between multi-agent systems and the world wide web. In: Proceedings of the 18th International Conference on Autonomous Agents and Multi-Agent Systems (AAMAS) (2019)

15. Cox, S., Little, C.: Time ontology in OWL. W3C Candidate Recommendation (2020)

16. Deitke, M., et al.: ProcTHOR: large-scale embodied AI using procedural generation. In: Proceedings of the 35th Annual Conference on Neural Information Processing Systems (NeurIPS) (2022)

17. Ferber, J., Müller, J.-P.: Influences and reaction: a model of situated multiagent systems. In: Proceedings of Second International Conference on Multi-agent Systems (ICMAS-96) (1996)

18. Fielding, R., Reschke, J.: Hypertext Transfer Protocol (HTTP/1.1): Semantics and Content, IETF RFC 7231 (2014). https://www.ietf.org/rfc/rfc7231

19. Fielding, R.T.: Architectural styles and the design of network-based software architectures. PhD thesis, University of California, Irvine (CA), USA (2000)

20. Gouaïch, A., Bergeret, M.: REST-A: an agent virtual machine based on REST framework. In: Proceedings of the 8th International Conference on Practical Applications of Agents and Multiagent Systems (PAAMS) (2010)

21. Guo, Y., Pan, Z., Heflin, J.: LUBM: a benchmark for OWL knowledge base systems. J. Web Semant. **3**(2-3) (2005)

22. Hagras, H., Callaghan, V., Colley, M., Clarke, G., Pounds-Cornish, A., Duman, H.: Creating an ambient-intelligence environment using embedded agents. IEEE Intell. Syst. **19**(6) (2004)

23. Haller, A., Janiwicz, K., Cox, S., Le Phuoc, D., Taylor, T., Lefrançois, M.: Semantic Sensor Network Ontology, W3C (2017). https://www.w3.org/TR/vocab-ssn/. Accessed 23 Sept 2020

24. Harth, A., Hose, K., Schenkel, R.: Linked Data Management. Chapman and Hall/CRC (2014)

25. Hartig, O., Bizer, C., Freytag, J.-C.: Executing SPARQL queries over the web of linked data. In: Bernstein, A., et al. (eds.) ISWC 2009. LNCS, vol. 5823, pp. 293–309. Springer, Heidelberg (2009). https://doi.org/10.1007/978-3-642-04930-9_19

26. Huberman, B.A., Clearwater, S.H.: A multi-agent system for controlling building environments. In: Proceedings of the 1st International Conference on Multiagent Systems (ICMAS) (1995)

27. IEA and UNEP: Global Status Report: towards a zero-emission, efficient and resilient buildings and construction sector (2018). https://www.unenvironment.org/resources/report/global-status-report-2018. Accessed 11 Dec 2019

28. Kaebisch, S., Kamiya, T., McCool, M., Charpenay, V., Kovatsch, M.: Web of Things (WoT) Thing Description. W3C Recommendation (2020)
29. Käfer, T., Harth, A.: Rule-based programming of user agents for linked data. In: Proceedings of the 11th International Workshop on the Web (LDOW) at the 27th Web Conference (WWW) (2018)
30. Käfer, T., Harth, A., Ciortea, A., Charpenay, V. (eds.): Proceedings of the All the Agents Challenge (ATAC) at the 20th International Semantic Web Conference (ISWC), vol. 3111. CEUR Workshop Proceedings. CEUR-WS.org (2022). http://ceur-ws.org/Vol-3111
31. Keppmann, F.L., Maleshkova, M., Harth, A.: DLUBM: a benchmark for distributed linked data knowledge base systems. In: Proceedings of the 16th International Conference on Ontologies, Databases, and Applications of Semantics (ODBASE) (2017)
32. Klapiscak, T., Bordini, R.H.: JASDL: a practical programming approach combining agent and semantic web technologies. In: Proceedings of the 6th International Workshop on Declarative Agent Languages and Technologies (DALT) (2008)
33. Kovatsch, M., Matsukura, R., Lagally, M., Kawaguchi, T., Toumura, K., Kajimoto, K.: Web of Things (WoT) architecture. W3C Recommendation (2020)
34. Mansour, E., et al.: A demonstration of the solid platform for social web applications. In: Proceedings of Posters & Demos at the 25th International Conference on World Wide Web (WWW). ACM (2016)
35. Morgan, J., et al.: Linked Data Objects. https://github.com/o-development/ldo/
36. Mulet, L., Such, J.M., Alberola, J.M.: Performance evaluation of open-source multiagent platforms. In: Proceedings of the 5th International Conference on Autonomous Agents and Multiagent Systems (AAMAS) (2006)
37. Ngonga Ngomo, A., García-Rojas, A., Fundulaki, I.: HOBBIT: holistic benchmarking of big linked data. ERCIM News **2016**(105) (2016)
38. O'Neill, E., Beaumont, K., Bermeo, N.V., Collier, R.W.: Building management using the semantic web and hypermedia agents. In: Käfer, T., Harth, A., Ciortea, A., Charpenay, V. (eds.) Proceedings of the All the Agents Challenge (ATAC 2021) Co-located with the 20th International Semantic Web Conference (ISWC 2021), Virtual Event, New York, USA, Oct. 26, 2021. CEUR Workshop Proceedings. CEUR-WS.org (2021)
39. Ogbuji, C.: SPARQL 1.1 Graph Store HTTP Protocol, W3C Recommendation (2013). http://www.w3.org/TR/sparql11-http-rdf-update/
40. Otto, B., Lohmann, S., Steinbus, S., Teuscher, A.: IDS Reference Architecture Model – Industrial Data Space, International Data Spaces Association (2018). https://www.fraunhofer.de/content/dam/zv/de/Forschungsfelder/industrialdata-space/IDS_Referenz_Architecture.pdf
41. Ricci, A., Ciortea, A., Mayer, S., Boissier, O., Bordini, R.H., Hübner, J.F.: Engineering scalable distributed environments and organizations for MAS. In: Proceedings of the 18th International Conference on Autonomous Agents and Multiagent Systems (AAMAS) (2019)
42. Schmid, S., Gui, Z., Freund, M., Wehr, T., Harth, A.: AEPIS: agent-enabled planning at scale. In: Proceedings of the 13th Knowledge Capture Conference (K-CAP). ACM (2025)
43. Schmid, S., Schraudner, D., Harth, A.: MOSAIK: an agent-based decentralized control system with stigmergy for a transportation scenario. In: Proceedings of the 20th Extended Semantic Web Conference (ESWC) (2023)

44. Schmidt, M., Görlitz, O., Haase, P., Ladwig, G., Schwarte, A., Tran, T.: FedBench: a benchmark suite for federated semantic data query processing. In: Aroyo, L., et al. (eds.) ISWC 2011. LNCS, vol. 7031, pp. 585–600. Springer, Heidelberg (2011). https://doi.org/10.1007/978-3-642-25073-6_37
45. Schmidt, M., Hornung, T., Lausen, G., Pinkel, C.: SP2Bench: a SPARQL performance benchmark. In: Proceedings of the 25th International Conference on Data Engineering (ICDE) (2009)
46. Stadtmüller, S., Speiser, S., Harth, A., Studer, R.: Data-Fu: a language and an interpreter for interaction with read/write linked data. In: Proceedings of the 22nd International Conference on World Wide Web (WWW) (2013)
47. Taelman, R., Van Herwegen, J., Vander Sande, M., Verborgh, R.: Comunica: a modular SPARQL query engine for the web. In: Vrandečić, D., et al. (eds.) ISWC 2018. LNCS, vol. 11137, pp. 239–255. Springer, Cham (2018). https://doi.org/10.1007/978-3-030-00668-6_15
48. UNEP and SKANSKA: Energy Efficiency in Buildings – Guidance for Facility Managers (2009). https://group.skanska.com/48dc86/globalassets/sustainability/environmental-responsibility/energy/unep_energy-efficbroch-final.pdf. Accessed 11 Dec 2019
49. Verborgh, R., Taelman, R.: LDflex: a read/write linked data abstraction for front-end web developers. In: Pan, J.Z., et al. (eds.) ISWC 2020. LNCS, vol. 12507, pp. 193–211. Springer, Cham (2020). https://doi.org/10.1007/978-3-030-62466-8_13
50. Wilensky, U., Rand, W.: An Introduction to Agent-Based Modeling: Modeling Natural, Social, and Engineered Complex Systems with NetLogo. MIT Press, Cambridge (2015)
51. A. Zimmermann: RDF 1.1: On Semantics of RDF Datasets, W3C Working Group Note (2014). https://www.w3.org/TR/rdf11-datasets/

In-Use

Energy Domain Ontology Matching with Large Language Model

Zhiyu Pan[1(✉)], Hanyang Li[1], and Antonello Monti[1,2]

[1] Institute for Automation of Complex Power Systems, RWTH Aachen University,
Mathieustr. 10, 52074 Aachen, Germany
`{zhiyu.pan,hanyang.li,amonti}@eonerc.rwth-aachen.de`
[2] Fraunhofer Institute for Applied Information Technology FIT,
Sankt Augustin 53757, Germany

Abstract. The rapid digitization and decentralization of modern energy systems have created vast amounts of heterogeneous data modeled with diverse ontologies. Ensuring semantic interoperability across these sources is essential but remains challenging due to domain-specific terminology and complex structural correspondences found in different ontologies. Traditional ontology matchers, which rely mainly on lexical or structural information, often struggle to capture the complex semantic. This paper presents a hybrid ontology matching framework designed for the energy domain. The approach combines a domain-pretrained transformer model with a large language model. Experiments on real-world energy ontologies show that the proposed method outperforms classical matchers, transformer-only approaches, and recent LLM-based systems. These results indicate that combining domain-specific encoders with prompt-guided LLM refinement offers an effective and scalable solution for ontology matching in a complex energy context.

Keywords: Ontology Matching · Large Language Models · Energy Domain

1 Introduction

As energy systems become increasingly digital, they generate large and heterogeneous datasets from smart meters, building automation, grid operations, forecasting services, and electricity markets. That information is encoded within distinct ontologies; for example, *SAREF4BLDG* and *SARGON*, which lack a shared semantic mapping. This misalignment undermines interoperability and reproducibility, so that even a basic query often requires manual reconciliation of classes and properties.

To address the semantic misalignment outlined above, ontology matching identifies semantically equivalent entities, including classes and properties, across heterogeneous ontologies. Classical matchers such as *LogMap* [12] and *AgreementMakerLight (AML)* [6] demonstrate that combining lexical similarity with

M. Acosta et al. (Eds.): ESWC 2026, LNCS 16550, pp. 343–359, 2026.
https://doi.org/10.1007/978-3-032-25159-6_18

structural reasoning yields alignments that are both scalable and logically consistent. Those works provide a strong methodological basis for ontology matching practices across domains.

However, the energy domain introduces challenges that complicate standard matching techniques. The vocabulary is highly specialized and technical, so terms such as *spinning reserve, distributed energy resource,* and *feeder reconfiguration* are domain-bound and often lack obvious lexical overlap. In addition, many useful correspondences are structural or composite. Several representative cases about ontology matching exist. First, a direct class-to-class correspondence holds; for example, *SARGON Flow meter* aligns with *SAREF4BLDG FlowMeter*. Second, a hierarchy and granularity mismatch occurs, where *SARGON Power_ meter* is subsumed by a more general class in *SAREF4BLDG*, such as *EnergyMeter* or *Meter*, indicating a subsumption rather than an equivalence relation. Third, object or contextual relations require reasoning over domains and ranges; for example, *SARGON in_ building* and *has_ zone* corresponding to *SAREF4BLDG hasSpace* and *isSpaceOf*. Finally, data properties such as *SARGON time_ stamp* align with *saref:hasTimestamp* in *SAREF*, showing that simple property-to-property correspondences coexist with structural ones. These patterns motivate matching strategies that combine domain knowledge with structural awareness rather than relying on lexical information alone.

In response to these challenges, recent advances in natural language processing, especially transformer-based language models, offer new opportunities. Models such as *BERT* [5], when fine-tuned on ontological text as in *BERTMap* [9], capture semantics beyond string similarity and support large-scale candidate discovery. At the same time, large language models have shown a strong capability for validating semantic correspondences through prompt-based reasoning, as demonstrated by *OLaLa* [10] and *Agent-OM* [17]. These developments show complementarity in performance: transformer encoders are effective for high-recall candidate generation, whereas large language models are well-suited for high-precision validation.

This paper targets ontology matching in the energy sector and proposes a two-stage hybrid framework that combines the strengths of transformer encoders and large language models. The contributions of this paper are as follows:

1. A two-stage hybrid ontology matching framework[1] tailored to energy informatics that integrates transformer driven candidate retrieval with LLM driven semantic validation.
2. An input scheme that encodes structural context, namely hierarchical relations and domain and range specifications, directly with the transformer to enhance recall and precision in candidate generation.
3. A comprehensive empirical evaluation of energy ontologies shows consistent gains compared to classical matchers (e.g., *LogMap, AML*), transformer only systems (e.g., *BERTMap*), and contemporary LLM-based methods (e.g., *OLaLa, Agent-OM*).

---

[1] Source code is available from Github https://github.com/zhiyupan/EnergyOM.

2   Related Work

Ontology matching techniques can be broadly categorized into three main approaches based on [17]: traditional knowledge-based methods, machine learning-based methods, and large language model(LLM)-based methods. Additionally, hybrid approaches have emerged to flexibly select multiple matching signals, such as lexical, structural, and textual descriptions. These methods offer a flexible framework that adapts matching strategies based on the characteristics of the input ontologies, effectively bridging the gap between machine learning and LLM-driven techniques. A comprehensive overview of these ontology matching paradigms and their core features is presented in Table 1. Furthermore, Table 2 summarizes the main limitations and distinctive characteristics of representative systems within each category. In the following subsections, we discuss each category in detail, highlighting representative systems, core methodologies, and key findings.

2.1   Traditional Knowledge-Based Ontology Matching

Traditional OM methods rely on knowledge-based resources (e.g., dictionaries, thesauri) to compute similarities between ontology entities [17]. They often integrate lexical similarity (e.g., entity names), structural information, and external resources, using fixed combination strategies. Representative examples include LogMap [12] [13], AML [6], ALIN [19]. These tools still serve as benchmark systems in the Ontology Alignment Evaluation Initiative (OAEI) [15].

In addition to limited flexibility, traditional knowledge-based matching systems exhibit two notable limitations: manual parameter tuning and feature sensitivity. First, parameters such as weights and thresholds in these systems are often manually tuned for each domain or dataset, which is time-consuming and may not generalize well across different domains. Second, as demonstrated by [3], the effectiveness of particular similarity features can vary significantly across domains, underscoring the limitations of a fixed heuristic approach.

Despite their limitations, knowledge-based systems remain integral to ontology alignment. Tools like AML and LogMap consistently performed at the top of OAEI competitions, only recently being challenged by machine learning and deep learning approaches.

2.2   Machine Learning-Based Ontology Matching

With the advancement of machine learning (ML) and natural language processing (NLP), an increasing number of OM systems have shifted from traditional knowledge-based approaches to data-driven methods. These methods typically learn feature representations and decision functions directly from data, enabling more flexible and adaptive alignment across diverse domains.

Among these, approaches based on pre-trained language models, particularly BERT [5] have emerged as effective models. Unlike traditional string-based matchers, BERT-based methods capture deep contextual semantic meaning beyond surface-level similarity [20], allowing for more robust performance in ambiguous or cross-domain matching scenarios.

BERTMap [9] is a notable system whose architecture combines BERT-based embedding with a candidate generation and repair module, offering both semantic depth and logical consistency. Their performance largely depends on textual features, such as class labels. In real-world ontologies, entities are embedded within inherently structured resources, forming rich networks of semantic relationships. Approaches that solely model textual similarity may overlook structural signals, especially in cases involving hierarchical alignment or complex, non-lexical correspondences.

To address this, recent methods like Matcha [7] incorporate both semantic and structural information. It evolved from AML's algorithms but with an updated architecture and ML enhancements. Matcha leverages transformer-based embeddings, particularly those generated by SentenceBERT, to calculate semantic similarity between entities. SentenceBERT [18] modifies the original BERT architecture by adding a pooling layer, enabling it to produce semantically meaningful sentence-level embeddings that are more suitable for similarity comparison tasks.

Limitation of BERT in Expert Domains. Inspired by the aforementioned advances, our approach adopts a similar embedding-based framework. However, we observe that standard BERT models often underperform in expert domains such as energy, due to their lack of exposure to domain-specific terminology and context. To address this, we employ EnergyBERT [16]: a domain-adapted variant of BERT fine-tuned specifically for the energy sector. This model better captures the semantic nuances of energy-related ontologies, resulting in more accurate and context-aware alignments.

Challenge and Improvement. Most ML-based systems require substantial amounts of labeled training data, which is often unavailable in specialized domains such as energy or biomedicine. Moreover, these approaches typically treat entities in isolation and may struggle with capturing nuanced semantic relationships or resolving ambiguous terms without contextual understanding. Challenges also remain in achieving robust cross-domain and multilingual generalization.

To address these limitations, recent work has explored LLM based methods. In the following section, we examine how LLMs are being applied to ontology matching tasks and the advantages they offer.

2.3 LLM-Based Ontology Matching

The newest paradigm in OM employs LLMs, including GPT-3/4, LLaMA, and DeepSeek, which are capable of leveraging contextual understanding to perform alignment more effectively. Unlike traditional or even BERT-based approaches, LLM-based methods can reason over rich textual descriptions and perform zero- or few-shot alignment with minimal supervision.

Early studies such as OLaLa [10] and LLMs4OM [8] adopt a purely prompt-based approach to ontology matching. In both systems, the task is framed as a binary classification problem: given a source and a target ontology entity, the LLM is prompted to determine whether the two refer to the same concept. Unlike earlier works, these methods integrate candidate generation with an embedding-based extractor, resulting in more targeted and context-aware prompting.

While these prompt-based systems represent an important step forward, they often struggle with complex alignment tasks that require multi-step reasoning, global consistency, or access to structured knowledge. Furthermore, LLMs are prone to hallucinations, particularly when operating with limited contextual information. They may generate responses that are syntactically well-formed but factually incorrect [11]. When relying solely on a single prompt to make binary (yes/no) decisions, LLMs lack internal mechanisms for verification or reasoning, which can result in false positives or false negatives during alignment.

To overcome these limitations, need for more structured, interactive frameworks, such as agent-based architectures, that can incorporate memory, reasoning, and external validation mechanisms to improve reliability. One such example is Agent-OM [17]. Agent-OM decomposes ontology matching into two coordinated LLM agents: a Retrieval Agent that extracts and indexes syntactic, lexical, and semantic information into a hybrid database, and a Matching Agent that performs multi-step reasoning using RRF (reciprocal rank fusion)-based [4] signal fusion, bidirectional candidate merging, and a secondary LLM validation step.

2.4 Dynamic Selection of Multiple Matching Strategy

One key limitation of fixed-strategy ontology matching systems lies in their dependence on accurate naming and well-structured ontologies. In scenarios where labels are noisy or class hierarchies are shallow, such methods often fail to produce reliable alignments.

To overcome this, we introduce a strategy-adaptive OM System, which dynamically combines multiple matching paradigms: name-based, hierarchy-based, and comment-based on ontology-specific features. The system evaluates the characteristics of the source and target ontologies and selects appropriate strategies accordingly. For example, ontologies with rich hierarchical organization benefit from structural matching, whereas flatter or sparse ontologies rely more heavily on lexical or annotation-based clues.

Similar ideas of strategy fusion have been explored in systems like AML [6] and Agent-OM [17], though these often use fixed rule-based fusion; in contrast, our OM-System uses ontology profiling indicators to inform the selection and weighting of alignment strategies, enabling fine-grained control and superior performance across benchmarks.

Table 1. Comparison of ontology matching methods and their features.

Method	Name	Structure	Description	Supervised	Hybrid	Method	Applications
LogMap	✓	✓	✗	✗	✓	Knowledge-based	Complex/Expert
AML	✓	✓	✓	✗	✓	Knowledge-based	Complex/Expert
ALIN	✓	✓	✓	✗	✓	Knowledge-based	Simple/Complex
BERTMap	✓	✓	✓	✓	✓	BERT-based	Complex/Expert
Matcha	✓	✓	✓	✓	✓	BERT-based	Complex/Expert
OLaLa	✓	✗	✓	✗	✓	BERT-based and LLM-based	Simple/Complex/Expert
LLMs4OM	✓	✓	✓	✗	✓	BERT-based and LLM-based	Simple/Complex/Expert
Agent-OM	✓	✓	✓	✗	✓	LLM-based	Complex/Expert
our system	✓	✓	✓	✗	✓	BERT-based and LLM-based	Simple/Complex/Expert

Table 2. Limitations and characteristics of various ontology matching systems.

Method	Hallucination	Label Dependence	Structure Dependence	Scalability	Automation Level
LogMap	None	High	High	High	High
AML	None	High	High	High	High
ALIN	None	High	High	Medium	Low
BERTMap	None	High	High	High	High
Matcha	None	High	High	High	High
OLaLa	High	High	Medium	Medium-High	High
LLMs4OM	High	High	Medium-High	Medium-High	High
Agent-OM	Medium	High	High	High	High
our system	Low	Low	Low	High	High

3   Methodology

This paper proposes a two-stage hybrid framework in Fig. 1 that combines the strengths of transformer encoders and large language models.

- **Stage 1**: High recall candidate generation using transformer encoders. To enrich contextual representation, each class is encoded with its label, comment, and local hierarchy, and each property is encoded with paraphrased structural context such as domain and range. This design introduces lexical and structural information at the beginning to maximize semantic coverage.
- **Stage 2**: Candidate filtering with a general purpose large language model using a single unified prompt. For each candidate pair, the model is asked only whether the two terms are semantically equivalent in the energy domain, with no labels, definitions, or structural context provided. The model returns an accept or reject decision that filters the candidate set. Restricting inference to the candidate pool and using a compact prompt reduces computational cost and the risk of hallucination.

This framework offers several advantages specific to the energy domain. First, domain pretraining encoders EnergyBERT resolve specialized terminology and symbol conventions. Second, encoding structural context such as hierarchy and domain and range improves semantic alignment, which is analogous to the refinements performed by logic based systems after initial matching. Third, restricting LLM inference to a curated candidate pool reduces false positives and enhances reliability while keeping inference costs tractable.

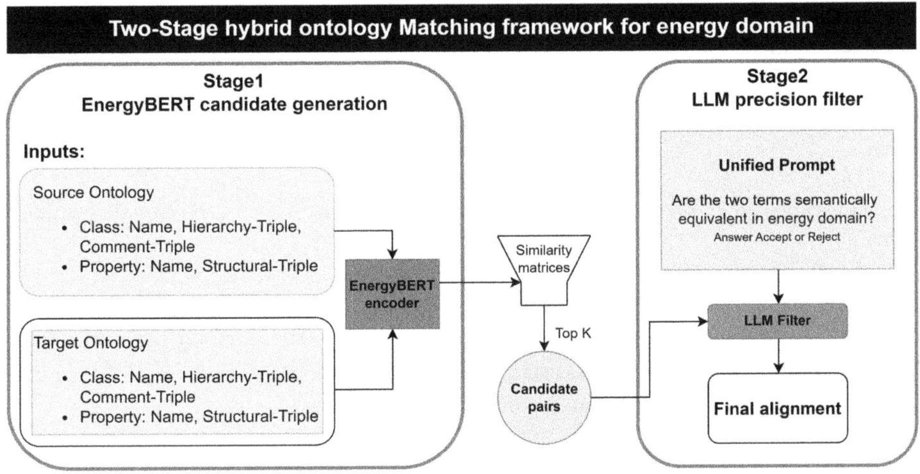

Fig. 1. Two Stage Hybrid Ontology Matching for Energy Domain.

We denote the source and target ontologies by $O_s = (C_s, P_s)$ and $O_t = (C_t, P_t)$, where C_s and C_t are the sets of classes and P_s and P_t are the sets of properties. A mapping M consists of class correspondences M_c and property correspondences M_p. M_c contains pairs (c_i^s, c_j^t) with $c_i^s \in C_s$ and $c_j^t \in C_t$, and M_p contains pairs (p_u^s, p_v^t) with $p_u^s \in P_s$ and $p_v^t \in P_t$.

Our framework contains four steps: (i) *Embedding* encodes all classes and properties in O_s and O_t with the encoder, yielding unified semantic and structural representations. (ii) *Mapping* constructs directional candidate correspondences by applying lexical and type aware blocking combined with neighborhood expansion. (iii) *Filtering* employs an LLM with ontology aware prompts to assess the semantic plausibility of each candidate and to refine the candidate set through re-ranking and pruning. (iv) *Fusion* formulates the final alignment as a constrained selection problem and jointly instantiates M_c and M_p under the chosen cardinality constraints.

Embedding Construction. We construct embedding input by feeding the encoder with (i) class names, (ii) hierarchy signals (parent/child relations and path encodings), (iii) textual comments, (iv) property names, and (v) structural information for properties (domain and range triples).

Cosine Similarity. The encoder produces one embedding per input type. For each class $x \in C$ we obtain embedding vectors $h_{\text{name}}(x)$, $h_{\text{hier}}(x)$, and $h_{\text{comm}}(x)$; for each property $x \in P$ we obtain $h_{\text{name}}(x)$ and $h_{\text{struct}}(x)$. All vectors are ℓ_2 normalised before comparison, and the cosine similarity for channel k is defined as

$$S_{\cos}^{(k)}(x, y) = \frac{h_k(x)^\top h_k(y)}{\|h_k(x)\|_2 \, \|h_k(y)\|_2}, \qquad k \in \{\text{name}, \text{hier}, \text{comm}, \text{struct}\}.$$

Similarity Score. For a candidate class pair (c_i^s, c_j^t) we aggregate those different similarities into a single score

$$S_c(c_i^s, c_j^t) = \lambda_1 S_{\cos}^{(\text{name})} + \lambda_2 S_{\cos}^{(\text{hier})} + \lambda_3 S_{\cos}^{(\text{comm})}, \quad \lambda_m \geq 0, \ \sum_{m=1}^{3} \lambda_m = 1.$$

For a candidate property pair (p_u^s, p_v^t) we combine name and structural similarity

$$S_p(p_u^s, p_v^t) = \mu_1 S_{\cos}^{(\text{name})} + \mu_2 S_{\cos}^{(\text{struct})}, \quad \mu_1, \mu_2 \geq 0, \ \mu_1 + \mu_2 = 1.$$

Here $S_{\cos}^{(\text{hier})}$ reflects coherence of parent and child paths derived from hierarchy triples, $S_{\cos}^{(\text{comm})}$ exploits textualised `rdfs:comment` triples, and $S_{\cos}^{(\text{struct})}$ encodes domain and range information from structural triples.

LLM Guided Filtering. This stage performs a simple yes or no verification rather than scoring. We first convert the candidates produced by EnergyBERT into structured dictionary pairs. Each dictionary pair is injected into a fixed, ontology aware prompt that asks the LLM to decide whether the pair is a correct match. The LLM returns a binary judgment (ACCEPT or REJECT). We keep only the accepted pairs and discard the rest. To improve stability, we optionally apply prompt paraphrases or few shot exemplars; we also perform simple post rules such as deduplication, type checks, and blacklist terms (e.g., "other", "misc") to avoid spurious matches. This filter reduces false positives and raises precision while intentionally allowing a moderate drop in recall.

Bidirectional Inference and Agreement. The matching is processed in both directions, producing $S_{s \to t}(x, y)$ and $S_{t \to s}(y, x)$ from the cosine based, fused scores. We aggregate them with a symmetric operator (e.g., mean, min, or geometric mean) and enforce agreement:

$$S_{\text{bi}}(x, y) = \text{Agg}\big(S_{s \to t}(x, y), \ S_{t \to s}(y, x)\big).$$

Final selection maximizes the bidirectional objective under mapping constraints (one to one or one to many), class hierarchy preservation, and property domain or range compatibility.

3.1   Ontology Class Matching

Figure 2 depicts the ontology class matching. Given source and target ontologies O_s and O_t, we construct three evidence lists on each side: (i) *Name List*, (ii) *Hierarchy Triple List*, and (iii) *Comment Triples List*. These lists feed three parallel matchers: name-based, hierarchy-based, and comment-based, whose outputs are fused to produce class level candidate pairs with high recall.

From O_s and O_t we extract: (a) **name** strings for each class; (b) **hierarchy triples** (e.g., $\langle c, \text{is a subclass of}, c' \rangle$); (c) **comment** triples about definitions or descriptions.

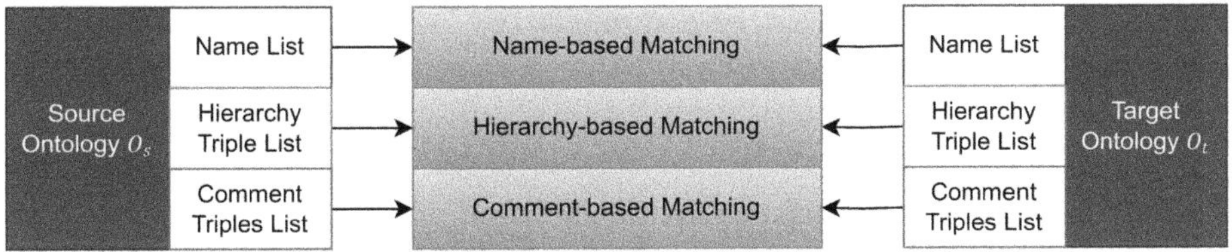

Fig. 2. Class Matching.

Name Based Matching. Each class name is encoded with the transformer model to obtain a name channel embedding h_{name}. For efficiency, we first form a blocked candidate set (e.g., via token overlap or prefix blocking). We then compute cosine similarities and, for each source class $c \in C_s$, retain the top k targets in C_t. This produces a similarity matrix $S \in \mathbb{R}^{|C_s| \times |C_t|}$ and a per source top k list. While effective, this stage relies only on name text and cannot capture hierarchy or annotations.

Hierarchy Based Matching. As shown in Fig. 3, the hierarchy based matches are $\mathcal{T} = \{(t_1, t_2, s)\}$, where each t_k is a string in the pattern `"Class c is a subclass of Class p"` and $s \in [0,1]$ is the cosine similarity computed. Only the pairs with similarity scores higher than the threshold are stored in the candidate set. If no threshold is provided, the upstream module uses the default $\tau = 0.7$.

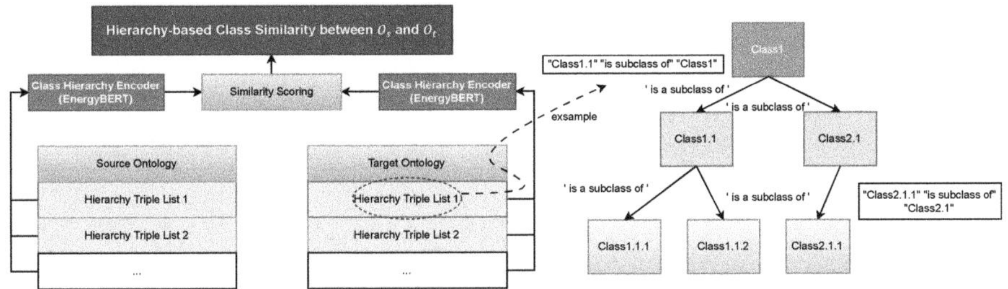

Fig. 3. Class hierarchy matching framework and triple construction.

As shown in Algorithm 1, for each $(t_1, t_2, s) \in \mathcal{T}$ that contains the substring `"is a subclass of"`, we parse

$$t_1 : (c_1, p_1), \qquad t_2 : (c_2, p_2),$$

by splitting on `"is a subclass of"` and stripping the leading prefix `"Class"` from both sides. We then generate exactly two role preserving candidates and discard cross role cases:

$$\underbrace{(p_1, p_2, s)}_{\text{parent} \to \text{parent}} , \qquad \underbrace{(c_1, c_2, s)}_{\text{child} \to \text{child}} .$$

After collecting all candidates, we deduplicate by keeping, for each ordered pair (u, v), the maximum similarity observed:

$$S^\star(u, v) = \max\{\, s \mid (u, v, s) \text{ was generated} \,\}.$$

The final hierarchy based class pairs are

$$\mathcal{M}_H = \{\, (u, v, S^\star(u, v)) \mid S^\star(u, v) \text{ defined} \,\}.$$

By construction, the procedure is role preserving (parent to parent and child to child only) and returns a single score per ordered pair obtained via max aggregation.

Algorithm 1: Role preserving combination

Input: Ontologies O_s, O_t; encoder model; upstream matches $\mathcal{T}$; threshold τ

1   $\mathcal{C} \leftarrow \emptyset$;
2   **foreach** $(t_1, t_2, s) \in \mathcal{T}$ **do**
3   **if** $s < \tau$ **then**
4        $\lfloor$ **continue**
5   **if** t_1 *does not contain* " *is a subclass of* " *or* t_2 *does not contain it* **then**
6        $\lfloor$ **continue**
7   parse t_1 into (c_1, p_1) by splitting on " `is a subclass of` " and removing a leading "`Class` ;
8   parse t_2 into (c_2, p_2) similarly;
9      $\lfloor$ $\mathcal{C} \leftarrow \mathcal{C} \cup \{(p_1, p_2, s), (c_1, c_2, s)\}$ #Parent to Parent and Child to Child only
10   Build a map $S^\star$ with the maximum score per ordered pair:
     $S^\star(u, v) \leftarrow \max\{\, s \mid (u, v, s) \in \mathcal{C} \,\}$;
11   **return** $\{\, (u, v, S^\star(u, v)) \mid (u, v) \in \text{keys}(D) \,\}$

Comment Based Class Matching. The comments for each class in the ontology are used to provide more semantic information. For each ontology, we extract class level comment triples and linearize each triple into a short sentence of the form "`ClassName: comment`" by concatenating the first and the third triple elements. This generates two sentence lists $S = \{s_i\}_{i=1}^{n}$ and $T = \{t_j\}_{j=1}^{m}$, where $s_i = $ `ClassName`$_i$: `comment`$_i$ and $t_j = $ `ClassName`$'_j$: `comment`$'_j$.

Encoding and Cosine Similarity. We compute embeddings for S and T and obtain a cosine similarity matrix using cosine similarity:

$$\sigma_{ij} = \text{cosine\_similarity}(S, T) \in [-1, 1]^{n \times m}.$$

Given a threshold τ (default 0.7), we keep sentence level matches

$$\mathcal{M}_{\text{sent}} = \{\, (s_i, t_j, \sigma_{ij}) \mid \sigma_{ij} \geq \tau \,\}.$$

Projection to Classes and Max Aggregation. Sentences are split at the delimiter ":" and the left segment is taken as the class name. Each sentence-level match (s_i, t_j, σ_{ij}) is thus projected to a class-level pair (c_i, d_j, σ). When multiple sentences map to the same (c, d), we retain the maximum similarity

$$S^\star(c, d) = \max\{\sigma \mid (c, d, \sigma) \text{ derived from } \mathcal{M}_{\text{sent}}\}.$$

The resulting comment-based candidates are

$$\mathcal{M}_{\text{com}} = \{(c, d, S^\star(c, d))\}.$$

This comment-based matching captures semantic cues present in textual descriptions, while max aggregation suppresses noise from less informative sentences.

3.2   Ontology Property Matching

This subsection introduces our property alignment workflow, which parallels class matching and leverages two types of semantic information: (i) a name based component that extracts lexical and semantic information from property labels, and (ii) a structure based component that exploits domain or range information and related structural constraints.

Name Based Matching. Each property label is encoded with EnergyBERT to obtain a name based embedding. We compute cosine similarities between all source to target pairs. While effective, this stage relies solely on label text and semantics and does not capture hierarchical relations or domain or range constraints.

Structure Based Matching (Data and Object). We encode each property as a short textual triple $\langle D, p, R \rangle$, where D is the domain, p is the property, and R is the range. For *object properties*, both D and R are classes (e.g., $\langle \mathsf{A}, \mathsf{Object\ property}, \mathsf{B} \rangle$). For *data properties*, R is a datatype (e.g., $\langle \mathsf{A}, \mathsf{Data\ property}, \mathsf{datatype} \rangle$). We textualize each triple using the pattern "Domain Property type Range", and compute pairwise cosine similarities between the resulting embeddings.

3.3   LLM-Based Filtering

This stage converts the unified candidate list produced upstream into final correspondences using a single large language model prompt. **Input** is a direction specific candidate list $\mathcal{C}$ of tuples (u, v, s), where u and v denote source and target items classes or properties and s is the aggregated similarity from prior modules. **Output** is a set $\mathcal{M} = \{(u, v)\}$ obtained by retaining only those pairs that the LLM deems correct. No ontology aware context is provided. For each $(u, v, \cdot) \in \mathcal{C}$, we extract only the two surface strings associated with u and v term names as they appear in their ontologies and feed them to a single, standardized prompt that asks for a yes or no equivalence judgment. The prompt is fixed across all pairs and requests a strictly binary output YES or NO to the question: "Are the following two terms semantically equivalent in the energy domain". The prompt incorporates few-shot examples—such as equating 'RoomTemperature' with 'IndoorTemperature'—to establish context. Furthermore, we integrated a Chain-of-Thought (CoT) directive, "For each pair, think step by step internally to verify equivalence", to enhance the model's reasoning capabilities.

4   Evaluation

This section offers a comprehensive empirical study of our ontology matching system in two settings: general domain ontologies and energy domain ontologies.

4.1   General Domain Experiments

Although our method is developed for energy-domain ontology alignment, it is important to verify that the approach also works beyond its target domain. We therefore evaluate the framework on the OAEI *Conference* [15] benchmark to test out-of-domain applicability under heterogeneous, non-energy schemas. For the general-domain experiments on *Conference*, we replace the domain-specific encoder with a general-purpose SentenceBERT while retaining DeepSeek as the LLM filter, because DeepSeek is open-source and offers an API at an affordable price. We use separate SentenceBERT thresholds for classes and properties: $\tau_{\mathrm{class}} = 0.80$, $\tau_{\mathrm{prop}} = 0.70$. A pair (u, v) is forwarded to the LLM only if $s(u, v) \geq \tau_{\mathrm{class}}$ when u, v are classes or $s(u, v) \geq \tau_{\mathrm{prop}}$ when they are properties. We use a higher class threshold (0.80) because class labels in the Conference ontologies are comparatively stable and benefit from stricter acceptance, whereas property labels are shorter and more paraphrastic (prepositions, verb–noun alternations), so a slightly lower threshold (0.70) preserves recall. Setting $k=1$ (top-1) is consistent with one-to-one mapping constraints and keeps the candidate pool small.

Experimental Procedure. We conduct ten controlled experiments (Exp1 – Exp10). Table 3 lists the configurations and reports the corresponding Precision, Recall and F_1. The experiments highlighted in the table are the configurations we carry forward as the representative results in the main comparisons.

We retain **Exp5** (Single, SentenceBERT + DeepSeek, Name + Hier.) and **Exp10** (Both, SentenceBERT + DeepSeek, Name + Hier.) as our representative settings because they isolate the effect of directionality while holding the encoder, evidence sources, and LLM constant. Among single-direction runs, Exp5 achieves the best precision–recall trade-off and the highest F_1. Exp10 is the bidirectional counterpart with reconciliation, delivering the best precision among the "Both" experiment while maintaining a competitive F_1. For clarity and comparability, we exclude (i) *no-LLM baselines* (Exp1/Exp2/Exp7/Exp8), which omit DeepSeek and thus serve only as ablations with lower precision;(ii)

Table 3. Configurations and performance of the 10 experiments. "Direction" = *Single*: $O_s \to O_t$ only, *Both*: bidirectional.

ID	Encoder-Model	LLM	Direction	Name	Hier.	Comm.	Precision	Recall	**F₁**
Exp1	SentenceBERT	–	Single	✓	✓	✓	51.20	61.95	54.21
Exp2	SentenceBERT	–	Single	✓	✓	✗	53.69	60.23	54.90
Exp3	EnergyBERT	–	Single	✓	✓	✗	14.04	61.14	22.00
Exp4	EnergyBERT	DeepSeek	Single	✓	✓	✗	48.39	55.29	47.52
Exp5	SentenceBERT	DeepSeek	Single	✓	✓	✗	81.45	45.49	**56.33**
Exp6	–	DeepSeek	Single	✓	✓	✗	44.11	**62.80**	50.08
Exp7	SentenceBERT	–	Both	✓	✓	✗	51.20	61.95	54.21
Exp8	SentenceBERT	–	Both	✓	✓	✓	32.56	45.51	36.42
Exp9	SentenceBERT	DeepSeek	Both	✓	✓	✓	31.86	46.13	36.60
Exp10	SentenceBERT	DeepSeek	Both	✓	✓	✗	**83.81**	41.40	53.04

the *domain-mismatch encoder* (Exp3, EnergyBERT), which falls outside the SentenceBERT setup used for the general-domain study; and (iii) *redundant DeepSeek variants* (Exp4/Exp6/Exp9), which underperform or rely on the Comment channel. Notably, the OAEI Conference ontologies have sparse/uneven `rdfs:comment` annotations, so adding the comment information introduces noise and degrades performance (cf. Exp8/Exp9); therefore, comment-based variants are not retained as principal results.

Table 4. Method performance on the OAEI Conference benchmark

Name	Prec	Rec	F_1	Name	Prec	Rec	F_1
ALIN	**88.00**	47.00	61.00	LogMapLt	73.00	50.00	59.00
AMD	87.00	43.00	58.00	LSMatch	**88.00**	42.00	57.00
GraphMatcher	76.00	**80.00**	**78.00**	Matcha	70.00	64.00	67.00
LogMap	81.00	58.00	68.00	PropMatch	83.00	8.00	15.00
TOMATO	61.00	50.00	55.00	SORBETMatch	78.00	64.00	70.00
OLaLa	56.00	66.00	61.00	Agent OM	69.55	61.70	64.21
Exp5	81.45	45.49	56.33	**Exp10**	83.81	41.40	53.04

Evaluation and Summary. The results compared with the benchmark are given in Table 4. Our two representative configurations **Exp5** and **Exp10**: Exp5 achieves **81.45%** Precision, 45.49% Recall, 56.33% F_1; Exp10 attains **83.81%** Precision, 41.40% Recall, 53.04% F_1. Compared with LLM-based baselines (e.g., OLaLa, Agent OM), our approach delivers higher precision at the cost of recall; relative to strong traditional/transformer systems (e.g., GraphMatcher, SOR-BETMatch), the strict top-1 gate and binary LLM filter trade coverage for precision, yielding lower F_1 on this general-domain benchmark.

Exp2 is exactly **Exp5** without DeepSeek; **Exp8** is **Exp10** without DeepSeek. These one-to-one pairs isolate the LLM's marginal effect under Single and Both, respectively (identical encoder/evidence/top-1/thresholds). As intended, adding DeepSeek substantially increases precision while reducing recall, with small net changes in F_1.

Why Both Direction Underperforms Single. Running $O_s \rightarrow O_t$ and $O_t \rightarrow O_s$ and then reconciling imposes mutual agreement, which removes asymmetric true positives and lowers recall. In addition, bidirectional processing increases the average input token length per LLM call (extra provenance/direction markers and longer labels in the reverse pass), enlarging prompt context. Empirically we observe that longer prompts make *DeepSeek* slightly more random `accept`/`reject` decisions occur more often even with temperature 0.0, so after reconciliation the bidirectional setting yields lower overall F_1 than the single-direction run.

Takeaways. The framework generalizes beyond energy ontologies and is competitive on precision in the general domain. When higher F_1 is required on Conference-like data, recall can be improved by relaxing similarity thresholds, increasing top-k before the LLM filter, or softening bidirectional reconciliation.

4.2   Energy-Domain Experiments and Results

We evaluate three representative energy ontologies and manually create the matching to serve as a benchmark dataset available in the Git repository for future domain specific research:

1. `SAREF4BLDG` [2]: ETSI SAREF building extension; interoperable device and space models with IFC4-inspired attributes and geolocation links.
2. `SARGON` [1]: General smart-energy ontology; unified vocabulary and relations spanning building and grid automation across assets and subsystems.
3. `SBEO` [14]: Smart building evacuation; geometry, network graph, devices, users, and context for localization and evacuation planning.

We adopt the same two-stage design as in the general-domain study, with an energy-specific encoder in Stage 1 and a binary LLM filter in Stage 2. We use EnergyBERT to embed classes and properties with the energy-aware textualization described in the methodology (labels, hierarchy for classes; domain–range and related structural cues for properties). Cosine similarity drives nearest-neighbor retrieval to form a high-recall candidate pool. Only candidates that pass the similarity gate are submitted to *DeepSeek* with a single fixed prompt asking for a yes/no decision on semantic equivalence in the energy context. The decoder temperature is set to 0.0 for deterministic outputs.

We use separate EnergyBERT thresholds for classes and properties: $\tau_{\text{class}} = 0.82$, $\tau_{\text{prop}} = 0.78$. EnergyBERT benefits from domain pretraining, which supports slightly higher thresholds without excessive recall loss. Properties typically have shorter labels and more paraphrastic variation, so τ_{prop} is set marginally lower than τ_{class}. We compare with three baseline systems, namely BERTMap, SORBETMatch, and AgentOM. BERTMap and SORBETMatch are transformer-only matchers. AgentOM serves as an LLM-centric ontology matching baseline.

Table 5 reports per-pair precision, recall, and F_1 for all methods. Our EnergyBERT+DeepSeek configuration achieves the highest F_1 on all three ontology pairs. On SAREF4BLDG-SARGON, our system with EnergyBERT reaches an F_1 of **83.50%**, substantially outperforming AgentOM (55.65%), BERTMap (49.02%), and SORBET (38.76%). The gap is even more pronounced on SAREF4BLDG-SBEO, where the transformer-only baselines (BERTMap at 6.90%, SORBET at 9.41%) exhibit very low precision, while AgentOM attains 40.00% and our method reaches **66.67%**. On the most heterogeneous pair, SARGON-SBEO, our system with EnergyBERT achieves **41.46%**, compared to 22.22% for AgentOM, 6.12% for BERTMap, and 3.04% for SORBET. Based on the results, our system using the general-domain SentenceBERT model also achieves, on average, better performance compared to other methods based on general-domain BERT models.

Table 5. Per-pair precision, recall, and F_1 for all methods (%).

Method	SAREF4BLDG–SARGON			SAREF4BLDG–SBEO			SARGON–SBEO		
	Prec	Rec	F_1	Prec	Rec	F_1	Prec	Rec	F_1
BERTMap	59.52	41.67	49.02	4.76	12.5	6.90	6.00	6.25	6.12
SORBET	36.23	41.67	38.76	5.79	25.00	9.41	2.01	6.25	3.04
AgentOM	58.18	53.33	55.65	55.56	31.25	40.00	**100.00**	12.50	22.22
Our System General	19.27	56.14	28.70	**100.00**	25.00	40.00	80.00	16.66	27.58
Our System Energy	**100.00**	**71.67**	**83.50**	81.82	**56.25**	**66.67**	35.42	**50.00**	**41.46**

5   Conclusion

This paper proposes a two stage hybrid ontology matching framework. A transformer encoder produces high recall candidate mappings, and a large language model then performs binary vetting to keep only the most plausible correspondences. The framework was evaluated on *Conference* and three energy ontologies. In the general domain, SBERT plus DeepSeek achieved strong precision. In the energy domain, the EnergyBERT to DeepSeek configuration reached the best F_1 on all three ontology pairs. Domain specific pretraining and structure-based matcher helped to disambiguate technical terms and device relations, while the binary LLM decision removed many mismatches. The method remained reproducible and cost aware: thresholds were fixed, prompts were single shot and shared across experiments.

There are several directions for future work. Scores from different evidence channels could be fused with a learning to rank model and combined with lightweight logical checks to preserve coherence. Additionally, the robustness of the LLM decision could be improved with self consistency or few shot examples. The evaluation can be extended to subsumption and compound correspondences.

Supplemental Material Statement:

– Source code is available at https://github.com/zhiyupan/EnergyOM.

Acknowledgments. The authors would like to thank the German Federal Government, the German State Governments, and the Joint Science Conference (GWK) for their funding and support as part of the NFDI4Energy consortium. Funded by the Deutsche Forschungsgemeinschaft (DFG, German Research Foundation) – 501865131.

Disclosure of Interests. The authors have no competing interests to declare that are relevant to the content of this article.

References

1. Sargon: Smart energy domain ontology. IET Smart Cities (2020). https://doi.org/10.1049/iet-smc.2020.0049
2. Saref4bldg: an extension of saref for the building domain. https://saref.etsi.org/saref4bldg/ (2025), Accessed 14 Oct 2025
3. Lécué, F.: Towards Constructive Evidence of Data Flow-Oriented Web Service Composition. In: Alani, H., Kagal, L., Fokoue, A., Groth, P., Biemann, C., Parreira, J.X., Aroyo, L., Noy, N., Welty, C., Janowicz, K. (eds.) ISWC 2013. LNCS, vol. 8218, pp. 298–313. Springer, Heidelberg (2013). https://doi.org/10.1007/978-3-642-41335-3_19
4. Cormack, G.V., Clarke, C.L.A., Büttcher, S.: Reciprocal rank fusion outperforms condorcet and individual rank learning methods. In: Proceedings of the 32nd International ACM SIGIR Conference on Research and Development in Information Retrieval (SIGIR 2009), pp. 758–759. ACM, Boston, MA, USA (2009). https://doi.org/10.1145/1571941.1572114
5. Devlin, J., Chang, M.W., Lee, K., Toutanova, K.: BERT: Pre-training of deep bidirectional transformers for language understanding. In: Proceedings of the 2019 Conference of the North American Chapter of the Association for Computational Linguistics: Human Language Technologies, Volume 1 (Long and Short Papers), pp. 4171–4186. Association for Computational Linguistics, Minneapolis, MN (2019). https://aclanthology.org/N19-1423
6. Faria, D., Pesquita, C., Santos, E., Palmonari, M., Cruz, I.F., Couto, F.M.: The agreementmakerlight ontology matching system. In: On the Move to Meaningful Internet Systems: OTM 2013 Conferences. Lecture Notes in Computer Science, vol. 8185, pp. 527–541. Springer, Cham (2013). https://doi.org/10.1007/978-3-642-41027-7_35
7. Faria, D., Silva, M.C., Cotovio, P., Ferraz, L., Balbi, L., Pesquita, C.: Matcha and matcha-dl results for oaei 2023. In: Proceedings of the 18th International Workshop on Ontology Matching (OM 2023). Athens, Greece (2023). http://ceur-ws.org/Vol-3534/, held in conjunction with ISWC 2023
8. Giglou, H.B., D'Souza, J., Engel, F., Auer, S.: LLMs4OM: Matching ontologies with large language models. arXiv preprint arXiv:2404.10317 (2024). https://arxiv.org/abs/2404.10317
9. He, Y., Chen, J., Antonyrajah, D., Horrocks, I.: Bertmap: a bert-based ontology alignment system. In: Proceedings of the Thirty-Sixth AAAI Conference on Artificial Intelligence (AAAI-22), pp. 5684–5691. AAAI Press (2022). https://doi.org/10.1609/aaai.v36i5.20510

10. Hertling, S., Paulheim, H.: Olala: Ontology matching with large language models. In: Proceedings of the 12th International Conference on Knowledge Capture (K-CAP 2023), pp. 131–139. ACM, Pensacola, FL, USA (2023). https://doi.org/10.1145/3587259.3627571

11. Ji, Z., et al.: Survey of hallucination in natural language generation. ACM Comput. Surv. **55**(12), 1–38 (2023). https://doi.org/10.1145/3571730

12. Jiménez-Ruiz, E., Cuenca Grau, B.: LogMap: Logic-Based and Scalable Ontology Matching. In: Aroyo, L., Welty, C., Alani, H., Taylor, J., Bernstein, A., Kagal, L., Noy, N., Blomqvist, E. (eds.) ISWC 2011. LNCS, vol. 7031, pp. 273–288. Springer, Heidelberg (2011). https://doi.org/10.1007/978-3-642-25073-6_18

13. Jiménez-Ruiz, E., Grau, B.C., Zhou, Y.: Logmap 2.0: towards logic-based, scalable and interactive ontology matching. In: Proceedings of the 4th International Workshop on Semantic Web Applications and Tools for the Life Sciences (SWAT4LS 2011), pp. 45–46. ACM, London, UK (2011). https://doi.org/10.1145/2166896.2166911

14. Khalid, Q., Fernandez, A., Lujak, M., Doniec, A.: Sbeo: Smart building evacuation ontology. Comput. Sci. Inf. Syst. (2022). final version and PDF available online

15. Ontology Alignment Evaluation Initiative: Oaei 2024: Ontology alignment evaluation initiative. https://oaei.ontologymatching.org/2024/ (2024). Accessed 26 May 2025

16. Pan, Z., Yang, M., Monti, A.: Schema matching based on energy domain pre-trained language model. Energy Inf. **6**(Suppl 1), 22 (2023). https://doi.org/10.1186/s42162-023-00277-0

17. Qiang, Z., Wang, W., Taylor, K.L.: Agent-OM: Leveraging LLM agents for ontology matching. arXiv:2312.00326 (2025)

18. Reimers, N., Gurevych, I.: Sentence-bert: Sentence embeddings using siamese bert-networks. In: Proceedings of the 2019 Conference on Empirical Methods in Natural Language Processing and the 9th International Joint Conference on Natural Language Processing (EMNLP-IJCNLP), pp. 3982–3992. Association for Computational Linguistics (2019). https://aclanthology.org/D19-1410/

19. da Silva, J., Revoredo, K., Baião, F.A., Euzenat, J.: ALIN: improving interactive ontology matching by interactively revising mapping suggestions. Knowl. Eng. Rev. **35**(e1), 1–22 (2020). https://doi.org/10.1017/S0269888919000249

20. Vaswani, A., et al.: Attention is all you need. In: Guyon, I., et al. (eds.) Advances in Neural Information Processing Systems (NeurIPS 2017). vol. 30. Curran Associates, Inc. (2017). https://papers.nips.cc/paper_files/paper/2017/hash/3f5ee243547dee91fbd053c1c4a845aa-Abstract.html

Context-Aware Visual Multi-turn Conversation Generation from Wikipedia and Wikidata

Basel Shbita[(✉)], Pengyuan Li, and Anna Lisa Gentile

IBM Research, San Jose, CA, USA
`{basel,pengyuan,annalisa.gentile}@ibm.com`

Abstract. Visual Question Answering (VQA) is a challenging task that demands not only accurate alignment between images and language, but also multi-step reasoning, contextual understanding, and the ability to incorporate external knowledge, especially in multi-turn settings where follow-up questions depend on previous dialogue. In this work, we present a novel framework for generating knowledge-grounded, multi-turn VQAs datasets that has been integrated into the IBM Granite-Vision development pipeline. The main novelty of our method is the generation of multi-turn conversations using large language models (LLMs), but grounded in knowledge-driven prompting: we leverage structured and unstructured knowledge sources from Wikipedia articles, associated images, and the Wikidata knowledge graph (KG). By combining both unstructured and structured knowledge sources, our approach advances VQA beyond shallow perception tasks toward more profound, knowledge- and entity-aware reasoning. We demonstrate the effectiveness of this approach by using it to fine-tune and evaluate existing vision-language models (beyond the Granite-Vision models), and share valuable insights about the complexity of the task and the nature of available benchmarks.

Keywords: knowledge graphs · vision language models · large language models · semantic web · semantic technologies · generative ai

1 Introduction

The Semantic Web community has long explored how structured, machine-interpretable representations such as RDF [3] knowledge graphs [10] and SPARQL [2] endpoints can support more explainable and verifiable intelligent systems. In this work we argue that the same principles are increasingly crucial for multimodal large language models (MLLMs) and vision language models (VLMs). Although these types of models have recently shown strong capabilities across tasks involving vision and language, such as image captioning, visual question answering (VQA), document understanding, and chart interpretation [1,4,5,9,16,23,27,32], they often lack access to the background knowledge needed to explain *why* something is relevant, *who* is depicted, or *how* objects and entities are related to each other.

M. Acosta et al. (Eds.): ESWC 2026, LNCS 16550, pp. 360–377, 2026.
https://doi.org/10.1007/978-3-032-25159-6_19

The majority of tasks that VQA models are currently tested against are limited in scope. The most popular benchmarks either focus on synthetic scenes, natural images with short factual queries, or narrow domains like science diagrams or charts [6,15,30,36]. These benchmarks often emphasize surface-level perception (e.g., object detection or scene description), leaving open the broader challenge of knowledge-grounded visual reasoning, where answering a question requires not only looking at an image but understanding its real-world context, entities, and relationships.

Real-world applications of VQA increasingly demand the ability to reason beyond what is explicitly visible, drawing on external, structured knowledge sources. For instance, answering a question about a historical artifact may require recognizing not just its appearance but also contextual knowledge about its historical significance and context, information often missing from current model outputs. Figure 1 illustrates this gap: while an open-source vision model (*Granite-Vision-3.3* [28]) gives a generic description, a more informative response would identify specific entities and offer factual, context-aware detail grounded in external knowledge. When paired with structured sources like Wikidata [31], and enriched by the contextual depth of Wikipedia articles, images become anchors for generating grounded and verifiable question-answer pairs at scale. This opens the door to vision-language systems that can reason not just about what they see, but about what it means.

In this work, we present a scalable pipeline for generating high-quality, multi-turn, knowledge-grounded visual QA data that can be used to generate both training data and more complex benchmarks for the task. We utilize images and article context from Wikipedia, aligned with structured Wikidata triples about entities mentioned in the articles. Our method builds on several key insights. First, not all knowledge triples are visually relevant or easy to ground, so we apply a combination of relation-based filtering, entity salience heuristics, and CLIP-based image-text alignment scoring to select high-quality inputs. Second, rather than relying solely on rule-based templates, we leverage prompting strategies with large language models (LLMs) to generate

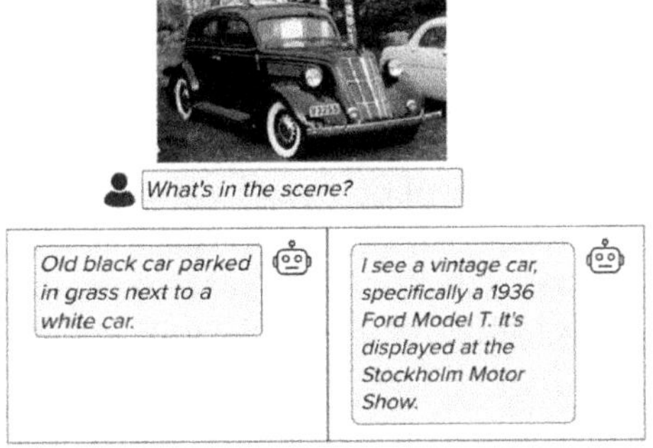

Fig. 1. Example of the gap between surface-level and knowledge-grounded visual question answering. When asked "What's in the scene?", a baseline model (*Granite-Vision-3.3*, left) provides a generic visual description. In contrast, the right-hand response demonstrates the kind of enriched, contextual answer we aim to enable, grounded in structured knowledge and capable of identifying specific objects like a 1936 Ford Model T.

diverse, fluent, and factually anchored QA pairs and multi-turn conversations. Specifically, we use *Granite-3.3-8B-Instruct* [8], an 8-billion parameter text-only language model with a 128K context window, fine-tuned for instruction-following and reasoning tasks, to synthesize multi-turn conversations from captions and verbalized triples. Its relatively small size makes it a practical choice for large-scale conversation generation, as it balances computational efficiency with sufficient generation quality for our needs. Since the model outputs are subsequently restructured and standardized, using a smaller model is feasible and allows for scalable deployment without the resource demands of larger LLMs. These conversations are then constructed in a chat-compatible format to support supervised fine-tuning of vision-language models.

We demonstrate our method by generating over half a million multi-turn image-text conversations, which we use to fine-tune two different vision-language models (*Granite-Vision-3.3* and *SmolVLM* [17]), spanning thousands of entities and relations across multiple languages. The resulting dataset enables the models to go beyond "what is seen" to answer "what is known," promoting multimodal reasoning grounded in both vision and structured knowledge. We believe this approach addresses a growing need for semantically rich and verifiable VQA and contributes to bridging the gap between perceptual grounding and factual understanding in VLMs. This capability directly strengthens Wiki-grounded VQA in deployed Granite-Vision models and addresses a growing need for encyclopedic multimodal reasoning in other state-of-the-art VLMs.

Our main technical contributions are as follows:

- **A scalable generation framework** for producing high-quality, multi-turn, knowledge-grounded visual question-answering data by aligning Wikipedia images and context with Wikidata triples.
- **An empirical analysis** of fine-tuning strategies, examining the training dynamics of vision-language models supervised with our generated data.
- **A comprehensive evaluation** across multiple VQA and multimodal benchmarks, demonstrating the effectiveness of our approach in improving knowledge-grounded reasoning and highlighting how Semantic Web technologies can be brought into practical multimodal training pipelines.

2   Related Work

Prior work on knowledge-grounded visual question answering (VQA) includes the usage of some popular datasets like LLaVA-OneVision [14], Cauldron [13], and Cambrian-10M [29] to train large vision-language models, offering diverse coverage across visual domains and instruction types. While these datasets support strong perceptual and general reasoning capabilities, they largely lack knowledge-grounded VQA content. In contrast, benchmarks like OK-VQA [18] highlight the importance of linking visual understanding with external knowledge. OK-VQA emphasizes questions that cannot be answered from the image alone, encouraging the development of systems that incorporate world knowledge and reasoning. A-OKVQA [24] extends this direction by offering a more diverse set of approximately 25,000 human-curated questions and multiple-choice answers that require commonsense and world knowledge, rather than direct querying of knowledge bases. This makes it a more realistic and challenging benchmark for evaluating multimodal reasoning in complex scenes. KVQA [25] similarly explores knowledge-based VQA, but focuses on multi-hop reasoning over large-scale knowledge graphs, particularly for entity-centric queries. Our work builds on this line of research by introducing a method to generate high-quality, multi-turn, knowledge-grounded VQA examples automatically. Instead of relying on human annotation, we leverage Wikidata triples and Wikipedia context to construct factual, entity-aware questions grounded in structured and visually-related inputs.

A closely related line of work is Encyclopedic VQA [21], which explicitly targets fine-grained, knowledge-intensive questions grounded in encyclopedic resources. This dataset introduces a large-scale collection (221k question-answer pairs and around 1M VQA samples) where each answer is supported by evidence in a controlled knowledge base derived from Wikipedia. The authors show that retrieval-augmented access to the underlying knowledge base is crucial to close this gap. Compared to Encyclopedic VQA, our work focuses on automatically generating *multi-turn* conversations rather than single-turn QA, and we explicitly operate on Wikidata enabling an explicit linking of conversational turns to Qnode-based entities and relations.

In the domain of synthetic and retrieval-augmented VQA, EchoSight [33] is a retrieval-augmented generation (RAG) framework that uses visual input to retrieve relevant Wikipedia articles to construct a joint vision-language relevance. This enhances grounding for LLM-based question answering, especially in tasks requiring fine-grained encyclopedic knowledge. Our work differs in that it focuses on generating training data for multi-turn, grounded QA rather than answering individual queries.

OVEN [12] introduces open-domain visual entity recognition by unifying image classification and QA datasets under a single label space grounded in English Wikipedia, covering a wide range of entity types and granularity. While OVEN focuses on large-scale visual classification and linking to Wikipedia pages, our work builds on this semantic space, along with structured Wikidata triples, to generate multi-turn QA data that emphasizes contextual reasoning over recognition. Another recent work on improving MLLM visual grounding through synthetic data includes EUCLID [35], which targets low-level geometric perception tasks such as accurately transcribing 2D visual structure. The authors demonstrate that synthetic high-fidelity visual descriptions, combined with a data curriculum and multi-stage training, significantly improve model performance on geometric understanding benchmarks. While this work focuses on geometric fidelity, our approach complements this line of work by targeting knowledge-grounded, multi-turn reasoning over real-world image and entity contexts.

3   Approach

In the scope of our context-aware visual conversation generation pipeline, we define *visual grounding* as the consistency of generated conversational responses with salient visual evidence in the image (objects, scene attributes, visible relations), while using Wikidata and Wikipedia as supporting context rather than a substitute for perception. Figure 2 illustrates our pipeline for generating multi-turn, knowledge-grounded conversations from image-text pairs aligned with structured Wikidata triples. We rely on existing entity-linkage in the Wikipedia-based Image Text (WIT) dataset [26], a large curated set of more than 37 million image-text associations extracted from Wikipedia articles. Our method integrates multiple components to construct high-quality training data for vision-language models. The process begins with an image-caption pair from the WIT dataset (highlighted in gray, left), where we use the Wikipedia page metadata to locate the source article and identify linked entities (Sect. 3.1). We resolve the discovered entities to QNodes, which serve as anchors for additional retrievals from Wikidata. This covers a wide range of triples capturing factual assertions such as class membership, taxonomic hierarchy, and descriptive attributes (for classes, instances, and literals). The retrieved triples undergo frequency- and heuristic-based filtering (Sect. 3.2) to discard overly generic or noisy relations (e.g., the `MeSH tree code` predicate in the accompanying example, shown in red). We then verbalize the remaining triples into natural language statements combined with the original image-caption to then prompt an LLM to generate coherent, grounded conversations (Sect. 3.3). We format the final conversations in a structured, chat-compatible format suitable for instruction-tuning VLMs (Sect. 3.4). Finally, and to ensure visual relevance, the generated outputs are further filtered, retaining only conversations closely aligned with the visual content (Sect. 3.5). As illustrated in Fig. 2, the resulting multi-turn conversation begins with a user query and progresses through factually grounded assistant responses, covering aspects like taxonomy, fruit type, and historical references, like in the example shown, and demonstrating how structured knowledge complements and extends beyond surface-level image understanding.

3.1   NER and KG Retrieval from Wikidata

The first stage of our pipeline involves extracting structured knowledge for each image-caption pair, corresponding to the "NER and KG Retrieval" module of Fig. 2. We begin by resolving the source Wikipedia article for each image in WIT using a rule-based URL extraction procedure. From the article URL, we obtain the associated Wikidata item (QNode) through the Wikipedia API[1], which uniquely identifies the underlying associated entity.

Once the QNode is identified, we retrieve all triples associated with it via SPARQL queries to the Wikidata endpoint[2] to retrieve all subject-predicate-object triples directly connected to that entity. These triples encode a broad range of encyclopedic facts such as class membership, geographic location, temporal attributes, and relationships to other entities, providing the structured context used to construct knowledge-grounded questions. At this stage we collect an unrestricted set of outgoing triples, without applying filtering or heuristics. This raw triple set forms the input to the filtering pipeline described next.

---

[1] https://www.wikidata.org/w/api.php.
[2] https://query.wikidata.org/.

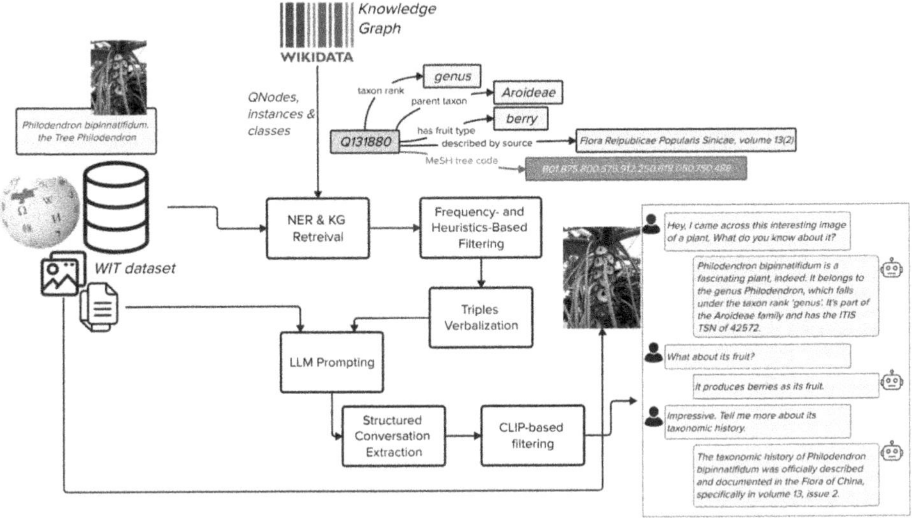

Fig. 2. Overview of our conversation generation pipeline. Left: image and caption input (gray) from the WIT dataset. Top right: retrieved Wikidata triples, color-coded by semantic relevance (orange: matched entity; yellow: relevant factual triples; red: low-salience or filtered triples). Bottom right: resulting multi-turn conversation, grounded in structured knowledge and initiated by a user query (green: user utterance; pink: assistant/model utterance). (Color figure online)

3.2   Filtering Triples

This stage corresponds to the "Frequency- and Heuristic-Based Filtering" and "Triples Verbalization" modules in Fig. 2. Before designing our filtering strategy, we conducted an initial analysis of predicate and object frequencies across a large sample of approximately 400k QNodes drawn from WIT. This analysis revealed that the majority of Wikidata facts associated with Wikipedia images concentrate around a small number of high-support semantic relations, summarized in Table 1.

The distributions in Table 1 reveal that a large portion of Wikidata facts associated with images fall into a few dominant semantic families: biographical (e.g., **human (Q5)**,

Table 1. Most notable frequent predicates and objects observed in our 400k-entity sample. These statistics inform our filtering strategy by highlighting relations that are common, human-interpretable, and often visually grounded.

Qnode	Label	Count
P31	instance of	846,144
P625	coordinate location	320,776
P106	occupation	165,531
P279	subclass of	123,876
P166	award received	55,738
Q5	human	267,990
Q16521	taxon	62,940
Q4830453	business	14,734
Q33999	actor	11,259
Q515	city	8,850

occupation (P106)), geographic (e.g., `city` (Q515), `coordinate location` (P625)), and taxonomic (e.g., `taxon` (Q16521), `subclass of` (P279)). Such relations tend to correspond well to visually grounded content in WIT images, making them ideal candidates for constructing informative multi-turn conversations.

Guided by our corpus-level analysis, we apply a single unified filtering stage that combines predicate frequency and administrative-property removal. Note that frequency-based filtering is a pragmatic, not optimal, heuristic: while it enables scalable data generation and noise control, it may amplify popularity bias and filter out precise, obscure relations where knowledge graphs typically add the most value compared to statistical models. Specifically, we keep only predicates with substantial support in the WIT-aligned corpus, discarding long-tail or overly narrow relations that rarely contribute meaningful visual grounding (less than 10 mentions) that often contain sparse, ambiguous relations that verbalize poorly in synthetic dialogue. To further clean the predicate space, we explicitly filter out administrative or low-value metadata properties whose labels match common identifier-related patterns such as ID, `identifier`, `number`, `code`, `username`, URL, or `website`. These properties are frequent in Wikidata but provide no conversational value and are not visually grounded, so they are removed regardless of their corpus counts. After predicate filtering, we verbalize the remaining triples by materializing the corresponding Wikidata labels for non-literal objects and converting each fact into a structured natural-language statement that can be reliably interpreted by the LLM. A full example showing verbalized triples for a detected entity is shown in Listing 1.2.

3.3   Multi-turn Conversation Generation

Aligned with the "LLM Prompting" module in Fig. 2, this stage transforms the factual textual snippets per image into multi-turn conversational format by forming the context for prompting an LLM to simulate a grounded conversation. We use a structured system prompt that instructs the model to generate a conversation initiated by a user asking about the image, where all assistant responses must reflect the provided factual context. The basis of the system prompt used to control the generation behavior is shown in Listing 1.1.

```
Generate a multi-turn conversation initiated by a user about an image. The
conversation should be grounded in the content of the image, which is
described by the caption below. Additionally, ensure that all factual
elements of the conversation are derived from or consistent with the
verbalized knowledge triples that follow.
```

Listing 1.1. System prompt used to guide the LLM in generating multi-turn, knowledge-grounded conversations from image captions and verbalized triples.

The result is a coherent multi-turn conversation in which the user asks questions about the image and the assistant provides contextually relevant, knowledge-informed answers. Because each verbalized fact is explicitly anchored in a Wikidata entity and the system prompt requires all responses to remain consistent with these triples, the LLM can combine and revisit facts across turns:

first identifying the primary object in the image, then elaborating on its historical background, geographic origin, semantic relations, or associated figures. In practice, this produces conversations where follow-up questions naturally refer back to previously mentioned entities, enabling the assistant to chain multiple knowledge graph facts while remaining grounded in the image caption and the provided structured context.

3.4 Restructuring to Chat Format

This step corresponds to the "Structured Conversation Extraction" module in Fig. 2. To prepare the generated conversations for use in instruction-tuning pipelines, we normalize the raw text into a consistent chat format. We first clean the LLM output by removing auxiliary markers (e.g., conversation tags, file separators, Markdown fences) and normalizing whitespace. Role markers from different prompting styles (such as "AI:", "assistant:", emoji-prefixed lines, or special tokens like `<|user|>`) are then unified via a series of regular-expression substitutions. The resulting text is split into turns by detecting these role labels; we discard outputs that do not contain at least one full user-assistant pair or that use unexpected roles. For valid cases, we construct a structured conversation by iterating over alternating role/content pairs and mapping them to a JSON-style schema and ensuring that the conversation begins with a user message and alternates properly between user and assistant turns.

This structured conversation is then formatted with the image metadata to produce a LLaVA-style [16] chat suitable for training VLMs. In addition to the chat content, we attach the originating Wikidata QNodes as separate metadata fields for each conversation. This creates an explicit bridge between the conversational supervision and the underlying RDF, enabling downstream use cases such as inspecting all conversations involving a given entity via SPARQL or joining model outputs with external Linked Data resources.

To illustrate the result of this extraction and normalization process, Listing 1.2 shows a complete example derived from an image associated with the QNode `Ishidori_Matsuri` (Q6080166). The example includes the detected entity, the filtered and verbalized triples used as factual grounding, the original image caption, and the final structured conversation produced after role normalization and turn extraction. This format represents the exact JSON-style schema consumed by our training pipeline and demonstrates how Wikidata-linked facts are preserved throughout the restructuring step.

```
Detected QNode:   Ishidori_Matsuri (Q6080166)

Image caption:    "Nishiyada-Machi's Ishidori Matsuri Cart"

Filtered triples (verbalized):
  - Ishidori Matsuri is a matsuri (festival).
  - It is located in Kuwana, Japan.
  - It takes place annually on the first Sunday in August.
  - It is inscribed on UNESCO's Representative List of the
    Intangible Cultural Heritage of Humanity.
  - It is part of the traditional Yama, Hoko, Yatai float festivals.
  - Its dedicated heritage entity is the Kuwana City Ishidori Hall.
```

```
Extracted conversation:
[ {role": "user", "content": [{"text": "I see this image is of the Ishidori
Matsuri Cart ... Can you tell me more about this event?"}]},
  {"role": "assistant", "content": [{text": "The Ishidori Matsuri ... is a
  significant cultural event held in Kuwana, Japan ..."}]},
  {"role": "user", "content": [{"text": "How often does this festival take
  place, and when is it usually held?"}]},
  {"role": "assistant", "content": [{"text": "This festival occurs annually
  on the first Sunday in August ..."}]},
  {"role": "user", "content": [{"text": "I noticed that this festival has an
  international recognition..."}]},
  {"role": "assistant", "content": [{"text": "It is included in UNESCO's
  Representative List of the Intangible Cultural Heritage of Humanity ..."}]}
  ]
```

Listing 1.2. Structured conversation after normalization and turn extraction.

3.5   Filtering via CLIP Embedding Similarity

After restructuring each conversation, we further ensure that the generated dialogue is meaningfully grounded in the visual content by applying a CLIP-based filtering step. This final stage corresponds to the "CLIP-Based Filtering" module in Fig. 2. We embed both the image and the initial user-assistant exchange using the ViT-B/32 CLIP [22] model[3], which embeds both images and text into a shared 512-dimensional space. We compute the cosine embedding distance between the image and text representations in the shared CLIP space and remove samples whose distance exceeds a fixed threshold of 0.7. This filters out conversations that, while structurally valid, diverge substantially from the visual content or rely too heavily on textual cues from the caption or knowledge triples. The threshold was determined empirically by analyzing the distance distribution over a subset of generated image-conversation pairs and manually inspecting examples near candidate cutoffs. In practice, pairs with distances above the mean (0.7) showed weak or generic visual grounding, whereas lower-distance pairs displayed strong alignment between the conversation and the content visible in the image. Restricting the dataset to samples below this threshold ensured that the final training data emphasized genuine multimodal grounding and reduced the risk of the model learning visually unsupported or hallucinated facts.

The dataset-wide distribution of image-text distances is shown in Fig. 3. The histogram reflects measurements across approximately half a million image-conversation pairs generated using our method, and is centered around a mean of roughly 0.7, indicating varying levels of semantic alignment. To better illustrate this variation, we highlight specific samples at either end of the distance spectrum. On the left (green), examples with low embedding distance exhibit strong alignment between visual and textual content. For instance, a user refers to the cover of an 1884 poetry book and asks about the poet, Isabella Valancy Crawford. The assistant provides a highly relevant response detailing her biography and literary significance, closely grounded in both the image and associated

---

[3] https://huggingface.co/sentence-transformers/clip-ViT-B-32.

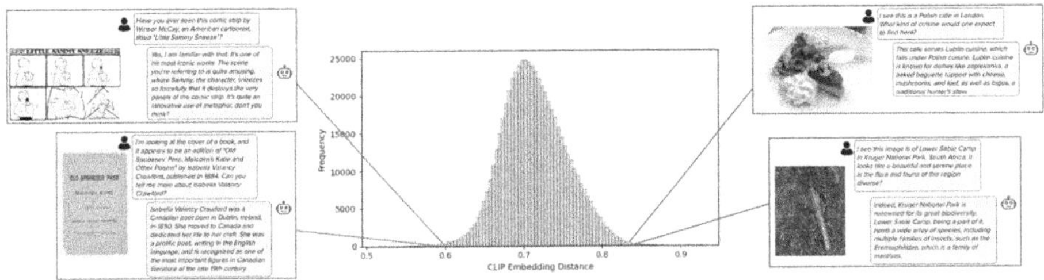

Fig. 3. Distribution of CLIP embedding distances between WIT images and their associated conversations. Examples on the left (green border) demonstrate strong semantic alignment between the visual and textual content, while those on the right (red border) exhibit weak or generic associations. Our CLIP-based filtering prioritizes samples with lower embedding distances to ensure visual grounding. Each illustrated conversation includes only the first two turns for simplicity, though the full conversations often continue for several additional turns. (Color figure online)

Wikidata content. On the right (red), high-distance samples show weaker or more tangential grounding. One example is the image of a mantis resting on a tree bark from Kruger National Park. While the assistant offers a biologically rich description of biodiversity in the region, the textual content does not clearly relate to the visual content of the image itself. These examples underscore the importance of CLIP-based filtering in identifying and removing noisy or overly generic image-text pairs.

4 Evaluation

We evaluate the impact and utility of our proposed generation framework across three dimensions: (1) the quality and grounding of the generated multi-turn visual conversations; (2) the training dynamics of vision-language models fine-tuned with our data; and (3) performance improvements on downstream VQA and multimodal reasoning benchmarks. We employ both qualitative and quantitative analyses to demonstrate the strengths of our approach and its practical contributions.

4.1 Training Dynamics

We analyze the training behavior of two vision-language models, *SmolVLM* and *Granite-Vision-3.3*, under various fine-tuning strategies, as shown in Figs. 4a and 4b, respectively. Each model is fine-tuned for a single epoch using full fine-tuning (FT) and Low-Rank Adaptation (LoRA) [11] for parameter-efficient tuning with and without CLIP-based visual filtering applied to the training data. The goal is to understand how tuning strategy and data quality affect convergence behavior. In Fig. 4a, *SmolVLM* demonstrates a clearer differentiation across fine-tuning strategies. Full fine-tuning (FT) achieves the lowest loss

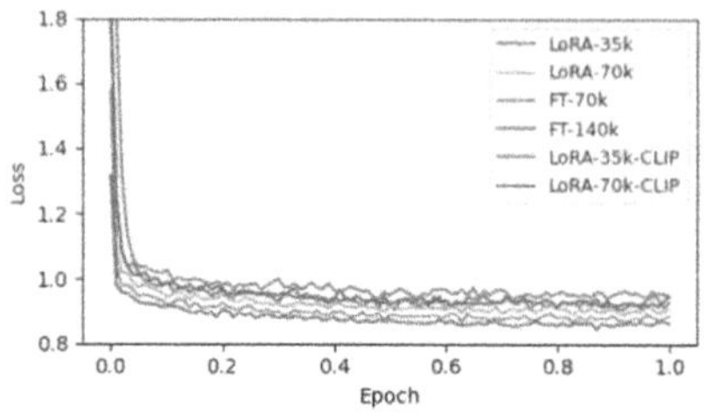

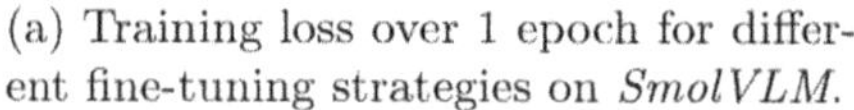

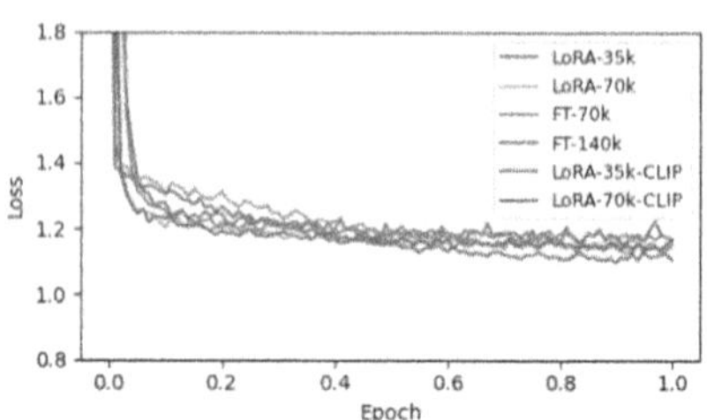

(a) Training loss over 1 epoch for different fine-tuning strategies on *SmolVLM*.

(b) Training loss over 1 epoch for different fine-tuning strategies on *Granite-Vision-3.3*.

Fig. 4. We compare full fine-tuning (FT) and LoRA-based tuning at different update scales (with and without CLIP-based filtered data). While full fine-tuning yields the lowest loss overall, LoRA-CLIP variants demonstrate more stable convergence early in training.

along the entire training epoch, reflecting the benefit of full parameter adaptation. Interestingly, LoRA with CLIP filtering converges faster than standard LoRA, indicating that improved visual grounding may help optimization in early steps. In Fig. 4b the loss curves are tighter across configurations, with FT still achieving lower loss compared to other methods but with smaller margins. LoRA variants benefit less visibly from CLIP-based filtering in this model, suggesting that *Granite-Vision-3.3* may already encode stronger visual-linguistic priors or be less sensitive to input noise.

Together, these results suggest that CLIP-based filtering enhances training stability and convergence, particularly in lighter models like *SmolVLM*. While full fine-tuning offers the best loss reduction, LoRA provides a viable and resource-efficient alternative, especially when coupled with high-quality, visually aligned data (the LoRA-CLIP variant). This general trend, along with the consistent convergence across configurations, supports the overall quality and usability of the generated dataset.

Key ablation findings motivate our pipeline design choices. Early experiments without the CLIP filtering step admitted more entity-only summaries that lacked strong visual grounding, resulting in models that memorized Wikipedia content without visual support. Removing the predicate-frequency filter similarly degraded performance by introducing sparse, noisy relations that verbalized poorly. These observations underscore the importance of each filtering stage in balancing coverage with quality.

4.2   Qualitative Effects on Model Behavior

To qualitatively assess the impact of grounding LLM responses in external structured knowledge, we compare outputs from a baseline multimodal model against our knowledge-enhanced and fine-tuned model. As shown in Fig. 5a, when asked *"What do you see in the image?"* about a vintage engraving of the University

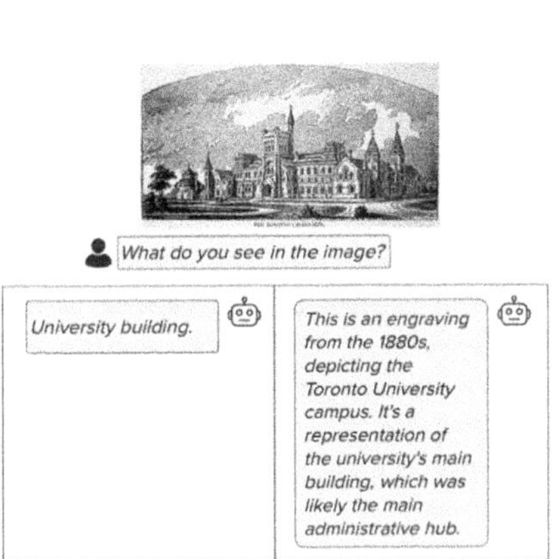

(a) A question about an image with a vintage engraving of the University of Toronto.

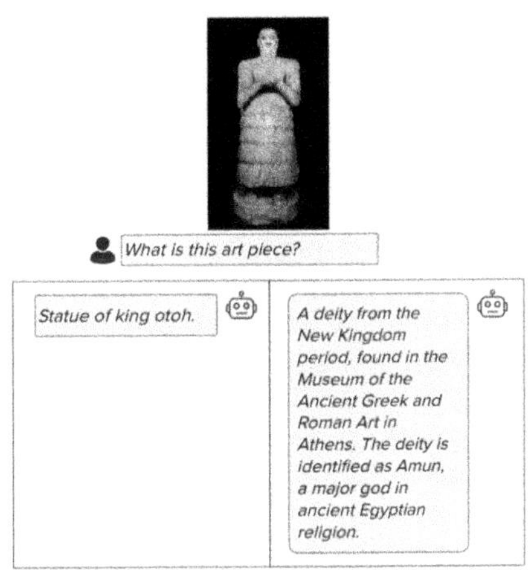

(b) A question for ancient statue identification.

Fig. 5. Example comparisons between a baseline vision-language model (left) and a model fine-tuned on our knowledge-grounded VQA data (right). The images were transformed from their training version to avoid memorization from training data.

of Toronto, the baseline model (left) produces a minimal response: *"University building"*. In contrast, the fine-tuned model (right) generates a richer, contextually grounded description: *"This is an engraving from the 1880s, depicting the Toronto University campus. It's a representation of the university's main building, which was likely the main administrative hub."* Although the source image is drawn from the training corpus, the one used in this example has been transformed (e.g., cropped and mirrored) to avoid any direct overlap or data leakage. This example highlights how even minimal knowledge grounding can significantly enhance semantic specificity and historical relevance in generated responses. Figure 1 shows another example, this time involving an ancient art artifact. When prompted with "What is this art piece?", the baseline model responds incorrectly with "Statue of king otoh.", a clearly erroneous guess. The fine-tuned model, however, produces an informative and factual output: "A deity from the New Kingdom period, found in the Museum of the Ancient Greek and Roman Art in Athens. The deity is identified as Amun, a major god in ancient Egyptian religion." This example demonstrates the factual correction and depth added by knowledge-grounded tuning, particularly for culturally or historically complex subjects and general knowledge present in Wikipedia and Wikidata.

4.3   Quantitative Evaluation on VQA Benchmarks

To assess the effectiveness of our generated data and fine-tuning strategies, we evaluate all model variants on a diverse suite of vision-language benchmarks.

These include general-purpose multimodal reasoning datasets (MMMU [34], MME [7]), document and chart understanding (DocVQA [20], ChartQA [19]), and knowledge-intensive VQA datasets such as A-OKVQA [24]. Our primary objective is not to maximize performance on general multimodal benchmarks, but rather to provide a scalable Semantic Web-grounded pipeline for generating knowledge-grounded, multi-turn visual conversations and to study how such supervision affects models. We include these benchmarks as stress tests to understand transfer effects and failure modes when introducing KG-grounded conversational supervision.

We consider seven configurations per model: an untuned baseline, LoRA with 35k and 70k randomly sampled image-text conversations, full fine-tuning (FT) using 70k and 140k samples, and LoRA using CLIP-filtered top-35k and top-70k samples to maximize visual relevance.

Table 2 presents the benchmark scores. The results show that LoRA methods generally preserve or modestly improve performance across several tasks, while full fine-tuning tends to degrade performance, particularly in benchmarks like MMMU, A-OKVQA, and DocVQA. This degradation in full fine-tuning is expected due to domain and format mismatch (Wikipedia/Wikidata-grounded conversational data differs significantly from general VQA formats), catastrophic forgetting with full parameter updates, and the inherent trade-offs of specializing a general-purpose model toward encyclopedic reasoning.

Table 2. Performance of baseline and tuned models on vision-language benchmarks.

		Granite-Vision-3.3 (3B)					SmolVLM (2B)				
		MMMU	$MME_{reasoning}$	DocVQA	$ChartQA_{human}$	A-OKVQA	MMMU	$MME_{reasoning}$	DocVQA	$ChartQA_{human}$	A-OKVQA
	Baseline	**0.372**	318.9	90.37	77.12	**0.847**	0.358	283.2	**79.73**	52.80	**0.771**
Tuning Method	LoRA 35k	0.371	338.2	**90.54**	75.68	0.834	0.361	235.0	68.28	46.72	0.733
	LoRA + CLIP 35k	0.370	350.4	90.44	**95.20**	0.840	**0.378**	230.7	64.79	47.20	0.749
	LoRA 70k	0.363	**353.9**	89.97	94.88	0.833	0.352	**318.9**	67.46	44.96	0.725
	LoRA + CLIP 70k	0.353	320.0	89.96	74.64	0.831	0.358	252.1	67.40	**82.56**	0.725
	FT 70k	0.303	227.8	6.93	12.72	0.412	0.350	227.9	45.43	58.48	0.663
	FT 140k	0.291	206.7	6.48	13.36	0.406	0.359	218.6	41.16	53.28	0.658

For *Granite-Vision-3.3*, the baseline achieves the best MMMU (0.372) and A-OKVQA (0.847) scores, while LoRA+CLIP70k and LoRA70k yield the best MME (353.9) and strong ChartQA performance (94.88 and 74.64). In *SmolVLM*, LoRA+CLIP35k outperforms all others on MMMU (0.378), with the LoRA+CLIP70k variant achieving a significant gain on ChartQA (82.56). Full fine-tuning, while yielding large losses in some benchmarks (e.g., DocVQA and A-OKVQA), shows moderate improvements in ChartQA, suggesting that it may lead to overfitting or knowledge drift in certain tasks. Although the overall MMMU score slightly favors the baseline, Table 3 offers deeper insight through per-category analysis. Fine-tuned models exhibit domain-specific improvements, with LoRA+CLIP35k showing the broadest gains, including *Clinical Medicine* (+17), *Computer Science* (+13.3), *Economics* (+11.7), and *Psychology* (+10).

LoRA35k similarly excels in *Architecture & Engineering* and *Math* (both +10.0), while full fine-tuning (FT140k) achieves isolated gains in *Physics* (+8.3)

Table 3. Category-wise performance change, in percentage (%), on MMMU benchmark relative to baselines across tuning strategies.

Tuning Method	Arch & Engr	Clinical Med	Comp Sci	Diagnostics	Economics	Electronics	Enrgy & Pwr	Finance	Geography	History	Math	Physics	Psychology
LoRA 35k	+10.0	-17.0	+6.7	+1.7	+8.3	+6.7	+8.3	-5.0	+10.0	+3.3	+10.0	+5.0	+0.0
LoRA+CLIP 35k	+3.3	+17.0	+13.3	+3.3	+11.7	+11.7	+10.0	-	+10.0	+6.7	+5.0	+5.0	+10.0
LoRA 70k	+5.0	-	+10.0	+3.3	+8.3	+3.3	+6.7	-3.3	+1.7	+5.0	+5.0	+5.0	+1.7
LoRA+CLIP 70k	+1.7	-	+10.0	+5.0	+6.7	+10.0	+10.0	-6.7	-1.7	+13.3	+10.0	+3.3	-
FT 70k	+6.7	-3.3	+13.3	-6.7	-1.7	+5.0	+8.3	-3.3	-1.7	-1.7	-1.7	+10.0	-11.7
FT 140k	+8.3	-13.3	+10.0	-5.0	-	+6.7	+3.3	-3.3	-	-3.3	-3.3	+8.3	-11.7

but suffers significant degradation in areas like *Psychology* (-11.7). These results suggest that CLIP-filtered LoRA tuning not only preserves general reasoning ability but also enables targeted knowledge enrichment across multiple domains. Part of the observed variance across categories stems from the inherent domain distribution in Wikidata, which favors encyclopedic, historical, and taxonomic knowledge, potentially leading to shifts or degradation in domains underrepresented in the training data (e.g., Finance). Overall, these findings demonstrate that our generated training data can effectively enhance downstream performance in knowledge-enriched tasks when paired with compute-efficient fine-tuning strategies like LoRA.

5 In-Use Scenario: Continuous Data Generation for Granite-Vision Post-training

Our framework is not only evaluated offline but is also employed as part of an operational data generation pipeline for the Granite-Vision [28] model family[4]. In this section, we describe how the system is integrated into a real model development workflow and how the generated conversations inform iteration on model design and training. By operating alongside established post-training tasks, the framework has become a practical component of the daily model development cycle and not merely an experimental prototype. Importantly, this system was introduced to directly combat catastrophic forgetting observed in production training pipelines when general VQA datasets overwrite encyclopedic knowledge. Rather than replacing retrieval-based approaches, our framework complements inference-time retrieval-augmented generation (RAG) paradigms in two key ways: (i) training-time KG grounding can reduce reliance on external calls at runtime, offering latency and operational benefits in production settings; and (ii) generating knowledge-grounded dialogues during training teaches models associations between visual entities and structured knowledge that are orthogonal to factual retrieval at inference.

---

[4] Available at https://huggingface.co/collections/ibm-granite/granite-vision-models.

5.1   Integration Into the Vision Post-training Pipeline

Within the Granite-Vision post-training stack[5], our pipeline runs as a data preparation stage that consumes Wikipedia/Wikidata resources and produces chat-formatted, knowledge-grounded image conversations. The output is ingested into the same infrastructure that handles other supervised fine-tuning data, so that vision engineers can combine or ablate our data source through configuration only. In practice, this means that new conversational data can be generated, validated, and incorporated into training runs within hours, enabling rapid iteration and targeted augmentation. Table 2 demonstrates that such integration not only preserves existing knowledge but actively restores important visual–encyclopedic grounding that general VQA tuning tends to erase. This also allows the post-training team to selectively generate conversations for specific domains or entities by issuing targeted SPARQL queries, effectively turning knowledge graph queries into actionable training data.

5.2   Benefits for Semantic Web-Driven Development

From an in-use perspective, an important advantage of our approach is that it keeps the entire data generation and training loop grounded in Semantic Web standards. Our experience so far suggests that aligning multimodal supervision with Semantic Web identifiers and infrastructure simplifies coordination between model developers and knowledge engineers. It also opens the door to future tooling that exposes model training artifacts (datasets, runs, evaluation results) directly as Linked Data that can be queried and audited using standard Semantic Web techniques. In effect, the pipeline positions Wikidata not only as a knowledge source but also as a control layer for guiding the scope and structure of multimodal supervision, offering a practical demonstration of how Semantic Web technologies can inform modern foundation-model development. Notably, the Granite-Vision training team has committed to incorporating our Wiki-grounded data into the next public release, reflecting operational confidence in the framework's ability to maintain long-term factual integrity. Given that state-of-the-art models such as GPT-5 and Qwen3-VL exhibit the same weakening of Wikipedia-derived knowledge, we expect this approach to benefit and be adopted by the broader VLM community as well.

6   Discussion and Future Work

Our work demonstrates the feasibility of generating grounded, multi-turn visual conversations at scale by aligning Wikipedia image-caption pairs with structured Wikidata knowledge. However, several key challenges remain. Ensuring that retrieved triples are both relevant and visually grounded is non-trivial, especially given the noisy and often sparse nature of Wikidata entries. While our

---

[5] Tutorial available at
   https://huggingface.co/learn/cookbook/en/fine_tuning_granite_vision_sft_trl

heuristic and frequency-based filtering reduces irrelevant relations, false positives still occur, which can propagate noise into the generated conversations. Additionally, maintaining fluency using LLMs without converging on repetitive patterns or rigid templates is difficult. Balancing visual cues with textual context across conversation turns also poses a challenge. Moreover, the reliance on entity-linked pages means that visually salient elements not explicitly mentioned in the article may be underrepresented in the retrieved triples, limiting the breadth of knowledge that the framework can incorporate.

While our work focuses on training-time KG grounding, we note that inference-time retrieval-augmented generation (RAG) represents a complementary paradigm that we view as orthogonal rather than competing. A hybrid approach combining both training-time supervision and inference-time retrieval could further improve conversational grounding and variation. Furthermore, incorporating topic-balancing filters during data construction could help mitigate unintended shifts in knowledge distribution, such as the degradation seen in finance-related performance (Table 3). This would allow for more controlled fine-tuning and generalization across domains. Additional evaluation on long-context image understanding would also strengthen the robustness of our approach. Finally, extending the framework to handle long-tail predicates more effectively, perhaps via curriculum learning or selective augmentation, could enhance the framework's ability to preserve rare but valuable knowledge.

7 Conclusions

We presented a scalable framework for generating multi-turn knowledge-grounded image-text conversations by aligning Wikipedia images and context with Wikidata triples. We demonstrate that this method can improve factuality in fine-tuned models, particularly when using parameter-efficient training strategies like LoRA. While benchmark performance does not consistently surpass strong baselines, qualitative analysis reveals promising improvements in terms of response grounding, suggesting that carefully curated multi-modal supervision can meaningfully contribute to grounded vision-language understanding and serve as a strong foundation for multi-modal synthetic data generation. Our method offers a scalable solution to enhancing VQA capabilities in production-level Granite-Vision models, with strong potential for external broader adoption.

Supplemental Material Statement: The use case is completely built upon publicly available resources: Wikipedia, Wikidata, and the Granite open-source vision-language models.

References

1. Chen, Z., et al.: Internvl: Scaling up vision foundation models and aligning for generic visual-linguistic tasks. In: Proceedings of the IEEE/CVF Conference on Computer Vision and Pattern Recognition, pp. 24185–24198 (2024)

2. Consortium, W.W.W., et al.: SPARQL 1.1 overview. Tech. rep., World Wide Web Consortium (2013)
3. Consortium, W.W.W., et al.: RDF 1.1 primer (2014)
4. Dai, W., et al.: NVLM: Open frontier-class multimodal LLMs. arXiv preprint arXiv:2409.11402 (2024)
5. Deitke, M., et al.: Molmo and pixmo: Open weights and open data for state-of-the-art multimodal models. arXiv e-prints pp. arXiv–2409 (2024)
6. Duan, H., et al.: Vlmevalkit: An open-source toolkit for evaluating large multi-modality models. In: Proceedings of the 32nd ACM International Conference On Multimedia, pp. 11198–11201 (2024)
7. Fu, C., et al.: MME-survey: a comprehensive survey on evaluation of multimodal LLMs. arXiv preprint arXiv:2411.15296 (2024)
8. Granite Team, I.: Granite 3.0 language models. URL: https://github.com/ibm-granite/granite-3.0-language-models (2024)
9. Grattafiori, A., et al.: The llama 3 herd of models. arXiv preprint arXiv:2407.21783 (2024)
10. Hogan, A., et al.: Knowledge graphs. ACM Comput. Surv. (Csur) **54**(4), 1–37 (2021)
11. Hu, E.J., et al.: Lora: Low-rank adaptation of large language models. ICLR **1**(2), 3 (2022)
12. Hu, H., et al.: Open-domain visual entity recognition: Towards recognizing millions of wikipedia entities. In: Proceedings of the IEEE/CVF International Conference on Computer Vision, pp. 12065–12075 (2023)
13. Kembhavi, A., Salvato, M., Kolve, E., Seo, M., Hajishirzi, H., Farhadi, A.: A diagram is worth a dozen images. In: European Conference on Computer Vision, pp. 235–251. Springer (2016)
14. Li, B., et al.: Llava-onevision: Easy visual task transfer. arXiv preprint arXiv:2408.03326 (2024)
15. Li, Z., Wu, X., Du, H., Liu, F., Nghiem, H., Shi, G.: A survey of state of the art large vision language models: Alignment, benchmark, evaluations and challenges. arXiv preprint arXiv:2501.02189 (2025)
16. Liu, H., Li, C., Wu, Q., Lee, Y.J.: Visual instruction tuning. Adv. Neural. Inf. Process. Syst. **36**, 34892–34916 (2023)
17. Marafioti, A., et al.: Smolvlm: Redefining small and efficient multimodal models. arXiv preprint arXiv:2504.05299 (2025)
18. Marino, K., Rastegari, M., Farhadi, A., Mottaghi, R.: OK-VQA: A visual question answering benchmark requiring external knowledge. In: Proceedings of the IEEE/CVF Conference on Computer Vision and Pattern Recognition, pp. 3195–3204 (2019)
19. Masry, A., Long, D.X., Tan, J.Q., Joty, S., Hoque, E.: Chartqa: A benchmark for question answering about charts with visual and logical reasoning. arXiv preprint arXiv:2203.10244 (2022)
20. Mathew, M., Karatzas, D., Jawahar, C.: Docvqa: A dataset for VQA on document images. In: Proceedings of the IEEE/CVF Winter Conference On Applications Of Computer Vision, pp. 2200–2209 (2021)
21. Mensink, T., et al.: Encyclopedic VQA: Visual questions about detailed properties of fine-grained categories. In: Proceedings of the IEEE/CVF International Conference on Computer Vision, pp. 3113–3124 (2023)
22. Radford, A., et al.: Learning transferable visual models from natural language supervision. In: International conference on machine learning, pp. 8748–8763. PmLR (2021)

23. Rahmanzadehgervi, P., Bolton, L., Taesiri, M.R., Nguyen, A.T.: Vision language models are blind. In: Proceedings of the Asian Conference on Computer Vision, pp. 18–34 (2024)
24. Schwenk, D., Khandelwal, A., Clark, C., Marino, K., Mottaghi, R.: A-okvqa: A benchmark for visual question answering using world knowledge. In: European Conference on Computer Vision, pp. 146–162. Springer (2022)
25. Shah, S., Mishra, A., Yadati, N., Talukdar, P.P.: KVQA: knowledge-aware visual question answering. In: Proceedings of the AAAI Conference on Artificial Intelligence, vol. 33, pp. 8876–8884 (2019)
26. Srinivasan, K., Raman, K., Chen, J., Bendersky, M., Najork, M.: Wit: Wikipedia-based image text dataset for multimodal multilingual machine learning. In: Proceedings of the 44th International ACM SIGIR Conference on Research and Development in Information Retrieval, pp. 2443–2449 (2021)
27. Team, G., et al.: Gemini 1.5: Unlocking multimodal understanding across millions of tokens of context. arXiv preprint arXiv:2403.05530 (2024)
28. Team, G.V., et al.: Granite vision: a lightweight, open-source multimodal model for enterprise intelligence. arXiv preprint arXiv:2502.09927 (2025)
29. Tong, P., et al.: Cambrian-1: A fully open, vision-centric exploration of multimodal LLMs. Adv. Neural Inform. Process. Syst. **37**, 87310–87356 (2024)
30. Tong, S., Liu, Z., Zhai, Y., Ma, Y., LeCun, Y., Xie, S.: Eyes wide shut? exploring the visual shortcomings of multimodal LLMs. In: Proceedings of the IEEE/CVF Conference on Computer Vision and Pattern Recognition, 9568–9578 (2024)
31. Vrandečić, D., Krötzsch, M.: Wikidata: a free collaborative knowledgebase. Commun. ACM **57**(10), 78–85 (2014)
32. Wang, P., et al.: Qwen2-vl: Enhancing vision-language model's perception of the world at any resolution. arXiv preprint arXiv:2409.12191 (2024)
33. Yan, Y., Xie, W.: Echosight: Advancing visual-language models with wiki knowledge. arXiv preprint arXiv:2407.12735 (2024)
34. Yue, X., et al.: Mmmu: a massive multi-discipline multimodal understanding and reasoning benchmark for expert AGI. In: Proceedings of the IEEE/CVF Conference on Computer Vision and Pattern Recognition, pp. 9556–9567 (2024)
35. Zhang, J., Liu, O., Yu, T., Hu, J., Neiswanger, W.: Euclid: supercharging multimodal LLMs with synthetic high-fidelity visual descriptions. In: Will Synthetic Data Finally Solve the Data Access Problem? (2025). https://openreview.net/forum?id=6DV6DCk8GS
36. Zhang, K., et al.: LMMS-EVAL: Reality check on the evaluation of large multimodal models (2024). https://arxiv.org/abs/2407.12772

A Wikidata-Based Workflow for Entity Reconciliation Strategies Evaluation: A Study on Early Modern Polish Personal Names

Luiz do Valle Miranda[(✉)][iD], Maciej Mozolewski[iD], Krzysztof Kutt[iD], and Grzegorz J. Nalepa[iD]

Department of Human-Centered Artificial Intelligence, Institute of Applied Computer Science, Faculty of Physics, Astronomy and Applied Computer Science, Jagiellonian University, prof. Stanisława Łojasiewicza 11, 30-348 Kraków, Poland
{luiz.miranda,m.mozolewski,krzysztof.kutt,grzegorz.j.nalepa}@uj.edu.pl

Abstract. Entity Reconciliation (ER) in Cultural Heritage often faces the problem of fragmented collections and inconsistent naming conventions. This paper presents a workflow for evaluating ER strategies, focusing on the complex case of early modern Polish personal names. Leveraging Wikidata's vast network of multilingual aliases, we construct a validation dataset of nearly 10,000 label–alias pairs. To address the specific variation of multi-part historical names, we introduce a novel component-wise coverage threshold strategy. Under this framework, we benchmark classical string metrics, such as Levenshtein and Jaro-Winkler, and phonetic algorithms like Beider-Morse and Daitch-Mokotoff. Our results demonstrate that supervised ensembles, particularly Random Forest and Gradient-Boosted Decision Trees, consistently outperform individual metrics. Comparison against a state-of-the-art multilingual Transformer (LaBSE) reveals that while the latter maximizes recall, it struggles with precision under limited data regimes and demands high computational resources. In contrast, tree-based classifiers offer a highly efficient alternative that remains transparent, explainable, and effective even when trained on small data samples. We conclude by discussing Wikidata's potential for scalable ER benchmarks.

Keywords: Wikidata · Linked Data · Entity Reconciliation · Cultural Heritage · SPARQL · Supervised Learning · Transformer-based Models

1 Introduction

The continuous process of data digitization and publication has led to the emergence of numerous digital collections, characterized by diverse structures and naming conventions. To effectively harness the value of this growing body of data, it is essential to integrate these collections and infer a common structure, including identifying correspondences across datasets. A fundamental challenge

M. Acosta et al. (Eds.): ESWC 2026, LNCS 16550, pp. 378–397, 2026.
https://doi.org/10.1007/978-3-032-25159-6_20

in this integration process is entity reconciliation (ER)—the task of identifying records across different datasets that refer to the same real-world object.

Workflows for ER can incorporate modules based on various frameworks, such as algorithm-based, graph-based, probabilistic, and learning-based approaches (see [7]). The choice of a suitable solution depends on several factors, including the structure of the datasets, the data types of the records to be linked, the availability of background knowledge in the relevant domain, and specific application requirements. A common scenario involves matching string-based information across heterogeneous datasets, with solutions independent of the data structure or domain-specific knowledge. In such cases, algorithm-based approaches—such as string similarity metrics or phonetic algorithms —are frequently employed due to their simplicity [4,6,11,21].

However, classical string metrics often fail to capture semantic equivalence when aliases share little lexical overlap [14]. To address this, Transformer-based language models emerged , mapping entity strings into dense vector spaces where similarity correlates with semantic proximity [13]. The Language-Agnostic BERT Sentence Embedding (LaBSE) [8] model is particularly promising for cultural heritage (CH), as its pre-training on over one hundred languages makes it robust to the multilingual variations, typical of early modern historical records.

A critical factor in deciding which algorithms to include in a reconciliation workflow for a data integration project is the evaluation of their performance on a representative validation dataset. While such a dataset can be manually constructed, this process is time-consuming and labor-intensive. Alternatively, synthetically introducing noise or perturbations into an existing dataset can simulate variation. However, the introduced errors may not accurately reflect the real-world causes of variation for the specific project, such as the specific heterogeneity in data sources or entry practices (see [4]).

This paper presents a workflow for evaluating strategies for ER, leveraging interoperable knowledge bases—specifically using Wikidata as a reference source. The case study addresses the challenge of matching personal names across heterogeneous datasets. Code corresponding to the analyses performed for this article can be found in https://github.com/luizdovalle2/ER-Wikidata-WF.

The remainder of the paper provides an overview of role of ER in the CH domain. Second, we present the context of our specific problem: matching names across Polish cultural heritage institutions, followed by the translation of our context into a Wikidata SPARQL query. We then present the evaluation of the algorithm-based ER strategies and propose a combination of different metrics using tree-based ensembles. Furthermore, we extend the evaluation to include a fine-tuned LaBSE model as a theoretical performance "upper bound". We investigate whether the semantic capabilities of such massive neural architectures justify their significantly higher computational costs and lower transparency compared to classical lexical methods. The paper concludes with a summary of findings and directions for future work.

2   Entity Reconciliation (not Only) in Cultural Heritage

A prominent method for reconciling personal names from different cultural heritage institutions is by matching independent resources to the same value in an authority file, like Virtual International Authority File (VIAF) and Library of Congress Name Authority File (LCNAF) [9,19] to name two of the most commonly used. While VIAF and LCNAF are centered around CH resources, Wikidata provides a more generalized, crowdsourced alternative that encompasses other domains of knowledge, that enables interoperability between diverse databases, allowing automated systems to recognize and analyze relationships across multiple sources [17,24].

Despite Wikidata's prominence as a LD platform—with over 119 million items as of November 2025[1]—its coverage exhibits significant disparities. These imbalances manifest in areas such as multilinguality (e.g., uneven representation of entities across languages) and niche domains (e.g., specialized cultural or scientific topics) [10,23]. In cases where personal records from different bases need to be reconciled without the possibility of matching to authority control resources, one alternative are algorithm-based workflows for ER.

Bryer et al. [4] conducted a comprehensive evaluation of algorithmic methods for reconciling book titles. With a dataset of 5.8 million records they analyzed the accuracy of reconciliation using string similarity metrics from several algorithms. However, one limitation against generalizing over Bryer et al.'s study is its reliance on a pre-processed dataset, which may not fully capture challenges in ER for other contexts. In contrast, [12] proposed an ER workflow using public Wikidata data, applying machine learning methods for familial network resolution. While their approach demonstrates a structured framework for ER experiments, it does not explicitly exploit the semantic relationships between entity labels and their aliases, a potentially valuable feature for improving disambiguation.

Thus, even when entities are absent from Wikidata, its expansive knowledge base can still enhance ER processes. While Wikidata has been utilized for constructing ER benchmark datasets, no prior work has specifically applied these techniques in CH domain—particularly to personal name headings. Their reconciliation is especially error-prone because variations in spelling, linguistic nuances, and alias relationships are fundamentally shaped by their source context. The institutional or cultural origin of a name pair—whether from library catalogs, archival records, or biographical databases—precisely determines the structure of these errors (e.g., translation discrepancies, or local cataloging conventions). Wikidata's structured data on alternative names and transliterations can provide a robust specialized dataset for experimenting and deciding on the best solution having a certain context for the target ER dataset. By experimenting with datasets capturing the context-dependent error patterns, a Wikidata-based ER pipeline can help decide on reconciliation methods that adapt to the specific challenges posed by different cultural heritage domains and applications.

---

[1] https://www.wikidata.org/wiki/Wikidata:Statistics, accessed November 30, 2025.

The landscape of ER evolved significantly with the transition from manual feature engineering to deep learning architectures. According to [14], deep neural models can automatically learn distributed representations of entity records, significantly outperforming traditional rule-based systems on textual data. This paradigm shift was accelerated by the introduction of pre-trained language models like BERT [5], which enable context-aware matching. Building on this, systems such as Ditto [13], fine-tune Transformer architectures to treat entity matching as a sequence classification task, achieving state-of-the-art results on standard benchmarks. However, the computational cost often limits their scalability. To address this bottleneck, [16] introduces Sentence-BERT (SBERT), a bi-encoder framework that derives fixed-size embeddings for input strings independently. This approach allows for the efficient computation of semantic similarity via cosine distance, making neural ER feasible for larger datasets while retaining much of the representational power of deep models.

In the context of CH , the challenge is further compounded by linguistic diversity. Early multilingual embedding models, such as LASER [3], enabled zero-shot transfer across languages, but have been superseded by more robust architectures. Specifically, the LaBSE model by Feng et al. [8] supports over 109 languages, aligning disparate labels (e.g., in Latin, Polish, and German) into a shared vector space. Recent studies [1,18] demonstrate that embedding-based methods significantly outperform statistical string alignment for cross-lingual ER in knowledge graphs like Wikidata. Nevertheless, applying these massive models to the nuances of historical personal names remains a non-trivial task [15], necessitating a rigorous evaluation of whether their semantic capabilities justify the resource overhead compared to classical ensembles.

3   Defining a Context: the CHExRISH Project

The Jagiellonian University (JU), Poland's oldest university and among the world's oldest continuously operating institutions, houses a diverse collection of CH artifacts connected to its more than 600 years history. Data concerning JU heritage resources are currently stored and analyzed separately in different units, including the Archive (AUJ), the Museum (MUJ), and the Library (BJ). Additionally, each unit models, stores and gives access to data in its own way.

The prototypical unification of these data in the Jagiellonian University Heritage Metadata Portal (JUHMP) with unified browsing and searching capabilities is one of the goals of the ongoing CHExRISH project[2]. A first prototype [20] has been implemented for 10 manually mapped records between the AUJ's Corpus Academicum Cracoviense[3] (CAC) and BJ's Jagiellonian Digital Library[4] (JDL).

The ongoing development of a second prototype requires more mappings. At this stage, records from CAC—an electronic database with around 67,000 records on students and graduates of the University of Kraków during the period

---

[2] https://chexrish.id.uj.edu.pl/.
[3] https://cac.historia.uj.edu.pl/.
[4] https://jbc.bj.uj.edu.pl/dlibra.

Mikołaj Kopernik (Copernicus) z Torunia, syn Mikołaja

First name:	Mikołaj
Surname:	Kopernik
Father's name:	Mikołaj
Place of origin:	Toruń
Place and date of birth:	Feb 19, 1473
Place and date of death:	May 24, 1543
Known events from period:	Feb 19, 1473 — the end of 1543 year

Education and academic degrees

#	Event type	Education stage/academic degree	Scientific discipline	Institution	Date
1	koniec studiów/nauki	student	medycyna	Uniwersytet Padewski	the beginning of summer 1503 — the end of summer 1503
1	początek studiów/nauki	student	medycyna	Uniwersytet Padewski	the beginning of August 1501 — the end of December 1501
1	immatrykulacja	student	prawo	Uniwersytet Boloński	after Jan 6, 1497
1	immatrykulacja	student	sztuki wyzwolone/filozofia	Uniwersytet Krakowski (Akademia Krakowska)	the beginning of winter semester 1491-1492 — the end of winter semester 1491-1492
1	koniec studiów/nauki	student	sztuki wyzwolone/filozofia	Uniwersytet Krakowski (Akademia Krakowska)	the beginning of summer 1495 — the end of autumn 1495

Functions/offices/roles

#	Event type	Function/office/role	Place/Institution	Date
1	wybór/mianowanie	kanonik	katedra we Fromborku	Aug 26, 1495 — the end of 1495 year
1	objęcie funkcji	kanonik	katedra we Fromborku	Oct 20, 1497
1	rezygnacja	scholastyk (koadiutor)	katedra we Wrocławiu	the beginning of 1538 year — the end of 1538 year
1	początek funkcji	scholastyk (koadiutor)	katedra we Wrocławiu	Jan 10, 1503

Works and books

Event type	Title	Shelf mark	Date	Source of information
autor	Nicolai Copernici Torinensis De Revolutionibus orbium coelestium libri VI (1543)		the beginning of 1543 year — the end of 1543 year	PSS, t. 14
autor	O szacunku monety (De aestimatione monetae)		the beginning of March 1522 — the end of March 1522	AM, t. II

Fig. 1. Excerpt of Nicolaus Copernicus' record from CAC.

1364-1780 (see Fig. 1 for an excerpt of the record for Nicolaus Copernicus)—are to be linked with ALMA, an integrated system for the management of around 9 million bibliographic records of BJ, and the MUJ system. The basis for the reconciliation process is the name and surname present in the personal records. The precision of the match can later be evaluated given temporal information, including date of birth, date of death and date of the early recorded event. The latter temporal properties are always present for every CAC record, however ALMA and MUJ records do not necessarily have such information.

Table 1. Samples of personal heading correspondences between CAC and ALMA.

CAC heading	ALMA heading
Marycjusz, Szymon	Maricjusz, Szymon
Najmanowicz, Jakub	Naymanowic, Jakub
Noskowicz, Jan	Noskowic, Jan
Ocko, Wojciech	Oczko, Wojciech
Pluciński, Franciszek Hieronim	Plucieński, Franciszek Hieronim
Rey, Mikołaj	Rej, Mikołaj
Śmieszkowicz, Wawrzyniec	Śmieszkowic, Wawrzyniec

While an exact match on names and surnames might seem sufficient for reconciling personal entities across databases, preliminary analysis reveals inconsistencies between CAC and ALMA. Table 1 demonstrates cases where the same individual is represented differently, preventing exact matching from correctly linking these records. To address this issue—where string representations diverge despite referring to the same entity—we evaluate different string similarity algorithms to enable more accurate fuzzy matching.

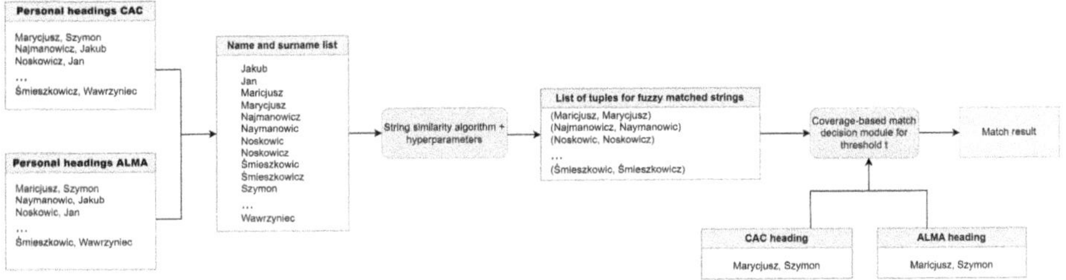

Fig. 2. String similarity-based reconciliation process.

The evaluation of the string similarity algorithms was based on the following reconciliation process. We (i) extract and normalize personal headings from both systems into comparable name–surname strings, (ii) compute pairwise similarities to generate candidate correspondences, and (iii) accept a reconciliation only when the evidence provided by the matched name components satisfies a coverage-based decision criterion. Figure 2 presents a diagram of the string similarity-based reconciliation process. Visible from the diagram are three factors that influence in the efficiency of the reconciliation process: the choice of string similarity algorithm, the hyperparameters and the decision threshold. Drawing from [4,12], we chose the following algorithms for evaluation:

- **Daitch-Mokotoff**: A phonetic algorithm designed to match surnames with similar pronunciation, especially in Slavic and Yiddish contexts.
- **Beider-Morse**: An improved phonetic matching algorithm that accounts for linguistic and geographic origins of names. Two hyperparameters are evaluated: `approx` and `exact`, which control the level of strictness in matching.
- **Levenshtein distance**: Measures the minimum number of single-character edits (insertions, deletions, substitutions) needed to transform one string into another. It is evaluated in both normalized and absolute versions, and with different values of n for determining whether a pair of strings is considered a match.
- **Jaro-Winkler similarity**: A string comparison metric that prefers matching prefixes and fewer transpositions, favoring short, similar strings. Different values of n are evaluated to determine whether a pair of strings qualifies as a match [22].

Since personal names often comprise multiple components (e.g., given names, surnames, prefixes), we introduce an additional coverage score in the matching workflow. It compares the proportion of matched components relative to the total number of components in both headings, i.e., the value is calculated using the Dice-Sørensen coefficient. Thus, the coverage threshold is a minimum required value of the coverage score that a pair of names must reach to be considered a match. For instance, consider the names *"Jan Crell"* and *"Jean Baptiste Corneille"* with the only match: *"Jan"* corresponding to *"Jean"*, while the other components do not match. Using the Dice–Sørensen coefficient, the coverage score is calculated as:

$$\text{Coverage} = \frac{2\,|\text{matched}|}{|\text{name1}| + |\text{name2}|} = \frac{2 \cdot 1}{2 + 3} = 0.4$$

The coverage score for this name pair is 0.4, and if the coverage threshold is set to 0.4 or lower, this pair would be considered a match. By using such a threshold, the workflow can effectively capture partial matches in multipart names while avoiding false positives from names that share few or no components.

4   Wikidata-Based Workflow for Entity Reconciliation Strategy Evaluation

As previously highlighted, Wikidata provides an extensive collection of alternative labels that can be retrieved through structured semantic queries. These queries enable the creation of curated datasets tailored for downstream processing in customized workflows. In our case, we developed a workflow in which a SPARQL query is formulated from the problem definition to retrieve a context-based dataset from Wikidata, multiple ER strategies are then instantiated over this dataset, and an evaluation routine compares their outputs to select the best-performing strategy. Figure 3 shows the complete workflow for the evaluation of the ER strategies.

4.1   The SPARQL Query

Having defined the context and the requisites for the ER strategies to be evaluated, the first step in the workflow is to define the query for the retrieval of the aliases' dataset. Listing 1.1 presents the defined query against Wikidata SPARQL endpoint[5].

Three requirements were predefined based on the context of JUHMP for the results of the query: (1) the retrieved names should pertain to individuals with a documented connection to Poland or its historical territories; (2) the person should have been active or influential within the timeframe of 1364–1780, ensuring relevance to the early modern period; (3) the record in Wikidata has to possess at least one main label and one alias.

---

[5] https://query.wikidata.org/.

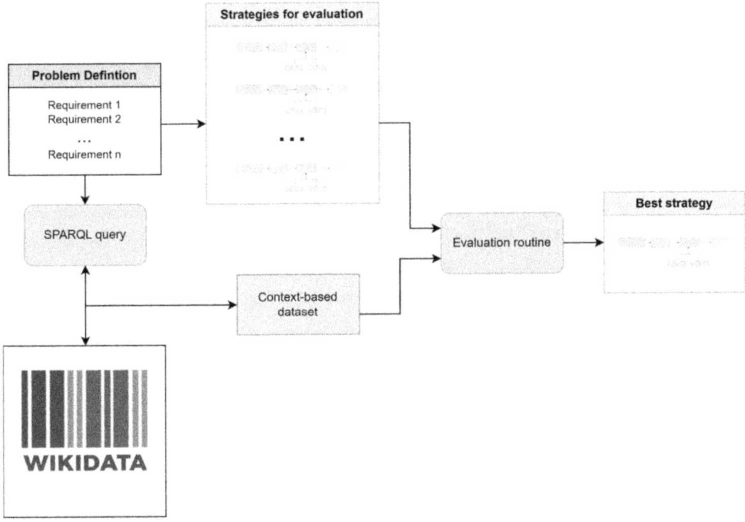

Fig. 3. Wikidata-based evaluation workflow.

For requirement (1), results were queried which contained either an identifier from the Polish Biographical Dictionary[6] (property *P8172*) or a provenance statement indicating the connection to the genealogical database "The
Great Genealogy"[7] (property *P1343*; here bound to *Q1485141*)., which document the descendants and elites of Central-Eastern Europe, including many
historical figures associated with academic, political, and cultural institutions.
Requirement (2) was implemented through a birthdate filter (property *P569*),
restricting results to individuals born between 1264 and 1770. This temporal
boundary ensures coverage of the relevant historical period while accounting for
potential variations in recorded birthdates. To satisfy requirement (3) the query
specifically requested both the primary label (rdfs:label) and at least one alternative label (skos:altLabel) for each entity. After removal of possible duplicates,
9793 pairs of main label and alias were included in the evaluation.

4.2 Evaluating Metrics and Results

We ran the pipeline from Fig. 3 for all the selected strategies and set of hyperparameters. Performance was quantified using three complementary measures:
precision, to gauge the accuracy of matches; recall, to assess coverage of true
positives; and a recall-weighted F_2 score, which gives twice as much importance
to recall as to precision when calculating the weighted harmonic mean. This
prioritization of minimizing false negatives over false positives is significant for

---

[6] https://www.psb.pan.krakow.pl/.

[7] https://wielcy.pl/.

```
prefix wdt:  <http://www.wikidata.org/prop/direct/>
prefix wd:   <http://www.wikidata.org/entity/>
prefix rdfs: <http://www.w3.org/2000/01/rdf-schema#>
prefix skos: <http://www.w3.org/2004/02/skos/core#>
prefix xsd:  <http://www.w3.org/2001/XMLSchema#>

SELECT ?person ?plLabel ?alias
WHERE {
  {
    ?person wdt:P31 wd:Q5 ;
            wdt:P8172 ?key .
  }
  UNION
  {
    ?person wdt:P31 wd:Q5 ;
            wdt:P1343 wd:Q1485141 .
  }

  ?person wdt:P569 ?birth ;
          rdfs:label ?plLabel ;
          skos:altLabel ?alias .

  FILTER(LANG(?plLabel) = "pl")
  FILTER(LANG(?alias) = "pl")
  FILTER (?birth >= "1264-01-01"^^xsd:dateTime
          && ?birth <= "1770-12-31"^^xsd:dateTime)
}
```

Listing 1.1. SPARQL query used for Wikidata.

alias matching in historical contexts, where missing valid connections (false negatives) carries greater disadvantages than including occasional incorrect matches (false positives)[8]. Coverage threshold was carefully calibrated for each strategy through empirical testing to balance sensitivity and specificity in our results. The comprehensive performance metrics for all evaluated strategies, including their optimal thresholds, are presented in Table 2.

As evidenced in Table 2, each approach demonstrates unique strengths at their respective optimal thresholds: normalized Levenshtein ($n = 0.3$) achieves the best balanced performance with an F_2-score of 0.676 at the coverage threshold 0.6, while Jaro-Winkler ($n = 0.2$) shows superior precision (0.579) at the coverage threshold 0.7. The purely phonetic Daitch-Mokotoff achieves very high recall (0.834), only slightly below normalized Levenshtein with $n = 0.4$ (recall 0.854), but at the expense of precision at the coverage threshold 0.6. These results suggest that a hybrid system could combine their strengths. Such an ensemble could potentially surpass individual method performance while maintaining the recall-precision balance critical for historical research, particularly when incorporating the coverage thresholds we established for component-wise matching. In the following sections we train tree-based ensembles on the coverage scores from all matching strategies. On the out-of-sample evaluation pool they achieve global F_2 scores that clearly exceed the best single-metric configuration.

---

[8] In our case we treat string matching as an early, high-recall stage in a multi-stage ER workflow, where false positives can be later easily filtered. The proposed workflow allows for easy configuration of the β value for the F_β calculation.

Table 2. Performance of individual string matching strategies on the out-of-sample evaluation pool (classifier training samples excluded).

ID	Matching strategy	Precision	Recall	F_2 score	Best coverage threshold
LEV-1	Levenshtein ($n = 1$)	0.378	0.705	0.601	0.6
LEV-2	Levenshtein ($n = 2$)	0.534	0.562	0.556	0.7
NLEV-0.1	Normalized Levenshtein ($n = 0.1$)	0.489	0.574	0.555	0.6
NLEV-0.2	Normalized Levenshtein ($n = 0.2$)	0.500	0.708	0.653	0.6
NLEV-0.3	Normalized Levenshtein ($n = 0.3$)	0.459	0.767	**0.676**	0.6
NLEV-0.4	Normalized Levenshtein ($n = 0.4$)	0.327	**0.854**	0.646	0.6
NJW-0.1	Normalized Jaro − Winkler ($n = 0.1$)	0.440	0.760	0.664	0.6
NJW-0.2	Normalized Jaro − Winkler ($n = 0.2$)	**0.579**	0.642	0.629	0.7
BM-A	Beider − Morse (approximate)	0.423	0.715	0.629	0.6
BM-E	Beider − Morse (exact)	0.443	0.671	0.608	0.6
DM	Daitch − Makotoff	0.289	0.834	0.605	0.6

4.3   Combining String Similarity Strategies

Coverage metric calculated for each label-alias pair served as our primary similarity measure for each strategy. Building on these individual scores, we explored the potential for machine learning classifiers to combine multiple matching approaches optimally. We considered several supervised learning alternatives, however ultimately we focused on tree-based ensembles. First, tree-based ensembles such as Random Forest and Gradient-Boosted Decision Trees typically provide strong predictive performance even when trained on relatively limited datasets, often outperforming more complex neural architectures in this regime. Second, despite their ensemble structure, these models are widely used in applied machine learning, remain comparatively intuitive, and offer improved interpretability through global feature importance estimates. Such importance profiles help reveal which features or matching strategies contribute most to overall classification performance (see [2]). This form of transparency is particularly valuable in semi-supervised entity resolution workflows, where understanding both general patterns and case-specific decisions is essential for effective evaluation and refinement of the model. Finally, both Random Forest and Gradient-Boosted Decision Trees are fast to train and evaluate and require only a modest number of hyperparameters, which makes them practical choices for iterative experimentation on large pools of candidate label − alias pairs.

In addition to the positive label-alias pair, we retrieved from the Wikidata results 71000 negative pairs. From this pool we randomly sampled a balanced subset of 625 positive and 625 negative pairs for training and cross-validation of the Random Forest classifier. A positive sample corresponds to a correct label-alias pair, while a negative sample corresponds to an incorrect label-alias pair. This balanced subset was then used in a stratified 5-fold cross-validation scheme. All remaining label-alias pairs that were not included in the balanced subset were reserved as an out-of-sample evaluation pool and were never used for fitting the classifiers, which prevents leakage between training and evaluation.

The classifiers were trained using all available metrics from the selected strategies, augmented by two additional features: the number of component strings in the main label and the number of component strings in the alias. The hyperparameters of both tree-based ensemble models were tuned by a Bayesian optimisation procedure over predefined search ranges using 3-fold cross-validation. Table 3 summarises the search spaces used for both models.

Table 3. Hyperparameter ranges used in Bayesian optimisation for ensemble models.

Parameter	Random Forest	Gradient-Boosted Decision Trees
Number of trees	100 – 1000	100 – 500
Maximum depth	5 – 30	3 – 10
Min. samples to split a node	2 – 20	–
Min. samples at a leaf	1 – 10	–
Feature subsampling	`sqrt, log2`	–
Learning rate	–	0.01 – 0.30 (log-uniform)
Subsample ratio (rows)	–	0.6 – 1.0
Subsample ratio (columns)	–	0.6 – 1.0
Class-weight factor (positive class)	–	1 – 6

In the inner loop we used the F_1 score as the optimisation objective, because it is directly supported by the optimisation library and provides a threshold-independent balance between precision and recall that correlates well with the F_2 score used in our final evaluation. During training of the Gradient-Boosted Decision Trees the underlying implementation minimizes logarithmic loss at each boosting step, as required by the library, while model selection across hyperparameter configurations still relies on cross-validated F_1. Since F_1 and F_2 are both harmonic means of precision and recall, with F_2 placing more weight on recall, we empirically observed that hyperparameter settings that perform well under F_1 also tend to perform well under the recall-weighted F_2 score adopted as our main evaluation measure.

After fitting a classifier in a given fold, we evaluated a small grid of decision thresholds on the predicted probabilities. The grid of thresholds was restricted to values between 0.35 and 0.50 in steps of 0.05, based on preliminary calibration experiments indicating that this interval offers a good trade-off between precision and recall: lower thresholds led to an unsustainable growth in false positives, while higher thresholds rapidly reduced recall. Importantly, the evaluation pool was only used for this final assessment of decision thresholds and was never used for model fitting or hyperparameter tuning, which ensures that there is no data leakage between training and evaluation.

For each threshold we computed two groups of metrics. First, on the balanced validation subset of the current fold we calculated the F_2 score and accuracy, which quantify how well the model separates positive and negative pairs under that decision rule. Second, we applied the same threshold to the large out-of-sample evaluation pool that contains all label – alias pairs never used in training or validation. In this case we treated the predicted positive pairs as a set and compared it with the reference set of correct label – alias pairs, obtaining set-based precision, recall and F_2 by counting true positives, false positives and false negatives at the pair level.

For each fold we selected the threshold that maximised the global F_2 score on the evaluation pool. This choice is consistent with our overall design goal of prioritising recall over precision in historical alias matching, that is, preferring to avoid missing valid connections even at the cost of accepting a limited number of spurious matches.

4.4 Results for Tree-Based Ensembles

The cross-validated results in Tables 4 and 5 show that both tree-based ensemble models achieve stable performance across the examined decision thresholds with the best trade-off on the out-of-sample evaluation pool obtained at threshold $\tau = 0.45$. XGBoost attains slightly higher global F_2 scores than the Random Forest at $\tau = 0.35$, 0.40 and 0.50, and generally offers marginally higher precision, whereas the Random Forest tends to achieve slightly higher recall. In applications that place a strong emphasis on avoiding missed aliases, operating at lower thresholds such as $\tau \in [0.35, 0.40]$ may therefore be preferable, at the cost of accepting more false positives. Conversely, if limiting the number of false positives is more important, a stricter threshold such as $\tau = 0.50$ yields the highest precision. As a default operating point, however, $\tau = 0.45$ provides a robust compromise that maximises global F_2 and is well-supported by both models.

Table 4. Random Forest classifier: cross-validated performance for different decision thresholds (mean $\pm$ standard deviation over 5 folds).

Metric	Decision threshold τ			
	0.35	0.40	0.45	0.50
Validation subset				
F_2 score	**0.887 $\pm$ 0.019**	0.880 $\pm$ 0.023	0.867 $\pm$ 0.020	0.862 $\pm$ 0.026
Accuracy	0.843 $\pm$ 0.011	0.850 $\pm$ 0.020	0.856 $\pm$ 0.011	**0.861 $\pm$ 0.019**
Out-of-sample evaluation pool				
Precision	0.483 $\pm$ 0.079	0.554 $\pm$ 0.082	0.649 $\pm$ 0.023	**0.683 $\pm$ 0.026**
Recall	**0.825 $\pm$ 0.020**	0.805 $\pm$ 0.023	0.775 $\pm$ 0.005	0.752 $\pm$ 0.012
F_2 score	0.718 $\pm$ 0.018	0.734 $\pm$ 0.016	**0.745 $\pm$ 0.007**	0.737 $\pm$ 0.009

Table 5. XGBoost (gradient-boosted tree classifier): cross-validated performance for different decision thresholds (mean ± standard deviation over 5 folds).

Metric	Decision threshold τ			
	0.35	0.40	0.45	0.50
Validation subset				
F$_2$ score	**0.894 ± 0.022**	0.882 ± 0.017	0.871 ± 0.027	0.865 ± 0.024
Accuracy	0.851 ± 0.012	0.860 ± 0.008	0.864 ± 0.014	**0.865 ± 0.016**
Out-of-sample evaluation pool				
Precision	0.493 ± 0.068	0.597 ± 0.057	0.656 ± 0.025	**0.682 ± 0.011**
Recall	**0.823 ± 0.018**	0.794 ± 0.013	0.770 ± 0.008	0.758 ± 0.010
F$_2$ score	0.722 ± 0.014	0.743 ± 0.011	**0.744 ± 0.006**	0.742 ± 0.007

Figure 4 presents receiver operating characteristic (ROC) curves for both models evaluated on the out-of-sample pool of label-alias pairs. They achieve high discriminative performance, with areas under the curve (AUC) of 0.937 for the Random Forest and 0.941 for XGBoost, which is consistent with the strong F$_2$ scores reported above. The curves are very similar, and the selected operating threshold $\tau = 0.45$ lies close to the optimal value of each ROC curve, confirming that this setting offers a reasonable balance between true positive and false positive rates.

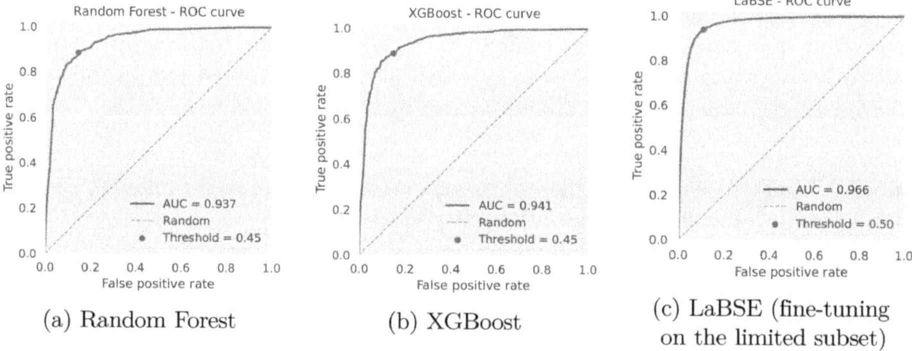

(a) Random Forest   (b) XGBoost   (c) LaBSE (fine-tuning on the limited subset)

Fig. 4. Receiver operating characteristic (ROC) curves for the two ensemble classifiers (a & b) and Transformer-based model (c) on the out-of-sample evaluation pool.

Across the examined thresholds, the best configurations of the Random Forest and XGBoost classifiers achieve global F$_2$ scores on the out-of-sample evaluation pool that clearly exceed those of any individual matching strategy reported in Table 2, confirming the benefit of combining complementary similarity features. Moreover, the optimal global F$_2$ scores of the two ensemble models at threshold

$\tau = 0.45$ are almost identical, which suggests that, for the current set of features, both models operate in a very similar performance regime. This indicates that substantial further improvements are more likely to come from enriching the similarity representation, for example with semantic or contextual features, rather than from switching to more complex classification architectures.

4.5 Results for the Transformer-Based Model

To assess the Transformer-based approach, we conducted the evaluation in two stages. First, to ensure direct comparability with the tree-based ensembles, we fine-tuned the LaBSE model using the exact same limited balanced subset of $1,250$ label–alias pairs (Table 6). In this restricted data regime, the semantic model demonstrates a clear advantage in identifying valid matches. At the optimal threshold of $\tau = 0.50$, LaBSE achieves a global F_2 score of 0.838 ± 0.003 on the out-of-sample evaluation pool, surpassing the best tree-based result (XGBoost) by approximately 0.09. This improvement is driven entirely by a massive increase in recall (0.938 ± 0.003 compared to 0.770 for XGBoost). It is worth noting that at this strict data limit, the neural model sacrifices some precision (0.587) compared to the tree ensembles (0.656) to maximise coverage. Nevertheless, the superior overall discriminative power is visually confirmed by the receiver operating characteristic curve in Fig. 4 (c), where the model reaches an AUC of 0.966.

Table 6. LaBSE: cross-validated performance (fine-tuning on the limited balanced subset) for different decision thresholds (mean $\pm$ standard deviation over 5 folds).

Metric	Decision threshold τ			
	0.35	0.40	0.45	0.50
Validation subset				
F_2 score	0.732 ± 0.021	0.769 ± 0.021	0.804 ± 0.008	$\mathbf{0.812 \pm 0.021}$
Accuracy	0.763 ± 0.018	0.814 ± 0.024	0.855 ± 0.012	$\mathbf{0.886 \pm 0.007}$
Out-of-sample evaluation pool				
Precision	0.383 ± 0.004	0.448 ± 0.005	0.517 ± 0.005	$\mathbf{0.587 \pm 0.005}$
Recall	$\mathbf{0.982 \pm 0.001}$	0.972 ± 0.002	0.958 ± 0.002	0.938 ± 0.003
F_2 score	0.748 ± 0.003	0.787 ± 0.003	0.818 ± 0.002	$\mathbf{0.838 \pm 0.003}$

In the second stage, we investigated the upper performance bound of the method by leveraging the full training dataset. We fine-tuned the model on 85% of the available data (reserving 15% for the final evaluation), allowing better adaptation to the nuances of historical entity resolution. Table 7 summarises these results. With sufficient data, the model exhibits a dramatic performance boost, achieving a global F_2 score of 0.912 ± 0.001 at a threshold of $\tau = 0.40$. This score is driven by an exceptional recall of 0.954 ± 0.004 combined with a

Table 7. LaBSE: cross-validated performance (fine-tuning on the full training dataset) for different decision thresholds (mean $\pm$ standard deviation over 5 folds).

Metric	Decision threshold τ			
	0.35	0.40	0.45	0.50
Validation subset				
F_2 score	0.898 ± 0.012	$\mathbf{0.901 \pm 0.010}$	0.897 ± 0.012	0.888 ± 0.013
Accuracy	0.944 ± 0.004	0.954 ± 0.003	0.960 ± 0.004	$\mathbf{0.962 \pm 0.004}$
Out-of-sample evaluation pool				
Precision	0.731 ± 0.007	0.775 ± 0.009	0.808 ± 0.008	$\mathbf{0.835 \pm 0.006}$
Recall	$\mathbf{0.967 \pm 0.004}$	0.954 ± 0.004	0.940 ± 0.002	0.919 ± 0.006
F_2 score	0.908 ± 0.003	$\mathbf{0.912 \pm 0.001}$	0.910 ± 0.002	0.901 ± 0.005

significantly improved precision of 0.775 ± 0.009. Compared to the initial tree-based baselines, the fully fine-tuned LaBSE model improves the F_2 score by approximately 0.17 and recall by over 0.18, while simultaneously maintaining significantly higher precision. This discriminative power is further evidenced by the area under the ROC curve (AUC), which reached 0.986 in this expanded data regime.

These findings indicate that while LaBSE is competitive on small datasets (prioritising recall), it is highly scalable and achieves high precision only when provided with larger training corpora. However, this performance comes with notable trade-offs. The LaBSE model requires significantly higher computational resources; for instance, the full fine-tuning process required 45 min and 42 s on dual NVIDIA RTX A5500 GPUs, whereas tree-based models train in seconds. Furthermore, the neural approach lacks the intrinsic interpretability offered by the Random Forest classifier.

4.6   Random Forest Under a Reduced Feature Set

Given the almost indistinguishable global F_2 performance of the two ensemble models, in the final step we treated the Random Forest as a representative classifier for a more detailed feature analysis. We were particularly interested in whether comparable performance could be maintained when the model is restricted to a smaller, carefully selected subset of similarity features, which would reduce computational costs and improve interpretability. In a subsequent experiment, we evaluated the classifier using a reduced feature set composed of one metric from each of four methods (Levenshtein, Jaro-Winkler, Beider-Morse, and Daitch – Mokotoff), along with the two component strings – based features. There were 24 different combinations of metrics. We used fixed 500 positive and 500 negative samples of label – alias pairs for training. This setup allowed us to assess the classifier's performance under constrained input conditions. If the model yields strong results using only a limited set of features, it suggests two key advantages: (1) for larger datasets, it may not be necessary to compute the full coverage metrics across all hyperparameters for each strategy—a process that can be computationally expensive and time-consuming; and (2) using fewer features enhances explainability, as the decision process becomes easier to trace and interpret, making it more straightforward to understand how each individual metric contributes to specific predictions.

Table 8. Performance of the Random Forest classifier for selected metric subsets on the out-of-sample evaluation pool.

Configuration	Precision	Recall	F_2 score
NLEV-0.3, NJW-0.2, BM-E, DM	0.602	0.785	**0.740**
NLEV-0.3, NJW-0.2, BM-A, DM	0.612	0.780	0.739
LEV-2, NJW-0.2, BM-A, DM	0.587	**0.787**	0.737
NLEV-0.2, NJW-0.2, BM-E, DM	0.590	0.785	0.736
NLEV-0.2, NJW-0.2, BM-A, DM	0.589	0.785	0.736
All metrics	**0.659**	0.755	0.733

Table 8 presents the results for the classifier using the five best-performing metric combinations alongside the classifier trained with all available metrics. It is evident that the trained Random Forest classifier significantly outperforms any individual matching strategy by effectively leveraging their complementary strengths. Furthermore, the comparison of different strategies for feature selection for the Random Forest classifier shows that performance across the top combinations is notably consistent, with only marginal differences in precision, recall, and F_2 scores. Interestingly, the classifier leveraging all metrics achieves a slightly higher precision but shows a small trade-off in recall compared to the selected combinations. The results suggest that a carefully selected subset of metrics can achieve comparable classification results. This reduction simplifies the model and enhances explainability without significant loss in performance.

5   Conclusion

This paper presented a Wikidata-based workflow for evaluating different ER strategies. We demonstrated how the richness and comprehensiveness of Wikidata can support the creation of dedicated datasets for context-aware evaluation. We reported performance metrics: precision, recall, and F_2 score for reconciliation pipelines based on various string similarity algorithms. Furthermore, we demonstrated that supervised ensembles, specifically Random Forest and Gradient-Boosted Decision Trees (XGBoost), consistently outperform individual metrics in terms of the F_2 score. Comparison with a fine-tuned multilingual Transformer (LaBSE) revealed that while semantic models maximize recall, tree-based classifiers offer a superior balance of precision, computational efficiency, and explainability, particularly under limited data regimes.

Importantly, we conceptualize string matching as an independent, early stage in a broader multi-stage ER workflow. Wikidata's semantic structure was instrumental in constructing a representative validation dataset, particularly by enabling temporal filtering via birth dates and a Poland-oriented selection through relevant metadata. At the same time, we acknowledge that additional Wikidata properties (e.g., place of birth, occupation, educated at) offer promising opportunities for precision refinement in later ER stages. However, because this paper focuses on string matching as a modular component, we intentionally abstracted from such additional filters to preserve generalizability across domains; integrating these semantic cues for multi-stage disambiguation remains an important direction for future work.

While we have demonstrated the advantages of combining multiple matching strategies through supervised learning, a longer-term study is necessary to assess the generalizability of this approach across different domains, entity types, and languages. This could be achieved by exploring larger datasets and extending the comparison to other semantic architectures, such as mBERT, XLM-R, or SoftTF-IDF. Future work should also focus on rigorous statistical validation to further quantify the reliability of observed differences, especially regarding the performance gap between lightweight ensembles and deep neural models. Further evaluation would also benefit from considering additional operational details, such as normalization of diacritics and historical variants, as well as scalability considerations. Such extended studies can also help to clarify the reason why certain combinations yield better results—highlighting the intricacies behind the complementary nature of different strategies and providing deeper insight into their respective contributions to classification performance. Another objective for future research is the development and release of a comprehensive benchmark dataset dedicated to historical entity reconciliation, potentially enriched with challenging negative samples and cross-lingual pairs.

Finally, this work underscores the value of Wikidata as a central resource for supporting the design, evaluation, and deployment of algorithms aimed at enhancing interoperability. Its multilingual, structured, and continuously updated nature makes it an ideal source of curated datasets for different domains, including, but not limited to, cultural heritage. Moreover, the open nature of Wikidata allows the reproduction of the dataset of label–alias pairs by executing the SPARQL queries provided in this study. This facilitates transparency and enables usage of the curated corpus as a benchmark dataset for future research on entity reconciliation in historical contexts, allowing for consistent evaluation of future solutions.

Supplementary Materials Statement Code and data supporting this study are available at: https://github.com/luizdovalle2/ER-Wikidata-WF

Acknowledgments. This publication was funded by a flagship project "CHExRISH: Cultural Heritage Exploration and Retrieval with Intelligent Systems at Jagiellonian University" under the Strategic Programme Excellence Initiative at Jagiellonian University. The research for this publication has been supported by a grant from the Priority Research Area DigiWorld under the Strategic Programme Excellence Initiative at Jagiellonian University. Contribution of Maciej Mozolewski for the research for this publication has been supported by a grant from the Priority Research Area (DigiWorld) under the Mark Kac Center for Complex Systems Research Strategic Programme Excellence Initiative at Jagiellonian University.

This publication benefited from the use of language models—gpt-4o, gpt-4o-mini, gpt-5.1, DeepSeek-V3 and Gemini 2.5/3.0 Pro—to support proofreading, enhance readability, and assist in code generation and refactoring. All generated text and code were reviewed and edited, and the authors take full responsibility for the publication's content.

Disclosure of Interests. The authors have no competing interests to declare that are relevant to the content of this article.

References

1. Amaral, G., Pinnis, M., Skadiņa, I., Rodrigues, O., Simperl, E.: Statistical and neural methods for cross-lingual entity label mapping in knowledge graphs. arXiv preprint arXiv:2206.08709 (2022). https://doi.org/10.48550/arXiv.2206.08709, https://arxiv.org/abs/2206.08709
2. Aria, M., Cuccurullo, C., Gnasso, A.: A comparison among interpretative proposals for random forests. Mach. Learn. Appl. **6**, 100094 (2021)
3. Artetxe, M., Schwenk, H.: Massively multilingual sentence embeddings for zero-shot cross-lingual transfer and beyond. Trans. Assoc. Comput. Linguistics **7**, 597–610 (2019)
4. Bryer, E., et al.: Analysis of clustering algorithms to clean and normalize early modern european book titles. In: Proceedings of the 2021 4th International Conference on Software Engineering and Information Management, pp. 106–112. ICSIM '21, Association for Computing Machinery, New York, NY, USA (2021). https://doi.org/10.1145/3451471.3451489
5. Devlin, J., Chang, M.W., Lee, K., Toutanova, K.: Bert: Pre-training of deep bidirectional transformers for language understanding. In: Proceedings of the 2019 Conference of the North American Chapter of the Association for Computational Linguistics: Human Language Technologies, Volume 1 (Long and Short Papers), pp. 4171–4186. Association for Computational Linguistics, Minneapolis, Minnesota (2019). https://doi.org/10.18653/v1/N19-1423
6. Doval, Y., Vilares, M., Vilares, J.: On the performance of phonetic algorithms in microtext normalization. Expert Syst. Appl. **113**, 213–222 (2018)
7. Enríquez, J., Domínguez-Mayo, F., Escalona, M., Ross, M., Staples, G.: Entity reconciliation in big data sources: a systematic mapping study. Expert Syst. Appl. **80**, 14–27 (2017)
8. Feng, F., Yang, Y., Cer, D., Arivazhagan, N., Wang, W.: Language-agnostic BERT sentence embedding. In: Proceedings of the 60th Annual Meeting of the Association for Computational Linguistics (Volume 1: Long Papers). pp. 878–891. Association for Computational Linguistics, Dublin, Ireland (2022). https://doi.org/10.18653/v1/2022.acl-long.62
9. Heng, G., Cole, T.W., Tian, T.C., and, M.J.H.: Rethinking authority reconciliation process. Cataloging Classification Quart. **60**(1), 45–68 (2022). https://doi.org/10.1080/01639374.2021.1992554

10. Kaffee, L.A., Piscopo, A., Vougiouklis, P., Simperl, E., Carr, L., Pintscher, L.: A glimpse into babel: An analysis of multilinguality in wikidata. In: Proceedings of the 13th International Symposium on Open Collaboration. OpenSym '17, Association for Computing Machinery, New York, NY, USA (2017).https://doi.org/10.1145/3125433.3125465
11. Koneru, K., Varol, C.: Metasoundex phonetic matching for English and Spanish. Global J. Enterprise Inf. Syst. **10**(1), 1–13 (2020). https://www.gjeis.com/index.php/GJEIS/article/view/266
12. Kouki, P., Pujara, J., Marcum, C., Koehly, L., Getoor, L.: Collective entity resolution in multi-relational familial networks. Knowl. Inf. Syst. **61**(3), 1547–1581 (2019)
13. Li, Y., Li, J., Suhara, Y., Doan, A., Tan, W.C.: Deep entity matching with pre-trained language models. Proc. VLDB Endowment **14**(1), 50–60 (2021). https://doi.org/10.14778/3421424.3421431
14. Mudgal, S., et al.: Deep learning for entity matching: A design space exploration. In: Proceedings of the 2018 International Conference on Management of Data (SIGMOD '18), pp. 19–34. Association for Computing Machinery, Houston, TX, USA (2018). https://doi.org/10.1145/3183713.3196926
15. Piotrowski, M.: Entity resolution for historical data. In: Proceedings of the 2021 Workshop on Computational Humanities Research (CHR 2021), pp. 12–25 (2021)
16. Reimers, N., Gurevych, I.: Sentence-BERT: Sentence embeddings using siamese BERT-networks. In: Proceedings of the 2019 Conference on Empirical Methods in Natural Language Processing and the 9th International Joint Conference on Natural Language Processing (EMNLP-IJCNLP), pp. 3982–3992. Association for Computational Linguistics, Hong Kong, China (2019). https://doi.org/10.18653/v1/D19-1410
17. Stinson, A., Fauconnier, S., Wyatt, L., ånäs, S., Darnell, J.: Why you should be paying attention to Wikidata and glam (2016). https://diff.wikimedia.org/2016/08/23/wikidata-glam/, accessed: 2024-06-21
18. Tang, X., Zhang, J., Chen, B., Yang, Y., Chen, H., Li, C.: BERT-int: a BERT-based interaction model for knowledge graph alignment. In: Proceedings of the Twenty-Ninth International Joint Conference on Artificial Intelligence (IJCAI 2020), pp. 3174–3180. International Joint Conferences on Artificial Intelligence Organization (2020). https://doi.org/10.24963/ijcai.2020/439
19. Tian, T.C., Cole, T., Yu, K.: Name and subject heading reconciliation to linked open data authorities using virtual international authority file and library of congress linked data service APIS. Lib. Res. Tech. Serv. **65**(4) (2021)
20. do Valle Miranda, L., Kutt, K., Nalepa, G.J.: CIDOC-CRM and the First Prototype of a Semantic Portal for the CHExRISH project. In: Proceedings of the Second International Workshop on Semantic Digital Humanities, co-located with the Extended Semantic Web Conference (ESWC 2025) (2025). forthcoming
21. Varada, V., Peris, C., Park, Y., Dipersio, C.: Using alternate representations of text for natural language understanding. In: Wen, T.H., Celikyilmaz, A., Yu, Z., Papangelis, A., Eric, M., Kumar, A., Casanueva, I., Shah, R. (eds.) Proceedings of the 2nd Workshop on Natural Language Processing for Conversational AI, pp. 1–10. Association for Computational Linguistics, Online (2020). https://doi.org/10.18653/v1/2020.nlp4convai-1.1
22. Wang, Y., Qin, J., Wang, W.: Efficient approximate entity matching using JARO-WINKLER distance. In: International Conference on Web Information Systems Engineering, pp. 231–239. Springer (2017)

23. Zhao, F.: A systematic review of wikidata in digital humanities projects. Digit. Scholarship Humanities **38**(2), 852–874 (2022). https://doi.org/10.1093/llc/fqac083
24. Zhu, L., Xu, A., Deng, S., Heng, G., and, X.L.: Entity management using wikidata for cultural heritage information. Catal. Classification Quart. **61**(1), 20–46 (2023). https://doi.org/10.1080/01639374.2023.2188338

LLM-Enhanced Semantic Data Integration of Electronic Component Qualifications in the Aerospace Domain

Antonio De Santis[1]([✉]), Marco Balduini[2,3], Matteo Belcao[2], Andrea Proia[4], Marco Brambilla[1], and Emanuele Della Valle[1]

[1] Politecnico di Milano, DEIB, 20133 Milano, Italy
{antonio.desantis,marco.brambilla,emanuele.dellavalle}@polimi.it
[2] Quantia Consulting, Milano, Italy
{marco.balduini,matteo.belcao}@quantiaconsulting.com
[3] Motus ml, Milano, Italy
[4] Thales Alenia Space, Roma, Italy
andrea.proia@thalesaleniaspace.com

Abstract. Large manufacturing companies face challenges in information retrieval due to data silos maintained by different departments, leading to inconsistencies and misalignment across databases. This paper presents an experience in integrating and retrieving qualification data for electronic components used in satellite board design. Due to data silos, designers cannot immediately determine the qualification status of individual components. However, this process is critical during the planning phase, when assembly drawings are issued before production, to optimize new qualifications and avoid redundant efforts. To address this, we propose a pipeline that uses Virtual Knowledge Graphs for a unified view over heterogeneous data sources and LLMs to enhance retrieval and reduce manual effort in data cleansing. The retrieval of qualifications is then performed through an Ontology-based Data Access approach for structured queries and a vector search mechanism for retrieving qualifications based on similar textual properties. We perform a comparative cost-benefit analysis, demonstrating that the proposed pipeline also outperforms approaches relying solely on LLMs, such as Retrieval-Augmented Generation (RAG), in terms of long-term efficiency.

Keywords: Knowledge Graphs · Large Language Models · Data Integration · Space Industry

1 Introduction

Dealing with data distributed across silos is a persistent challenge for large manufacturing companies, where different departments maintain systems that evolve independently over time. This fragmentation leads to inconsistencies and heterogeneity, making information retrieval a challenging task. This is particularly

© The Author(s), under exclusive license to Springer Nature Switzerland AG 2026
M. Acosta et al. (Eds.): ESWC 2026, LNCS 16550, pp. 398–415, 2026.
https://doi.org/10.1007/978-3-032-25159-6_21

pronounced in industries like aerospace, where products are highly complex and are produced in low volumes. As a result, most of the data is manually recorded in spreadsheet-like databases, which are prone to errors and heterogeneity.

The experience presented in this article focuses on the problem of qualification of components, i.e., ensuring that a produced item consistently meets the expected quality and reliability standards for its use in different scenarios. More precisely, we consider the qualification of electronic components (e.g., capacitors, resistors) used as part of electronic boards that are installed on Thales Alenia Space's satellites. Given a certain component, determining whether it is qualified according to space agencies regulations and the type of qualification is challenging due to the need to cross-reference two distinct and heterogeneous data sources. However, accessing past qualifications is essential for optimizing qualification planning and preventing unnecessary redundancies. To address this use case, we adopted a hybrid approach that enables semantic data integration and access across heterogeneous sources by combining Semantic Web technologies and Large Language Models (LLMs). Our solution leverages Virtual Knowledge Graphs (VKGs) to provide a unified semantic view over heterogeneous sources while allowing symbolic queries through SPARQL and utilizes LLMs to enhance data cleansing, handle unstructured textual content, and support similarity-based retrieval via vector search.

Structure of the Work. The rest of this paper is structured as follows. Section 2 presents a review of related work. Section 3 describes our motivating use case. Section 4 explains our proposed methodology, while a comparison with a RAG-based approach and an analysis of the solution's potential impact are provided in Sect. 5. Section 6 discusses the deployment experience and the lessons learned, while Sect. 7 concludes the paper with directions for future work.

2 Related Work

Ontology-Based Data Integration. Virtual Knowledge Graphs (VKGs) [25] have already been utilized for data integration in various industrial settings, as they allow unified access to different data sources via a virtual semantic layer (i.e., an ontology). For instance, in the aerospace domain, VKGs have been applied to integrate and query test reports data of Printed Circuit Boards (PCBs) at Thales Alenia Space [9]. Statoil has also adopted a similar approach utilizing the *Ontop* [22] framework to integrate diverse data sources [16], significantly benefiting geologists involved in oil and gas exploration and production. VKGs and Ontop have also been used to integrate Bosch manufacturing data for analyzing the production pipeline of electronic components [15]. Similarly, Ontop has been used for a global manufacturing company, enabling semantic integration of heterogeneous data sources, facilitating tool management and energy consumption monitoring [21]. In this paper, we expand on this body of applied research by including LLMs in the integration pipeline and adapting these technologies to solve a concrete industrial challenge.

Knowledge Engineering using LLMs. In recent years, Large Language Models (LLMs) have advanced significantly, particularly with the introduction of models such as OpenAI's GPTs [20,26]. LLMs have found applications in numerous knowledge engineering tasks, such as knowledge extraction [8,9], ontology engineering [7,12,23,28], and information retrieval [10,18]. Regarding applied research in real-world scenarios, LLMs have been increasingly integrated into domain-specific knowledge engineering workflows as they enhance the accessibility and usability of heterogeneous data and achieve automation and efficiency while incorporating human oversight to ensure precision. For instance, in the healthcare and life sciences domain, LLMs have been leveraged for both ontology learning [5,13] and knowledge graph population [6]. Similarly, in the scholarly domain, LLMs have been employed to semi-automatically populate and validate metadata for academic conferences in Wikidata [19] as well as for scholarly ontology generation [2]. LLMs have also played a role in the cultural heritage domain, automatically populating a cultural tourism knowledge graph with human oversight [11]. Regarding industrial settings, LLMs have already been used at Thales Alenia Space for automating extraction and validation of textual test reports of PCBs, achieving significant time-saving [9]. Siemens has also explored LLM-guided ontology term definition generation, where LLMs assist in formulating precise definitions for domain-specific terms [4]. Additionally, automated ontology generation has been used to enable zero-shot defect identification in manufacturing by leveraging structured domain knowledge [27].

3   Motivating Use Case

In this section, we explore the motivations behind our case study. We then describe the structure and formats of Thales Alenia Space's qualification data and the different types of qualifications we are aiming to retrieve.

Context and Motivation. In the aerospace sector, mission success and safety depend heavily on the quality and compliance of electronic assemblies, such as capacitors, resistors, and integrated circuits, used in satellites and other spacecraft. Each of these components must adhere strictly to rigorous qualification standards defined by aerospace regulatory frameworks established by entities such as the European Space Agency (ESA) and the European Cooperation for Space Standardization (ECSS) [1].

Furthermore, the qualification of electronic components involves extensive and resource-intensive verification processes, including mechanical, thermal, and electrical tests, to ensure that each component meets the demanding conditions of space environments. Given these high costs associated with qualification processes, it is of strategic importance to try to reuse existing qualifications whenever possible. Consequently, access to previous qualifications is essential during the planning phase, when assembly drawings are issued before production, to optimize the planning of new qualifications and avoid the proliferation of new ones while ensuring compliance with stringent aerospace standards.

Qualification Data. The main challenge addressed in this paper is that the retrieval of qualification information within the company requires cross-referencing two different internal databases:

- A *Product Lifecycle Management Database (PLM-DB)*, which comprehensively lists all electronic components utilized by Thales Alenia Space in the production of PCBs, along with their technical characteristics.
- A *Qualification Catalog (QC)*, which tracks electronic components that have either completed or are undergoing a qualification process. The status of the qualification can be *Closed, Ongoing, Failed,* or *Obsolete*.

Both databases, but QC in particular, are manually maintained and updated by designers and technologists, resulting in numerous inconsistencies, heterogeneities, and potential data-entry errors. Furthermore, essential information is often hidden within free-text columns, making the task of accurately cross-referencing components between PLM-DB and QC highly complex and time-consuming. Before the deployment of the solution described in this paper, Thales Alenia Space designers managed this challenge by manually searching through web forms of both databases, significantly increasing the risk of oversight and inefficiency given that PLM-DB contains tens of thousands of entries while QC's size is in the thousands. Some examples of entries from the two databases, featuring a small subset of columns, are provided in Table 1.

In the PLM-DB, the combination of "Part Number", "Package Code", "Sub-package Code", and "Manufacturer" identifies a certain entry, while other columns contain various characteristics such as "Family", "Pitch", "Dimension", and many others. On the other hand, for QC, an entry is identified by a "Qualification Number", while other columns contain a subset of component characteristics, the status of the qualification, and a column "Notes" which contains other information in free-text format. In the vast majority of cases, the "Part Number (PN)" is buried within a textual note in this column, as there is no dedicated column for PN in QC. An example of text contained in "Notes" is the following: *"R1 (pn P3333333) double component soldered on double pad, soldered in HS (Hot Soldering) with a stand-off of 0.2–0.3 mm on polyimide"*. Another example of heterogeneity is the inconsistent representation of manufacturer names, which might appear as "ABC", "ABC Corp", "ABC Inc.", or other variations.

Given these structural discrepancies and different naming conventions, retrieving previous qualifications is challenging, as it requires integrating and querying these databases. Furthermore, even when structural differences are addressed through a local integration view, the data remains difficult to merge due to quality issues. For instance, the heterogeneity in the "Manufacturer" column would hinder the effective use of standard join operations.

Qualification Types and Requirements. Qualifications obtained for specific components can potentially be applied to other components, provided they adhere to established aerospace standards and demonstrate comparability. This comparability includes similarities in critical attributes such as package design,

Table 1. Examples of records from the PLM-DB and QC.

Product Lifecycle Management Database (PLM-DB)

Part Number (PN)	Package Code	Subpackage Code	Manufacturer	Family	Pitch
P1111111	FP1	a1	ABC	Hybrid	1.27
P2222222	C2	x2	XYZ	Capacitor	2.2
P3333333	R1	a3	ABC	Resistor	1.92

Qualification Catalog (QC)

Qual. Number	Package Code	Subpackage Code	Manufacturer	Qual. Status	Notes
qc1	FP1	a1	ABC Corp	Closed	...
qc2	C2	x2	XYZ Inc.	Closed	...
qc3	R1	a3	ABC Inter.	Ongoing	...

terminal finish, thermal characteristics, assembly processes, and intended operating conditions. Specifically, our case study includes three types of qualifications under ECSS guidelines (Sect. 13.8 in [1]):

- *Direct qualification*: Achieved when the attributes "Part Number", "Package Code", "Subpackage Code", and "Manufacturer" in the PLM-DB correspond exactly with the respective attributes in QC, indicating that the component in question has previously undergone the full formal qualification process.
- *Qualification by similarity*: Achieved when the attributes "Package Code", "Subpackage Code", and "Manufacturer" in the PLM-DB precisely match those present in QC, however, the "Part Number" is different. In this scenario, the component is not identical to the originally qualified unit but exhibits sufficiently comparable physical, electrical, and manufacturing characteristics to justify extending the qualification.
- *Alternative qualification*: The specific criteria for alternative qualification are dependent on the component type and always require detailed technical review and validation by qualified experts[1]. For instance, Flat Packages (FP) require that the "Package Code" and "Pitch" match between PLM-DB and QC, that the "Pin Dimension" is within a range of ± 5 microns from the original qualified component, and that the assembly process is also equivalent.

Considering this, the requirements for a qualification retrieval tool must include the capability to automatically identify direct and by-similarity qualifications, as well as to propose alternative qualification candidates, which must subsequently be reviewed and validated by a designer or technologist.

Data Confidentiality. The exemplary data shown in the paper is not disclosed in its original form to protect Thales Alenia Space's confidentiality. We modified free-text descriptions, codes, and numerical values, ensuring the structure and syntax remained intact without disclosing any confidential information.

---

[1] This type of qualification serves more as a suggestion rather than a definitive answer.

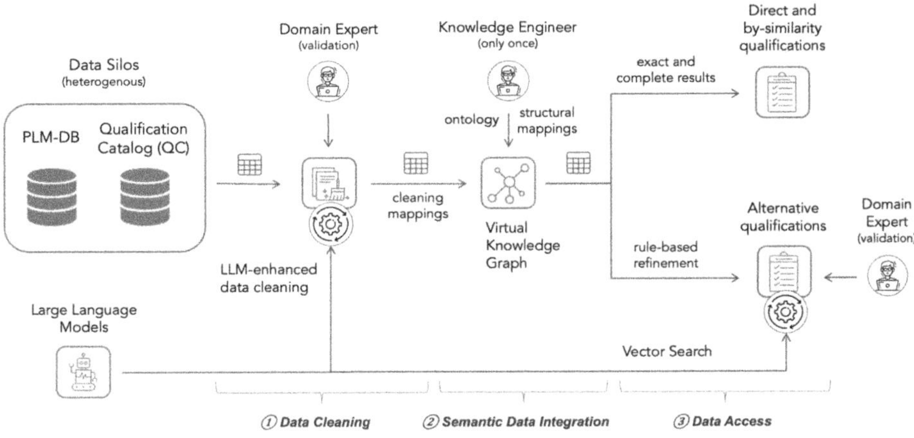

Fig. 1. Overview of the proposed LLM-enhanced semantic integration pipeline. The approach integrates two heterogeneous data silos (PLM-DB and QC) by combining LLMs and VKGs. LLMs assist in semi-automated data cleaning, including normalization and extraction of key fields, while knowledge engineering is performed once to create a semantic layer over the data. Direct and by-similarity qualifications are retrieved via exact symbolic queries over the VKG, whereas alternative qualifications are retrieved through vector search and refined using rules based on the component type and finally validated by a domain expert.

4   Methodology

This section describes the methodology we adopted for implementing our qualification retrieval tool. This process, which is outlined in Fig. 1, is structured into three main phases, each executed in close collaboration with domain experts from Thales Alenia Space:

- *Data Cleaning:* LLMs are used to streamline data cleansing by addressing heterogeneities and inconsistencies.
- *Semantic Data Integration:* A one-time knowledge engineering process is performed to define the semantic model and mappings to integrate heterogeneous data sources into a unified VKG.
- *Data Access:* The retrieval of direct and by-similarity qualifications is then achieved via SPARQL queries. Database records are also vectorized to allow retrieval of alternative qualifications via vector-based search.

Data Cleaning. The first phase focuses on data cleansing to address heterogeneities and structural differences between the two databases, with particular attention to the columns required for the automatic retrieval of both direct and by-similarity qualifications. This includes standardizing column names and resolving inconsistent naming conventions in the "Manufacturer" column. Additionally, a new column for the "Part Number (PN)" must be added to QC. While column names can be easily standardized manually, addressing the data entries

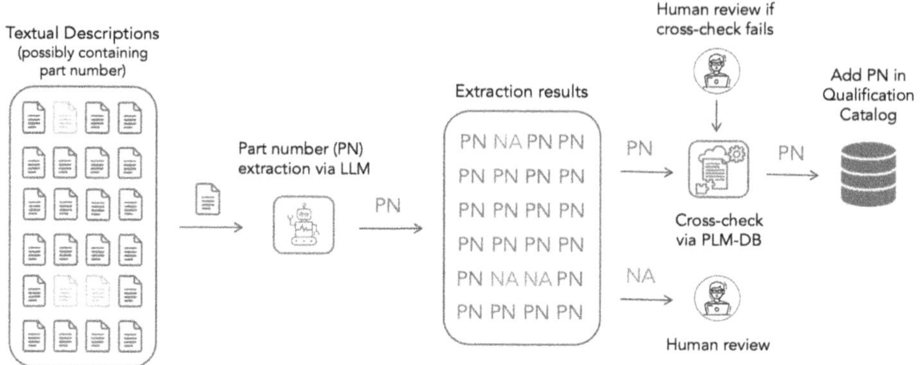

Fig. 2. Overview of the pipeline for semi-automated extraction of Part Numbers (PN) from textual descriptions using LLMs. Textual entries, potentially containing Part Numbers, are first processed by an LLM that automatically extracts candidate PNs. These candidates are subsequently cross-checked against the PLM-DB to validate correctness. Entries that cannot be verified through this step are flagged and sent to human domain experts for manual review, including those for which a candidate PN could not be extracted (indicated as NA). Successfully validated PNs are then added to QC.

within the "Part Number" and "Manufacturer" columns follows a semi-automated approach, combining LLMs with human-in-the-loop validation to ensure correctness. We used GPT-OSS-120B for this purpose, running on-premise to ensure data security and compliance.

For the "Manufacturer" column, we ask the model to generate cleaning rules in a semi-automated workflow. All unique manufacturer names are first extracted from the two databases. Then, using a few-shot prompt, the LLM is asked to infer normalization rules (i.e., merging variations of the same name). A simple example of such a rule can be:

$$\{\texttt{ABC Corp, ABC, ABC Inc., ABC International}\} \longrightarrow \texttt{ABC}$$

The model is also prompted to format the response as JSON so that a cross-check can be performed on-the-fly to ensure that none of the names are hallucinated and that the LLM is not ignoring any of the provided names. These rules are subsequently reviewed and refined by domain experts. Their feedback can also be incorporated to enrich the prompting context; for example, in cases where country-specific branches of the same manufacturer must remain distinct, this information can be included in the prompt to guide the model accordingly.

In our case study, out of 466 unique manufacturer names extracted from the dataset, GPT-OSS-120B generated 50 normalization rules. Of these, 9 (approximately 18%) required further refinement by domain experts. Furthermore, in a few notable instances, the model correctly identified and merged manufacturer names that were lexically dissimilar but referred to the same company, typically due to historical name changes or corporate acquisitions. Cleaning rules are stored in a separate SQL table without modifying the original manufacturer name, as it may potentially be relevant for tasks other than qualification retrieval. This table is later used to automatically generate cleaning mappings for the VKG. Despite the LLM-based cleaning being far from perfect, the semi-automated approach still proved more efficient than conducting the process entirely manually. For instance, the rule refinement process was completed in a single 2-hour workshop with company stakeholders, whereas, based on our prior experience, asking them to perform the same cleaning entirely manually would have required at least several days of work.

Considering the "Part Number", it is not explicitly listed in QC and therefore must be extracted from unstructured textual descriptions within the "Notes" column and added as a stand-alone column. The overall process is summarized in Fig. 2. The extraction is performed automatically using GPT-OSS-120B and a few-shot prompt. Once a candidate PN is identified, it is cross-referenced with the PLM-DB to verify both its existence and the fact that it is actually the one that underwent the qualification in question. This latter check is done by comparing the "Package Code", "Subpackage Code", and "Manufacturer" columns. However, a human-in-the-loop still remains necessary to handle cases where the cross-check fails or the LLM fails to extract a PN from the textual description. In practice, manual intervention was mainly required because, in approximately 2% of qualifications, the PN is not present in the description and must be looked up in the qualification documents. In a handful of additional cases, the PN was present but the model failed to comply with the requested output format, so the PN was added manually on-the-fly. Overall, validating PNs for all qualifications required only a few hours of expert time whereas performing the same extraction and verification fully manually for thousands of qualifications would have required several weeks.

Semantic Data Integration. Central to the VKG paradigm is the development of an ontology to model the concepts, entities, and relationships relevant to electronic components in PLM-DB and their qualification data. This ontology serves as the foundation for constructing a VKG using the Ontop framework. Figure 3 provides a high-level overview of the main classes and properties used. The ontology for this use case is not particularly complex, with the two primary entities being `PLMDB_COMPONENT` and `QUALIFICATION_CARD`. The former represents components from PLM-DB, and exposes the core attributes required for retrieval as data properties such as `componentPn`, `manufacturerName`, `pkgCode`, and `spkgCode`, together with other physical characteristics (e.g., `leadFinish`, `rawMaterial`). Similarly, a `QUALIFICATION_CARD` includes the corresponding identifiers (`qualificationPn`, `manufacturerName`, `pkgCode`, `spkgCode`) and qualification metadata such as `status`, `qualificationType`, and `qualificationDesc`.

The ontology is then linked with the underlying relational databases by defining a series of mappings. To handle the heterogeneity of the manufacturer name,

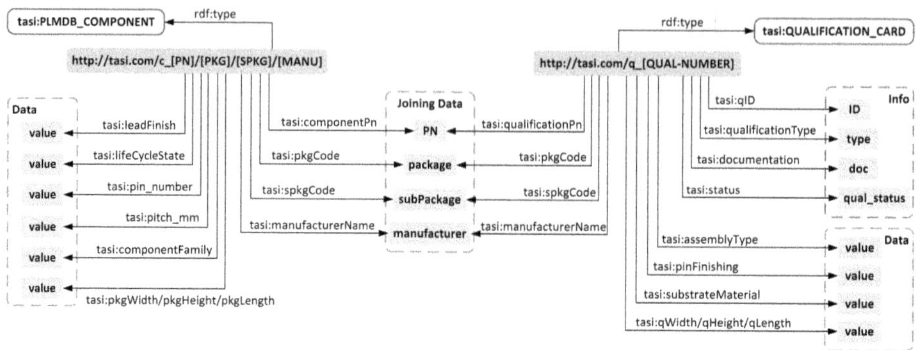

Fig. 3. High-level overview of the VKG semantic model for qualification retrieval. It shows the two main entities (**PLMDB_COMPONENT** and **QUALIFICATION_CARD**) and the core join attributes (PN, manufacturer, package and subpackage) used for direct and by-similarity matching. Additional domain-specific properties are omitted for readability.

we define an Ontop Lens[2], which performs a virtual normalization by joining the original database with the SQL table containing the LLM-generated cleaning rules. This ensures that all variants referring to the same manufacturer are mapped to a canonical name. The Ontop Lens can directly be used in a mapping as shown in Listing 1. The target of the mapping is a URI that identifies the qualification within the ontology, while the source is a SQL query that fetches the data from the virtual view generated by the Lens. When the mapping is executed, the **manufacturerName** is filled with the cleaned manufacturer name, allowing us to enable access to an RDF graph without materializing or modifying the underlying data sources.

```
@prefix : <http://tasi.com/> .
@prefix tasi: <http://tasi.com#> .

mappingId Qualification_Manufacturer
target :q_{number} tasi:manufacturerName {canonical_manufacturer_name}.
source SELECT number, canonical_manufacturer_name
       FROM lenses.qualification_manufacturer
```

Listing 1. The mapping for the manufacturer of the **QUALIFICATION_CARD**.

---

[2] An Ontop Lens is a relational view defined at the level of Ontop instead of at the level of the underlying database, specified upon a relational table or other lenses. Further details at: https://ontop-vkg.org/guide/advanced/lenses.html.

Data Access. With the VKG in place, structured access to qualification data is achieved directly through SPARQL queries over the ontology. The query is dynamically rewritten into executable SQL by Ontop. The SQL answer is then translated back to RDF or SPARQL result format, avoiding physically materializing the PLM-DB and QC into a knowledge graph. We adopt this approach to extract both direct and by-similarity qualifications, as their identification relies on strict matching rules. An example of a SPARQL query for extracting direct qualifications is shown in Listing 2. The query binds a selected component's part number (`componentPn`) and checks for the existence of a `QUALIFICATION_CARD` with the same `qualificationPn`, as well as identical `pkgCode`, `spkgCode`, and `manufacturerName`. Additional metadata, such as qualification status, type, and documentation, is also extracted using the optional clause ensuring resilience to missing values in the underlying data. The methodology for qualifications by similarity is equivalent but without binding the `qualificationPn` to the `componentPn`.

In contrast, alternative qualifications require a more flexible retrieval strategy, as they are not based on fixed attribute identity. Since the criteria for this type of qualification vary depending on the component type and are based on unstructured data (e.g., the description of the assembly process), one approach is to use symbolic queries to filter qualifications based on the available structured data and have the designer (i.e., a domain expert) manually review each candidate. This process, however, can further be optimized by leveraging an LLM's capability to process unstructured data and rank such candidates based on their relevance. A natural approach to implement this is through a vector-based retrieval mechanism. Both PLM-DB components and qualification entities are converted to JSON with their relevant attributes (e.g., assembly description, mounting type, manufacturer, package dimensions, and others) and are embedded into vector representations using `multilingual-e5-large-instruct`, an open-source language embedding model. Finally, given a PLM-DB component, a SPARQL query is first used to filter qualifications that are not compliant with the criteria and then the embeddings of the resulting qualification candidates are ranked by cosine similarity to the embedding of the component to be subsequently reviewed by the designer. This step is only done when no direct or by-similarity qualifications were found by their respective queries.

```
PREFIX rdf: <http://www.w3.org/1999/02/22-rdf-syntax-ns#>
PREFIX tasi: <http://tasi.com#>


SELECT ?generic_pn ?package_nbr ?subpackage_nbr ?manufacturer_name
       ?conf_coating ?conf_substrate ?conf_mounting
       ?q_number ?q_description ?q_status ?q_type ?q_documents
WHERE {
  ?c a tasi:PLMDB_COMPONENT; tasi:componentPn "{selected_value}";
     tasi:pkgCode ?package_nbr;
     tasi:spkgCode ?subpackage_nbr;
     tasi:manufacturerName ?manufacturer_name .
```

```
  ?qc a tasi:QUALIFICATION_CARD; tasi:qualificationPn "{selected_value}
    ";
    tasi:pkgCode ?package_nbr;
    tasi:spkgCode ?subpackage_nbr;
    tasi:manufacturerName ?manufacturer_name .
OPTIONAL {
    ?qc tasi:qID ?q_number; tasi:status ?q_status;
        tasi:documentation ?q_documents;
        tasi:qualificationDesc ?q_description;
        tasi:qualificationType ?q_type;
        tasi:conformalCoating ?conf_coating;
        tasi:substrateMaterial ?conf_substrate;
        tasi:assemblyType ?conf_mounting .
    ?c tasi:componentGenPn ?generic_pn .
  }
}
```

Listing 2. SPARQL query for retrieving directly qualified components and their respective qualification data for a given Part Number (PN).

5   Evaluation and Impact

This section presents a benchmarking experiment assessing the performance of a RAG-based system as an alternative to our pipeline. We then provide a cost-benefit analysis, highlighting both the strengths and limitations of each method across different scales of use.

Comparison with a RAG-based approach. Given the increasing popularity of Retrieval-Augmented Generation (RAG) in enterprise settings [3,14], it is natural to consider whether such an approach could serve as a viable alternative to our more structured pipeline. RAG combines pre-trained language models with a retrieval mechanism that fetches relevant information from a vector database at query time, appending them to the model's prompt to enrich its answers with external knowledge [18]. This allows LLMs to answer questions based on up-to-date and domain-specific data, even if that data was not seen during training. These types of systems offer a very quick deployment, making them especially appealing in industrial scenarios where timelines are constrained. This is because they can operate directly on heterogeneous and unstructured data and thus do not require schema alignment, data cleaning, or ontology engineering.

Based on this idea, we implemented a RAG pipeline tailored to the qualification retrieval task. Both the PLM-DB (containing electronic components) and QC (containing qualification records) were converted into JSON objects and embedded using `multilingual-e5-large-instruct`. Given a component from PLM-DB, the system ranks QC entries based on cosine similarity and selects the top 200 most similar records. These 200 entries are then added as context into a prompt for the LLM, which is asked to identify and classify qualification matches according to our predefined types. Specifically, for direct and by-similarity qualifications, we use the definition described in Sect. 3, while for alternative qualifications, we consider as a rule that only the "Package Code" and "Manufacturer" must match. To evaluate the performance of the RAG approach, we conducted a benchmarking experiment using the results of our deployed VKG+LLM pipeline as ground truth. We selected a representative subset of 675 components from the PLM-DB and retrieved their qualification matches using both methods. In this subset, 17.48% of components have never been qualified, while the remaining components have on average 0.63 direct qualifications, 7.98 by similarity, and 2.23 alternative qualifications. The system's performance was measured using standard metrics such as Precision, Recall, F1-score, and Intersection over Union (IoU). The results using two different open-source LLMs are reported in Table 2. In general, RAG using the GPT model achieves some promising results in our use case, particularly for direct and by-similarity qualifications, with F1-scores of 92.07% and 93.71%, respectively. However, for alternative qualifications, the performance is noticeably lower, with an F1-score of 61.88% and IoU of 59.64%. The variant based on Mistral 3 24B performs substantially worse, with an overall F1-score of 13.06%. This could be due to the model's difficulty in handling a large context window with hundreds of qualifications, which is a challenging task for smaller LLMs. While performing a model call for each component-qualification pair might improve accuracy, we did not consider this possibility as it is computationally infeasible, taking more than one hour per component.

Regarding the applicability of this approach in a real industrial scenario, it is evident that while the RAG-based method can serve as a suggestion tool for designers, it still necessitates a human-in-the-loop component for manual verification. A possible approach would be to provide designers with the 200 most similar entries to manually review and verify the qualifications suggested. This would still be an improvement compared to manual search since we found that 78.5% of qualifications for a component are found within the top 50 results, 93.4% within the top 100, and 99.8% within the top 200. However, whether this

Table 2. Performance of the RAG-based approach with different LLMs for direct, by-similarity, and alternative qualifications.

Model	Qual. Type	Precision	Recall	F1-score	IoU
GPT-OSS-120B	Direct	0.947	0.896	0.921	0.959
	Similarity	0.969	0.907	0.937	0.907
	Alternative	0.516	0.773	0.619	0.596
	Overall	0.866	0.886	0.876	0.846
Mistral 3 24B	Direct	0.055	0.654	0.101	0.287
	Similarity	0.152	0.306	0.203	0.143
	Alternative	0.024	0.384	0.046	0.131
	Overall	0.081	0.335	0.131	0.237

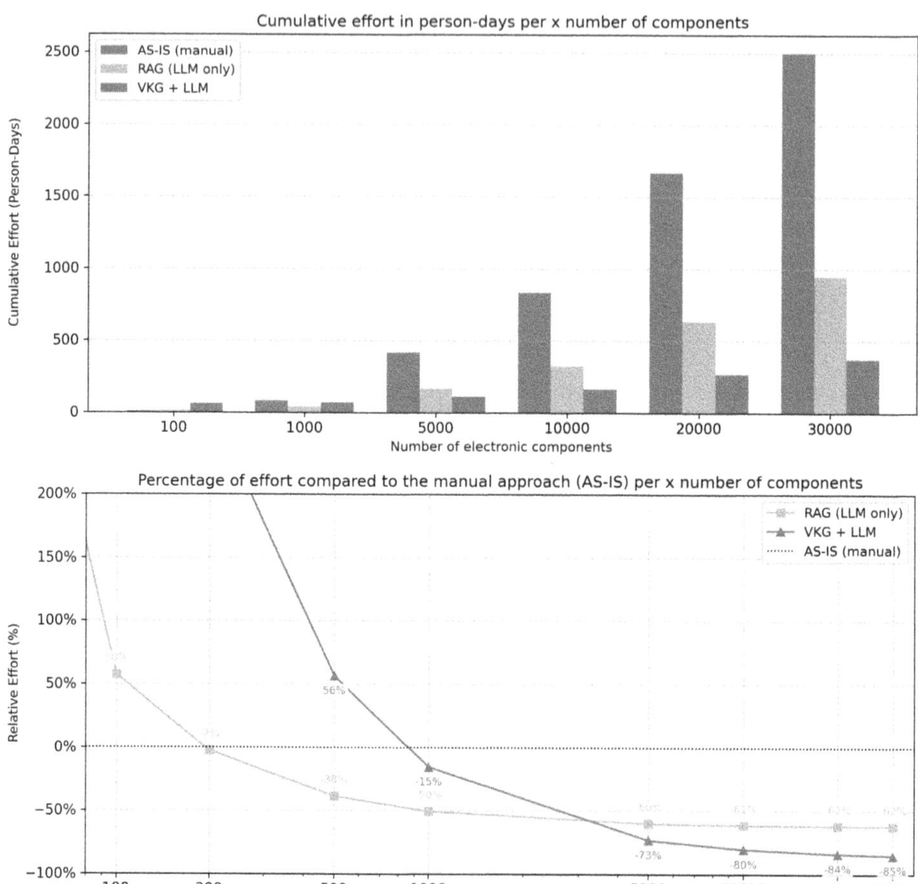

Fig. 4. Cumulative effort in person-days required for each approach as a function of the number of electronic components, considering system setup time and human-in-the-loop validation. The top figure shows the absolute effort of the three pipelines, while the bottom figure shows the relative effort compared to the manual (AS-IS) baseline. The VKG+LLM pipeline requires higher setup time but becomes significantly more efficient at scale, reducing effort by over 70% beyond 5000 components.

is the best long-term strategy from a business perspective is discussed in the subsequent paragraph.

Cost-Benefit Analysis. While the RAG-based approach requires minimal engineering effort and was implemented in just a few days, our structured pipeline involved several weeks of collaboration with domain experts for data cleaning, ontology modeling, and implementation. To quantify the trade-offs between approaches, we developed a cost model that estimates the effort required to extract qualification manually, using RAG with the GPT model, or our VKG+LLM pipeline. For the manual procedure (AS-IS), there are no upfront costs, but each PLM-DB component requires an average of 40 person-minutes to manually review qualifications and check for compliance. RAG took only about 10 person-days to design and implement, but it still relies on humans to validate LLM answers and revert to manually search within ranked qualifications in case these answers are incorrect. This reduces the average effort to roughly 15 person-minutes per component. Regarding the VKG+LLM approach, it took approximately 60 person-days to set up, but direct and by-similarity qualifications are automated, and a manual review is required only in case of alternative qualifications, resulting in an overall average of about 5 person-minutes per component.

The plots in Fig. 4 show the absolute and relative effort (y-axis) over an increasing amount of components (x-axis). The AS-IS approach is advantageous for only up to 200 components, after which the RAG approach becomes the most convenient, offering a decent trade-off between setup and validation effort. Still, it is eventually overtaken by the VKG+LLM pipeline, which, despite its higher upfront time investment, reduces cumulative effort significantly, offering over 80% savings at the scale of 10000 components.

While the VKG+LLM approach is more efficient at higher scales (i.e., in the long term), it is also interesting to see that LLMs alone were able to provide a good effort reduction even with a small upfront investment. This emphasizes their usefulness for rapid prototyping (i.e., developing proofs-of-concept for large-scale solutions). As the performance of LLMs continues to improve, we expect their break-even point with respect to structured approaches like VKG+LLM to shift further toward higher scales. However, the increased computational requirement would introduce new cost dimensions such as hardware and energy consumption, which for simplicity are not considered in this analysis.

6 Uptake and Lessons Learned

This section describes our deployment experience at Thales Alenia Space and discusses the broader organizational and strategic implications. We then reflect on the main lessons learned working on this case study.

Deployment Details. The tool was successfully deployed in a production-ready state within an operational aerospace industrial environment, representing a significant milestone in the transition from research to practice. The system is currently in active use, with authenticated access granted to approximately 50 users across different departments such as AT (Advanced Technologies), M&P (Materials and Processes), IPE (Industrial Product Engineering), DESIGN, and ENG (Engineering). These groups represent key functional areas involved in component qualification, evaluation, and integration. Before deployment, a structured training program was conducted that involved more than 50 individuals from the same departments. The training aimed to familiarize users with the application interface and clarify its role in supporting the identification and comparison of qualification cases across different component families.

This training phase was instrumental in fostering an early understanding of the system's capabilities and aligning expectations regarding its use. Currently, the application is in adoption and under continuous evaluation to assess whether it can be integrated into existing workflows and decision-making processes.

A broader European-scale deployment is being considered and a feasibility study is being discussed to explore this possibility. The study should address several critical dimensions, including data source heterogeneity across institutions, interoperability with existing repositories, access control models for inter-organizational usage, and the technical infrastructure needed to support secure and scalable deployment across multiple industrial actors.

Furthermore, the deployment has initiated strategic conversations with stakeholders at the European level, including the European Space Agency (ESA), regarding the potential development of a shared service to consolidate qualification knowledge. This initiative aims to facilitate the reuse of qualification experiences across the space industry, opening pathways toward greater cross-organizational collaboration and mutualization of noncompetitive technical assets.

Lessons Learned. Several key lessons emerged from the experience presented in this paper. In real-world KG implementation, data cleaning proved to be as crucial and even more time-consuming than data modeling. Without thorough data cleaning, even a well-designed semantic model cannot support effective querying due to underlying data inconsistency. Leveraging LLMs significantly streamlined this process, substantially reducing manual efforts in correcting heterogeneities and enabling knowledge engineers to focus primarily on modeling tasks. Furthermore, another interesting insight was the effectiveness of integrating symbolic queries with embedding-based methods. This allowed the system to maximally exploit both structured and unstructured data and address complex retrieval scenarios. Collaboration with domain experts and iterative development also proved to be essential for validating the ontology as well as fine-tuning the LLM prompts used in the data cleaning phase, ensuring that the LLM answers align with the desired outcome.

Another important lesson was that RAGs can be very valuable for rapidly testing proofs of concept, given the speed and ease with which these systems can be implemented. Still, they are less suitable for long-term use, especially in critical domains such as aerospace, where accuracy is paramount and errors carry high costs. In such scenarios, the initial advantage provided by quick setup is offset by the continual need for human oversight and real-time correction of system outputs. On the other hand, semantic technologies have been shown to be efficient and scalable to tens of thousands of components which corresponds to the scale of our retrieval task.

Finally, many of these lessons are not limited to the electronic components domain. The benefits of semi-automated data cleaning, integration of symbolic and vector-based methods and using RAGs for rapid prototyping are all potentially applicable across other industrial qualification processes and data integration challenges in general. For instance, within Thales Alenia Space, we plan to implement a similar methodology for mechanical qualifications, which exhibit similar issues related to fragmented and unstructured data.

7 Conclusion and Future Work

This paper presents a novel semantic data integration pipeline that enhances the VKG paradigm through the use of LLMs for semi-automated data cleaning and to enable the support of vector-based search. The pipeline was developed to address a concrete aerospace use case involving the retrieval of electronic component qualifications from heterogeneous data silos. It is currently deployed and actively used at Thales Alenia Space, demonstrating considerable operational benefits, and is under evaluation for broader adoption at the European level. Comparative analysis also highlighted that the proposed approach is more efficient and scalable than purely LLM-based methods, such as RAG, which may be better suited for proofs of concept or short-term solutions.

In future work, we plan to explore the applicability of hybrid workflows in which LLMs also support knowledge engineering tasks such as ontology design and mapping generation. While this remains an active area of research, it is gaining increasing attention from the scientific community [12,17,23,24,28] and may offer a promising direction to mitigate the main weakness of the VKG+LLM framework, which was its high initial setup effort compared to a RAG baseline.

Supplemental Material Statement: The code used in some of the experiments, including all the prompts used for the LLMs, is made available to enhance reproducibility and potential reuse for future research. The data itself and the implementation of the deployed tool cannot be shared as they are part of confidential material belonging to Thales Alenia Space. The available resources can be accessed in our GitHub Repository at https://github.com/Antonio-Dee/tasi-qual.

Acknowledgments. Antonio De Santis's doctoral scholarship is funded by the Italian Ministry of University and Research (MUR) under the National Recovery and Resilience Plan (NRRP), by Thales Alenia Space, and by the European Union (EU) under the NextGenerationEU project. Conference attendance was partially supported through a Young Researcher and Innovator Conference Grant under COST Action CA23147 GOBLIN - Global Network on Large-Scale, Cross-domain and Multilingual Open Knowledge Graphs, supported by COST (European Cooperation in Science and Technology, https://www.cost.eu.

References

1. Ecss-q-st-70-61c: High reliability assembly for surface mount and through hole connections. Tech. rep., European Cooperation for Space Standardization (ECSS) (2022), published 8 April 2022

2. Aggarwal, T., Salatino, A., Osborne, F., Motta, E.: Large language models for scholarly ontology generation: An extensive analysis in the engineering field (2024). https://arxiv.org/abs/2412.08258

3. An, Z., Ding, X., Fu, Y.C., Chu, C.C., Li, Y., Du, W.: Golden-retriever: High-fidelity agentic retrieval augmented generation for industrial knowledge base (2024). https://arxiv.org/abs/2408.00798

4. Bischof, S., Filtz, E., Parreira, J.X., Steyskal, S.: LLM based guided generation of ontology term definitions. In: Meroño Peñuela, A., et al. (eds.) The Semantic Web: ESWC 2024 Satellite Events, pp. 133–137. Springer Nature Switzerland, Cham (2025)

5. Bouchouras, G., Bitilis, P., Kotis, K., Vouros, G.A.: LLMs for the engineering of a parkinson disease monitoring and alerting ontology. In: GeNeSy'24: First International Workshop on Generative Neuro-Symbolic Artificial Intelligence, co-located with ESWC 2024. Hersonissos, Crete, Greece (2024). https://ceur-ws.org/Vol-3749/genesy-02.pdf

6. Caufield, J.H., et al.: Structured prompt interrogation and recursive extraction of semantics (spires): a method for populating knowledge bases using zero-shot learning. Bioinformatics 40(3), btae104 (02 2024). https://doi.org/10.1093/bioinformatics/btae104

7. Crum, E., et al.: Enriching ontologies with disjointness axioms using large language models. In: Proceedings of the 2nd Workshop on Knowledge Base Construction from Pre-Trained Language Models (KBC-LM 2024) Co-located with the 23rd International Semantic Web Conference (ISWC 2024), pp. 1–12 (2024). https://ceur-ws.org/Vol-3853/paper1.pdf

8. Dagdelen, J., et al.: Structured information extraction from scientific text with large language models. Nat. Commun. (2024). https://doi.org/10.1038/s41467-024-45563-x

9. de Santis, A., et al.: Integrating large language models and knowledge graphs for extraction and validation of textual test data. In: Demartini, G., et al. (eds.) The Semantic Web - ISWC 2024, pp. 304–323. Springer Nature Switzerland, Cham (2025)

10. Edge, D., et al.: From local to global: A graph rag approach to query-focused summarization (2025). https://arxiv.org/abs/2404.16130

11. Fan, Z., Chen, C.: Cupe-kg: Cultural perspective–based knowledge graph construction of tourism resources via pretrained language models. Inform. Process. Manage. 61(3), 103646 (2024). https://www.sciencedirect.com/science/article/pii/S0306457324000062

12. Fathallah, N., Das, A., De Giorgis, S., Poltronieri, A., Haase, P., Kovriguina, L.: Neon-GPT: a large language model-powered pipeline for ontology learning. In: Extended Semantic Web Conference, ESWC2024. Hersonissos, Greece (2024)

13. Fathallah, N., Staab, S., Algergawy, A.: Llms4life: Large language models for ontology learning in life sciences. (2024)

14. Heredia Álvaro, J.A., Barreda, J.G.: An advanced retrieval-augmented generation system for manufacturing quality control. Adv. Eng. Inform. 64, 103007 (2025). https://www.sciencedirect.com/science/article/pii/S1474034624600658X

15. Kalaycı, E.G., et al.: Semantic integration of Bosch manufacturing data using virtual knowledge graphs. In: Pan, J.Z., et al., (eds.) ISWC 2020. LNCS, vol. 12507, pp. 464–481. Springer, Cham (2020). https://doi.org/10.1007/978-3-030-62466-8_29

16. Kharlamov, E., et al.: Ontology based access to exploration data at statoil. In: Arenas, M., Corcho, O., Simperl, E., Strohmaier, M., d'Aquin, M., Srinivas, K., Groth, P., Dumontier, M., Heflin, J., Thirunarayan, K., Staab, S. (eds.) ISWC 2015. LNCS, vol. 9367, pp. 93–112. Springer, Cham (2015). https://doi.org/10.1007/978-3-319-25010-6_6
17. Laurenzi, E., Mathys, A., Martin, A.: An LLM-aided enterprise knowledge graph (ekg) engineering process. Proc. AAAI Symposium Series **3**(1), 148–156 (2024)
18. Lewis, P., et al.: Retrieval-augmented generation for knowledge-intensive NLP tasks. In: Larochelle, H., Ranzato, M., Hadsell, R., Balcan, M., Lin, H. (eds.) Advances in Neural Information Processing Systems. vol. 33, pp. 9459–9474. Curran Associates, Inc. (2020)
19. Mihindukulasooriya, N., Tiwari, S., Dobriy, D., Nielsen, F.Å., Chhetri, T.R., Polleres, A.: Scholarly wikidata: population and exploration of conference data in wikidata using LLMs. In: Alam, M., Rospocher, M., van Erp, M., Hollink, L., Gesese, G.A. (eds.) Knowledge Engineering and Knowledge Management, pp. 243–259. Springer Nature Switzerland, Cham (2025)
20. OpenAI: Gpt-4 technical report (2024). https://arxiv.org/abs/2303.08774
21. Petersen, N., Halilaj, L., Grangel-González, I., Lohmann, S., Lange, C., Auer, S.: Realizing an RDF-based information model for a manufacturing company - a case study. In: d'Amato, C., et al., (eds.) The Semantic Web - ISWC 2017, pp. 350–366. Springer International Publishing, Cham (2017)
22. Rodríguez-Muro, M., Kontchakov, R., Zakharyaschev, M.: Ontology-based data access: *Ontop* of databases. In: Alani, H., et al., (eds.) ISWC 2013. LNCS, vol. 8218, pp. 558–573. Springer, Heidelberg (2013). https://doi.org/10.1007/978-3-642-41335-3_35
23. Shimizu, C., Hitzler, P.: Accelerating knowledge graph and ontology engineering with large language models. J. Web Semantics **85**, 100862 (2025)
24. Val-Calvo, M., et al.: Ontogenix: leveraging large language models for enhanced ontology engineering from datasets. Inform. Process. Manage. **62**(3), 104042 (2025)
25. Xiao, G., Ding, L., Cogrel, B., Calvanese, D.: Virtual knowledge graphs: an overview of systems and use cases. Data Intell. **1**(3), 201–223 (06 2019). https://doi.org/10.1162/dint_a_00011, https://doi.org/10.1162/dint_a_00011
26. Ye, J., et al.: A comprehensive capability analysis of GPT-3 and GPT-3.5 series models (2023)
27. Yhdego, T.O., Wang, H.: Automated ontology generation for zero-shot defect identification in manufacturing. IEEE Trans. Autom. Sci. Eng. (2025). https://doi.org/10.1109/TASE.2025.3537463
28. Zhang, B., et al.: Ontochat: a framework for conversational ontology engineering using language models. In: Meroño Peñuela, A., et al. (eds.) The Semantic Web: ESWC 2024 Satellite Events, pp. 102–121. Springer Nature Switzerland, Cham (2025)

Towards Context-Aware Search: Dynamic Facet Generation in Digital Libraries

Mutahira Khalid[1]([✉])[ID], Mohamad Yaser Jaradeh[2][ID], Sören Auer[1,2][ID], and Markus Stocker[1][ID]

[1] TIB Leibniz Information Centre for Science and Technology, Hannover, Lower Saxony, Germany
{Mutahira.Khalid,Auer,Markus.Stocker}@tib.eu
[2] L3S Research Center, Leibniz University Hannover, Hannover, Lower Saxony, Germany
Yaser.Jaradeh@l3s.de

Abstract. Academic search engines are essential resources for researchers, enabling the discovery of new work and access to relevant literature. Typically, researchers rely on keyword-based search combined with static filters such as publication date, paper type, and citation count. While useful, these static filters operate on metadata and fail to capture the content of research papers. Platforms such as the Open Research Knowledge Graph (ORKG), which aims to structure research findings, and ORKG ASK, an advanced question-driven search system designed to retrieve precise, contextually relevant information from a large corpus of scholarly articles, also face this limitation. To address the shortcomings of static filtering, we introduce Smart Filters, a novel feature integrated into both ORKG and ORKG ASK, enabling context-aware exploration. Smart Filters adopt a neuro-symbolic approach that combines Knowledge Graphs and Large Language Models to dynamically generate semantic facets from paper content. Evaluation results demonstrate that Smart Filters significantly enhance the user search experience and improve content discovery efficiency, doing so more intuitively and effectively.

Keywords: Faceted Search · Dynamic Facets · Digital Library · Academic Search engine · Knowledge base · Large Language Model · Neuro-symbolic AI

1 Introduction

The volume of scientific publications has grown exponentially over the past decades, with millions of new articles being published each year [12]. Digital libraries and academic search platforms such as Google Scholar [11], CORE [18], and Semantic Scholar [4] have become indispensable tools for finding and accessing this vast body of knowledge. Despite their ubiquity and usefulness, finding relevant content remains a significant challenge. The most valuable insights are

© The Author(s), under exclusive license to Springer Nature Switzerland AG 2026
M. Acosta et al. (Eds.): ESWC 2026, LNCS 16550, pp. 416–434, 2026.
https://doi.org/10.1007/978-3-032-25159-6_22

often buried deep within massive research collections, making it increasingly difficult for users to find the information they need.

To support the discovery of relevant research, most digital libraries offer keyword-based search, combined with faceted search and metadata-level filters such as publication year, document type, research field, and citation count. Faceted search [28] is one of the most commonly used techniques to help users refine results efficiently. However, facets are typically static and do not reflect the content of research articles. This limitation results in imprecise or overly broad filtering, hindering users from easily finding the most relevant work. Although various dynamic facet generation methods have been proposed in recent years, they have yet to be adopted in mainstream digital libraries.

In this work, we address this gap through the development of Smart Filters, a standalone microservice for dynamic, content-aware facet generation. Smart Filters are built upon a neuro-symbolic approach called KBLLM (Knowledge Base and Large Language Model) [17], which combines structured knowledge representations with LLM reasoning to generate semantically meaningful facets at runtime. Unlike static filtering mechanisms, Smart Filters dynamically analyze retrieved content and produce contextual filters aligned with the user's query. In this work, context refers to the content of the retrieved research papers—primarily titles and abstracts—together with the user's query.

A central contribution of this paper is not only the development of a production ready microservice, but also its integration into two operational scholarly platforms: the Open Research Knowledge Graph (ORKG)[1] [5,26] and ORKG ASK[2] [24]. ORKG structures scholarly knowledge into machine-readable representations (currently hosting 54,000 papers, 1,700 comparisons, and 900,000 research resources), while ORKG ASK provides question-driven access to over 77 million scholarly articles. We designed and implemented an API that enables seamless communication between Smart Filters and these platforms, allowing dynamic facet generation over heterogeneous and variably sized content sets.

The integration process required addressing several practical and engineering challenges, including heterogeneous data representations, response-time constraints, scalability across varying result sizes, and maintaining real-time performance suitable for interactive search environments. Smart Filters operate on demand, triggered by the user, and generate semantically rich facets without introducing significant computational overhead.

To evaluate the effectiveness of Smart Filters, we conducted two user studies assessing its quality and usability within ORKG and ORKG ASK. The first study evaluated the quality of the generated facets using parameters such as relevance, coverage, clarity, redundancy, and specificity. The second study evaluated the usability and perceived usefulness of Smart Filters by examining how they enhance the user's search and exploration experience. Together, these evaluations provide a comprehensive understanding of Smart Filters' impact on improving content discoverability and user interaction in scholarly search systems.

---

[1] https://orkg.org/.
[2] https://ask.orkg.org/.

The paper is organized as follows. Section 2 reviews relevant literature while Sect. 3 elaborates on our methodology. In Sect. 4, we rigorously evaluate our approach, followed by a discussion of its limitations and a summary of our findings in Sect. 5.

2   Related Work

Faceted search is a core interaction paradigm in digital libraries (e.g., Google Scholar, Semantic Scholar, IEEE Xplore, CORE, PubMed, ACM Digital Library, Scopus, ORKG, ORKG ASK), helping users refine and navigate large scholarly collections. Traditional systems rely on static, metadata-based facets—such as publication year, content type, authorship, venue, or subject categories—which limit query-specific exploration. While platforms like Semantic Scholar, CORE, and IEEE Xplore offer richer faceting (e.g., research field, author affiliations, conference names), and ORKG ASK adds content-based topic clusters, facets remain predefined and independent of user queries. They do not dynamically adapt to the content retrieved for a given question, limiting the depth and personalization of the search experience. This lack of query-sensitive, content-driven facet generation underscores the need for runtime, dynamic facet generation—a capability that remains underexplored, mainly in current digital libraries. Such an approach would enable semantically rich, context-aware filtering tailored to the content of search results rather than to generic metadata fields. This motivates the development of Smart Filters.

Dynamic facet generation is inherently complex because it involves identifying meaningful facet values and mapping them to appropriate facet categories to form coherent facet – value pairs. A variety of methods based on semantic web technologies, natural language processing (NLP), machine learning (ML), and statistical techniques [1, 10, 13–15, 20, 23] have been explored to address these challenges. Knowledge graphs and ontologies—particularly Wikidata—have been widely utilized to extract facet values and build semantic hierarchies [10, 14, 20]. For example, Facetedpedia [20] generated query-specific facets for Wikipedia articles, while another study [14] used GeoNames to support geographical filtering in ORKG research comparisons. Several studies have combined NLP and Semantic Web resources to improve facet extraction and organization. Such methods typically integrate entity recognition, tagging, or linking tools with external knowledge bases to enhance the semantic relevance of generated facets [7, 8]. Topic modeling techniques like Latent Dirichlet Allocation (LDA) [2, 6, 21, 29] have also been used to uncover latent topics and support domain-specific facet generation (Table 1).

Other research has focused on clustering semantically related facet values using similarity measures or hybrid approaches. For instance, FacetX [1] combined clustering and topic modeling to generate structured facets across multiple domains. Recent work has extended this direction by experimenting with modern learning paradigms, such as sequence labeling, autoregressive generation, multi-label classification, and large language models (LLMs), for web

Table 1. Related work on dynamic facet generation approaches.

Reference	Methods	Use case	Online
[1]	Statistical	Job, movie, and recipe search	✗
[2]	Statistical	Ratings (Movie, Yahoo!, etc.)	✗
[7]	Symbolic + NLP	Web search results	✗
[10]	Symbolic	Films and novels	✗
[14]	Symbolic	Research papers	✓
[15]	Statistical + ML	Bibliographical search	✗
[19]	Statistical	Medical journals	✗
[20]	Symbolic	English articles	✗
[21]	Statistical + NLP	Research and news papers	✗
[23]	Statistical + Symbolic	IT support, Virtual Assistant	✗
[19]	Statistical	Medical journals	✗
[25]	Machine Learning + NLP	Web search	✗

search – based facet generation [25]. Incorporating user feedback has also proven valuable for refining facet quality, as demonstrated by feedback-driven language modeling approaches [3]. More recently, hybrid systems have emerged that combine symbolic methods (e.g., Wikidata-based reasoning) with neural representations (e.g., topic or entity embeddings) to enhance faceted exploration in applications such as virtual assistants [23].

While existing approaches have significantly advanced dynamic facet generation, only a limited number—such as [14,15,19,21]—focus on academic or research-oriented content. However, many of these methods, including those developed for other domains, exist only as research prototypes and are not publicly accessible. Although the approach proposed for ORKG comparisons [14] has been implemented in an interactive, end-user-facing system, it remains restricted to a specific use case and lacks scalability to larger digital libraries. Overall, most existing work remains confined to prototype-level or offline evaluations, with few systems transitioning into fully deployed, user-integrated applications. This gap underscores the ongoing challenge of translating theoretical or model-level innovations into practical, scalable solutions that enable real-time, user-driven exploration across diverse domains.

For a structured overview and exploration of related work on dynamic facet generation approaches, we refer readers to the related ORKG comparison[3] [16].

3   Methods

This section describes the development of the Smart Filters API and methodology along with challenges faced for integrating Smart Filters into the ORKG and

---

[3] https://doi.org/10.48366/R720446.

ORKG ASK platforms. The integration pipeline is designed to enable dynamic, context-aware filtering based on the semantic understanding of research content. Figures 2 and 4 illustrate the system architecture and the service integration flow within each platform.

3.1 Smart Filters API

The Smart Filters API builds upon the KBLLM (Knowledge Base and Large Language Model – based) approach [17]. While the earlier work established the conceptual foundation of combining symbolic knowledge bases with neural language models for dynamic facet generation, our contribution in this paper is the implementation and deployment of a fully operational API based on this methodology. This hybrid neuro-symbolic approach leverages structured knowledge bases to extract meaningful entities and employs a Large Language Model to group them into semantically relevant facet – value pairs. In our implementation, DBpedia, accessed via DBpedia Spotlight, is used as the external knowledge base for facet-value and entity recognition. At the same time, GPT-4o served as the LLM responsible for inferring and grouping entities into relevant facets. This combination enables the Smart Filters API to produce high-quality, semantically grounded filters suitable for real-world research exploration scenarios.

As shown in Fig. 1, the Smart Filters API takes paper metadata (titles and abstracts), paper IDs, and the user's query as input. DBpedia Spotlight first processes titles and abstracts to extract candidate facet values (entities or concepts), with each value mapped to its source paper for context and frequency computation. These values, together with the user query, are then sent to GPT-4o, which infers appropriate facets and returns structured facet – value pairs. Finally, the tracked paper IDs are reattached to each pair, producing facet – value – paper ID triples that are returned to the user.

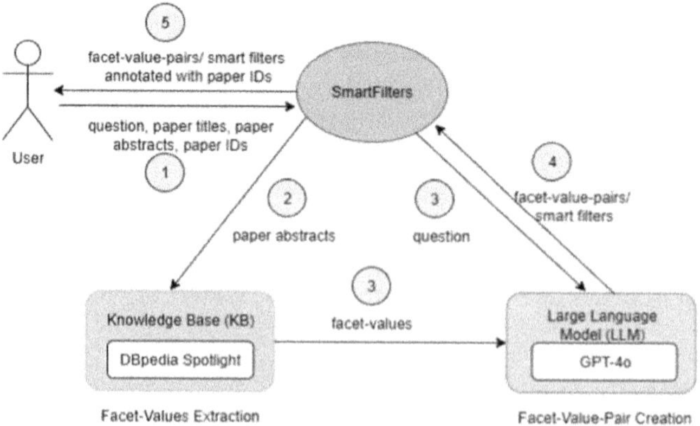

Fig. 1. Workflow of the Smart Filters API.

DBpedia as a Knowledge Base and GPT-4o as Large Language Model.
Before selecting DBpedia, we evaluated several established approaches to facet-value extraction, including statistical methods (e.g., TF-IDF), graph-based techniques (e.g., TextRank), and embedding-based models (e.g., KeyBERT). While these methods provided some insights into term importance, semantic similarity, or centrality, they exhibited notable limitations when applied to abstracts. Statistical techniques struggled with variations in terminology and phraseology across diverse research topics, while graph-based and embedding approaches sometimes failed to capture domain-specific concepts consistently. Pre-trained deep learning models, though powerful, also faced challenges in generalizing across different research areas. These limitations highlighted the need for a knowledge-based solution that could provide structured and semantically meaningful entities.

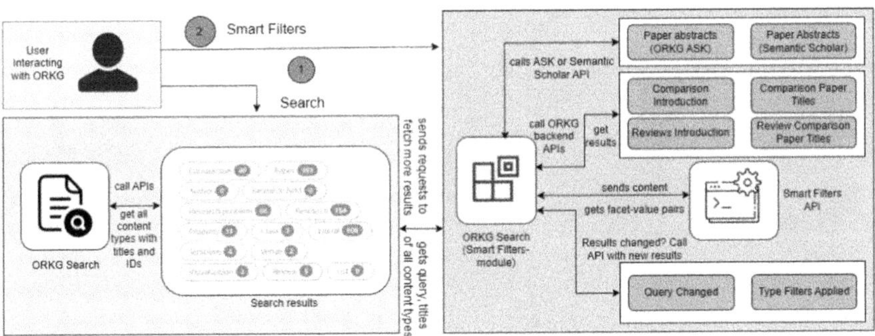

Fig. 2. Smart Filters workflow in ORKG: from query to dynamic facets.

DBpedia, accessed via DBpedia Spotlight [9,22], emerged as the most suitable choice for our dynamic facet generation system. As a Wikipedia-derived knowledge base, DBpedia provides high-quality, curated information on entities across virtually any research domain, making it domain-agnostic and highly adaptable. DBpedia Spotlight offers additional advantages: it performs reliable entity recognition and disambiguation, links entities to their corresponding DBpedia URIs, and enables frequency and context tracking across documents. While other knowledge bases, such as Wikidata, were considered, DBpedia Spotlight's combination of entity extraction, disambiguation, and URI mapping made it the practical choice for generating coherent facet-value pairs.

Before adopting GPT-4o, we initially experimented with Mistral-7B as the underlying LLM for generating facet – value pairs. Although Mistral-7B produced coherent outputs, its outdated knowledge often resulted in inaccuracies, especially for rapidly evolving research topics or domain-specific terminology. GPT-4o, with a larger parameter size and more up-to-date training data, consistently delivered higher accuracy, stronger semantic alignment, and greater coherence. Its improved reasoning capabilities enable more reliable interpretation of entities extracted by DBpedia Spotlight, yielding clearer, more meaningful facet – value pairs. Importantly, the Smart Filters API is modular: both the knowledge base (KB) and the language model (LLM) can be swapped to accommodate domain-specific needs. This enables the framework to be easily adapted to domains such as medicine, chemistry, and physics, supporting domain-aware facet generation and significantly enhancing the search experience across specialized scholarly ecosystems.

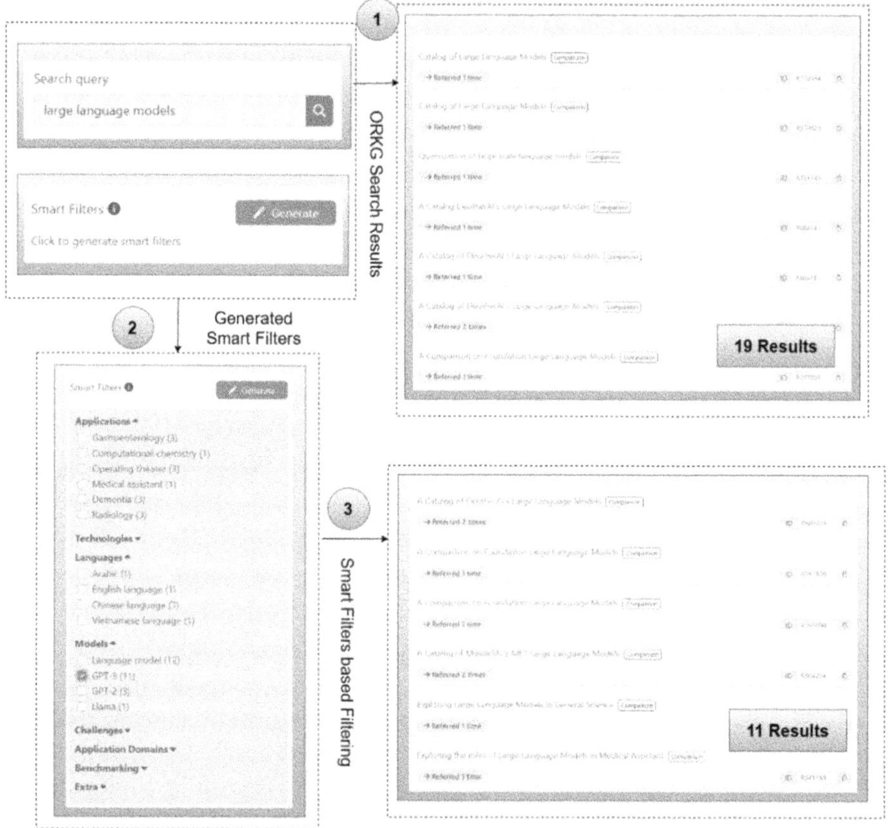

Fig. 3. Smart Filters in ORKG illustrated for query "Large Language Models."

3.2 Integration with Open Research Knowledge Graph (ORKG)

The Open Research Knowledge Graph (ORKG) is a collaborative infrastructure for representing scholarly knowledge in a structured, machine-readable form. Unlike traditional digital libraries that primarily store documents, ORKG models research contributions as structured entities such as comparisons, reviews, research papers, datasets, and other research resources. Its search functionality is keyword-based and supports filtering by metadata attributes such as content type, class, or observatory. However, as in many digital libraries, these filters are static and metadata-driven, limiting content-aware exploration. To address this limitation, we integrated Smart Filters into ORKG to enable semantic, context-aware filtering. This integration required addressing several design and engineering challenges related to scalability, interaction dynamics, and the hierarchical structure of ORKG content.

Scalability and Runtime Constraints. A single query in ORKG can return a large and heterogeneous result set. For example, the query "Large Language Models" yields dozens of comparisons, hundreds of papers, and numerous additional research resources. Applying dynamic facet generation globally across all retrieved results would be computationally expensive and impractical for an interactive system, as processing the full result set would significantly increase latency and compromise responsiveness. To maintain real-time usability, Smart Filters operate only on the content displayed at runtime (i.e., the top-15 visible results), ensuring bounded computational cost while preserving contextual relevance. Furthermore, dynamic facet generation is highly sensitive to changes in the result set: interactions such as modifying the query, applying static filters, excluding items, or loading additional results can invalidate previously computed facets and require recomputation. To balance responsiveness, computational cost, and facet quality, Smart Filters were implemented as an on-demand microservice rather than a continuously running facet generator. Facet generation is triggered only when the user explicitly clicks the "Generate" button and remains inactive during the initial search. Any change in the query, application of static filters, or interaction such as "Load more" resets Smart Filters to their default state, ensuring that computational resources are utilized only when users explicitly request semantic filtering.

Hierarchical Content Handling:A second challenge arises from the hierarchical nature of ORKG content. Different content types embed multiple layers of information: A *review* contains a title, an introduction, and multiple embedded comparisons. A *comparison* includes a title, an introduction, and multiple associated research papers. A *research paper* contains metadata such as a title and a third-party retrieved abstract as well as structured descriptions of research contributions. Processing all nested content exhaustively would lead to prohibitive computational overhead. Therefore, we designed a content selection strategy that balances semantic richness with efficiency: For *reviews*, we use the review title, its introduction, and the paper titles contained in embedded comparisons. For *comparisons*, we use the comparison title, its introduction, and the titles of associated papers. Since ORKG does not store abstracts or full-text content, for *research papers* we retrieve abstracts via the Semantic Scholar API and the ORKG ASK API. For other content types, we extract only titles.

Integration Workflow: Figure 2 illustrates the detailed workflow within ORKG, while Fig. 3 shows a representative integration example. When a user submits the query "Large Language Models", the ORKG backend returns initial results containing titles, content types, and ORKG identifiers. Suppose the user selects the content type Comparison, resulting in top-15 comparisons. Upon clicking the Generate button, additional ORKG backend APIs are invoked to retrieve enriched content according to the content selection strategy described above. The collected query, extracted content, and corresponding ORKG IDs are then sent to the Smart Filters API. The API returns a set of dynamically generated facet – value pairs along with their frequencies, indicating how often each facet value appears across distinct instances of the selected content type. Example facet categories include *Applications*, *Languages*, *Models*, *Challenges*, and *Benchmarking*. If the user selects the facet value "GPT-3" for the facet *Models*, the interface filters results to the subset in which GPT-3 appears in the comparison title, introduction, or associated paper titles.

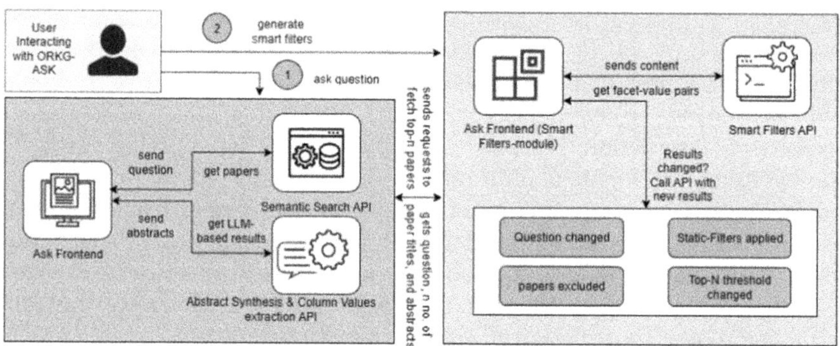

Fig. 4. Smart Filters workflow in ORKG ASK: from query to dynamic facets.

3.3 Integration with **ORKG ASK**

ORKG ASK is an advanced question-driven search system designed to retrieve relevant scholarly articles in response to natural language research queries. The system hosts a vast collection of approximately 77 million research articles, out of which 19.1 million (24.8%) contain full-text content. Given the scale and diversity, enabling meaningful, context-aware filtering became a key goal that motivated the integration of the Smart Filters service. Similar to the ORKG integration, this process required addressing some design and engineering challenges, primarily related to scalability, runtime responsiveness, and interaction dynamics within a large-scale question-answering environment.

Scalability and Runtime Constraints. Applying dynamic facet generation globally across the ORKG ASK corpus is computationally infeasible due to the scale of more than 77 million articles. Processing all retrieved results would introduce prohibitive latency and prevent interactive use. ORKG ASK displays the top-5 papers by default to ensure fast initial response times; however, generating semantic facets from only five papers provides insufficient contextual diversity for meaningful filter generation. To balance semantic richness with computational efficiency, Smart Filters operate on a configurable subset of retrieved results,

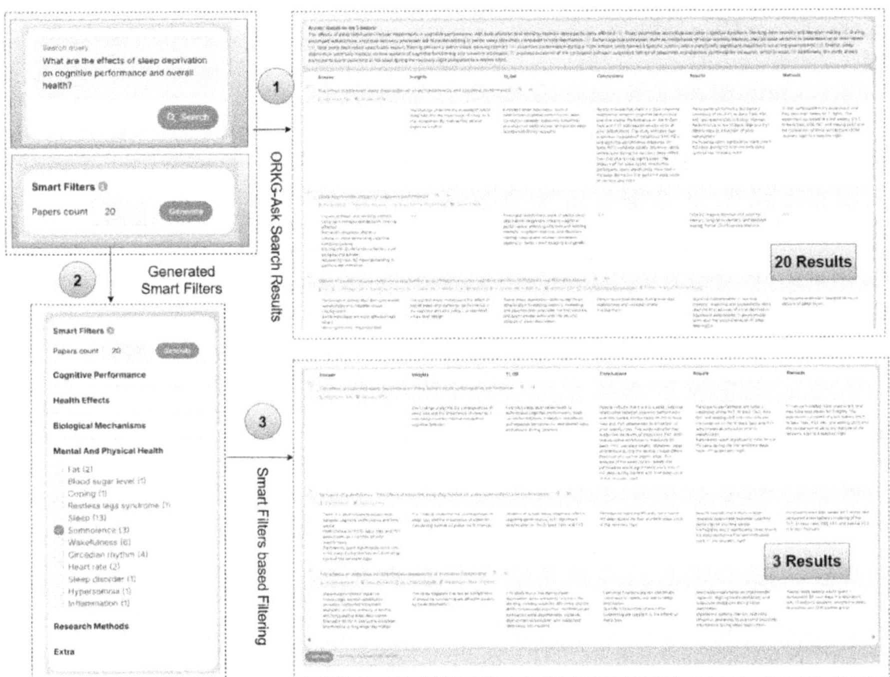

Fig. 5. Smart Filters in ORKG ASK illustrated for the query "What are the effects of sleep deprivation on cognition performance and overall health."

using the top-20 papers by default. Users can adjust this range between 5 and 100 papers in increments of five, allowing them to trade off between performance and facet coverage depending on their exploration needs. Furthermore, dynamic facet generation is highly sensitive to user interactions: modifying the research question, applying static filters, or loading additional papers alters the underlying result set and invalidates previously computed facets. Automatically recomputing facets after every interaction would significantly increase system load and degrade responsiveness. To address this, Smart Filters were implemented as an on-demand microservice, where facets are generated only when the user explicitly clicks the "Generate" button. Any interaction that changes the result set resets Smart Filters to an empty state, ensuring efficient resource utilization while preserving interactive performance.

Content Selection and Processing: Unlike ORKG, ORKG ASK primarily retrieves research papers rather than heterogeneous structured entities. To support dynamic facet generation in ORKG ASK, the Smart Filters service utilizes the titles and abstracts of retrieved research papers, as full-text access is limited. This approach allows the system to provide semantically rich filtering options while maintaining computational efficiency.

Integration Workflow: Figure 4 illustrates the Smart Filters workflow in ORKG ASK, while Fig. 5 presents a representative usage example. When a user submits a question—for example, "What are the effects of sleep deprivation on cognitive performance and overall health?"—the ORKG ASK backend retrieves the most relevant papers and displays the top results. Smart Filters initially appear in an empty state. Upon clicking the "Generate" button, the system collects the user's question, paper titles, abstracts, and corresponding ORKG ASK identifiers and sends them to the Smart Filters API. The API returns dynamically generated facet – value pairs together with frequency counts representing how often each value appears across the selected papers. Example facets include *Cognitive Performance, Health Effects, Biological Mechanisms*, and *Mental and Physical Health*. Selecting a facet value such as "Somnolence" filters the results to papers containing that concept, enabling precise, content-driven exploration of large scholarly collections.

4   Evaluation

To evaluate Smart Filters, we conducted two structured user studies focusing on their quality and usability. **Protocol:** Participants first received a brief introduction to faceted search and the Smart Filters feature. They then completed evaluation tasks using the ORKG and ORKG ASK interfaces. In the quality assessment study, 40 participants examined retrieved papers and rated the generated facet – value pairs using six predefined criteria. In the usability study, 20 participants performed two information-seeking tasks comparing the baseline system (without Smart Filters) and the enhanced system (with Smart Filters), followed by survey questions on ease of use, usefulness, satisfaction, and overall usability. In total, 60 different participants took part in the evaluation, and all responses and interaction metrics were recorded for analysis. The two study components—quality and usability—are detailed below.

4.1   Quality Assessment

To evaluate the quality of Smart Filters generated by the Smart Filters API, we conducted a structured quality assessment survey using data from ORKG ASK. The goal was to assess how well the automatically generated facet – value pairs reflected the content of the retrieved papers. We employed six evaluation criteria—Relevance, Coverage, Clarity, Redundancy, Specificity, and Completeness—with survey questions aligned to these dimensions (see Table 2).

We selected a random sample of 20 research questions from ORKG's demo repository *(e.g., How does climate change impact biodiversity?)*. For each question, the top 10 retrieved papers from ORKG ASK were included directly in the survey interface, using their titles and abstracts, so participants could assess them without navigating to external sources. Smart Filters were generated for each paper set, and the resulting facet – value pairs were presented in textual form (**Facet:** Facet values), supplemented with a visual mock-up mirroring their

Table 2. Survey questions and response scales for quality assessment.

Survey Question	Response Scale
Q1: Overall, how relevant do you find the generated facets for the question? (Relevance)	1 (Not at all relevant) to 5 (Highly relevant)
Q2: Do the facet – value pairs cover important concepts for this topic? (Coverage)	1 (Very poor coverage) to 5 (Comprehensive)
Q3: How clear and understandable are the facet labels and values? (Clarity)	1 (Not clear at all) to 5 (Very clear)
Q4: Are multiple facet – value pairs overlapping in meaning? (Redundancy)	1 (Strongly disagree) to 5 (Strongly agree)
Q5: The generated facet – value pairs were too broad. (Specificity)	1 (Strongly disagree) to 5 (Strongly agree)
Q6: The generated facet – value pairs were too narrow. (Specificity)	1 (Strongly disagree) to 5 (Strongly agree)
Q7: Did you notice any facet – value pairs irrelevant or unrelated to the topic (except the Extra facet)? (Irrelevance)	Free-text response
Q8: Based on the 10 titles and abstracts you read, did you feel important facets or concepts were missing? (Missing Information)	Free-text response
Q9: Rate the overall quality of filters.	1 (Worst) to 10 (Best)

appearance in the ORKG ASK interface. Participants were instructed to read the paper titles and abstracts, inspect the generated Smart Filters, and provide an overall quality judgment for each research question using the six criteria, rather than evaluating individual facets separately.

To obtain a neutral, diverse, and domain-relevant participant pool, we conducted the quality assessment study on Prolific[4], a widely used platform for academic user studies. Prolific enabled us to recruit participants with specific educational backgrounds and domain expertise relevant to our research questions. Each of the 20 selected questions was mapped to an academic discipline (e.g., Psychology, Social Sciences, Health and Welfare), and participants were filtered accordingly. Each question was deployed as an individual survey, with two independent responses collected per survey[5]. In total, we gathered 40 responses, ensuring disciplinary relevance, topic coverage, and multiple perspectives on the quality of the generated Smart Filters.

Overall, respondents generally found the facet – value pairs highly relevant (4.42) to the questions, with good coverage (3.98) and clear articulation (4.08). The specificity of the facet – value pairs was perceived as balanced—neither too narrow (2.25) nor overly broad (3.25). However, a slight tendency toward broader

---

[4] https://www.prolific.com/.

[5] Participants were compensated £3.50; sessions lasted approximately 15 – 20 min.

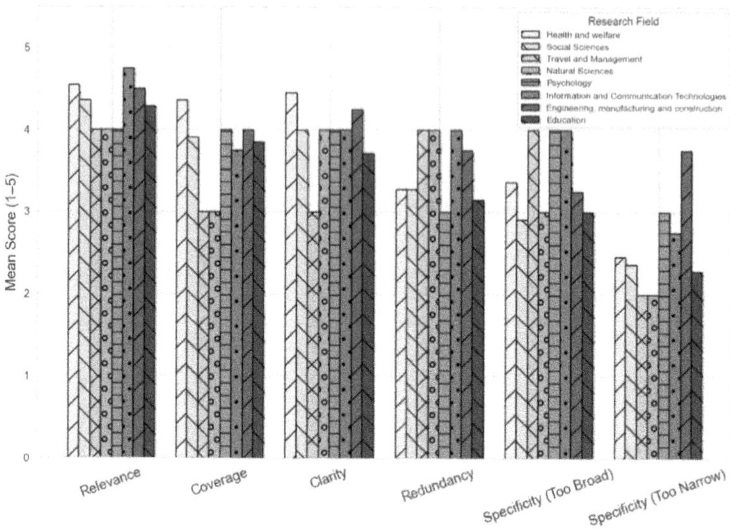

Fig. 6. Quality assessment results for relevance, coverage, clarity, redundancy, and specificity.

terms was observed, likely due to the use of DBpedia as a source. As an online encyclopedia, DBpedia may sometimes suggest terms that are more general and not always specific to academic research. Redundancy was noted as relatively high, with an average score of 3.4. While redundancy is typically undesirable, in this context it resulted from intentionally prompting the language model to associate certain facet-values with multiple facets when appropriate. This design choice leads to some values appearing under more than one facet. Although this increases redundancy, we consider it a necessary trade-off to enhance the completeness of the facet – value pairs representation. Figure 6 presents the survey results based on research fields. The final rating for Smart Filters asked via Survey Question 9 was 7.67/10.

4.2   Usability Assessment

The usability study was designed around two interactive tasks, allowing participants to directly compare the experience of using a traditional faceted search system (System A, without Smart Filters) versus the enhanced system (System B, with Smart Filters). The aim was to assess how effectively and efficiently users could complete information-seeking tasks under both settings.

Task 1 involved predefined research queries for both ORKG and ORKG ASK. For ORKG, we used general topics such as "information extraction" and "large language models", while for ORKG ASK, questions were included, such as: "How does exposure to natural environments impact cognitive restoration and creativity?" and "What are the social implications of widespread adoption of autonomous vehicles?" Participants identified facet-values (e.g., loca-

Table 3. Survey questions and response scales for usability assessment.

Survey Question	Response Scale
Q1: How easy was it to complete the task (e.g., finding geographical regions)?	1 (Very difficult) to 5 (Very easy)
Q2: How easy was it to complete the task (e.g., finding main topics)?	1 (Very difficult) to 5 (Very easy)
Q3: How useful did you find this system for your search?	1 (Not useful at all) to 5 (Extremely useful)
Q4: How satisfied were you with your search experience?	1 (Not satisfied at all) to 5 (Extremely satisfied)

tions, methods, applications) from the returned content. The task did not require deep domain expertise; for example, for "Wind energy assessment," participants extracted all geographical regions mentioned. Each participant completed Task 1 using System A and then System B, with a 5-minute limit per system. Task 2 allowed participants to choose a research query or question and identify the main topics or facets from the retrieved content, simulating an exploratory search. Both systems were used in sequence, again with a 5-minute limit per system. We measured the number of facet-values identified, task completion time, perceived ease, and System Usability Scale (SUS) scores. Participants also rated overall usefulness, satisfaction, and system preference. Table 3 lists the survey questions.

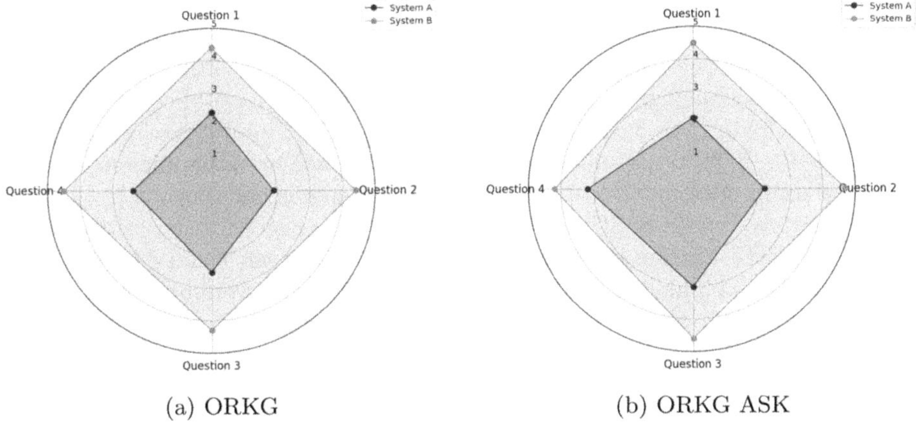

(a) ORKG   (b) ORKG ASK

Fig. 7. Usability assessment of System A vs. System B for task ease, usefulness, and user satisfaction in ORKG and ORKG ASK.

For ORKG ASK, each query consistently included 20 retrieved papers, whereas in ORKG participants could explore all available results. We conducted

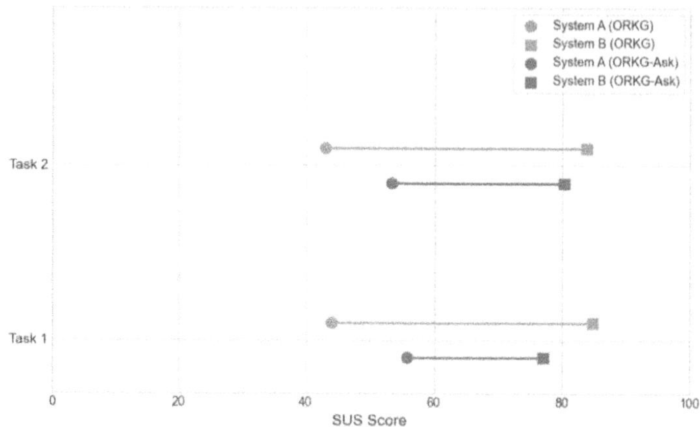

Fig. 8. System Usability Scores for Tasks 1 & 2, comparing Systems A & B.

20 usability surveys—ten for ORKG and ten for ORKG ASK—with participants primarily consisting of researchers, PhD students, and post-docs, most of whom search for scholarly literature several times per week. The 20 participants performed both Tasks 1 & 2.

In ORKG, System B (with Smart Filters) substantially outperformed System A across both objective and subjective measures.For Task 1, participants identified an average of 19 facet values in 30.70 s using System B (SD = 13.31, 95% CI [21.18, 40.22]), whereas with System A they identified 11 values in 253.20 s on average (SD = 73.15, 95% CI [200.87, 305.53]). For Task 2, participants extracted an average of 16 topics in 23.50 s with System B (SD = 7.09, 95% CI [18.43, 28.57]), whereas with System A they extracted 9 topics in 179.30 s on average (SD = 102.97, 95% CI [105.64, 252.96]). Subjective usability ratings as illustrated in Fig. 7a showed a consistent advantage for System B. Ease-of-completion was significantly higher for System B in both Task 1 ($t = -4.74$, $p = 0.0011$) and Task 2 ($t = -6.23$, $p = 0.00015$). System B also achieved significantly higher perceived usefulness ($t = -5.01$, $p = 0.00073$) and higher satisfaction ($t = -4.16$, $p = 0.0024$). System-level SUS scores as illustrated in Fig. 8 further reinforce this trend. In ORKG, System B achieved SUS = 84.75 for Task 1 and 83.75 for Task 2, compared to 44 and 43 for System A. These differences were highly significant (Task 1: $t = -7.49$, $p < 0.001$; Task 2: $t = -7.07$, $p < 0.001$).

For ORKG ASK, System B (with Smart Filters) enabled participants to identify an average of 10 facet values in 36.50 s for Task 1 (SD = 14.15, 95% CI [26.38, 46.62]), whereas with System A they identified 5 values in 285.00 s on average (SD = 47.43, 95% CI [251.07, 318.93]). In Task 2, participants identified an average of 13 topics in 39.00 s using System B (SD = 11.97, 95% CI [30.44, 47.56]), whereas with System A they identified 6 topics in 267.00 s on average (SD = 45.72, 95% CI [234.30, 299.70]). As shown in 7b, System B consistently achieved higher ease-of-completion, usefulness, and satisfaction scores. The differences were statistically significant for Task 1 ease-of-completion ($t = -5.13$,

$p = 0.00062$), Task 2 ease-of-completion ($t = -7.32$, $p = 0.000045$), and search usefulness ($t = -4.00$, $p = 0.0031$). Search satisfaction also reached significance, though with a smaller effect size ($t = -2.37$, $p = 0.0418$). In terms of SUS, System B achieved scores of 77.0 for Task 1 and 80.25 for Task 2, compared to 55.75 and 53.25 for System A. These differences approached or reached statistical significance (Task 1: $t = -2.03$, $p = 0.073$; Task 2: $t = -2.51$, $p = 0.0335$), indicating that Smart Filters substantially improved perceived usability in ORKG ASK.

As shown in Figs. 7a, 7b, and 8, System B consistently outperformed System A across ease of completion, usefulness, satisfaction, and SUS scores, while drastically reducing task times from minutes to seconds. The improvement was more pronounced for ORKG, where SUS differences between systems were substantial, whereas in ORKG ASK the gap was smaller. This is likely due to ASK's design, which already aids users by summarizing topics and extracting key elements such as insights, conclusions, and TL;DRs from abstracts. Nevertheless, System B's time efficiency made it the preferred choice in both cases, with majority of participants consistently favoring it over System A.

Computational Performance and Cost Analysis: We also evaluated the computational cost and runtime of the Smart Filters generation for 20 queries with an average of 20 papers (titles and abstracts) per query. Across the evaluated queries, the total pipeline time—combining DBpedia Spotlight entity linking and LLM-based facet extraction—averaged 10.06 s per query (SD = 2.55 s, 95% CI [8.75, 11.38]), with execution times ranging from 5.36 s to 14.51 s. In terms of monetary cost, generating Smart Filters for the evaluated papers required a total of \$0.0556, corresponding to an average cost of \$0.0033 per paper (0.33 cents). The cost distribution showed low variability (SD = \$0.0005, 95% CI [\$0.0030, \$0.0035]), indicating that the pipeline operates with stable and predictable costs across queries.

5   Discussion and Conclusion

Our evaluation indicates that Smart Filters efficiently generate dynamic, context-aware facets in near real-time. Predictions from the LLM can vary slightly across repeated queries, leading to minor differences in facet labels. Implementing caching for facet – value pairs could improve stability for repeated queries. Currently, Smart Filters support a single-level hierarchy: facets contain values, but these values cannot themselves act as sub-facets. Expanding to multi-level hierarchies would enable more structured exploration. User feedback from the Prolific study also highlighted a desire to retrieve methodological details and greater control over facet generation, such as specifying desired facet types. Addressing these aspects will require additional knowledge sources and more interactive, user-driven customization.

Beyond technical performance, our system demonstrates practical value and clear potential for adoption. Smart Filters have been deployed in two large-scale scholarly platforms—ORKG and ORKG ASK—which together serve a substantial user base. In 2025, ORKG attracted 57,724 visiters, and ORKG ASK recorded 117,432 visits and overall more than 730,906 unique questions. With Smart Filters now integrated into both platforms, these users can directly benefit from improved semantic exploration and context-aware facet suggestions.

Leveraging DBpedia Spotlight as the knowledge base enables Smart Filters to generalize across multiple research domains, addressing limitations of other facet – value extraction techniques that often fail to disambiguate entities or scale beyond specific fields. As a large, publicly available knowledge graph, DBpedia provides vast, structured knowledge that enhances the semantic richness and relevance of generated facet – value pairs. This lightweight, modular service can be readily adapted for integration into additional TIB services (e.g., TIB Portal) and external platforms such as Google Scholar and Semantic Scholar, where static facets currently constrain exploratory search.

We proposed Smart Filters, a dynamic, context-aware, neuro-symbolic, faceted search approach for scholarly digital libraries that generates facets and values directly from retrieved content rather than relying solely on metadata. Through quality and usability assessments, users rated the facet – value pairs as relevant, clear, and comprehensive. In usability tests, Smart Filters reduced task completion times from several minutes to seconds, while substantially improving System Usability Scores—from 44 to 84.75 (Task 1) and 43 to 83.75 (Task 2) in ORKG, and from 55.75 to 77.0 (Task 1) and 53.25 to 80.25 (Task 2) in ORKG ASK. All participants preferred the Smart Filters – enabled system. Smart Filters can be generated efficiently, with an average runtime of 10.06 s per query and a cost of $0.0033 per paper. Overall, these results demonstrate that Smart Filters can be generated efficiently and at very low cost, making the approach practical for large-scale deployment in digital libraries.

Acknowledgments. We thank the frontend development team (Allard Oelen, Kheir Eddine Farfar) for integrating the implemented features into the user interface, and Manual Prinz from the backend team for deploying the API. Their technical support was invaluable in completing this work. This work was co-funded by the German Ministry of Education and Research (BmBF) for the project KISSKI AI Service Center (01IS22093C) and by DFG for project NFDI4DataScience (460234259).

Code and Data Availability. The Smart Filters service is deployed and operational on both ORKG and ORKG ASK. The API is publicly accessible at https://gitlab.com/TIBHannover/orkg/smart-filters/-/tree/Smart-Filters-going-live?ref_type=heads. Frontend integrations for ORKG are available at https://gitlab.com/TIBHannover/orkg/orkg-frontend, and for ORKG ASK at https://gitlab.com/TIBHannover/orkg/orkg-ask/frontend. The scientific findings presented in this article are published in reproducible form in the TIB Knowledge Loom at https://doi.org/10.82209/6krg-xg29 following the Knowledge Loom method [27].

Declaration of AI Technologies. During the preparation of this manuscript, generative AI tools were used to support language editing and to assist with small programming tasks (e.g., code formatting and debugging). The authors reviewed and validated all outputs, and remain fully responsible for the content of the paper.

References

1. Affolter, R., Weiler, A.: FacetX: Dynamic Facet Generation for Advanced Information Filtering of Search Results. In: EDBT/ICDT Workshops (2020). https://ceur-ws.org/Vol-2578/
2. Agarwal, D., Chen, B.C.: fLDA: matrix factorization through latent dirichlet allocation. In: Proceedings of the Third ACM International Conference on Web Search and Data Mining, pp. 91–100. ACM (2010). https://doi.org/10.1145/1718487.1718499
3. Allan, J.: Relevance feedback with too much data. Proceedings of the 18th Annual International ACM SIGIR Conference on Research and Development in Information Retrieval - SIGIR '95 (1995). https://doi.org/10.1145/215206.215380
4. Allen Institute for AI: Semantic Scholar. https://www.semanticscholar.org/. Accessed: 2025-08-18
5. Auer, S., et al.: Improving Access to Scientific Literature with Knowledge Graphs. Bibliothek Forschung und Praxis **44**(3), 516–529 (2020. https://doi.org/10.1515/bfp-2020-2042
6. Blei, D.M., Ng, A.Y., Jordan, M.I.: Latent Dirichlet Allocation. J. Mach. Learn. Res. **3**(Jan), 993–1022 (2003). https://www.jmlr.org/papers/v3/blei03a.html
7. Chang, W., Koh, J.-L.: Dynamic Facet hierarchy constructing for browsing web search results efficiently. In: Ali, M., Kwon, Y.S., Lee, C.-H., Kim, J., Kim, Y. (eds.) IEA/AIE 2015. LNCS (LNAI), vol. 9101, pp. 293–304. Springer, Cham (2015). https://doi.org/10.1007/978-3-319-19066-2_29
8. Cui, P.: A Tighter Analysis of Set Cover Greedy Algorithm for Test Set. Combinatorics, Algorithms, Probabilistic and Experimental Methodologies p. 24–35 (2007). https://doi.org/10.1007/978-3-540-74450-4_3
9. Daiber, J., Jakob, M., Hokamp, C., Mendes, P.N.: Improving efficiency and accuracy in multilingual entity extraction (2013). http://dx.doi.org/10.1145/2506182.2506198
10. Feddoul, L., Schindler, S., Löffler, F.: Automatic facet generation and selection over knowledge graphs. In: Acosta, M., Cudré-Mauroux, P., Maleshkova, M., Pellegrini, T., Sack, H., Sure-Vetter, Y. (eds.) SEMANTiCS 2019. LNCS, vol. 11702, pp. 310–325. Springer, Cham (2019). https://doi.org/10.1007/978-3-030-33220-4_23
11. Google: Google Scholar. https://scholar.google.com/. Accessed: 2025-08-18
12. Hanson, M.A., Barreiro, P.G., Crosetto, P., Brockington, D.: The strain on scientific publishing. Quan. Sci. Studies **5**(4), 823–843 (2024)
13. Harth, A.: VisiNav: A system for visual search and navigation on web data. J. Web Semantics **8**(4), 348–354 (2010). https://doi.org/10.1016/j.websem.2010.08.001
14. Heidari, G., Ramadan, A., Stocker, M., Auer, S.: Leveraging a federation of knowledge graphs to improve faceted search in digital libraries. In: International Conference on Theory and Practice of Digital Libraries, pp. 141–152. Springer (2021). https://doi.org/10.1007/978-3-030-86324-1_18
15. Hoon, G.K., Wei, T.C.: Flexible facets generation for faceted search. In: Proceedings of the First EAI International Conference on Computer Science and Engineering, pp. 399–401. EAI (2017). https://doi.org/10.4108/eai.27-2-2017.152348
16. Khalid, M., Auer, S., Stocker, M.: Comparison on dynamic facet generation approaches (2024). https://doi.org/10.48366/R720446
17. Khalid, M., Auer, S., Stocker, M.: A neuro-symbolic approach for faceted search in digital libraries (2024). https://doi.org/10.3233/faia240620

18. Knoth, P., Zdrahal, Z.: Core: Three access levels to underpin open access. D-Lib Mag. **18**(11/12) (2012). https://doi.org/10.1045/november2012-knoth
19. Latha, K., Veni, K.R., Rajaram, R.: AFGF: An Automatic Facet Generation Framework for Document Retrieval. In: 2010 International Conference on Advances in Computer Engineering, pp. 110–114. IEEE (2010). https://doi.org/10.1109/ACE.2010.63
20. Li, C., Yan, N., Roy, S.B., Lisham, L., Das, G.: Facetedpedia: dynamic generation of query-dependent faceted interfaces for Wikipedia. In: Proceedings of the 19th International Conference on World Wide Web, pp. 651–660. ACM (2010). https://doi.org/10.1145/1772690.1772757
21. Mei, Q., Shen, X., Zhai, C.: Automatic labeling of multinomial topic models. In: Proceedings of the 13th ACM SIGKDD International Conference on Knowledge Discovery and Data Mining. ACM (2007). https://doi.org/10.1145/1281192.1281246
22. Mendes, P.N., Jakob, M., Garcia-Silva, A., Bizer, C.: DBpedia spotlight: shedding light on the web of documents. In: Proceedings of the 7th International Conference on Semantic Systems. ACM (2011). https://doi.org/10.1145/2063518.2063519
23. Mihindukulasooriya, N., et al.: Dynamic faceted search for technical support exploiting induced knowledge. In: Pan, J.Z., Tamma, V., d'Amato, C., Janowicz, K., Fu, B., Polleres, A., Seneviratne, O., Kagal, L. (eds.) ISWC 2020. LNCS, vol. 12507, pp. 683–699. Springer, Cham (2020). https://doi.org/10.1007/978-3-030-62466-8_42
24. Oelen, A., Jaradeh, M.Y., Auer, S.: Orkg ask: a neuro-symbolic scholarly search and exploration system (2024). https://doi.org/10.48550/ARXIV.2412.04977
25. Samarinas, C., Dharawat, A., Zamani, H.: Revisiting open domain query facet extraction and generation. In: Proceedings of the 2022 ACM SIGIR International Conference on Theory of Information Retrieval, pp. 43–50 (2022). https://doi.org/10.1145/3539813.3545138
26. Stocker, M., et al.: FAIR scientific information with the Open Research Knowledge Graph. FAIR Connect **1**(1), 19–21 (2023). https://doi.org/10.3233/FC-221513
27. Stocker, M., et al.: Rethinking the production and publication of machine-readable expressions of research findings. Sci. Data **12**(1) (2025). https://doi.org/10.1038/s41597-025-04905-0
28. Tunkelang, D.: Faceted search. Springer Nature (2022)
29. Zhu, J., Ahmed, A., Xing, E.P.: MedLDA: maximum margin supervised topic models for regression and classification. In: Proceedings of the 26th Annual International Conference on Machine Learning, pp. 1257–1264. ACM (2009). https://doi.org/10.1145/1553374.1553535

An Ecosystem of Knowledge Graph-Based Web Services for Greenhouse Gas Assessment

Ekaterina Aymon[1,4], Ivan Kostanjevec[1,4], Jan Grau[2,3], Alexander Kirsten[2], Benjamin Pocklington[1], Alessandro Giugno[3], Daniel Lachat[2], Kimberly Garcia[3], Didier Beloin-Saint-Pierre[2], and Jean-Paul Calbimonte[1,4(✉)]

[1] University of Applied Sciences and Arts Western Switzerland HES-SO, Sierre, Switzerland
`benjamin.pocklington@hevs.ch`
[2] EMPA Swiss Federal Laboratories for Material Science and Technology, Delémont, Switzerland
`{jan.grau,alexander.kirsten,daniel.lachat,`
`didier.beloin-saint-pierre}@empa.ch`
[3] University of St. Gallen, St. Gallen, Switzerland
`{alessandro.giugno,kimberly.garcia}@unisg.ch`
[4] The Sense Innovation & Research Center, Lausanne, Switzerland
`jean-paul.calbimonte@hevs.ch`

Abstract. Technological and societal changes today are accompanied by a growing need to address the challenges posed by rising greenhouse gas (GHG) emissions. While national and international efforts are made to improve accountability for these emissions and to ensure accurate carbon footprint calculations, these efforts require substantial intervention of experts and the use of highly heterogeneous data sources and methodological standards. In this paper, we present the WISER ecosystem for the integration, exchange, computation, and elaboration of GHG assessments, based on ontologies and Knowledge Graphs to model not only GHG data sources but also the assessment standards that guide their use and application. Specifically, the WISER ecosystem has been applied to multiple real use-cases, of which we focus on: GHG assessments of industrial/production sites, GHG-informed procurement for ICT in cities, and decarbonization of data centers. These use cases have been developed and validated in collaboration with external industry partners as well as the ICT departments of Swiss municipalities.

Keywords: Knowledge Graph · Digital Ecosystem · GHG assessment

1 Introduction

The evolution of human industrial activity, transportation, and energy use has significantly increased greenhouse gas (GHG) emissions, leading to the worrying effect of climate change [13, 15]. It is therefore helpful to develop GHG assessment

M. Acosta et al. (Eds.): ESWC 2026, LNCS 16550, pp. 435–452, 2026.
https://doi.org/10.1007/978-3-032-25159-6_23

methods and tools that support the reliable evaluation of these emissions and their linkages to human activities, so that we may identify pathways to slow the pace of climate change [16].

In essence, GHG assessments can be differentiated into three categories: those that account for atmospheric gas concentrations, those that account for annual GHG emissions, and those that model GHG emissions across the full life cycles of human activities [1,5]. The assessments can be based on direct or indirect measurements, observations, or modeled from scientific and environmental knowledge. To perform these assessments, it is important to follow standards that provide a well-defined framework for the accounting of GHG emissions from various anthropogenic sources [1]. Different perspectives exist in this regard. At the geopolitical level, guidelines exist to produce annual country-level inventories of GHG emissions, which may inform macroeconomic decisions and political actions. Other standards refer to organizational or product-level carbon footprints, linked to direct and indirect GHG emissions across their value chains.

Navigating these standards [8], as well as the different data sources and methods used for GHG assessments, is not straightforward (see Sect. 2) and currently requires substantial support from domain experts, limiting access to reliable GHG evaluations. This situation does not align with the current available resources of companies, public organizations, and SMEs, which need simpler tools and methods to facilitate this task while maintaining high levels of transparency, consistency, and comprehensiveness.

In this work, we present the WISER ecosystem of digital services, called WIDS (Web of Interoperable Digital Services). This ecosystem aims to provide decentralized web services and applications that facilitate the creation and realization of GHG assessments. Within its structure, the WIDS deliver various features such as the harmonization of GHG data source formats, verification of alignment between standards and studies, and the usage of different computational carbon footprint tools.

The main contributions of this work are: (i) the WISER ecosystem infrastructure, including its core API, which provides access to sustainability data and domain-expert knowledge on procedures and standards by connecting to ontologies and semantic models; (ii) the implementation of the WIDS, which includes methods for integrating heterogeneous knowledge and enabling the construction of advanced GHG assessment applications; and (iii) the implementation, deployment and validation of the WIDS for three main use cases, with direct involvement of public and private partners invested in GHG assessments of manufacturing sites, ICT use in cities, and electricity use in data centers.

2   Challenges in GHG Assessments

Several challenges arise in assessing GHG emissions from human activities [1]. We focus on those related to the data needed for these assessments, as well as the variations in standards and methodologies.

2.1   Heterogeneity Between and Within GHG Databases

Different types of data can be used to assess GHG emissions from products, services, or systems, depending on the selected assessment standard. For example, in the context of the well-known GHG protocol, Scopes 1 and 2 respectively account for direct emissions and indirect emissions from used and purchased energy [21]. These value chains typically require data from internal measurements or estimates for Scope 1, and energy mixes, as well as usage data for Scope 2. For Scope 3, which refers to indirect life-cycle GHG emissions across the value chain, the data may come from different sources with varying degrees of alignment with the rules of standards, as well as varying levels of availability. Among these data sources, Life Cycle Assessment (LCA) databases [10] are of primary importance, as they describe GHG emissions associated with the inputs and outputs of different products, services, and activities of a company or organization along value chains. These may include detailed emissions information for a reference product per functional unit of mass or volume, enabling the calculation of complex emission analytics. Certain LCA data sources capture data for an extensive range of services, activities, and industries in different geographic locations, such as ecoinvent[1] [2], or the GLAD repository [11], while others are specialized in specific sectors, as in the case of Plastics Europe [19] or worldsteel [22].

Given the degree of heterogeneity between LCA databases and their different models, their datasets cannot be seamlessly combined for a consistent assessment of different parts of value chains. The differences are not limited to the format or structure of the data, but also to various assumptions, modeling decisions, and uncertainty, which are the origin of inconsistencies among these data sources. Differences may lie, for example, in assumptions about the allocation rules for GHG emissions from multi-product systems, or in differences in granularity. In addition, different strategies for modeling future evolutionary scenarios may be important, as these models may evolve substantially over the life cycle. Thus, it is common for multiple versions of these datasets to be released over the years, and it is essential to maintain consistent choices for combining them reliably.

2.2   Diversity in GHG Assessment Standards

As mentioned earlier, different assessment standards are proposed to evaluate GHG emissions for a given system or value chain [1]. Such standards set out a series of rules and requirements for conducting GHG assessments, but may differ substantially depending on, among other factors, their goals and scope [8].

Standards defined by organizations, such as the GHG Protocol Consortium [21], offer rule sets complemented by recommendations, as well as specific vocabulary and reporting structures, to enable broader assessments, with the goal of generalizability and applicability across a wide range of companies across domains. Some sectors and consortia seek to address these differences

---

[1] https://ecoinvent.org/.

by publishing more specific guidelines as supplements to their general assessment standards (e.g., the Agricultural Guidance [20] as a supplement to the Corporate Standard). However, the lack of clarity regarding mandatory and voluntary considerations in standards and on how to validate them leaves room for interpretation. This fact, combined with heterogeneity between and within LCA databases, may lead to inconsistencies and incoherent application of the standards. This suggests that there is a clear, common structure to all GHG assessment standards and that evaluations should typically proceed in a similar manner. With that said, the devil is in the details, and differences in standards prevent straightforward, consistent comparisons across assessments, making the exchange of GHG knowledge between organizations very challenging. Here are some examples of aspects that impede consistent comparison.

- Differences in system boundaries: If two organizations or value chains do not consider activities at the same level of comprehensiveness, there is a high probability that some GHG sources will be neglected in one of the assessments [17]. This means that comparisons between assessments will not be made on equal grounds. This can happen if the standards leave the decision to include or exclude certain parts of a value chain to the assessing party.
- Differences in how a system is modeled: The web of human activities worldwide is complex, and it can be modeled in many ways. For instance, one can assume that a company uses the electricity available in the region where it operates (location-based modeling), leading to a specific representation of its associated GHG emissions. On the other hand, it is also reasonable to link a company's electricity use to the energy sources they pay for, which may yield a completely different picture if substantial amounts of renewable energy are purchased with certificates (market-based modeling) [9].
- Differences in accepted background data sources: Database providers may use different sources of information to evaluate the scope 3 GHG emissions of various activities. These sources of information can vary in detail, use different modeling choices, and be updated at different times. Furthermore, database providers must use complex models that are rarely fully documented, which often prevents consistency checks across the entire background databases [18].
- Differences in the formats and vocabularies: The many organizations behind the various existing standards are connected to different regions of the world and sectors of the economy, which makes it easy to understand why they use different formats and naming conventions for their disclosure requirements, for instance. While easily understandable and very difficult to change, this nevertheless hinders a clear and efficient comparison between assessments.

In this landscape, the assessment work of companies is challenging since they need to comply with various standards, and need to find the proper guidance matching their system(s) and activities. This situation is even more complex to navigate for organizations operating internationally with diverse portfolios of products, services, and activities.

2.3   Procedures for GHG Assessment

Regarding the actual implementation of the GHG assessment, which will use the standards and data described previously, the limitations become clearer, and their consequences are evident in current approaches. We analyze them through the different steps that constitute this process:

- Goal and scope definition. For this step, a comprehensive description of the system, including its activities, products, and services, is crucial to guarantee transparency and reproducibility. This includes, but is not limited to, specifying boundaries, scope, and modeling choices. However, these choices need to be consistent with those made by data creators who inform databases of GHG emission factors (e.g., LCA databases). Given the multiple, heterogeneous data sources and different assumptions and modeling approaches across standards, the involvement of LCA experts is required to ensure the development of consistent models for the assessed value chains.
- Information gathering and modeling. In this step, detailed information is gathered from the GHG data sources identified in step 1, namely internal measurements and their corresponding emissions, as well as external data from LCA databases and other sources providing GHG emission factors for energy use and complex value chains. At this point, it is necessary to integrate different data sources and verify that they comply with the chosen standard and its requirements, which requires extensive domain knowledge.
- Carbon footprint calculation. Using the previously gathered information, all necessary aggregations will be performed in accordance with the rules of the chosen standards, resulting in a calculated mass of GHG emissions, typically expressed as carbon dioxide equivalent (e.g., kg CO2-eq.). In many cases, these calculations are still performed in spreadsheets, but professional tools are available to aid in this task. These calculation tools are helpful only if experts have conducted consistency checks when creating models of value chains. In other words, they depend on the reliability of their inputs to offer standard-compliant results.
- Carbon footprint reporting. This stage will take the aforementioned steps to construct a data-informed narrative that presents the assessment findings, in accordance with the chosen standards' disclosure requirements. The produced reports will thus need to include the necessary information to provide transparency, comparability, and compliance with the regulations of the country, the sector or other regulatory bodies that the assessing party is obliged to. Current reporting strategy often lack the means to provide clear traceability that links the carbon footprint results with the data sources, including their different versions, assumptions, modeling decisions, etc.

3   The WISER Ecosystem

To address the previous challenges in GHG assessments, this work presents the WISER initiative[2], which aims to streamline the processes and tasks undertaken

---

[2] WISER initiative: https://wiser-climate.com/.

by organizations to perform GHG assessments and carbon footprint calculations, while significantly reducing the required knowledge to carry them out.

More concretely, WISER proposes a digital ecosystem of Web services and applications that enable the integration of existing GHG assessment databases and standards while guarding consistency and transparency. In particular, the WISER ecosystem provides the following features:

- Harmonization of different LCA data models and data sources through a common ontological model.
- Integration of heterogeneous LCA databases through a common interface—the WISER API—based on the ontology model.
- Modeling of GHG assessment standards within a Knowledge Graph, integrated into the WISER API.
- Consistency in the usage of different GHG assessment standards and the integrated data sources.
- Provision of API connections for decentralized computational tools.
- Access of third-party systems to the WISER services, to create user-facing dashboards and applications.

These components are schematically represented in Fig. 1, where examples of LCA databases are shown in the upper-right corner and may include API-fronted data sources (e.g., ecoinvent [2]) accessible through various interfaces (e.g., REST). The WISER Knowledge Graph represents different smaller interconnected KGs, of which the most prominent are: knowledge pertaining to rules and guidelines within GHG standards (e.g., GHG Protocol, or ISO 14064-1), and also those fully embedding LCA information specific to a certain domain (e.g., the KBOB[3] data for the Swiss building sector). The WISER KG also incorporates other linked datasets, like Geonames [6], to provide geographical context for the assessments and databases, among others. The WISER ecosystem also interconnects computational tools that can provide additional calculations based on the activities/services LCA data. At the center of this ecosystem lies the WISER API, which, apart from providing the infrastructure to interconnect the aforementioned components, allows establishing authentication and authorization policies for these resources. The decentralized services that integrate the WIDS can operate autonomously while also interoperating with the rest of the WISER ecosystem through the API's semantic interfaces. Within this ecosystem, applications in specific domains can use parts or all of these features to develop their own user interfaces for very concrete tasks, such as GHG-aware procurement or GHG emission forecasting.

Overall, the WISER ecosystem targets the following aspects in GHG assessments, it (i) provides consistency of GHG assessments, by respecting rules of specific standards; (ii) enhances comprehensiveness of assessments by including relevant direct/indirect emissions linked to activities in different LCA data sources; (iii) allows transparency and recency of information, through versioning and provenance of all the information encoded in the KG and the original data

---

[3] KBOB: https://www.kbob.admin.ch/de/oekobilanzdaten-im-baubereich.

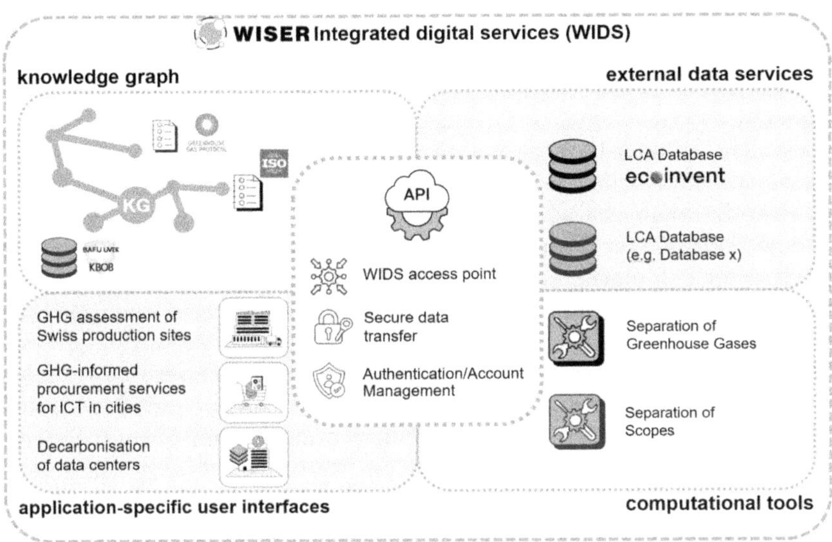

Fig. 1. The WISER ecosystem of integrated digital services: centered on the WISER API for interconnection and the WISER KG for data harmonization, the ecosystem includes access to heterogeneous data sources, GHG standards, and computational tools. Domain-specific applications connect to the WIDS for specialized tasks.

sources; and (iv) allows extensibility by permitting other services and applications to be connected to the WISER ecosystem through its API.

To bootstrap the WISER initiative, a consortium has been formed comprising academic, industrial, and institutional partners. This initiative has been supported by Innosuisse Flagship's funding and includes the active participation of companies and the ICT department of a city in Switzerland[4]. These consortium members have actively participated in the construction and the validation of the different components of the WISER ecosystem.

4 WISER Architecture and Implementation

To design the WISER architecture and its components, we investigated the main pain points and concrete problems faced by professionals working on GHG assessment within the WISER consortium. We provide a summary of the main design principles as follows:

- Uniform GHG data accessibility: Different LCA databases (sources of GHG emission factors) can be accessed (e.g., ecoinvent, UVEK, etc.) through an API (e.g., ecoinvent API) or as data resources in different formats. The WISER API provides access to the information from these different databases in terms of the ontologies supported by WISER.

---

[4] WISER industrial and academic members: https://wiser-climate.com/partners/.

- Fine-grained access control: Access rights to databases conform to stakeholders' permissions (e.g., paid and/or restricted access according to a license). The WISER API ensures that the enforced restrictions on data access are respected.
- Assessment standards representation: Metadata about the requirements of assessment standards are made available through the WISER API. This includes the main characteristics of these standards (e.g., the GHG Protocol) and their potential compatibility with existing databases and the selected activities for a potential assessment. This information is integrated within the WISER Knowledge Graph and is represented using explicit semantic relationships that allow reasoning and querying.
- Homogeneous activities/processes: Activities (or processes) of value chains are listed in the environmental data sources previously mentioned, and the WISER API is accessible through a unique model and format. Search and filtering are possible by category, location, and other activity characteristics. These relationships are accessible through specific APIs or embedded directly in the WISER Knowledge Graph.
- Unique identification: Entities in the WISER API have a unique identifier, i.e., URI that can help to name them across databases and versions, even though each database may have its own identifier (e.g., hashes, URNs).
- Access to GHG emission factors: The WISER API provides access to GHG emission factors of LCA databases. They may be linked to different impact methods and characterization factors. In WISER, the focus is on the climate change impact of all GHG represented in kg of CO2-eq.
- Filtering options for location/region are made available in the WISER API. Certain databases are specific to an area or country. Within a single database (e.g., ecoinvent), datasets for a process or activity may exist for different countries or regions.
- Computation capabilities: LCA calculations can be made available through the API, although actual implementation strategies may vary. For instance, computational results can be generated by invoking existing libraries such as Brightway2 [14] and then the outcomes can be stored and integrated within the WISER KG using the same semantic model.

4.1   WISER Implementation

The WIDS proposed by the WISER initiative addresses the challenges of supporting integrated digital GHG assessment procedures that check for alignment between GHG data sources and assessment standards, proposing a digital ecosystem of open-source Web services. Concretely, the WIDS are implemented as independent Web Services that can be consumed and/or used by external applications in a configurable manner, as building blocks depending on specific needs. The architecture of how these services are structured and interconnected responds to the principles listed above. In the following, we first describe the WISER API implementation and then the different types of WIDS services currently integrated into the platform (Fig. 2).

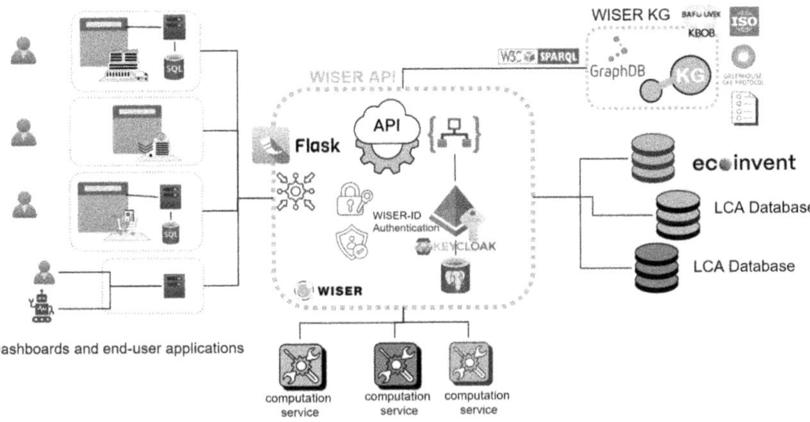

Fig. 2. Implemented and deployed architecture of the WIDS.

WISER API Implementation. The current version of the WISER API has been implemented in Python, based on the Flask framework. The source code is available for project partners on Github. Although it is planned to be open-sourced, at this stage its usage requires an agreement with the WISER consortium, given the intellectual property restrictions of the participating industrial partners. The WISER API exposes data from the heterogeneous GHG data source using a unified semantic model, and provides access to the WISER KG (see Sect. 4.2). The API uses an authentication mechanism based on the OAuth 2.0 protocol, implemented with the Keycloak framework and a self-hosted service. The WISER API is currently deployed and accessible on a public address[5].

The rest of the WIDS can be classified in four main groups, those dedicated to: (i) the management of external and heterogeneous LCA data sources, (ii) the management of GHG assessment knowledge with a Knowledge Graph (KG), (iii) the provision of computation services, and (iv) the interaction with end-users through specialized dashboards and other front-ends for GHG assessment.

WIDS for Heterogeneous Data Sources. Goal 1: Use and compare different data sources for various contexts. These WIDS can access data from various data managers/organizations. Data sources include primarily LCA databases like ecoinvent or UVEK, but it can potentially reach other relevant sources in the wider sustainability domain. These data sources conform to different schemas and data structures, therefore requiring a process of homogenization. For example, data sources like ecoinvent use de-facto standards like Ecospold [12]. This can be compatible to a certain degree with other standards like ILCD [4], but requires an explicit formalization and mappings between both models. We use ontologies to semantically represent all incoming data sources and map key concepts such as Activities, Classifications, Operational Boundaries, and GHG Emission Factors.

---

[5] WISER API docs: https://docs.wiser.ehealth.hevs.ch/.

WIDS for GHG Assessment Knowledge. Goal 2: Align value chain models and related data to different GHG assessment standards in a consistent way. GHG assessment standards require verification of compatibility among different aspects, including LCA methods, datasets, regions, and other assessment details. The selection of these parameters is often part of the expertise offered by the LCA expert preparing a GHG assessment report. The WISER ecosystem integrates the information and logic needed to ensure consistency in how GHG assessment standards are used. This information is formalized in the WISER Knowledge Graph (KG), which is made accessible through the WIDS. The KG is currently deployed with GraphDB, and serves data following a unified ontology model inspired from Ecospold or ILCD. Sample queries are detailed in Sect. 4.2. Other services in the WISER ecosystem can access the Knowledge Graph using a dedicated API built specifically for that purpose.

WIDS for Computational Tools and APIs. Goal 3: Connect various computational tools from different organizations with Application Programming Interfaces (APIs) to enable more useful and representative evaluations. Computations for GHG assessments can use specialized services and tools for this type of task, such as Brightway [14]. These tools may work in various ways, using different parameter sets and assumptions that might be difficult to master. For this reason, in WISER, the WIDS can invoke and perform these computations, integrating their outcomes into the WISER KG, so that other applications can reuse them according to their needs.

WIDS for Expert Users Apps and Services. Goal 4: Bring the GHG information from environmental experts to businesses with *easy-to-use* Web services. Although different client applications can access the WIDS, they are not intended for end users, e.g., LCIA experts or sustainability employees preparing assessment reports. For end-users, specialized applications can connect to the rest of the WIDS via the WISER API. These Apps typically implement domain-specific features, as shown in the use cases in Sect. 5.

4.2   Knowledge Graph Interfaces

One of the key components of the WISER ecosystem is its Knowledge Graph, which includes different–yet interconnected–subgraphs containing specific, well-defined subdomains. In particular, we have integrated the requirements of different assessment standards into the KG, as well as LCA databases. These resources are also linked to classification terminologies and geographic entities, which are essential for discerning which elements to filter or include in a GHG assessment. In the following, we present snippets of queries used to manage these resources, illustrating the use of the KG (prefixes and details omitted for brevity).

The snippet below shows a simple request for all operational boundaries linked to a given assessment standard available in the WISER KG. As we will see later, domain-specific applications (e.g., dashboards) can use this feature to allow selection of a specific standard (e.g. GHG Corporate Protocol [21], GRI

Standard [7], ISO 14064-1, etc.) and operational boundary (e.g., Scope 1, Scope 3 upstream or downstream, etc.) for a GHG assessment. Regardless of their heterogeneous origin, all of the standards are typed as `wiser:Framework`, and similarly for operational boundaries. As stated above, we provide unique URIs to identify each framework (e.g., `wiser: GHG-CP-LOC` for the GHG Corporate Protocol-Location-based) and potentially connect them to other Web resources.

```
1   SELECT DISTINCT ?framework ?operationalBoundary ?operationalBoundaryLabel
2   WHERE {
3       ?framework a wiser:Framework.
4       ?framework rdfs:label ?frameworkLabel.
5       ?operationalBoundary a wiser:OperationalBoundary.
6       ?operationalBoundary wiser:hasFramework ?framework.
7       ?operationalBoundary rdfs:label ?operationalBoundaryLabel.
8   }
```

Listing 1.1. Query snippet for GHG standards and operational boundaries.

Similarly, one can identify different categories linked to a framework and operational boundary. This can help maintain consistent use of categories relevant to a given standard and within a specific scope. Notice that categories may vary across assessment standards; therefore, the WISER KG provides the necessary mappings in this case. The query snippet below allows filtering these categories, given the assessment standard and assessment scope.

```
1   SELECT DISTINCT ?category ?categoryLabel  WHERE {
2       ?framework a wiser:Framework.
3       ?category a wiser:Category.
4       ?category wiser:hasOperationalBoundary ?operationalBoundary.
5       ?category wiser:hasFramework ?framework.
6       ?category rdfs:label ?categoryLabel.
7   }
```

Listing 1.2. Query snippet linking GHG standards to categories.

Finally, the following snippet shows a query that searches for activities using different criteria, including country codes, classification, activity search terms, the LCA database to search, and its version (the WISER ecosystem may use multiple versions at the same time). This will allow finding different activities that match the search criteria, even if they are in very different (e.g., in terms of format and sectoral coverage) sources such as Plastics Europe or UVEK.

```
1    SELECT DISTINCT ?act ?uuid ?referenceProduct ?activity ?location ?dataset ?version
         WHERE {
2        ?act a wiser:Activity; wiser:UUID ?uuid; wiser:Geography ?geo .
3        ?geo wiser:geographyTerm ?location .
4        FILTER(?location IN {?location_codes})
5        ?act wiser:activityName ?activityName .
6        ?activityName rdfs:label ?name .
7        ?act wiser:classification ?class .
8        ?class a ?classSubType .
9        ?classSubType wiser:subClassificationOf ?classType .
10       FILTER(?classType IN {?classifications}) .
11       BIND(?name AS ?referenceProduct)
12       BIND(?name AS ?activity)
13       FILTER (CONTAINS(LCASE(?name), ?search_term))
14       ?act wiser:wiserInfo ?wiserInfo .
15       ?wiserInfo wiser:dataset ?dataset; wiser:version ?version .
```

```
16       FILTER (?dataset = ?dataset)
17       FILTER (?version = ?dataset_version)
18 }
```

Listing 1.3. Query snippet filtering activities in an LCA database.

5 Domain-Specific Use-Cases

The key services of the WISER ecosystem (WIDS) have been deployed and tested through *user-friendly* interfaces in collaboration with different companies and public institutions that are members of the WISER consortium, such as the ICT department of a Swiss city. In this paper, we present three examples of use cases. The three use cases touch different aspects in which GHG assessments can be of primordial importance in the scope of recent regulatory developments related to climate change. They target different users through dedicated dashboards and tools in the following domains:

- Dashboard for Manufacturing Sites: This dashboard offers a user-friendly interface, presenting and dealing with the GHG assessments of manufacturing sites that connect with global value chains.
- Dashboard for ICT Procurement and Planning: This dashboard aggregates and displays information about GHG emissions from ICT equipment in terms of procurement and usage.
- Dashboard for Electricity Uses in Data Centers: Addressing the environmental impact of data centers' electricity use on behalf of any time-flexible and energy-intensive operation, this dashboard focuses on detailed GHG assessments of energy use based on electricity mix data with a high temporal resolution, as well as forecasting electricity mixes in European countries (Switzerland and neighboring countries).

The front-end dashboards communicate with the WISER REST API, serving as a bridge between end users and the backend infrastructure. The API backend, in turn, interfaces with the Knowledge Graph, facilitating the verification of alignment between of rules and vocabularies of standards and GHG data. Simultaneously, the WISER API backend communicates with the external ecoinvent API and other data sources (in the Knowledge Graph), enriching the ecosystem with additional environmental data and formats.

The separation of front-end and backend components is a fundamental design principle. The dashboards serve as user interfaces that interact with the backend through the API. This separation ensures a streamlined user experience while maintaining the robustness and scalability of the backend infrastructure.

For all three dashboards, the GHG assessment procedure is split into three main support steps. These steps include: (i) assessment definition; (ii) user data input management, and (iii) reporting definition. This three-step process breaks down the complex path to analyzing GHG assessments into easily understandable chunks for the user, while maintaining strong alignment with the typical structure of GHG assessment procedures found in most assessment standards.

5.1   GHG Assessments for Manufacturing Sites

This use case centers on GHG assessments in the manufacturing domain for companies with multiple production sites and complex global value chains. Within the WISER initiative, an industrial partner has actively participated in the co-development and evaluation of this prototype. The main goal of this prototype is to provide a Web service that streamlines GHG assessments of manufacturing sites in Switzerland with the following key features: (i) giving guidance on the choice of GHG emission factors that align with a selected assessment standard; (ii) semi-automated translation of the terminology and structure of the models when a different GHG assessment standard is selected for the same assessment; (iii) searching and filtering of activity datasets and their related GHG emission factors (see Fig. 3) from different LCA databases based on their suitability and compliance with the selected assessment's context (e.g., standard, region, location); (iv) report creation with auto-generated elements based on disclosure requirements of the selected standard or from calculations made by the computational tools of the WIDS.

The three support steps of the dashboard help complete key aspects of any GHG assessment, including system definition, description of modeling choices, carbon footprint calculation, and creation of tables and figures for the report. After validating the dashboard with a leading Swiss industrial partner, we demonstrated how it helps reducing the effort of creating GHG assessments that meet the requirements of the selected assessment standard. Thanks to the WISER Knowledge Graph, the tool is able to guide and validate the creation of the assessment. It also provides a consistent structure that we considered reusable for other domains, and as such we used it as a basis for the other two use-cases. In the future, it could also be adapted with low effort to support the GHG assessment of other systems, products, or services if new GHG assessment standards are included in the Knowledge Graph.

5.2   GHG-Informed Procurement for ICT Equipment Used in Cities

The main goal of this use-case is to provide a quick, easy-to-use tool that informs on carbon footprint values of ICT equipment for a more sustainable procurement procedure by city officials. The dashboard can also help to evaluate and communicate on the decarbonisation potential of different ICT development scenarios within a selected time horizon. Key features that support city officials with simpler GHG assessment of ICT equipment in this example are: (i) semi-automated connection between GHG datasets of a standard-compliant database, and their related GHG emission factors, with inventories of ICT equipment provided by the ICT department; (ii) restricted list of options for system definition based on a Swiss-compliant definition of the GHG assessment standard that are in the Knowledge Graph; (iii) possibility to define and calculate a timeline of GHG emissions from the buying and decommissioning of ICT equipment within a specified time horizon. This is done by separating information on the carbon

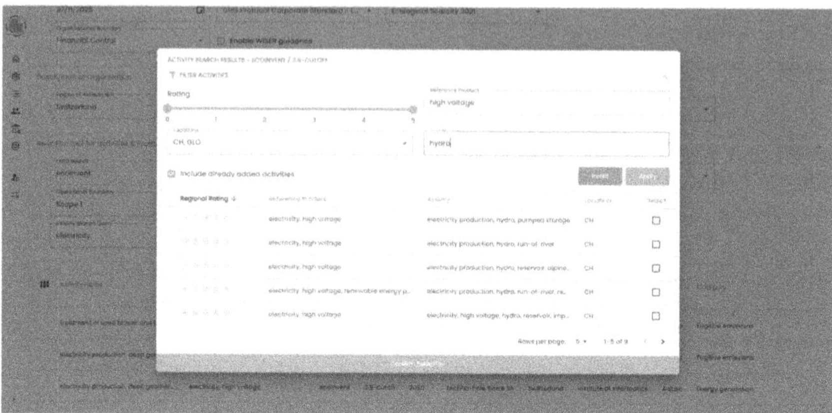

Fig. 3. Activity search for the manufacturing use case, connecting external databases (e.g. ecoinvent) for a specific GHG standard, enabled by the WISER KG and API.

footprint values of ICT equipment into GHG emissions from manufacturing and from waste management.

A validation phase was implemented in partnership with the ICT department of a major Swiss city, and therefore the dashboard was designed to tackle the key implementation concerns raised by the city officials. This sets a well-defined scope for the data that can be used in the assessment (i.e., Swiss context), aligning with the need for support from domain experts in defining the limits that the city should respect to provide reliable assessments that can be shown to the public. In Fig. 4, the carbon footprint is shown for different types of ICT equipment, based on GHG Emission factor data from ecoinvent and user-controlled consumption data. Thanks to the semantic information integrated through the WISER API, different data sources can be consulted in order to perform the assessment and the calculations, providing actionable data for decision-making.

5.3 GHG Emissions from Electricity Use in Data Centers

The main goal of this application is to provide a web service that delivers detailed, temporally differentiated GHG emission values for electricity use in data centers. The key feature of this dashboard is related to the GHG evaluation for electricity use. More specifically, it enables the connection of standard-compliant GHG emission factors from different LCA databases with external information on electricity production from various European Countries to provide temporal profiles of variation in the past (historical), present (annual), and near future (forecasting). The main challenge in the development of this dashboard was the required effort to map the connections between LCA databases and external sources of information on electricity production (i.e., ENTSO-E [3]) since the data sources use different levels of aggregation in describing energy sources.

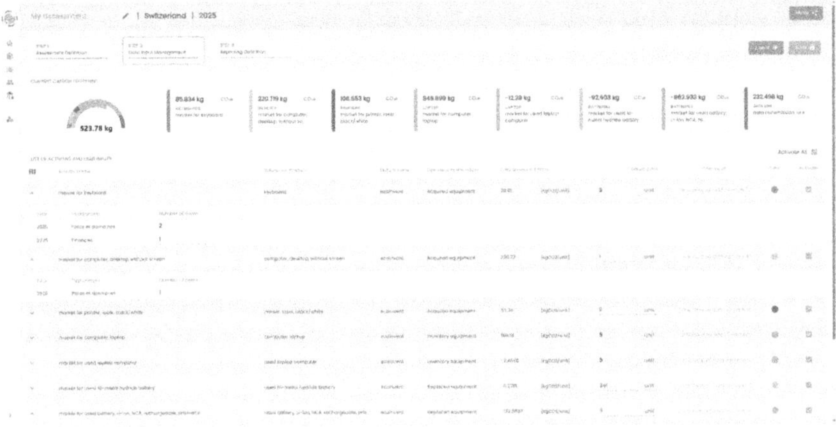

Fig. 4. Data input management and carbon footprint for a GHG assessment in the use case of ICT in Swiss cities.

Thanks to the WISER ecosystem, data center managers have access to detailed temporal profiles of *GHG intensity* that can drive automated changes in electricity consumption (e.g., the computation period) to reduce the overall carbon footprint of data center activities. In the future, the Knowledge Graph could also support the provision of electricity GHG profiles, using diverse sources of GHG emission factors that better align with the requirements of assessment standards for specific European countries. This would simplify evaluation across different assessment contexts relevant to international organizations. This, however, will require further domain-expert work on the connection between the many available LCA databases and the ENTSO-E model of electricity production. In Fig. 5, the carbon footprint profile of the Swiss electricity mix is shown

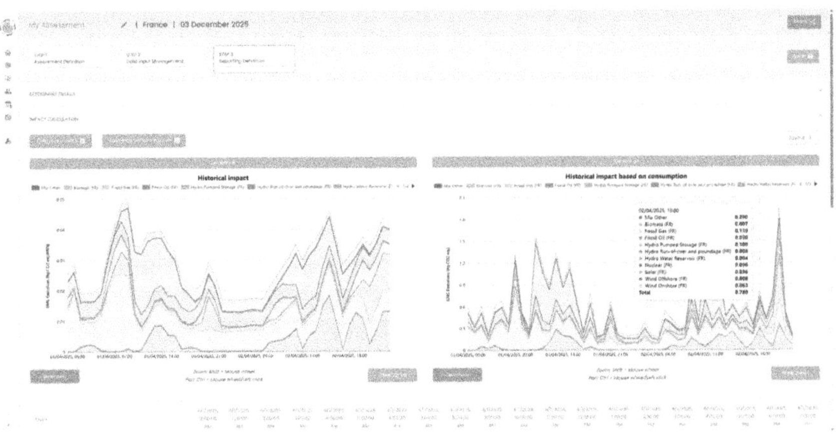

Fig. 5. Carbon footprint profile of the Swiss electricity mix with detailed GHG emissions data for different energy sources.

for a selected period of time and with detailed information on the key sources of GHG emissions between different energy sources.

6   Conclusions

We have presented the WISER interoperable digital services, their underlying architecture and a concrete implementation of its core components. We have demonstrated how this digital ecosystem can connect different services and stakeholders of the GHG assessment procedure in an integrated manner to simplify the sharing of domain-expert knowledge, so that future GHG assessments can be aligned with different assessment standards. A fundamental aspect of the WISER ecosystem is its reliance on semantic models and ontologies that enable the integration of heterogeneous LCA data sources, while also encoding the dependencies, restrictions and relationships of GHG standards. We have also showcased the importance of the WISER Knowledge Graph, and the ability of the WISER API to integrate the multiple input sources used by GHG assessment professionals.

Moreover, we have shown initial adoption by domain practitioners in different sub-domains, with active participation of companies, an organization in the sector of data center management, and an ICT department in a Swiss city. The implementation of the WISER showcases the usage of semantic data, in this case especially through the WISER Knowledge Graph, for the integration of inherently diverse data sources. Specifically, we have co-constructed and validated the WISER platform in three use cases: one on GHG assessment in manufacturing sites, another on ICT procurement, and finally one on data center decarbonization with detailed information on the carbon footprint of electricity. All use cases have been conceived and evaluated in direct collaboration with these stakeholders, and will continue to be developed in the context of the WISER initiative and its extended membership.

One of the key lessons learned from this experience is the importance of including an interdisciplinary team including GHG and sustainability experts, ontology engineers, computer scientists, and end-users. The participation of the latter has been crucial for the validation of the ecosystem, as it forced the rest of the team to solve specific issues, relying on the available data and the WISER capabilities. Another important takeaway is the importance of extensibility, given the very wide range of potential users of this ecosystem. Being applicable to multiple domains, it has been crucial to focus on loosely coupled components working as an ecosystem, rather than providing a centralized framework.

The WISER ecosystem will continue to evolve through the collaboration of stakeholders from specific application domains, together with environmental, information and computer scientists. These synergies are expected to develop in the future, including the addition of other use cases (e.g., wind energy, hydropower, citizen-driven carbon footprint, product carbon footprint). We also anticipate the inclusion of AI-driven assistance, and the tighter integration of GHG assessment computational pipelines.

Acknowledgments. Supported by the Innosuisse Flagship grant PFFS-21-72: Web of Interoperable Digital Services for Knowledge on Decarbonisation Pathways.

References

1. Atmaca, A.: Chapter twenty-two - understanding carbon footprint: impact, assessment, and greenhouse gas emissions. In: Rahimpour, M.R., Makarem, M.A., Meshksar, M. (eds.) Advances and Technology Development in Greenhouse Gases: Emission, Capture and Conversion, pp. 497–516. Elsevier (2024). https://doi.org/10.1016/B978-0-443-19231-9.00015-6
2. Ecoinvent: Ecospold2 (Nov 2021). https://ecoinvent.org/the-ecoinvent-database/data-formats/ecospold2/
3. ENTSO-E: European network of transmission system operators for electricity (2025). https://www.entsoe.eu/
4. European Commission. Joint Research Centre. Institute for Environment and Sustainability.: International Reference Life Cycle Data System (ILCD) Handbook :general guide for life cycle assessment : detailed guidance. Publications Office, LU (2010). https://data.europa.eu/doi/10.2788/38479
5. Franchetti, M.J., Apul, D.: Carbon footprint analysis: concepts, methods, implementation, and case studies. CRC press (2012)
6. GeoNames: GeoNames (2023). https://www.geonames.org/. Accessed 01 Dec 2025
7. Global Reporting Initiative: Gri stnadards. https://www.globalreporting.org/standards/ (2025)
8. Jia, J., Axelsson, K., Chaudhury, A., Taylor, E.: A rapid review of GHG accounting standards. SSRN Electron. J. (2023). https://doi.org/10.2139/ssrn.4523132
9. Johnathon, C., Agalgaonkar, A.P., Planiden, C., Kennedy, J.: A proposed hedge-based energy market model to manage renewable intermittency. Renew. Energy **207**, 376–384 (2023)
10. Klöpffer, W., Grahl, B.: Life cycle assessment (LCA): a guide to best practice. John Wiley & Sons (2014)
11. Life Cycle Initiative: Global LCA data network (GLAD). https://www.lifecycleinitiative.org/resources-2/global-lca-data-network-glad-2/ (2023)
12. Meinshausen, I., Müller-Beilschmidt, P., Viere, T.: The EcoSpold 2 format—why a new format? Int. J. Life Cycle Assess. **21**(9), 1231–1235 (Sep 2016). https://doi.org/10.1007/s11367-014-0789-z
13. Mukherji, A., et al.: Synthesis report of the IPCC sixth assessment report (AR6) (2021)
14. Mutel, C.: Brightway: an open source framework for life cycle assessment. J. Open Source Softw. **2**(12), 236 (2017)
15. NASA: The effects of climate change (Nov 2022). https://climate.nasa.gov/effects/
16. Nations, U.: The sustainable development goals report 2022 (2022). https://unstats.un.org/sdgs/report/2022/
17. Palermo, V., Bertoldi, P., Crippa, M., Franco, C., Monforti-Ferrario, F., Pisoni, E.: Uncovering divergences and potential gaps in local greenhouse gases emissions accounting and aggregation. Curr. Res. Environ. Sustain. **8**, 100263 (2024)
18. Pan, Y., Zhang, L.: Data-driven estimation of building energy consumption with multi-source heterogeneous data. Appl. Energy **268**, 114965 (2020)
19. Plastics Europe: Eco-profiles for determining environmental impacts of plastics (2025)

20. Protocol, G.G.: Ghg protocol agricultural guidance. https://ghgprotocol.org/agriculture-guidance
21. Ranganathan, J., Corbier, L., Bhatia, P., Schmitz, S., Gage, P., Oren, K.: The GHG protocol corporate accounting and reporting standard. https://ghgprotocol.org/corporate-standard (2004)
22. World Steel Association: Climate action data collection (2025). https://worldsteel.org/climate-action/climate-action-data-collection/

VRTI Knowledge Graph Explorer: Adoption, Use and Impact

Alex Randles[1]([✉])[iD], Lucy McKenna[1][iD], Lynn Kilgallon[2][iD], Peter Crooks[2][iD], and Declan O'Sullivan[1][iD]

[1] ADAPT Centre, School of Computer Science and Statistics, Trinity College Dublin, Dublin, Ireland
alex.randles@adaptcentre.ie
[2] Department of History, Trinity College Dublin, Dublin, Ireland

Abstract. The Virtual Record Treasury of Ireland (VRTI) was established in 2016 as an initiative to reconstruct the Public Record Office of Ireland that was lost in the civil war fire in 1922. The initiative has become a nationwide effort to re-discover, digitize, reconstruct and catalogue the lost archives of the subsequent years. Leveraging information technologies, historians have constructed the one of the largest historical archives in Ireland, contributing to preserving the heritage of a nation. The VRTI has implemented a data-centric approach to exploiting the Knowledge Graphs (KG) capabilities for organizing, cataloguing and sharing knowledge across different resources. However, interacting with a KG requires the creation of complex and time-consuming queries which limits the uptake of the information. Therefore, it was decided to design an intuitive user interface to facilitate exploration of the Knowledge Graph for Irish History (KGIH). The KGIH is a five star linked open data resource, the first of this kind to exist for Irish historical research. The VRTI-KG explorer provides a straightforward method for casual users and researchers to interact with the KG and apply it in real-world use cases. This paper presents the uptake and impact of the explorer and outlines recent enhancements designed to provide historians with more control over the information presented on the interface.

Keywords: User Interface · Digital Humanities · User Uptake

1 Introduction

The Virtual Record Treasury of Ireland (VRTI)[1] is a digital reconstruction of the Public Record Office of Ireland (PROI), which was destroyed during a fire in the 1922 Irish civil war, damaging records dating back seven centuries. A century later the Beyond 2022 [5] project was launched with the objective to create an openly accessible digital resource containing the historical documents. As a result the VRTI was created and published, which contains over 350,000

---

[1] https://virtualtreasury.ie/.

M. Acosta et al. (Eds.): ESWC 2026, LNCS 16550, pp. 453–469, 2026.
https://doi.org/10.1007/978-3-032-25159-6_24

records and 250 million words of searchable Irish history. Complementing these records, is the Knowledge Graph for Irish History (KGIH), which contains over 2.9 millions statements, detailing notable people and places and their interconnections. The KGIH [21] is a five star linked open data resource, the first of this kind to exist for Irish historical research. The project involves a high-level of interdisciplinarity research between historians (collecting information) and computer scientists (digitising and linking information). The use of semantic web technologies provides a method to interlink this information, resulting in the discovery of previously unknown historical relationships. The Knowledge Graph (KG) is available as an open, free and permanently accessible resource, which allows researchers to use the data in real-world use cases.

Given the nature of the content in the KGIH, it was decided to create a user interface to allow casual users and researchers to interact seamlessly with the extensive data available in the KG. The VRTI-KG explorer[2] [19] is an interface designed to facilitate exploration of the KGIH and associated VRTI archival records. The tool bridges the gap between non technical users and semantic web technologies. The explorer was implemented as a web based application that uses a combination of technologies to interact with the KG and visual the data on the user interface. The explorer was released as a freely accessible web application, promoting the core principles of open linked data. The explorer provides a straightforward mechanism to search over 70,000 distinct resources. In addition, over 4,000 treasury records have been directly linked to resources in the KGIH, allowing users to explore the involvement of the resource in historical documents hosted on the VRTI website dating back centuries. The interdisciplinary collaboration involved in the development of the explorer identified that historians required direct control over certain information presented on the KGIH explorer interface because of their extensive knowledge of historical material. It was hoped that integrating functionality to allow historians to configure interface components independently would result in a more efficient editorial process and decrease workload for the knowledge engineers in the team.

The explorer was released at the end of June 2025 and in the five month period between then and the start of December 2025 it has received over 11,000 unique visits, with consistent engagement from both casual users and researchers. The users have originated from 93 different countries, with over 60% of the visitors based outside of Ireland, which demonstrates worldwide impact of the application. In addition, the explorer has received over 46,000 user engagements, over 60,000 page views and over 12,000 keyword searches. In addition, the explorer has been used as an educational and research resource for those interested in history and semantic web technologies. The KGIH and explorer were also awarded two gold awards at the 2025 Irish Digital Media Awards in recognition of its significant impact on digital engagement and innovation in online platforms.

This paper presents the uptake and impact of the explorer, along with the lessons learned during the design, development and deployment of the explorer. In addition, recent enhancements to the explorer are discussed. These

---

[2] https://kg.virtualtreasury.ie/.

enhancements are designed to facilitate linking of VRTI records to the KGIH and add features to support independent editing of interface components by historians. This paper is structured as follows. Section 2 describes the KGIH. Section 3 includes an overview of the VRTI-KG explorer. Section 4 describes the uptake of the explorer by casual users and researchers. Section 5 describes recent updates to the explorer. Section 6 discusses related work. Section 7 discusses next steps and concludes the paper.

2   Background

The first version of the KGIH [5] was published in 2022. The VRTI ontology[3] was developed by extending the CIDOC-CRM [4] ontology, which was designed to represent knowledge associated with the cultural heritage domain. Since then the KG has been refactored to improve the schemas and introduce controlled vocabularies. The current version of the KG contains the following:

- 10,653 people from different historical time periods.
- 66,771 places structured in a nested hierarchy, where smaller places are container within larger geographical areas.
- 2.9M triples representing historical statements and their interconnections.

The KGIH interlinks to other existing KGs, such as Wikidata [20] and DBpedia [2], which enrich the semantic context of the data and enable users to explorer historical entities within a broader network of open linked data. The schemas of the project are designed to be transferable, allowing them to be applied to uplift similar information in other research projects. The person and places schemas are currently being applied in the digital humanities European Research Council project called VOICES [16], which aims to represent the voices of women in early modern Ireland. These people have been uplifted into (and where appropriate interlinked with existing people and places) in the existing KGIH. A dedicated event schema has been created to provide historians with a structured framework to represent information related to historical events, such as wars or depositions. It supports multiple levels of granularity level to model events of different importance. A nested hierarchy of sub-events allows historians to organise events of different granularity, and importance and interlink associated information. The previous version of the KG was browsed and visualised using the following out of the box tools: LodView [3], Ontodia [15] and OSCAR [6]. LodView supports browsing of resources by URI, however, it does not include built-in search functionality. OSCAR supports simple free-text search, however, it does not support multi-faceted search and presents search results with a basic table view. Ontodia supports visualisation of the KG in graph form, however, it requires users to be familiar with graph terminology, such as instances and classes. Due to these limitations, it was decided to develop a bespoke tool to support multi-faceted exploration of the KGIH.

---

[3] https://www.w3id.org/virtual-treasury/ontology#.

3 KG Explorer

This section provides summary of the main functionality of the explorer that has been previously described in more detail in [19]. The users can search the KG with a simple free-text search or using an advanced search[4] with multiple search filters, including drop downs, checkboxes and timelines. Figure 1 presents a screenshot of a sample search results view.

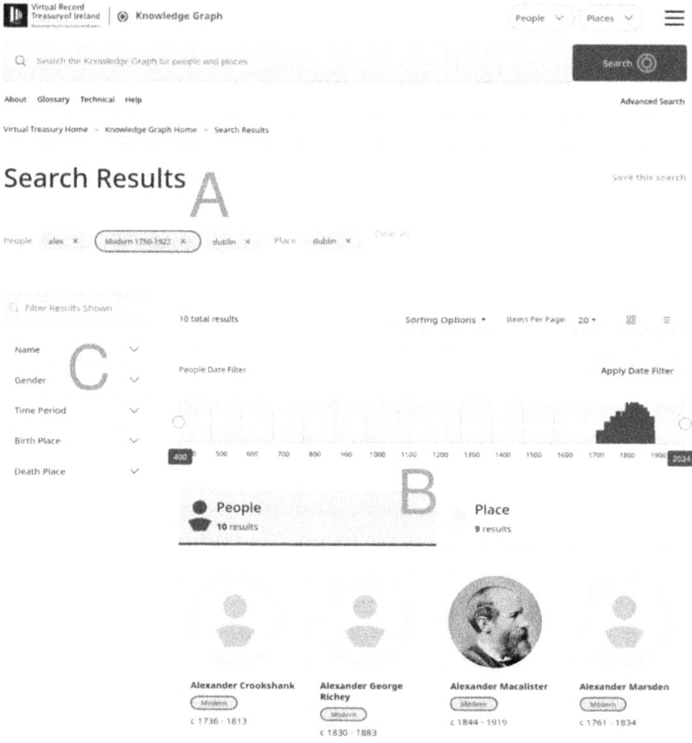

Fig. 1. Search results view

This view shows results for people named Alex in the modern time period who were born in Dublin and places named Dublin (Fig. 1 - A). Each result contains a snippet of information (B) for each result, including their name, image and dating information. The results can be sorted by attributes, such as surname or birth date. The results for people and places are grouped in separate tabs (B). A timeline visualises (B) the time periods of people and allows filtering by dates. Additional filters (C) can be applied to the initial result set to further filter them. A user is redirected to an entity card when a result is selected. An entity card

---

[4] https://kg.virtualtreasury.ie/advanced-search.

summarises the information in the KG related to a particular resource. Figure 2 presents a screenshot of an entity card of a person[5] in the KGIH.

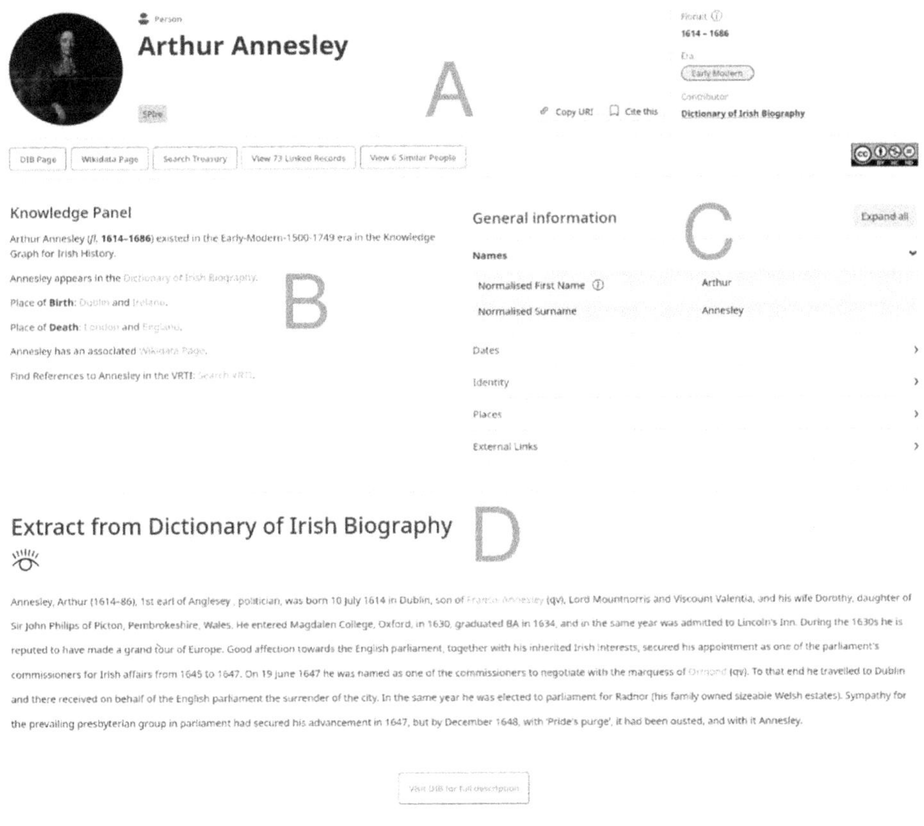

Fig. 2. Entity card of person in KGIH

The entity cards include the name of the person, basic dating information and contributor of data (Fig. 2 - A). Links to other KGs, such as Wikidata [20] and DBpedia [2] are included when available. Natural language statements (B) describe the attributes of the person, such as birth and death place. A set of tables (C) present categorised information related to the person. In addition, the entity card includes a snippet of information (D) that is dynamically fetched from the Dictionary of Irish Biography (DIB)[6] page associated with the person.

A place entity card[7] contains similar information to the person entity card. However, it includes a map that displays the geospatial information associated with the place, including the centre point and boundary. The map includes

---

nearby places that can be selected to view their entity card. In addition, a place entity card includes a listing of people related to the place.

4   Uptake and Impact

This section describe how the explorer has been received and the influence it is having since release. Three main categories of users have emerged from the publication of the explorer, which are public, education and research.

4.1   General Public

Figure 3 presents the user metrics from the top ten countries who visited the explorer most frequently since release. Google analytics[8] was integrated into the explorer to collect data on website traffic and interactions from users.

	Country	↓ Active users	New users	Engaged sessions	Engagement rate
	Total	11,833 100% of total	11,684 100% of total	10,376 100% of total	62.88% Avg 0%
1	Ireland	4,673 (39.49%)	4,627 (39.6%)	4,710 (45.39%)	64.34%
2	United Kingdom	2,676 (22.61%)	2,662 (22.78%)	2,238 (21.57%)	59.32%
3	United States	2,179 (18.41%)	2,167 (18.55%)	1,700 (16.38%)	59.21%
4	Australia	686 (5.8%)	687 (5.88%)	609 (5.87%)	67.89%
5	Canada	408 (3.45%)	408 (3.49%)	340 (3.28%)	65.01%
6	New Zealand	135 (1.14%)	134 (1.15%)	107 (1.03%)	57.84%
7	Netherlands	107 (0.9%)	105 (0.9%)	78 (0.75%)	61.9%
8	Spain	106 (0.9%)	105 (0.9%)	90 (0.87%)	63.38%
9	France	95 (0.8%)	93 (0.8%)	80 (0.77%)	74.07%
10	Germany	86 (0.73%)	86 (0.74%)	72 (0.69%)	64.86%

Fig. 3. Overview of users from top ten countries

The explorer has received over 11,684 unique visits between June and December of 2025, with consistent engagement from both casual users and researchers. The users have originated from 93 different countries with over 60% of the visitors based outside of Ireland, which demonstrates worldwide impact of the application. 10,376 engaged sessions indicates most users meaningfully interacted. 63,538 page views indicate significant interest and repeated interaction from users. The explorer has received 46,143 user engagement events (e.g. scroll,

---

[8] https://support.google.com/analytics.

click), which measures how many time users actively engaged with the site. 12,280 keyword searches have been completed using the explorer, which indicates active engagement and meaningful use of the search features. The explorer averaged 76 visits per day during this period, which demonstrates the continued impact of the application. Users on average viewed 5 or more pages and engaged with the site for approximately 1.5 min, which indicates a high level of user engagement and a relatively low bounce rate. In addition to the analytics, the explorer was awarded with two gold awards at the 2025 Irish Digital Media Awards in recognition of its significant impact on digital engagement and innovation in online platforms. The VRTI website contains over 4,000 links to the the KGIH, which provides a method for users of the site to become aware of the KG entities associated with records and collections. Vice versa, the KG explorer also allows users to visit records or collections on the VRTI website associated with a KG person or place. In conclusion, the metrics indicate meaningful engagement by public users from a global audience, with users repeatedly interacting with the site to discovery new information.

4.2 Research

The KG explorer has been adopted by researchers in the fields of digital humanities and computer science as a discovery tool for identifying information and relationships from Ireland's lost history. Historians within the VRTI project use the explorer to find KG entities which are associated with archival records, dating back centuries. The historians can link these entities by associating the KG URI with the metadata of the record. Over 4,000 records have been linked with KG resources by historians in the VRTI. In addition, the entity cards for each KG resource contain a citation feature, which allows researchers to copy formatted citations and reference the KGIH in their studies.

The KGIH infrastructure and the KG Explorer has also been adopted by the European Research Council VOICES [16] project for their work in uncovering the hidden of lives of women in the Early Modern period. Their data was uplifted using the existing KGIH schemas and the data was made available to search using the explorer.

The explorer is being used in a a PhD project at Trinity College Dublin to create a framework [14] to address the initial exploration problem of interfaces, which is defined as a set of barriers non-technical users face when first attempting to explorer complex KGs. The approach proposes using curated questions to guide new users in the beginning of their interaction with the KGIH by suggesting starting points for their exploration. Curated questions have been integrated into the explorer by the PhD researcher to investigate the potential benefits to user interaction. In addition, the questions provide users with an understanding of what types of questions can be answered by data in the KG.

4.3   Education

The explorer has become a educational resource for digital humanities and computer science modules at Trinity College Dublin. The explorer is used by students in the Knowledge and Data Engineering module to provide insights into how KG data can be visualised and presented on user interfaces to provide data driven interaction. In addition, the explorer provides links to the VRTI ontology and SPARQL endpoint, along with sample queries that students can use to practice and understand KG modelling and querying.

The explorer is being used by students in undergraduate and MPhil Irish history modules to investigate connections between KG resources and archival records, supporting their research through interactive data exploration. In addition, the links in the KG to other sources provide useful contextual information.

The uptake of the explorer by casual users and researchers demonstrate the real-world value of the application, demonstrating that it meets practical needs for everyday use, while offering the depth for more rigorous research.

5   Recent Updates to Explorer

The following new features have been integrated into the explorer since the evaluation [19] that was completed on the initial version. The updates emerged in response to the results of the evaluation, in addition to internal testing by the VRTI team. The updated features have been implemented, being tested and will shortly be made available in the version available to the public.

5.1   Linking VRTI Records with Entities in the KGIH

The archival records in the VRTI are stored in relational databases, while people and places mentioned in these records are stored in the KGIH. Establishing a process to link records with KG resources is intended to meaningfully enrich both data sources.

5.2   Linking KG Resources and Archival Records

A bespoke data ingestion pipeline has been developed to allow historians to inject KG URIs into the metadata of VRTI records. First, the historians use the explorer to find the URI of the desired KG resource. Then, the URI is copied from the explorer and uploaded into the metadata of the record, along with the role (e.g. creator, sender and editor) of the resource. To date, the historians have created over 4,000 links between archival records and KG resources. The entity cards for people and places include a section which show the links to the appropriate VRTI record or collection (Fig. 4).

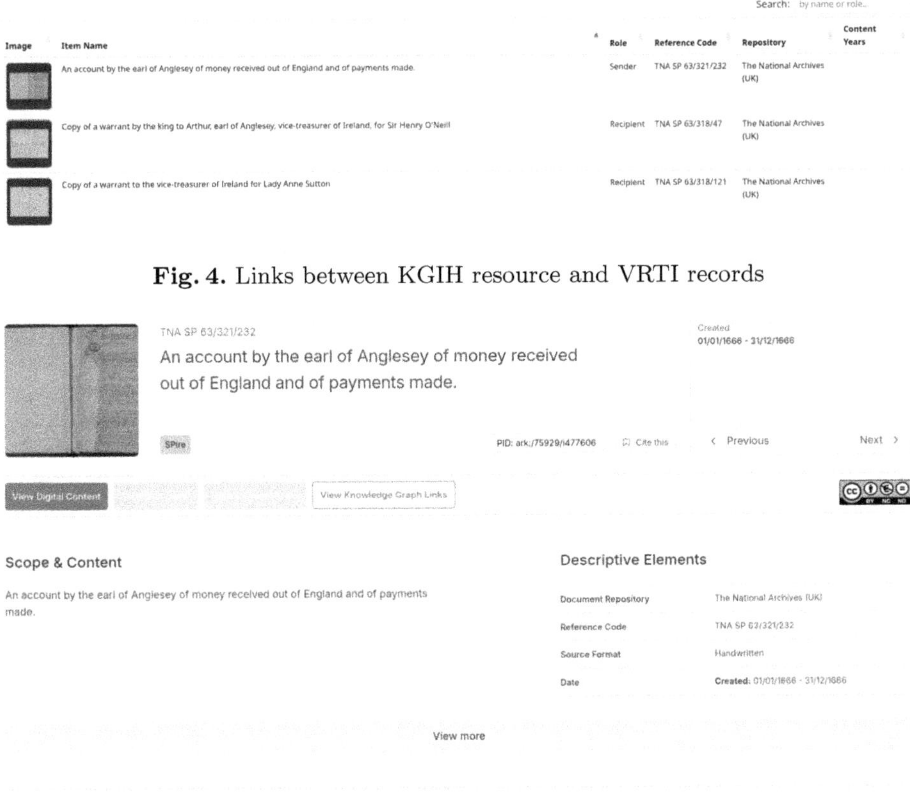

Fig. 4. Links between KGIH resource and VRTI records

Fig. 5. VRTI record linked to KGIH Resource

Figure 5 presents a record[9] related to a person in the KGIH. This record is a handwritten letter from 1666 and shows a person in the KGIH who is defined as the sender of a letter. The users can click on the URI shown, which redirects them to the entity card of the person in the KG explorer.

The implementation of this feature will increase the usage of the explorer by exposing relevant information to users of the VRTI website and allowing them to easily navigate into the KGIH, and vice versa.

---

[9] https://virtualtreasury.ie/item/TNA-SP-63-321-232.

5.3   Curated Collections

The VRTI website contains curated collections[10] that group together archival records with similar sources. For instance, the Census Reports, 1821–1891 collection contains all the Census Commissioners' Report and statistical tables published in association with the Census of Ireland for 1821, 1831 and 1841. It was identified that there is a need for the KG explorer to facilitate search and visualisation of KG resources linked with records in these collections. Figure 6 presents an overview of the components designed to support interaction with collections. The search method involves the users navigating through a carousel[11] (Fig. 6 - A) with the details of each curated collection. Then, the user selects a collection and is redirected to the search results view (B) with all the people that have been associated with records in the collection. Then, the user can select a person to view their entity card (C), which contains a listing of the collection records related to the person.

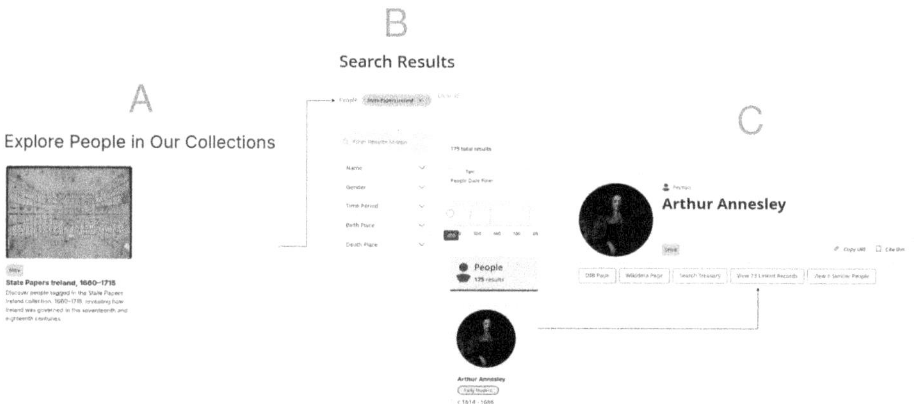

Fig. 6. Searching curated collections

A process was developed to visualise records in collections on interactive geospatial maps. The process involves retrieving all place URIs associated with the records in the collection. A SPARQL query is executed to find the associated geospatial information from the KG used to plot markers on the map. Figure 7 presents the map[12] designed to facilitate geospatial discovery of records in the census curated collection and links to places in the KG.

The places are plotted on a map using a heat map feature (Fig. 7 - left screenshot) to visually indicate the concentration of places in different regions across Ireland. The map includes the centre point and boundary of each place, which is

---

[10] https://virtualtreasury.ie/curated-collections.

[11] https://kg.virtualtreasury.ie/#kg-collections.

[12] https://kg.virtualtreasury.ie/collection-map/census-gleanings.

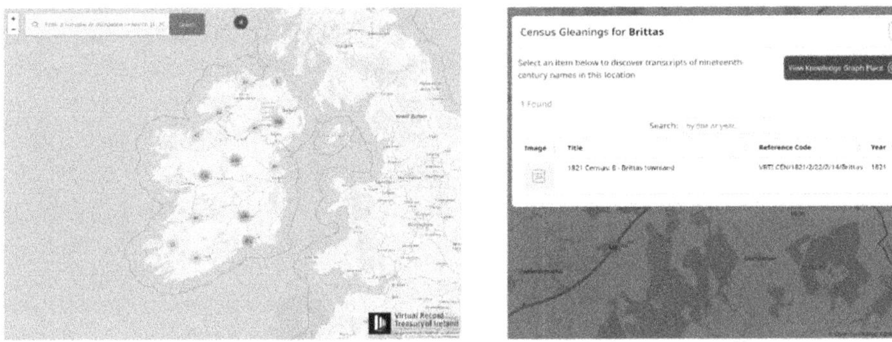

Fig. 7. Geospatial map visualisation of curated collection

retrieved from the geospatial data in the KG. A pop-up (right screenshot) with a table listing records related to a place appears once a marker is selected. Then, a user can select a record or place to view their entity card in the explorer. The map supports keyword search of the records in the collection and toggles the visibility of relevant markers when a search is completed.

5.4   Configurability of Interface

The interdisciplinary collaboration between historians and knowledge engineers during the development of the explorer identified a requirement to allow historians to independently update interface components.

5.5   Spreadsheet Configuration

It was decided to use spreadsheets to enable the configuration process, as the historians in the VRTI are familiar with them and already use them during data curation and uplift into the KGIH. The spreadsheets contain schemas used to support editing of the glossary, frequently asked questions and featured entities. Figure 8 presents a screenshot of the spreadsheet and resulting interface view for featured people[13].

The spreadsheet contains columns to populate the carousel items shown for the featured people, such as the name, time period, dating information, textual description and link to an image. In addition, the KG URI of the person is used to redirect the user to their entity card when a person is selected. The spreadsheet contains a script which notifies the explorer when a cell is edited and it then retrieves the latest version. The input from historians to the configuration spreadsheets is validated using data validation features available in the google spreadsheets that we use.Âă This spreadsheet allows historians to easily present notable people on the interface when certain related events occur throughout the year, such as a birth anniversary or political milestone. The historians found this feature to be useful and it resulted in decreased workload for computer scientist.

---

[13] https://kg.virtualtreasury.ie/people-homepage.

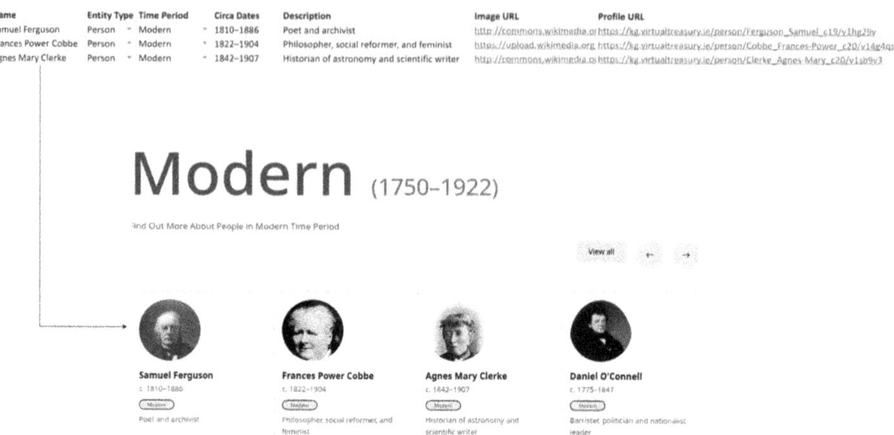

Fig. 8. Spreadsheet configuration of featured people on the explorer

5.6   KG Views

Historians required functionality to group together related resources, such as those within an event or time period. A KG view in the explorer facilitates the creation and presentation of related resources in the KGIH. A bespoke process was designed to enable historians to create views on the explorer independently to the technical team. Initially spreadsheets were experimented with to create views, however, it was decided to develop instead a dedicated web interface (that will be incorporated into our KG editor [18]) to improve data management as the number of views grows.

An interface component was integrated into the explorer which includes a form that allows historians to create and edit views. First basic metadata (Fig. 9 - A) is input when creating a view. This metadata includes the view name, description, associated image and related existing views. Then, the group of resources (B) are associated with the view by either copying an explorer search results link with the required filters (e.g. all women in the early modern time period) or by adding individual URIs. Featured URIs can be included, which highlight notable resources in the view (C). The input from this form is then translated into a view homepage on the explorer (D). The users can search through the resources in the view and filter the original group by other attributes to find relevant information. A carousel (E) is used to display featured URIs and their images. Clicking a featured URI redirects to their entity card. Finally, the views homepage includes a carousel (F) with all of the related views, allowing users to easily navigate to their homepages. The views interface includes user input validation to ensure that historians do not enter incorrect values for different attributes. For instance, regular expressions are used to ensure that users enter URIs that correctly match the URI structure of the KGIH. The views are currently being tested internally and will be made available to the public in July 2026.

Fig. 9. Creation of KG Views

6   State of Art

The following approaches are designed to allow users engage with KGs in the cultural heritage domain. Most prominent amongst them are the implementations that have used the SampoUI framework [9], which is a customisable framework for creating semantic portals from SPARQL endpoints.

AcademySampo [13] provides a portal and LOD service published on the Linked Data Finland platform. The portal facilitates the search of 28,000 biographical descriptions of students of the University of Helsinki between 1640–1852 and 1853–1899. The KG was created from data extracted from texts and database structures and includes information beyond the persons study, including their career and relatives.

BiographySampo [7] created a KG by automatically extracting information from 13,100 textual Finnish biographies with links to 16 external data sources. The user interface contains a collection of seven perspectives, representing entities and relationships to support the prosopographical research method. These perspectives include search filtering of target groups and data-analytic and visualisations of these groups.

WarSampo [11] KG contains information about Finland in Second World War. The data contains 94,676 records about soldiers who were killed in action. In addition, it contains metadata of all known 4,200 Finnish prisoners of war and 5,611 notable individuals from other data sources, such as Wikipedia. The WarSampo Portal is a tool for simple data analysis of people in the registers. The portal includes nine perspectives on the events, people and places in the registers. SampoUI is a powerful out of the box tool, however, it is not automatically intuitive for non-technical users who are unfamiliar with the domain specific ontology terms.

Other interesting examples are also worth highlighting. MuseumFinland [8] transformed collections in Finish museums from heterogenous formats into a KG. The KG is represented using seven domain ontologies related to locations, objects and actors. The KG contains cultural artifacts and metadata associated with historical sites. The KG was implemented using a tool called OntoViews, which includes reasoning engines and search engine. The search engine (Ontogator) provides multi-faceted search functionality of the KG and expresses search results as an RDF-XML tree. OntoViews includes faceted search which reflects the ontology, which requires users to be familiar with their structure. In addition, users must be familiar with XML to understand the search results.

ResearchSpace [17] is a platform developed by the British Museum for developing digital humanities KGs. The platform supports collaborative building and refinement by researchers. In addition, the platform supports data enrichment and semantic search. The enrichment component provides a platform for community semantic authoring. The search component formulates a query based on a pattern or piece of text, where the researcher creates the SPARQL queries used to guide the exploration. As a result users must be familiar with SPARQL queries to search data using ResearchSpace.

The Vienna History Wiki [12] is a project that created a KG for the History of Vienna using the Semantic MediaWiki (SMW). SMW provides an interface that facilitated open-source collaboration by the city government and general public of Vienna. The KG contains people, places and objects from the city of Vienna. SMW includes a semantic search interface to retrieve information from the Vienna History Wiki. Users can search the data using semantic conditions, such as matching categories, page ID and Wikidata ID. The results can be viewed as a table or exported in formats, such as CSV and JSON. In addition, an API is available to facilitate querying by applications. The SMW search tool has a learning curve that requires users to be familiar with the syntax of the search conditions.

Europeana [10] is a cultural heritage portal that contains metadata from over 4,000 libraries and museums across Europe. A subset of the data is available as a KG represented using the Europeana data model and can be retrieved using the API and SPARQL endpoint. However, the interface searches the database rather than the data in the KG.

The VRTI-KG explorer provides visualisations of data from the KGIH in various formats other than tabular, including timelines and geospatial maps. The explorer facilitates search using plain-text search terms and does not use a specialised syntax. In addition, it does not require users to be familiar with the terms in the VRTI ontology.

7   Future Work and Conclusions

It is planned to publish the source code of the explorer as an open-source resource to enable researchers to reconfigure it to search other KGs. The explorer can be customised to interact with a specified SPARQL endpoint along with associated queries to retrieve relevant information. A Wikidata property proposal[14] has been proposed to link resources in Wikidata with corresponding resources in the KGIH. The unique alphanumeric associated with each KGIH resource will be linked to Wikidata to facilitate transversal of links between both graphs. Finally, curated questions will be integrated into the interface of the explorer to guide new users when beginning their exploration of the KG. The questions will also provide an understanding of what types of questions can be answered from the data available in the KG.

The KGIH contains over 2.9 millions statements and over 70,000 distinct resources published as open linked data for others to use. However, retrieving useful information from the KG requires the creation of complex and time-consuming SPARQL queries. The approach described in this paper is designed to provide a straightforward method to allow non-technical users to easily retrieve information from the KG and use it in real-world use cases. The public release of the explorer has identified key insights when non-technical users are interacting with semantic web technologies. The explorer has been utilised by casual users, along with educational and research users. Over 11,000 unique visits to the explorer from 60% global audience, indicates interest from worldwide users. Over 60,000 page views indicates active engagement and repeated interaction from users. The over 12,000 keyword searches indicates active exploration. It is hoped the approach improves the uptake of linked data, particularly in the cultural heritage domain. In the next period, it is also hoped to see how appropriately configured [1] machine learning algorithms can be utilised to allow for more extensive analysis over the data, and the KG Explorer will need to be modified to accommodate appropriate user interaction. Finally, it is hoped the design considerations and lessons learned can be used by researchers when designing and deploying interfaces, which interact with semantic web technologies.

Acknowledgments. Virtual Record Treasury of Ireland (VRTI) is funded by the Government of Ireland, through the Department of Tourism, Culture, Arts, Gaeltacht, Sport and Media, under the Project Ireland 2040 framework. The project is also partially supported by the ADAPT Centre for Digital Content Technology under the SFI Research Centres Programme (Grant 13/RC/2106_P2).

Disclosure of Interests. The authors have no competing interests to declare that are relevant to the content of this article.

Use of Generative AI. The authors did not use Generative AI tools in the preparation of this manuscript.

Supplementary Material Statement. The VRTI-KG explorer can be accessed at https://kg.virtualtreasury.ie/.

---

[14] https://www.wikidata.org/wiki/Wikidata:Property_proposal/
Knowledge_Graph_for_Irish_History_ID.

References

1. Assem, H., Xu, L., Buda, T.S., O'Sullivan, D.: Machine learning as a service for enabling internet of things and people. Pers. Ubiquit. Comput. **20**(6), 899–914 (2016)
2. Auer, S., Bizer, C., Kobilarov, G., Lehmann, J., Cyganiak, R., Ives, Z.: Dbpedia: A nucleus for a web of open data. In: International Semantic Web Conference, pp. 722–735. Springer (2007)
3. Camarda, D.V., Antonuccio, A.: Lodview: Iri dereferencer, rdf to html. GitHub repository (2017). https://github.com/LodLive/LodView, available at https://github.com/LodLive/LodView
4. CIDOC CRM Special Interest Group: Definition of the cidoc conceptual reference model, version 7.1.3. https://cidoc-crm.org/Version/version-7.1.3 (February 2024). Accessed April 2025
5. Debruyne, C., Munnelly, G., Kilgallon, L., O'Sullivan, D., Crooks, P.: Creating a knowledge graph for Ireland's lost history: Knowledge engineering and curation in the beyond 2022 project. ACM J. Comput. Cult. Heritage **15**(2), 1–25 (2022)
6. Heibi, I., Peroni, S., Shotton, D.: Oscar: a customisable tool for free-text search over sparql endpoints. In: International Workshop on Semantic, Analytics, Visualisation, pp. 121–137. Springer (2017)
7. Hyvönen, E., et al.: Biographysampo - publishing and enriching biographies on the semantic web for digital humanities research. In: European Semantic Web Conference, pp. 574–589. Springer (2019)
8. Hyvönen, E., et al.: Museumfinland—finnish museums on the semantic web. J. Web Semant. **3**(2), 224–241 (2005). https://doi.org/10.1016/j.websem.2005.05.008. https://www.sciencedirect.com/science/article/pii/S157082680500017X, selcted Papers from the International Semantic Web Conference, 2004
9. Ikkala, E., Hyvönen, E., Rantala, H., Koho, M.: Sampo-UI: a full stack Javascript framework for developing semantic portal user interfaces. Semantic Web **13**(1), 69–84 (2022)
10. Isaac, A., Haslhofer, B.: Europeana linked open data - data.europeana.eu. Semantic Web **4**, 291–297 (2013). https://api.semanticscholar.org/CorpusID:27103052
11. Koho, M., Rantala, H., Hyvönen, E.: Digital humanities and military history: analyzing casualties of the warsampo knowledge graph. In: Proceedings of the 6th Digital Humanities in the Nordic and Baltic Countries Conference (DHNB 2022). CEUR-WS. org (2022)
12. Krabina, B.: Building a knowledge graph for the history of Vienna with semantic Mediawiki. J. Web Semantics **76**, 100771 (2023)
13. Leskinen, P., Rantala, H., Hyvönen, E.: Analyzing the lives of finnish academic people 1640–1899 in nordic and baltic countries: Academysampo data service and portal. In: Proceedings of the 6th Digital Humanities in the Nordic and Baltic Countries Conference (DHNB 2022). CEUR-WS. org (2022)
14. McNamara, C., Hederman, L., O'Sullivan, D.: Mind the gap: LLMs, competency questions, and the non-technical user in the humanities domain. In: First International Workshop on Users and Knowledge Graphs@ SEMANTiCS 2025 (2025)
15. metaphacts: Ontodia: data diagramming library (2017). https://github.com/metaphacts/ontodia, gitHub repository, archived September 26, 2024
16. Ohlmeyer, J.: Voices: Life and death, war and peace, c.1550–c.1700: Voices of women in early modern Ireland. Research project, Trinity College Dublin (2023). https://voicesproject.ie/, eRC Advanced Grant 101097003

17. Oldman, D., Tanase, D.: Reshaping the knowledge graph by connecting researchers, data and practices in researchspace. In: The Semantic Web – ISWC 2018: 17th International Semantic Web Conference, Monterey, CA, USA, October 8–12, 2018, Proceedings, Part II, pp. 325–340. Springer-Verlag, Berlin, Heidelberg (2018). https://doi.org/10.1007/978-3-030-00668-6_20
18. Randles, A., McKenna, L., Kilgallon, L., Yaman, B., Crooks, P., O'Sullivan, D.: Evaluating the knowledge graph editor of the virtual record treasury of ireland. In: Fu, B., Lambrix, P., Li, H., Nunes, S., Pesquita, C. (eds.) Proceedings of the 9th International Workshop on the Visualization and Interaction for Ontologies, Linked Data and Knowledge Graphs co-located with the 23rd International Semantic Web Conference (ISWC 2024), Baltimore, USA, November 12, 2024. CEUR Workshop Proceedings, vol. 3773, pp. 62–77. CEUR-WS.org (2024). https://ceur-ws.org/Vol-3773/paper5.pdf
19. Randles, A., McKenna, L., Kilgallon, L., Yaman, B., Crooks, P., O'Sullivan, D.: The knowledge graph explorer for the virtual record treasury of ireland. In: Proceedings of VOILA 2024: The 9th International Workshop on the Visualization and Interaction for Ontologies, Linked Data and Knowledge Graphs co-located with the 23rd International Semantic Web Conference (ISWC 2024). CEUR-WS. org (2024)
20. Vrandečić, D., Krötzsch, M.: Wikidata: a free collaborative knowledgebase. Commun. ACM **57**(10), 78–85 (2014)
21. Yaman, B., McKenna, L., Randles, A., Kilgallon, L., Crooks, P., O'Sullivan, D.: Digital prosopographical information in the virtual record treasury of Ireland's knowledge graph for irish history. In: Proceedings of the 1st International Workshop of Semantic Digital Humanities (SemDH) Co-Located with the 21st Extended Semantic Web Conference (2024)

A Graph-Based RAG System for Enhanced Information Gathering in Local Newsrooms

Reshmi Pillai[1,2]([✉]) [iD], Antske Fokkens[2] [iD], and Wouter van Atteveldt[2] [iD]

[1] University of Amsterdam, Amsterdam, The Netherlands
r.pillai@uva.nl
[2] Vrije University Amsterdam, Amsterdam, The Netherlands

Abstract. Newsrooms handle a plethora of incoming stories. Gathering relevant background information in a time-efficient way is a core challenge faced by journalists. We propose a knowledge graph-based RAG system to suggest background information relevant to an incoming news story. We use Large Language Models to construct a knowledge graph leveraging multiple web-based information sources, from which the background information is retrieved upon query. Using human annotation and specifically designed metrics, we evaluate and compare the retrieved information using different LLMs. We further report the evaluation done by the target users of the system (journalists in a local newsroom). Thus, the research provides implementation and evaluation of a novel application of knowledge-graph-based RAG for the retrieval of background information for a news story.

Keywords: Retrieval Augmented Generation · Knowledge Graphs · Newsrooms · Information gathering

1 Introduction

While navigating the paradigm shift induced by artificial intelligence, newsrooms with a local focus often find themselves at a disadvantage, due to knowledge and resource constraints. At the same time, local-focused journalism is essential for the public sphere, as audience interests or concerns often focus on what is happening in their local communities and not solely on global or national issues [15] [26]. Even more importantly, local media have the task of connecting people in a geographical context with a sense of belonging, to strengthen social cohesion in an area [12] and its continued existence is crucial to ensure the diversity of voices in the media landscape of any country.

Following the introduction of ChatGPT by OpenAI in November 2022, there has been a surge in interest in applying generative AI techniques to knowledge-centric domains. In the discussions about generative AI in newsrooms, its potential in the throughput (report-writing) stage has been vastly explored. In comparison, the possible contributions of generative AI in the input stage, by gathering and processing incoming information and building knowledge, is rather

© The Author(s), under exclusive license to Springer Nature Switzerland AG 2026
M. Acosta et al. (Eds.): ESWC 2026, LNCS 16550, pp. 470–487, 2026.
https://doi.org/10.1007/978-3-032-25159-6_25

overlooked. However, looking more closely at the journalistic process, information gathering is at its center [1]. Stories in other news media and earlier reports often become an important source of information for journalists [11]. Algorithmic tools are identified to have the potential to support information gathering, but there is a lack of appropriate techniques to fulfill these functions effectively [9].

In the context of local newsrooms, the key question is to find background information about locally relevant entities involved in the incoming news story. This information is significant in two contexts: 1) The story itself is about a global or national event, but the journalist seeks a connection to a person or organization within the locality to make the news more relatable to the audience, e.g., a newly elected Prime Minister of the country had previously studied in a local school 2) The story is of local relevance, and the journalist seeks connection to other events or persons in the locality, e.g., A local school is currently on the verge of closing due to the decrease in the number of students, but had won a district sports event five years earlier.

To find connections such as these, traditional information gathering tools, such as web search, are not sufficient. Retrieval Augmented Generation (RAG) systems, which combine Large Language Models (LLM) with retrieval from a reference text-based database, have the possibility of addressing the unmet needs of supplying background information to journalists, but it has not been explored so far in applied research. The evaluation of RAG systems is still an upcoming research area, and the existing performance metrics do not reflect the specific requirements of the task of retrieval of news background information. Thus, we identify clear research gaps in the problem area, the solution of graph-based RAG, and performance metrics.

Given this background, we investigate the following research question:

RQ: *How can we design and implement a knowledge base and retrieval system to provide relevant background information about a news story?*

This work was done in collaboration with Rijnmond,[1] a regional news media broadcaster in the Netherlands, to ensure the real-life relevance of the problem of background information generation and the applicability of the proposed solution.

We built a graph structure of locally relevant entities and their relationships, using a dataset of past news reports extracted from the Rijnmond website, national news outlets in the Netherlands, and Wikipedia pages of the 14 municipalities that fall under the focal area of this news organization. As a next step, we implemented a RAG system that retrieves relevant information for an incoming news story, based on this graph knowledge base. We defined and implemented the metrics **context relevance**, **entity coverage**, and **novelty**, to measure the performance of the RAG in the news background information task. Performance metrics of multiple implementations of the RAG system using different LLMs were reported and analyzed. We also evaluated the LLMs for accuracy of entities and relationships extracted from the text sources, compared with human-coded annotations. Finally, we involved the end-users of the system (journalists in a

---

[1] https://www.rijnmond.nl/nieuws.

local newsroom) and gathered their evaluation on relevance, serendipity, usefulness, diversity, and factual correctness of the retrieved background information.

The main contribution of this research is the design, implementation, and evaluation of a retrieval-generation pipeline using the knowledge base to provide relevant background information for journalists about news stories. Our work was developed and evaluated in collaboration with a regional newsroom with the aim of supporting journalists in their daily background research. Our contribution lies in demonstrating how established knowledge-graph construction methods and retrieval-augmented generation can be adapted, integrated, and evaluated in a real professional workflow. The resulting application reveals how semantic technologies can address concrete information-seeking challenges and deliver measurable benefits in a domain where timely access to contextual knowledge directly affects the quality of public information.

2 Related Work

2.1 Artificial Intelligence (AI) and Journalism

There has been a great deal of interest in applying AI-based techniques in newsroom processes, especially after the introduction of ChatGPT by OpenAI in november 2022. The approach to AI applications in the newsroom has been dichotomical: on one end, there is enthusiasm in embracing these techniques; on the other end, there is caution and fear about the risks of AI adoption [20].

The impact of AI adoption or the lack thereof in newsrooms has been studied in the context of media organizations focusing on various geopolitical regions (South Africa, Pakistan, Europe) [17,18,24] and scopes (national/local) [3]. Technology-oriented research work often describes novel or adopted methods relevant for journalistic tasks such as audio transcriptions, headline generation or routine event report writing [5]. AI-based techniques have also been used in the news gathering stage to identify 'newsworthy' events and leads from heterogeneous sources like press releases and social media. Surpassing the hype around AI and related technologies, and driven by the idea that AI systems manifest values their designers attempt to build into them [2], discussions are now oriented towards what AI should do rather than what AI can do in news processes [14].

2.2 Structured Knowledge in Newsrooms

There have been attempts to incorporate knowledge representations such as Resource Description Frameworks (RDF) [16] into the news making process, thus enhancing search and retrieval. Using LSTM, RDF triples were used to generate automated news reports on football matches [7]. There have also been knowledge graph representations of events based on Wikipedia [22] and news data [29]. In another interesting application area, journalistic knowledge graphs have been enhanced with the possibility of extracting news angles [28] and local connections of events [27]. While these works propose ontologies and pipelines to represent and extract knowledge for newsrooms, they base their knowledge building on encyclopedic resources. In practice, the entries for many locally relevant entities (e.g., local politicians, locations, regional organizations) in such data sources are often non-existent or have sparse information. In addition, they often lack quantitative evaluation and analysis of the performance of such proposed systems. Our work addresses these gaps by leveraging data in local news reports and implementing performance metrics fine-tuned for the background information retrieval task.

2.3 Synergy of LLMs and Knowledge Graphs

Knowledge graphs and LLMs are shown to have complementary strengths and weaknesses. Even the most advanced LLMs are challenged in the representation of factual knowledge [25]. The structured and decisive nature of knowledge graphs can address hallucination, a known weakness of LLMs. On the other hand, LLMs are capable of efficiently extracting knowledge from large amounts of data, thus easing the construction process of knowledge graph. Synergized solutions unifying LLMs and knowledge graphs can mutually enhance each other through knowledge representations and explainable reasoning [19]. We propose using this synergy in a RAG pipeline.

2.4 RAG Evaluation

The evaluation of RAG systems is especially tricky. An approach commonly considered is to evaluate the RAG system based on question-answering. However, this may not be representative of the system's performance if it is used for a different purpose. Also, such evaluation strategies are dependent on the availability of human-annotated question-answer pairs, which may not be the case in domain-specific tasks. Driven by the motivation for reference-free metrics, RAGAS [8] put forward three aspects: faithfulness of the retrieved answer to the context, relevance of the answer with respect to the given question, and relevance of the retrieved context to the given question. ARES [23], another RAG evaluation framework, is based on these three metrics, but makes use of the combination of a small human-annotated set of data points and the recently proposed statistical metric of prediction-powered inference (PPI), to provide tightly bound confidence intervals for the system scores. In eight different datasets, ARES is shown to outperform RAGAs and a GPT-3.5 judge on faithfulness and relevance metrics. Combining RAGs with LLMs has not only been shown to enhance functionalities but also result in undesirable features. Retrieval Augmented Benchmark (RGB) [6] is proposed as a news corpus for RAG evaluation, based on noise robustness, negative rejection, information integration and counterfactual robustness. Through experiments with 5 different models in 2 languages (English and Chinese), these features are found to be challenging for RAGs in LLMs.

These existing RAG evaluation metrics do not directly reflect the performance of a RAG system to retrieve background information for a news story. Hence, while motivated by the design and implementation of these metrics, we introduce new metrics for the evaluation of the RAG system in our work.

3   Methodology

3.1   Development of the Usecase

We reached the problem definition through a series of sprints involving employees of the regional news media collaborator, Rijnmond, following the design thinking framework [4,21]. The motivation behind this paradigm choice was to follow a bottom-up approach and bring on the domain expertise of the journalists while defining the problem of background information retrieval and designing the knowledge base and retrieval architecture. We used the three phases, *empathize*, *define* and *ideate*, of the design thinking process to closely interact with the people working in the newsroom, management and technology profiles in the news organization. Through these sessions, we identified, mapped, and prioritized challenges faced in the newsmaking process, facilitated clear definitions of these problems, and extracted storyboards depicting the journalists' vision of the desirable solution. This bridging between the research problem and the domain expertise motivated our research through the real-life relevance of the problem.

3.2   Solution Pipeline

As shown in Fig. 1, the core component of the solution pipeline is a graph representation of the locally relevant entities and their relationships. We use a collection of past news reports and Wikipedia pages related to the Rijnmond-Rotterdam area to create the graph representation. The envisioned user of the RAG system is a journalist who queries the system for background information about an incoming local news story. An LLM renders the background information retrieved from the graph and presents it to the journalist, who can decide to further persuade this information to write a news report on it. We additionally use five baselines for comparison: background information retrieval methods based on TF-IDF, BERT-based embeddings using sentence-transformers/LaBSE and LLM-only generation using GPT-3.5, GPT-4o and Mistral-7B.

3.3   Data Collection

To build the graph that can serve as the knowledge base of the RAG, we required information about the region of the local news media, which in the context of this research is Rotterdam and neighboring towns in the Netherlands. Using newsplease [10], an open-source Python package, we collected 894 randomly chosen news reports from the official online news portal of Rijnmond, the regional news broadcaster. The oldest of these news reports is from 20th July 2022 and the latest is from 16th April 2024. The median length of the articles is 437.0 with minimum length of 152 and maximum of 7876 characters.

In addition, using the same python package, we collected news stories from the national media outlets in the Netherlands published during the same time window. We filtered the collected news reports for the occurrence of Rotterdam and related keywords to ensure relevance for local information. Including this,

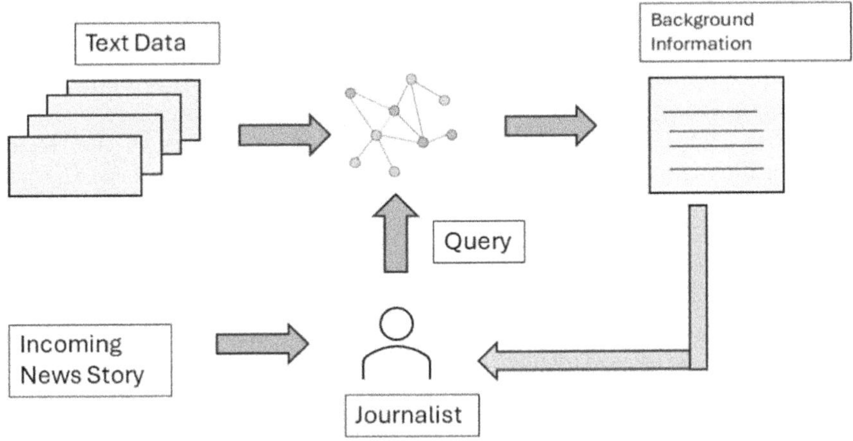

Fig. 1. Framework to retrieve background information for an incoming news story

there were 1200 news reports in the dataset. We also collected Wikipedia pages about Rotterdam, the other 14 municipalities that comprise the Rijnmond area in the Netherlands, and 11 locally relevant entities.

3.4   Construction of the Graph and Information Retrieval

We used LangChain document loaders to access and process text data from news reports and Wikipedia pages. These documents were merged and split into tokens using LangChain libraries. The chunk size was kept as 512 and chunk overlap as 24. We used three different LLMs (GPT-3.5-turbo, GPT-4o, Mistral 7B)[2] to create graph representations of the text data and evaluated these different implementations according to the metrics outlines in Sect. 3.5.

The graph follows an explicit schema for nodes, implemented via Neo4j node labels, which constrain the allowed entity types. In contrast, relation types are kept flexible and based on verbs to accommodate the heterogeneous sources. Semantic consistency is ensured through constrained entity typing and normalization of relation labels. To reduce duplication in the knowledge graph, we perform entity resolution in different stages. First, we normalize entity labels by lowercasing, stripping punctuation, and removing generic modifiers (e.g., *gemeente* (municipality), *provincie* (provence)). Then, we compute the character-level string similarity using Jaro–Winkler. Entity pairs whose similarity exceeds a threshold are treated as coreferent and merged using a union-find clustering scheme, after which all edges are rewired to a canonical node per cluster. This procedure effectively collapses spelling variations and surface-form variants (e.g., *Mark Rutte, M. Rutte, minister-president Rutte*) into single entities while preserving graph structure. We initially considered LLMs to be used for entity

---

[2] At the time of experimentation, GPT-3.5-turbo was a supported model, and although now deprecated, we report results for reproducibility and historical comparison.

matching; we opted for string similarity for better reproducibility and transparency in the pipeline. This allows us to attribute improvements specifically to the Knowledge Graph extraction step rather than conflating multiple LLM-driven components.

The news story from the journalist is presented as a query to the retriever, which identifies the entities present in the news story and retrieves the neighboring nodes of each entity node and the corresponding relationships. This is presented as the context to the LLM component which generates a natural language response to the query, summarizing the information retrieved from the graph. The retriever is not implemented as an LLM; instead, it relies on structured queries over the constructed graph. Only the relevant subgraph retrieved for a given query is provided to the generator, rather than the full graph.

Additionally, we implemented five baselines for extended comparisons: 1) a naive TF-IDF based model: using cosine similarity to compare the vectors of the incoming story and stored documents, this baseline method retrieves contextually relevant text spans as background information for the given story. 2) a BERT-based embeddings model using sentence-transformers/LaBSE. All documents in the corpus are encoded and indexed using FAISS for efficient similarity search. For each incoming news item, we encode the input query and retrieve the top-k most similar texts using cosine similarity. We then pass these retrieved documents into the generation component using the same LLM prompting setup as our graph-based RAG. 3) LLM-only Generation (No Retrieval) using GPT-3.5-turbo, GPT-4o, Mistral 7B. These baselines removed the retrieval stage and The LLM is prompted directly with the news item and asked to generate background information purely from parametric memory. This provides a naive comparison illustrating how much the RAG and baseline retrieval methods meaningfully contribute.

3.5   Evaluation

We perform a three-part, extensive evaluation of the framework: 1) analysis of the graph representation of the text data 2) assessment of the retrieved background information 3) qualitative evaluation by journalists.

1. Evaluation of the graph constructed from text data We evaluate the performance of LLMs in extracting the graph representation of entities and relationships from text data. For entities, we chose a random subset of 200 news reports and 4 Wikipedia pages and extracted entities from these using the flair/ner-dutch-large model. This model was chosen for its near-human performance (F1 score 95%) in entity extraction task.[3] The entities extracted using this model were further verified by a human coder. Considering this as the gold-standard, we report the accuracy of the LLMs in extracting the entities from the same text.

For the evaluation of relationship extraction from text, we annotated, with the help of a human coder with native-level proficiency in Dutch, relationships

---

[3] https://dataloop.ai/library/model/flair_ner-dutch-large/.

between 105 entity pairs extracted from 30 randomly chosen news text passages. We compared the relationships extracted using the LLMs with these human annotations and report the accuracy scores.

2. Evaluation of the retrieved background information

Secondly, we evaluated the performance of the framework in the task of retrieving background information. As mentioned in Sect. 2.4, existing RAG evaluation metrics do not reflect the performance of a system specifically for this purpose. We implement three performance metrics: **context faithfulness** (similar to RAGAs [8]), **entity coverage** and **novelty** all defined below.

Context Faithfulness: Since the RAG system is used to provide background information to be used to write news reports, it is especially important to ensure that it is factually correct and devoid of hallucinations. Given IS as the collection of information statements generated by the RAG for a given news story NS, using an LLM judge (LLaMA-3.1-70B-instruct) we evaluate whether each of the information statements can be verified with the context $C(NS)$. The LLM judge was prompted to identify the verifiability with a short reasoning. A human coder manually validated this judgment, thus supporting validity while considerably reducing the cognitive/annotation effort. With VS being the subset of IS which can be verified with the context, the metric context faithfulness is defined as:
$$\text{faithfulness} = \frac{|VS|}{|IS|}$$

Coverage: A news story will have several entities present in it such as persons, locations etc. Ideally, there should be background information provided by as many entities mentioned in the news story as possible. We define coverage as the fraction of the entities mentioned in the news story which are present in the background information.

Given the news story NS and background information statements IS with $E(NS)$ being the set of entities mentioned in the news story, and $E(IS)$ being the set of entities mentioned in the information statements, we define entity coverage as:
$$\text{coverage} = \frac{|E(NS) \cap E(IS)|}{|E(NS)|}$$

Novelty: In the coverage metric, we looked at the extent to which the entities mentioned in the original news story are covered in the background information generated by the system. At the same time, the system should also add to the information already present in the news story. The background information provided should not be merely a paraphrasing of the news story; it should contain new entities connected to the ones in the original news story so that the journalist can potentially find leads to pursue and develop the report.

Given the news story NS and background information statements IS with $E(NS)$ being the set of entities mentioned in the news story, and $E(IS)$ being the set of entities mentioned in the information statements, we define novelty as:
$$\text{novelty} = \frac{|E(IS) - E(NS)|}{E(IS)}$$

3. Qualitative evaluation by journalists In addition to these quantitative metrics, we evaluated the quality of the background information retrieved for given news stories with journalists of a regional newsroom. Four journalists from the regional news broadcaster Rijnmond participated in this evaluation. We provided the annotator with 125 background information inputs (2 or 3 for each of 45 fictitious news stories). The journalists evaluated the generated background information in scales of 1 (minimum score) to 5 (maximum score), for relevance, serendipity, usefulness and diversity. The key annotation instructions given to the journalists were as follows:

Relevance : Is the background information connected to the topic or content of the news story?

Serendipity : Is the background information new and serendipitous to the journalist?

Usefulness : Is the background information useful for the journalist in the context of the news story?

Diversity : Does the background information include perspectives that are typically not covered?

This evaluation was done in a workshop physically attended by all four journalists and the first author.

Though provenance was preserved by storing explicit links from retrieved graph nodes back to their originating source documents (including document identifiers), this was not included in the evaluation. In an envisioned usage, this information should be accessible to the journalists allowing them to inspect, verify and contextualize the background information before use.

4   Results

4.1   Knowledge Representation of Locally Relevant Entities

We constructed the graph representation of entities extracted from local news reports and Wikipedia pages, using multiple LLMs. For demonstration, we include here the visualization of the graph created by OpenAI model GPT-4o, using the graph visualization library, yFiles, in Fig. 2). We observed entity nodes with several and sparse connections in the graph.

Two of the nodes with most connections are marked with red circles. One such node represents the football club of Feyenoord. The other node marked with the red circle represented the city of Rotterdam. Given the context of the dataset, this aligns with intuitive expectations. Another example is a node representing Dutch national Volleyball player *Twan Wiltenburg*, showing the information present in the news reports and Wikipedia data about the sports organizations he has associated with. It should be noted that the Volleyball organizations shown in this figure are not located in the Rotterdam area or nearby towns. *ZAKSA Kędzierzyn Koźle* is a Volleyball club in Poland, *Narbonne Volley* is in France and *Orion Doetinchem* is in the east of the Netherlands. However, they are still present in the graph since *Twan Wiltenburg* is originally from Zuidplas (a municipality near to Rotterdam); and the Wikipedia page about this area is included in our dataset for graph creation. If an incoming news story mentions the French volleyball club of Narbonne Valley, our retrieval system can find its connection to the Dutch volleyball player.

We observed that the graph represents a comprehensive network of interconnected entities and relationships. It comprises of 5,510 nodes and using analytical metrics, we noted the structural characteristics:

Degree Distribution: Most nodes (5,497) fall within the in-degree range of 0–49, while only a handful of nodes have higher in-degrees, including key hubs like "Rotterdam" (419), "Feyenoord" (209), and "Rijnmond" (120). The average in-degree is 2.81, suggesting a relatively sparse graph, while a few highly connected nodes dominate interactions.

Central Nodes: Nodes with the highest in-degree and centrality metrics, such as "Rotterdam" and "Feyenoord," act as pivotal entities, connecting various topics and communities. "Rotterdam" leads in both betweenness centrality (0.1123) and eigenvector centrality (0.3958).

Clustering and Community Structure: With an average clustering coefficient of 0.5865, the graph exhibits moderate local clustering, indicating that neighbors of a node are likely to be interconnected. The Louvain community detection algorithm identifies 34 communities, ranging from the largest (563 nodes) to the smallest (2 nodes), highlighting both broad thematic clusters and niche groups.

Sample Communities: - Large communities, such as Community 0 (563 nodes), include entities like "Feyenoord" and "Fortuna Sittard," representing major topics or regions. - Smaller communities, like Community 4 (2 nodes), focus on highly specific or isolated subjects.

Table 1. Accuracy of extraction of entities and relationships from text sources.

Model	entities	relationships
GPT-3.5-Turbo	0.912	0.873
GPT-4o	0.952	0.925
Mistral-7B	0.901	0.871

The results of the evaluation of graph representation created by the three LLMs using annotations of entity and relationships are given in Table 1. We find GPT-4o to have a better performance in terms of accuracy but we do not find the difference to be statistically significant for entity extraction, using Pairwise Wilcoxon test. For relationship extraction, GPT-4o's performance is significantly different ($p < 0.05$) from the other two models.

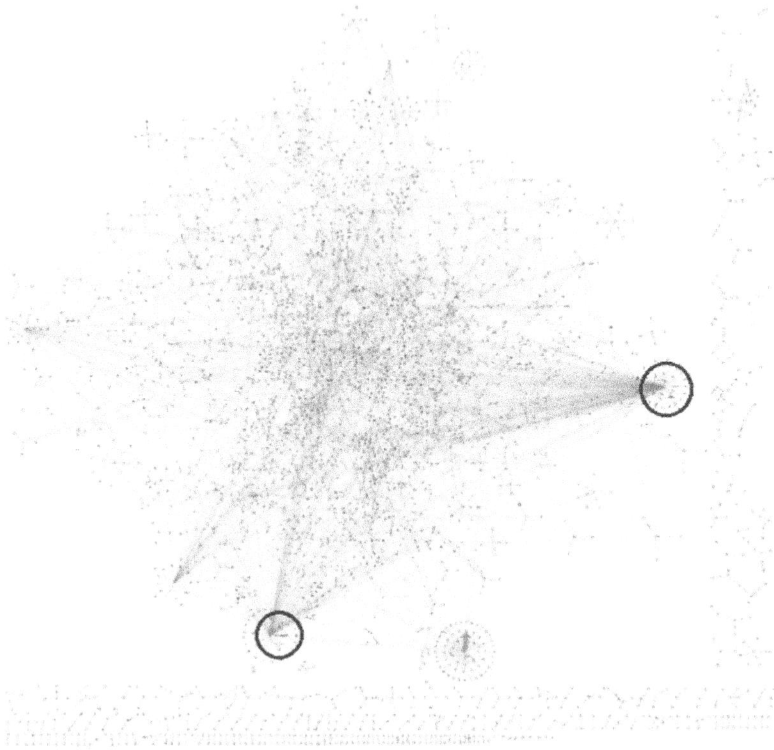

Fig. 2. Visualization of the entity relationship graph created from the Wikipedia pages and news reports, using GPT-4o

4.2 Performance of Background Information Retrieval

In this section, we present the performance of the RAG systems with the three different LLMs, GPT-3.5-turbo-0125, GPT-4o and Mistral 7B. We wanted to find out whether the additional information from the past news reports can add to the performance of the background information generation task. For each LLM, we measure the performance 1) only the Wikipedia pages as data (wi) and 2) past news reports and Wikipedia pages (wi+ne). The results are presented in Table 2.

We observe that the RAG systems with graph created from both past news reports and Wikipedia pages perform better in terms of entity coverage and novelty than those created from only Wikipedia pages. This shows that the addition of data sources indeed improves the quality of the generated background information, in alignment with our intuitive expectation. We also see GPT-4o and Mistral-7B perform better in terms of faithfulness though within these models, the data variants had similar faithfulness scores.

We also see the relatively low performance with respect to entity coverage. This should be due to the relatively smaller dataset we built our knowledge graph

Table 2. Performance of RAG systems with different LLMs and data sources.

Model	Faithfulness	Entity Coverage	Entity Novelty
GPT-3.5-Turbo(wi)	0.732	0.525	0.493
GPT-3.5-Turbo(wi+ne)	0.745	0.671	0.542
GPT-4o(wi)	0.833	0.745	0.512
GPT-4o(wi+ne)	0.834	0.793	0.581
Mistral 7B(wi)	0.830	0.512	0.504
Mistral 7B(wi+ne)	0.831	0.645	0.573
tf-idf(wi+ne)		0.335	0.287
bert(wi+ne)		0.443	0.327
llm-only-GPT3.5(wi+ne)		0.423	0.422
llm-only-GPT4o(wi+ne)		0.467	0.434
llm-only-Mistral-7B(wi+ne)		0.385	0.402

on. But we also see that by including news data, the entity coverage is improved for all models; this indicates potential improvement of the RAG systems by adding further data sources about locally relevant entities.

Using pairwise Wilcoxon test, we find the superior performance of LLM-graph RAG models in comparison with all baselines as statistically significant for both entity coverage and novelty ($p < 0.01$). We also find the better performance of the models with both news and Wikipedia data in comparison with only using the Wikipedia data is also statistically significant.

4.3   Qualitative Evaluation by Journalists

Four journalists working in the Rijnmond local broadcaster newsroom participated in the qualitative evaluation of the retrieved background information. They evaluated, on a scale of 1 to 5, the relevance, serendipity, usefulness and diversity factors of the retrieved background information snippet. Additionally, they provided comments on whether the retrieved background information was factually correct (with options completely true/mostly true/partially true/completely false/I don't know). They also gave overall comments on the possible adoption of such a system in their workflow. The key results from this evaluation are given in Table 3.

There was no background information which was found to be completely false or mostly false by journalists. Averaging across the coders, for 32.4% of the background information, the journalists could not comment on the factual accuracy of the background information ('I don't know' option). Among the rest, 41.5% was completely true and 26.1% was mostly true. Although the qualitative evaluation was conducted on a relatively small subset, this indicates that the background information retrieved by the graph-RAG system is found to be (largely) factually

Table 3. Average scores given by journalists for the background information (in a scale of 1 to 5)

Metric	Coder1	Coder2	Coder3	Coder4
Relevance	4.2	3.8	4.1	4.3
Serendipity	3.5	4.5	4.3	3.7
Usefulness	3.6	3.7	4.5	4.2
Diversity	3.2	3.4	3.1	3.2

correct by the journalists. An example of mostly true background information is given below (translated from Dutch to English).

News: Vogelklas Karel Schot saves rare birds near Europoort Rotterdam
Background information: Vogelklas Karel Schot, located in Rotterdam, is a wildlife sanctuary that cares for injured and orphaned birds The journalist pointed out that while the organization Vogelklas Karel Schot mainly cares for birds and hedgehogs, it could not be termed as a care center for wild animals and hence not a wildlife sanctuary.

The journalists also added comments on the applicability of such a system to their workflow. One journalist particularly appreciated the background information providing contact persons mentioned in earlier related news reports, which might give ideas for a portrait news story with the mentioned person. Another journalist mentioned that this background information was "useful for a first quick check for previous news stories and factual hooks for the story to continue on". However, he also mentioned that he would like to be 'surprised a little more often' with serendipitous background information.

5   Discussion

The results demonstrate the applicability of a graph structure of locally relevant entities and their interrelationships, and the retrieval of the background information for an incoming news story from this graph. Our work aligns with recent graph-based RAG approaches in using structured graphs to support retrieval. The key distinction is our focus on journalistic background generation and the use of LLM-based graph construction from news text, rather than relying on pre-existing knowledge graphs. This supports the retrieval of locally relevant information and the preservation of the journalistic agency by choosing the data sources. We also see that adding information sources improves the performance of the RAG system.

Revisiting the use cases we had envisioned in Sect. 1, we see concrete and relevant examples from the knowledge graph. We note the presence of international entities such as football clubs in the graph (e.g., the example of Narbonne Valley in Sect. 4.1), with local connections (e.g., A volleyball player from Zuidas having played for Narbonne Valley in the past). For the second use case, we look at the example of the node representing the person Nataniel 'Tati' Gomes. This hairdresser is famous as the 'inventor' of the 'kapsalon', a popular food item in The Netherlands. The graph also shows him as a member of the Cape Verdeans migrant community. If given a story about this migrant community, the RAG system will thus fetch Nataniel Gomes as background information, presenting an interesting connection to a local news story.

A key challenge in our work was the evaluation of the system in the absence of human-annotated ground truth. Due to the distinct nature of the task, the quality of the background information generated could not be evaluated using an external dataset. Hence, we designed and implemented evaluation metrics specifically for this task.

While envisioning the usage of this system, we were driven by a human-in-the-loop approach, maintaining the agency with the journalist in the news information gathering process. We envision this system as a useful tool for, and not a replacement, of the human journalist. The journalist is responsible for judging whether the background information generated by the system is worth pursuing further and verifying its accuracy. In our evaluation setup, the journalists reported positively on the factual correctness of the retrieved information. Nevertheless, in the adoption of such a system into journalistic workflow, the journalist should always verify the background information before using it in news reports, and provenance information or sources should be provided to facilitate this.

We reflect here on the ethical aspects of our work. The system relies on previous news reports and Wikipedia pages as information sources, and the inaccuracies or bias present in these sources might also be reflected by the system. The transparency of information sources is a core journalistic value. When deploying this framework in a journalistic organization, journalists should clearly understand that the information suggested is machine-generated.

The applicability of this system beyond local newsrooms is another point to discuss. Based on our experience, the system components that are most specific to local journalism are the definition of locality (e.g., geographic scope, community issues) and the curation of locally relevant knowledge sources. In contrast, the core architecture - LLM-based extraction, graph construction, and retrieval - is domain-agnostic and could be transferred to other newsroom settings (e.g., national or thematic desks) by adapting the underlying knowledge sources and relevance criteria.

Limitations We identify three key limitations of our work:

Data sources: the purpose of this research is to demonstrate graph RAG as a potential technique to generate background information. Hence, our data sources were not exhaustive; in real-life local newsrooms, there are other important data sources such as social media, press releases, websites of emergency services which are regularly monitored by journalists. We report statistically significant better performance of all considered models with a higher number of data sources, and more sources can be incorporated in future work.

Scalability: The scalability of the system is to be explored more in the context of local newsrooms. While the TF-IDF approach is computationally efficient and scales linearly with the number of documents, LLM-based models are more resource-intensive, with complexity increasing significantly for tasks like embedding generation and similarity computation, especially for large document collections. We would like to further explore the possibility of language models much smaller in resource requirements and compare their performance according to the metrics considered in this work. As indicative latency observations: at 500 articles (approximately 2000 nodes), average query latency was around 400 ms. For the full corpus of 1,200 articles (5,510 nodes), end-to-end query latency remained below 700 ms. These figures are only indicative observations rather than a formal scalability evaluation.

Scope: We defined the scope of this research in the context of local newsrooms. The application of RAG-based systems is most successful in areas that require domain-specific knowledge. Since the audience interest for local news media typically revolves around locally relevant entities such as persons, locations, and organizations, it was a sensible step to develop RAG systems using information sources which contain relevant information. Such an approach will be much more challenging in the case of newsrooms with a broader scope. Also, making use of multi-hop connections in graph representations [13] is a promising direction that we have not explored in our current study.

Bias, Temporality: News is constantly evolving. We demonstrate the applicability of background information retrieval from text sources, but we only consider static data sources. To enhance usability in a newsroom, it would be helpful to incorporate social media sources and other highly dynamic information sources in the input data. At the same time, the bias present in the input data sources could propagate to the graph representation and hence the background information. It further emphasizes the role of the journalist: we reiterate the principle of human-in-the-loop in the design of this system. The graph-RAG presents potentially relevant information to journalists; and complements (not replaces) their own expertise and ethical guidelines through its ability to retrieve information from large data sources. Through the factual correctness inputs from the journalists, we observe the lack of hallucinations' or completely false background information. However, this needs further exploration and potentially evaluation in a larger sample to validate this observation.

Evaluation: While we assess whether the resulting natural-language background information is relevant and useful through quantitative and user-centric measures as described in Sects. 4.2 and 4.3, we did limited experiments to evaluate the extraction of entities and relations (Sect. 4.1). Our work focuses on and reflects the practical utility of the extracted entities and relations for the intended journalistic task, and future work could potentially evaluate the entity-relationship extraction step more closely. Also, we note that we do not separate the evaluation of the retriever from the evaluation of the generator in RAG. In our setting, the primary goal is to assess end-to-end usefulness for journalistic background generation. We therefore treat retrieval and generation as a coupled pipeline and note component-level evaluation as future work.

6   Conclusion

Journalists in local newsrooms try to find locally relevant background information for forthcoming news stories. We implemented a graph-based retrieval augmented generation system for providing natural language suggestions for background information to journalists. We used real-world data of 25 Wikipedia pages and 1200 past news reports to construct a graph and implement a retrieval system using different LLMs to extract background information from this knowledge base. Since existing RAG evaluation metrics do not address the specific requirements of a news information retrieval task, we introduced three performance metrics as a novel contribution. We evaluated the combinations of 3 different LLMs and 2 datasets on these performance metrics. We evaluated the background information with human-annotated dataset of relationship between entity pairs and qualitative evaluation with four local newsroom journalists.

We identify clear scope for future work. Information is dynamic, especially in the context of newsrooms. The proposed system at the moment draws its data from existing news reports and snapshots of Wikipedia pages and can be developed to incorporate dynamic sources of locally relevant data, such as social media streams and updates in emergency service websites (e.g., 112.nl in the Netherlands). We also intend to explore less resource-intensive language models to construct the knowledge representations and extract background information. As suggested by the journalists in the evaluation session, it will be a helpful extension to provide the source document together with the background information snippet to enhance the context for the journalists.

This work illustrates the value of combining knowledge graph representations with RAG in an applied setting that carries significant societal importance. Journalism, and particularly local journalism, operates under strong time constraints while being expected to provide accurate, contextual, and well-connected information. By developing and evaluating our system in close collaboration with an active newsroom, we show how widely studied semantic technologies can be operationalized in ways that meaningfully support practitioners rather than exist only at the conceptual or experimental level.

Our experience highlights several insights that emerge only when such technologies are put into use. Constructing a graph from heterogeneous sources such as historical reports and Wikipedia reveals practical challenges related to entity consistency that are rarely visible in controlled research settings. Likewise, aligning retrieval signals with journalists' expectations exposes important considerations around evaluation metrics (e.g., faithfulness, novelty, and local relevance) that do not appear in generic IR benchmarks. Overall, this work demonstrates that established graph-based RAG approaches, when thoughtfully adapted, can provide real and positive impact in journalistic information ecosystem.

Supplementary Materials The code used in the experimental setup is provided at https://github.com/rgopala/news-rag and is publicly available. Since the data consists of copyrighted news articles, access to the data can be discussed through direct requests to the corresponding author.

Generative AI Statement We used AI assistance in writing this paper through the Writefull plugin in Overleaf to support language refinement, such as improving phrasing, reducing repetition, and enhancing clarity. The authors would like to clarify that AI assistance was not used for content generation, but for refinement of existing text. The authors did not, at any point, accept AI-assisted content in the article without direct human supervision. Every suggestion generated by Writefull was carefully considered by the authors and accepted only based on contextual judgment.

Acknowledgments. This work is supported by the Volkswagen Foundation under the "Artificial Intelligence and the Society of the Future" program (grant number: 9B390). The authors would also like to thank: Prof. Dr. Yael de Haan, University of Groningen, for input and guidance throughout the design thinking process; Lieneke Grolle, facilitator of the design thinking sprints; the management and journalists of the regional broadcaster Rijnmond, Rotterdam.

References

1. Berkowitz, D.: Reporters and their sources. In: The handbook of journalism studies, pp. 165–179. Routledge (2019)
2. Broussard, M.: Artificial unintelligence: How computers misunderstand the world. MIT Press (2018)
3. Broussard, M., Diakopoulos, N., Guzman, A.L., Abebe, R., Dupagne, M., Chuan, C.H.: Artificial intelligence and journalism. J. Mass Commun. Quart. **96**(3), 673–695 (2019)
4. Brown, T.: Design thinking. harvard business review p. 1 (2008)
5. Caswell, D., Dörr, K.: Automating complex news stories by capturing news events as data. J. Pract. **13**(8), 951–955 (2019)
6. Chen, J., Lin, H., Han, X., Sun, L.: Benchmarking large language models in retrieval-augmented generation. In: Proceedings of the AAAI Conference on Artificial Intelligence, vol. 38, pp. 17754–17762 (2024)
7. Cremaschi, M., Bianchi, F., Maurino, A., Pierotti, A.P., et al.: Supporting journalism by combining neural language generation and knowledge graphs. In: CLiC-it (2019)
8. Es, S., James, J., Anke, L.E., Schockaert, S.: Ragas: automated evaluation of retrieval augmented generation. In: Proceedings of the 18th Conference of the European Chapter of the Association for Computational Linguistics: System Demonstrations, pp. 150–158 (2024)
9. de Haan, Y., Kruikemeier, S., Lecheler, S.: Searching, selecting, and verifying: Online journalistic sourcing strategies. Tijdschrift voor Communicatiewetenschap **52**(2), 150–176 (2024)
10. Hamborg, F., Meuschke, N., Breitinger, C., Gipp, B.: news-please-a generic news crawler and extractor. In: ISI (2017)

11. Hertzum, M.: How do journalists seek information from sources? a systematic review. Inform. Process. Manage. **59**(6), 103087 (2022)
12. Hess, K., Waller, L.: Local journalism in a digital world: Theory and practice in the digital age. Bloomsbury Publishing (2017)
13. Huang, Z., Li, S., Cheng, G., Kharlamov, E., Qu, Y.: Micron: making sense of news via relationship subgraphs. In: Proceedings of the 28th ACM International Conference on Information and Knowledge Management, pp. 2901–2904 (2019)
14. Lin, B., Lewis, S.C.: The one thing journalistic ai just might do for democracy. Digit. J. **10**(10), 1627–1649 (2022)
15. Meijer, I.C.: What does the audience experience as valuable local journalism?: Approaching local news quality from a user's perspective. In: The Routledge companion to local media and journalism, pp. 357–367. Routledge (2020)
16. MILLER, E.: An introduction to the resource description framework. Bull. American Society Inform. Sci. **25**(1), 15–19 (1998)
17. Munoriyarwa, A., Chiumbu, S., Motsaathebe, G.: Artificial intelligence practices in everyday news production: The case of South Africa's mainstream newsrooms. J. Pract. **17**(7), 1374–1392 (2023)
18. Noor, R., Zafar, H.: Use of artificial intelligence in Pakistani journalism: navigating challenges and future paths in tv newsrooms. J. Asian Develop. Stud. **12**(3), 1638–1649 (2023)
19. Pan, S., Luo, L., Wang, Y., Chen, C., Wang, J., Wu, X.: Unifying large language models and knowledge graphs: a roadmap. IEEE Trans. Knowl. Data Eng. (2024)
20. Peña Fernández, S., Meso Ayerdi, K.: News in the hybrid media system. Servicio Editorial de la Universidad del País Vasco/Euskal Herriko (2023)
21. Rowe, P.G.: Design thinking. MIT press (1991)
22. Rudnik, C., Ehrhart, T., Ferret, O., Teyssou, D., Troncy, R., Tannier, X.: Searching news articles using an event knowledge graph leveraged by wikidata. In: Companion Proceedings of the 2019 World Wide Web Conference, pp. 1232–1239 (2019)
23. Saad-Falcon, J., Khattab, O., Potts, C., Zaharia, M.: Ares: An automated evaluation framework for retrieval-augmented generation systems. In: Proceedings of the 2024 Conference of the North American Chapter of the Association for Computational Linguistics: Human Language Technologies (Volume 1: Long Papers), pp. 338–354 (2024)
24. Sirén-Heikel, S., Kjellman, M., Lindén, C.G.: At the crossroads of logics: Automating newswork with artificial intelligence–(re) defining journalistic logics from the perspective of technologists. J. Am. Soc. Inf. Sci. **74**(3), 354–366 (2023)
25. Sun, K., Xu, Y.E., Zha, H., Liu, Y., Dong, X.L.: Head-to-tail: How knowledgeable are large language models (LLM)? aka will llms replace knowledge graphs? arXiv preprint arXiv:2308.10168 (2023)
26. Swart, J., Peters, C., Broersma, M.: The ongoing relevance of local journalism and public broadcasters: Motivations for news repertoires in the netherlands. Participations: J. Audience & Reception Stud. **14**(2), 268–282 (2017)
27. Tessem, B., Gallofré Ocaña, M., Opdahl, A.L.: Construction of a relevance knowledge graph with application to the local news angle (2023)
28. Tessem, B., Opdahl, A.L.: Supporting journalistic news angles with models and analogies. In: 2019 13th International Conference on Research Challenges in Information Science (RCIS), pp. 1–7. IEEE (2019)
29. Vossen, P., et al.: Newsreader: using knowledge resources in a cross-lingual reading machine to generate more knowledge from massive streams of news. Knowl.-Based Syst. **110**, 60–85 (2016)

Knowledge-Enhanced Multimodal Retrieval over Cultural Heritage Knowledge Graphs

Xuemin Duan[1]([✉]) [iD], Federico D'Asaro[2,3] [iD], Ruben Peeters[1] [iD],
Giacomo Blanco[2] [iD], Tommaso Monopoli[2] [iD], Giuseppe Rizzo[2] [iD],
and Anastasia Dimou[1] [iD]

[1] KU Leuven – Flanders Make@KULeuven – Leuven.AI, Leuven, Belgium
`{xuemin.duan,ruben.peeters,anastasia.dimou}@kuleuven.be`
[2] LINKS Foundation, Turin, Italy
`{giacomo.blanco,tommaso.monopoli,giuseppe.rizzo}@linksfoundation.com`
[3] Politecnico di Torino, Turin, Italy
`federico.dasaro@polito.it`

Abstract. The widespread digitization of cultural heritage (CH) collections has enabled institutions to make vast artefacts accessible through knowledge graphs (KGs). Yet existing retrieval systems predominantly rely on keyword-based search with semantic enrichment via linking to controlled vocabularies, while only a few systems support visual information search, and none support queries involving relational metadata constraints such as temporal ranges, or combinations of metadata and visual information. Although advanced approaches, including multimodal retrieval and knowledge reasoning, have emerged in research, few have been deployed in operational CH environments, creating a substantial gap between academic innovations and real-world practice. In this paper, we present a knowledge-enhanced multimodal retrieval system that operates over KG, integrating a domain-adaptive fine-tuned CLIP model with an LLM-based Text2SPARQL and unifying their results through a fusion strategy. We deployed the system as a backend API integrated into a CH platform and conducted a comprehensive evaluation through quantitative metrics and qualitative user studies with museum curators. Results demonstrate that our domain-adaptive CLIP achieves high retrieval accuracy, while Text2SPARQL effectively further enhances precision through knowledge reasoning. Our work provides practical experience into the benefits and limitations of these technologies in production environments, informing broader adoption across CH institutions.

Keywords: Information Retrieval · Cultural Heritage · Multimodal · CLIP · Large Language Model · Text2SPARQL · Knowledge Graph

M. Acosta et al. (Eds.): ESWC 2026, LNCS 16550, pp. 488–506, 2026.
https://doi.org/10.1007/978-3-032-25159-6_26

1 Introduction

The widespread digitization of cultural heritage (CH) worldwide has transformed how museums, libraries, and archives preserve and provide access to collections, with Knowledge Graphs (KGs) emerging as a critical infrastructure for organizing these artefacts [6]. The CH community has increasingly adopted KG technologies to model artefacts, with visual content of images and rich structured metadata describing properties, including creator, creation period, and provenance [17], providing a foundation for sophisticated retrieval.

Despite this rich foundation, users of CH retrieval systems face significant challenges in accessing digitized artefacts due to limitations in supported information types in queries. Users may seek artefacts based on textual visual content (e.g., *"paintings with yellow flowers"*), metadata constraints (e.g., *"paintings by Van Gogh in 1889"*), or both combined (e.g., *"yellow flower paintings by Van Gogh in 1889"*). Current CH retrieval systems predominantly rely on keyword-based search for metadata, with some semantic enrichment to controlled vocabularies [9,24,27,31–33], while only a few systems support visual information via textual query to image retrieval [28,30], and none support knowledge reasoning to handle metadata relationships such as temporal range queries.

Given these limitations, existing systems struggle to effectively process queries involving both relational metadata constraints and visual information. Although advanced approaches have emerged in research, including multimodal retrieval models [18,20,29], with CLIP [29] gaining particular traction in CH [7,8], and Large Language Model (LLM)-based Text2SPARQL systems [15] for knowledge reasoning, few of these have been deployed in operational CH environments [37]. This deployment gap leaves CH institutions without sufficient practical evidence of these technologies' benefits and limitations under real-world conditions. Moreover, the absence of published technical details and evaluations on retrieval performance of such deployed systems hinders the broader adoption of these technologies across the CH community.

To address these challenges, we developed and deployed a knowledge-enhanced multimodal retrieval system over a KG that integrates digitized artefacts from three pilot museums (MoMu [23], Olympic [25], and Aquileia [10]) with CH entities from Wikidata [36], serving both museum curators and the public. The system integrates two complementary modules: a domain-adaptive CLIP model retrieves artefacts through semantic matching between user queries and image-text pairs generated from the KG, while an LLM-based Text2SPARQL module enhances these retrieval results through knowledge reasoning over the KG. Their results are fused through a weighted combination, enabling the system to handle queries containing visual content information, metadata information, or both, with knowledge-enhanced retrieval performance. The system is deployed as a backend API[1] and integrated with a public CH platform[2]. All datasets[3], models[4], and experimental code[5] presented in this paper are open-source.

We evaluated the system through quantitative metrics and user studies with museum curators. Specifically, we systematically assessed individual module performance and fusion results on a dataset of synthetic user queries, complemented by qualitative user studies. Results demonstrate that knowledge enhancement effectively improves retrieval precision beyond the already high accuracy achieved by the multimodal retrieval module. User study discussions further provide insights into current limitations of CH retrieval systems and future directions.

The remainder of this paper is structured as follows: Sect. 2 presents user requirements, data foundations, and system architecture; Sect. 3 describes domain-adaptive CLIP fine-tuning and evaluates multimodal retrieval performance; Sect. 4 details the Text2SPARQL approach and evaluates retrieval effectiveness over the KG; Sect. 5 presents a fusion strategy and system deployment, evaluates knowledge-enhanced multimodal retrieval, and discusses user studies; Sect. 6 discusses deployment lessons and practical implications; Sect. 7 reviews related work; and Sect. 8 concludes work with future directions.

2   System Overview and Data Foundations

This section provides an overview of our system. Section 2.1 identifies user requirements. Section 2.2 introduces the KG and our approach to constructing training and evaluation data. Section 2.3 presents the system architecture.

2.1   User Requirements

Users of CH retrieval systems search for artefacts using diverse types of information in their textual queries. Queries may specify visual content observable in artefact images, such as depicted objects, colors, or artistic style (e.g., *"paintings with yellow flowers"*); artefact metadata such as creator, creation time, or label (e.g., *"paintings by Van Gogh in 1889"* or *"paintings titled 'Still Life: Vase with Fifteen Sunflowers'"*); or combinations of both, requiring retrieval over both visual features and metadata constraints.

Running Example. To illustrate how our system addresses these diverse scenarios, we use the following query as a running example throughout this paper: *"yellow flower paintings by Van Gogh between 1887 and 1889"*. It exemplifies a challenging scenario by combining visual content (*"yellow flower"*) which requires multimodal retrieval with relational metadata constraints (*"Van Gogh"*, temporal range *"between 1887 and 1889"*) which requires knowledge reasoning.

---

[1] https://reevaluate.cs.kuleuven.be/search/docs.

[2] https://reevaluate.cs.kuleuven.be/search.

[3] https://huggingface.co/datasets/xuemduan/reevaluate-image-text-pairs.

[4] https://huggingface.co/xuemduan/reevaluate-clip.

[5] https://github.com/REEVALUATE/knowledge_enhanced_multimodal_retrieval.

2.2 KG and Image-Text Pairs Generation

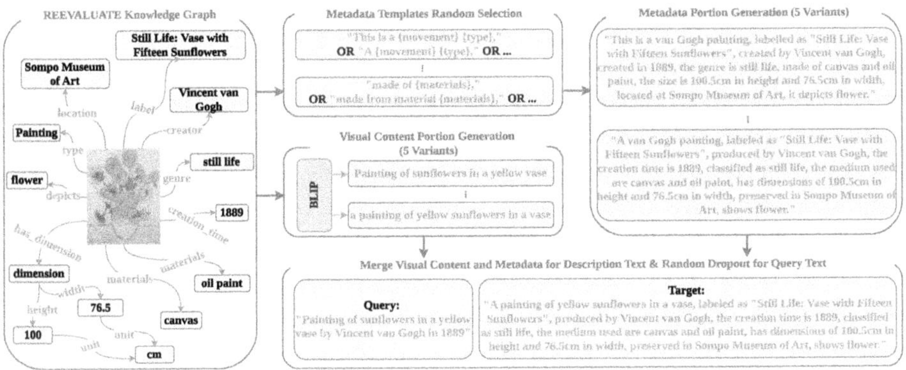

Fig. 1. The description texts and query texts generation in image-text pairs from KG.

Our retrieval system operates over a KG [4,5] that integrates digitized arte-facts from three pilot museums (MoMu [23], Olympic [25], and Aquileia [10]) with CH entities from Wikidata to enable exploration of local collections within a broader cultural context. The KG contains 147,497 artefacts with linked single images so far, spanning diverse object types, including paintings, sculptures, medals, musical instruments, textiles, mosaics, and documents. Each artefact is associated with various structured metadata fields, including object type, label, creator, creation time, material, location, genre, movement, and rights.

To support queries containing visual and metadata information through multimodal retrieval, we constructed and publicly released a dataset(see footnote 3) of image-text pairs generated from the KG. It consists of artefact images paired with descriptive texts generated from KG metadata and image visual content, along with synthetically generated query texts simulating user queries. The artefact images and descriptive texts serve as retrieval targets in deployment, while the query texts, with varying specificity, are exclusively used for model training and evaluation. Our generation methodology of the texts in this dataset is illustrated in Fig. 1, producing both description texts and query texts.

To introduce diversity and avoid overfitting to fixed patterns when generating description and query texts, we employ variation and stochastic sampling throughout the process. We first generated the visual content portion of description texts by applying the BLIP (Bootstrapping Language-Image Pre-training) [20] captioning model (blip2-opt-2.7b) to produce five textual variants per image (e.g., for our running example target artefact generated: *"painting of sunflowers in a yellow vase"*), capturing visual characteristics such as depicted objects, colors, and artistic style. In parallel, we generated the metadata portion by pre-defining five template variants per metadata field (e.g., *"made of {material}"* vs. *"the materials are {material}"*), randomly sampling templates across available fields, and instantiating them with KG values. For each artefact, we randomly sampled one variant from each portion and concatenated them (content first) to form the final description text. Similarly, we generated query texts by randomly sampling from both portions, but applied stochastic field dropout to simulate varying query specificity, with higher dropout rates for optional metadata fields than for core fields like object type and creator.

At the time of our data preparation, the KG contained 43,500 artefacts, from which we generated 43,500 image-text pairs. These were split into training (80%), validation (10%), and test (10%) sets using stratified sampling by object type to ensure balanced representation across artefact categories. Our current deployed retrieval system operates over the expanded KG with 147,497 artefacts.

2.3   System Architecture

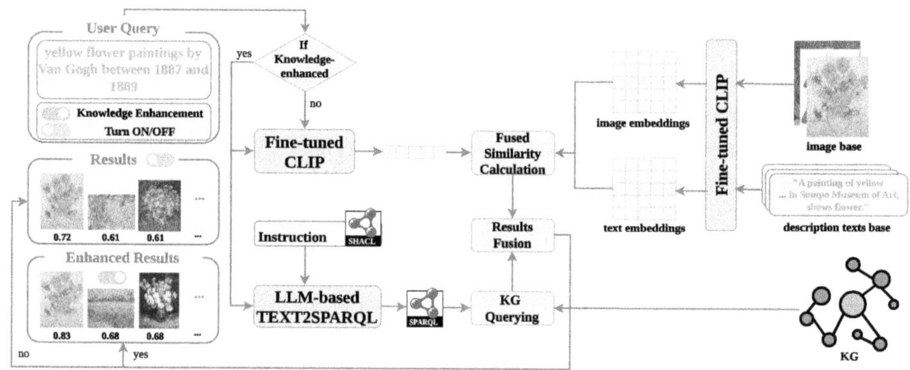

Fig. 2. System architecture including multimodal retrieval by fine-tuned CLIP, and user-toggleable knowledge enhancement via LLM-based Text2SPARQL.

To support queries with diverse information types (i.e., visual content and metadata) and those requiring knowledge reasoning (e.g., temporal range), our system architecture (Fig. 2) is designed to consist of two core modules: domain-adaptive fine-tuned CLIP model that supports queries containing visual and metadata information through multimodal retrieval, and LLM-based Text2SPARQL that enhances retrieval results through knowledge reasoning. We adopt CLIP [29] (Contrastive Language-Image Pre-training) as our multimodal encoder to encode images and texts, as it has demonstrated strong performance and gained particular traction in CH [7,8]. While CLIP performs semantic matching over image and text embeddings to retrieve relevant artefacts from visual content and metadata information, it cannot resolve relational metadata constraints such as temporal ranges, necessitating knowledge reasoning to refine retrieval results. To address this, we integrate Text2SPARQL, which translates natural language queries into executable SPARQL [16] queries on KG, to perform knowledge reasoning. The results from CLIP's ranked retrieval and Text2SPARQL's retrieved artefact set are combined through weighted fusion to produce the final ranking.

Since Text2SPARQL introduces additional latency, the system provides a toggleable knowledge enhancement option: when disabled, queries are processed by CLIP alone; when enabled, CLIP and Text2SPARQL operate in parallel with their results fused. Users can balance precision and response time as needed.

When knowledge enhancement is disabled, queries are processed solely by CLIP, as illustrated in the Fig. 2. For our running example *"yellow flower paintings by Van Gogh between 1887 and 1889"*, CLIP first encodes the query into an embedding vector. During system deployment preparation, we used our fine-tuned CLIP model to precompute embeddings for images and their description texts of all KG artefacts, storing them in an H5 file for efficient retrieval. The query embedding is then matched against these precomputed embeddings through both text-to-text similarity (query vs. description texts) and text-to-image similarity (query vs. images). These two similarity scores are combined to produce the final ranking. As shown in the middle left result panel (Fig. 2), while the top-ranked artefact correctly matches our target painting, the second and third Van Gogh paintings were created in 1890, violating the temporal constraint. This illustrates CLIP's limitation, where the embedding-based semantic matching cannot handle relational metadata constraints.

When knowledge enhancement is enabled, queries are processed by both CLIP and Text2SPARQL modules in parallel, as illustrated in Fig. 2. For our running example, CLIP generates the ranked results as described above, while Text2SPARQL translates the query into an executable SPARQL query that encodes precise metadata constraints (e.g., *"creation time"* ≥ 1887 AND ≤ 1889) over the KG. This SPARQL query retrieves a set of artefacts satisfying the structured constraints. The ranked results from CLIP and the retrieved artefact set from Text2SPARQL are then integrated through weighted fusion to produce the final ranking. As shown in the bottom left result panel (Fig. 2), not only does the top one match our target, but the second and third paintings have creation times of 1888 and 1887 respectively, both satisfy the temporal constraint, demonstrating the benefit of introducing knowledge enhancement.

3   Domain-Adaptive CLIP for Multimodal Retrieval

To enable multimodal retrieval over CH artefacts, we employ CLIP [29] as the unified embedding model for both **text-to-image (T2I)** and **text-to-text (T2T)** retrieval tasks. Given a textual query (e.g., query text in Fig. 1), T2I retrieval computes similarity between the query embedding and artefact image embeddings, while T2T retrieval computes similarity between the query embedding and description text embeddings. The image-text pairs constructed in Sect. 2.2 enable CLIP fine-tuning on both tasks. Section 3.1 describes our evaluation metrics and baseline performance, and Sect. 3.2 presents our domain-adaptive fine-tuning strategy and score fusion mechanism on both tasks.

Table 1. The retrieval performance of baselines and fine-tuned CLIP on the test set.

Model	T2I					T2T				
	R@1	R@5	R@10	MRR	MR	R@1	R@5	R@10	MRR	MR
CLIP ViT-B/32 pretrained	17.08	33.70	42.04	25.50	280.72	51.75	64.53	68.80	57.86	143.23
CLIP ViT-L/14 pretrained	39.36	62.64	72.07	50.25	23.40	69.38	81.95	85.86	75.22	19.32
CLIP ViT-L/14 fine-tuned	**50.97**	**75.89**	**83.24**	**62.18**	**15.61**	**92.00**	**98.11**	**99.06**	**94.73**	**1.93**

3.1 Evaluation Methodology and Baselines

We conduct experiments on the 43,500 image-text pairs. The **training set** is used for fine-tuning, the **validation set** for hyperparameter selection, and the **test set** for final performance reporting. We assess model performance on two tasks: T2I, where the model ranks all images by similarity to a query text, and T2T, where the model ranks all description texts by similarity to the query.

We evaluate retrieval effectiveness using standard metrics. Let N denote the number of artefacts in the evaluation set, and r_i denote the rank of the ground-truth for query i. **Recall@K** measures the percentage of queries for which the correct item appears within the top K retrieved results, where Recall@K $= \frac{1}{N}\sum_{i=1}^{N} \mathbb{1}[r_i \leq K]$. **Mean Reciprocal Rank** is computed as MRR $= \frac{1}{N}\sum_{i=1}^{N} \frac{1}{r_i}$, emphasizing early retrieval of relevant results. **Mean Rank** is computed as MR $= \frac{1}{N}\sum_{i=1}^{N} r_i$, with lower values indicating better performance. We report Recall@1, Recall@5, Recall@10, MRR, MR for both T2I and T2T tasks.

To establish baseline performance, we evaluate two pretrained CLIP variants, ViT-B/32 and ViT-L/14 [29], which differ in model scale. Both models are evaluated on T2I and T2T tasks without domain-specific fine-tuning. Baseline performances are reported in Table 1. ViT-L/14 consistently outperforms ViT-B/32 across all metrics on both tasks. On T2I, ViT-L/14 achieves MRR of 50.25% compared to ViT-B/32's 25.50%; on T2T, it achieves MRR of 75.22% compared to 57.86%. We therefore select ViT-L/14 as the base model for fine-tuning to leverage its stronger baseline performance. Our system requires a unified encoder for both visual and textual modalities to ensure similarity scores from T2I and T2T tasks reside in the same embedding space, enabling direct score fusion. This constraint necessitates using a language-image pre-training model, CLIP, rather than specialized text embedding models, which lack visual encoding capabilities.

3.2 Fine-Tuning Strategy and Results

We fine-tune both the vision and text encoders of ViT-L/14 to adapt to our CH domain. Similar to the original CLIP model [29] and relevant models [8], we minimize InfoNCE loss [26] during contrastive learning on T2I and T2T (Fig. 3). The losses are combined through weighted summation: $\mathcal{L}_{\text{total}} = \alpha_{T2I} \cdot \mathcal{L}_{\text{T2I}} + \beta_{T2T} \cdot \mathcal{L}_{\text{T2T}}$, where $\mathcal{L}_{\text{T2I}}$ and $\mathcal{L}_{\text{T2T}}$ are losses per task. Based on validation performance, we set $\alpha_{T2I} = 0.7$ and $\beta_{T2T} = 0.3$ to prioritize visual-language

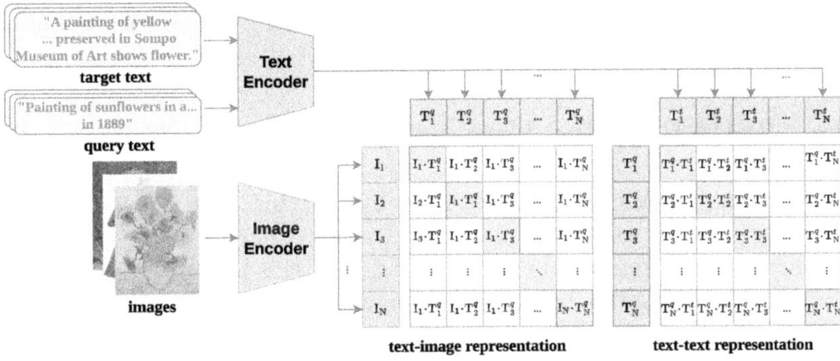

Fig. 3. Fine-tuning architecture. Images, query texts, and description texts are encoded into a unified embedding space with joint contrastive loss.

alignment while leveraging textual descriptions for additional supervision. The model is fine-tuned for 20 epochs using the AdamW optimizer with learning rate 5e-6 on one H100 GPU, with early stopping based on validation MRR.

Our domain-adaptive fine-tuning substantially improves CLIP performance on both tasks, as shown in Table 1. On T2I, MRR increases from 50.25% to 62.18%. On T2T, MRR increases from 75.22% to 94.73%. Both improvements validate the effectiveness of domain adaptation for CH artefact retrieval. However, the substantially higher absolute performance on T2T (94.73% vs. 62.18% of MRR) reflects the lexical overlap between synthetic query texts and description texts in our dataset, even after applying dropout to queries (Sect. 2.2). This suggests that T2T performance is inflated relative to its expected contribution in real retrieval scenarios, where user queries may diverge more from the synthetic description texts than our evaluation setup captures.

To produce unified rankings that leverage both retrieval tasks, we fuse T2I and T2T similarity scores through a weighted combination: $\mathrm{score}_{\mathrm{CLIP}}(q, a) = w_{\mathrm{T2I}} \cdot \mathrm{sim}_{\mathrm{T2I}}(q, I_a) + w_{\mathrm{T2T}} \cdot \mathrm{sim}_{\mathrm{T2T}}(q, D_a)$ where q is the query, a is an artefact, I_a is its image, and D_a is its description text. Since both tasks operate within the same embedding space, their similarity scores are directly comparable.

Table 2 reports the performance under different fusion weights. We explored fusion weights via grid search over $\{0.1, 0.2, \ldots, 0.9\}$ on the validation set. The search identified optimal weights of $w_{\mathrm{T2I}} = 0.1$ and $w_{\mathrm{T2T}} = 0.9$ (MRR 94.76%), heavily favoring T2T. This outcome aligns with T2T's substantially higher absolute performance (94.73% MRR) compared to T2I (62.18% MRR), reflecting the similarity between our synthetic queries and description texts despite dropout regularization. However, real user queries are likely to exhibit greater variance and may diverge more substantially from description texts, suggesting that the 0.1/0.9 weights optimize for our specific evaluation setup rather than general retrieval scenarios. We therefore adopt balanced weights of $w_{\mathrm{T2I}} = 0.5$ and $w_{\mathrm{T2T}} = 0.5$ (MRR 91.18%) for deployment and subsequent evaluation, providing a more conservative baseline that can be refined as real user queries and interaction signals accumulate through continued system deployment.

4 Text2SPARQL for Knowledge-Enhanced Retrieval

While CLIP enables multimodal retrieval by matching queries with visual and textual information, its embedding-based approach cannot resolve metadata relationships such as temporal ranges. To address this limitation, we introduce a Text2SPARQL module for enhancing retrieval results via knowledge reasoning. Section 4.1 describes our methodology and Sect. 4.2 analyzes its performance.

Table 2. Retrieval performance of fine-tuned CLIP and knowledge enhancement via Text2SPARQL. Weights (w_{T2I}, w_{T2T}) combine CLIP's T2I and T2T scores; weights (α_{CLIP}, β_{SPARQL}) combine CLIP results with Text2SPARQL results.

Model	w_{T2I}	w_{T2T}	α_{CLIP}	β_{SPARQL}	R@1	R@5	R@10	MRR	MR
CLIP fine-tuned fused	0.1	0.9	–	–	92.07	98.07	99.03	94.76	1.78
Text2SPARQL enhanced CLIP	0.1	0.9	0.8	0.2	**96.37**	**99.13**	**99.56**	**97.63**	1.41
CLIP fine-tuned fused	0.5	0.5	–	–	86.90	96.53	98.46	91.18	1.83
Text2SPARQL enhanced CLIP	0.5	0.5	0.8	0.2	95.03	98.87	99.43	96.68	**1.37**

4.1 LLM-Based Text2SPARQL Pipeline

Our Text2SPARQL module is inspired by the multi-stage pipeline of Sparnatural AI [11,13]. It leverages a Mistral LLM [1] to generate SPARQL queries from natural language via an intermediate JSON representation [14]. Although Sparnatural was originally designed for browser-based visual query building with SHACL-based [19] configuration, we adopted this pipeline design and reimplemented its logic to meet the specific requirements of our CH domain and Python-based backend deployment. The Text2SPARQL pipeline (Fig. 4) transforms natural language queries into executable SPARQL through four stages: (1) LLM-based JSON generation with entity placeholders, (2) entity linking via SPARQL lookups to resolve placeholders to entity IRIs, (3) JSON-to-SPARQL conversion, and (4) query execution against the KG. Our implementation adapts this approach through: a SHACL configuration defining the ontological structure of our KG, modified instruction prompts [12] incorporating our SHACL shapes and CH-specific examples, an implemented entity linking for named entity disambiguation, and a reimplemented JSON-to-SPARQL conversion in Python following Sparnatural AI's translation logic.

LLM-based JSON Generation with Placeholders. The first stage employs an LLM (Mistral-small, temperature 0.1) to translate user queries into structured

JSON representations that encode query constraints. Rather than directly generating SPARQL, we adopt Sparnatural's intermediate JSON format [14], which represents subject-predicate-object patterns with type constraints in a predefined hierarchical structure. This JSON is then converted to SPARQL via deterministic translation rules, minimizing syntax errors compared to direct LLM-based SPARQL generation. The instruction prompt includes: (1) task guidelines and JSON structure requirements, (2) SHACL shapes that we created based on the ontology of our KG, specifying classes and properties, and (3) few-shot examples. Since the SHACL configuration provides schema information but not individual entity URIs, the LLM generates placeholder URIs for named entities. For example, given our running example query, the LLM produces a JSON, as the fragment shown in Fig. 4, encoding the constraints on object type (*painting*), creator (with label "*Van Gogh*" and placeholder URI), and creation time (temporal range between *1887* and *1889*).

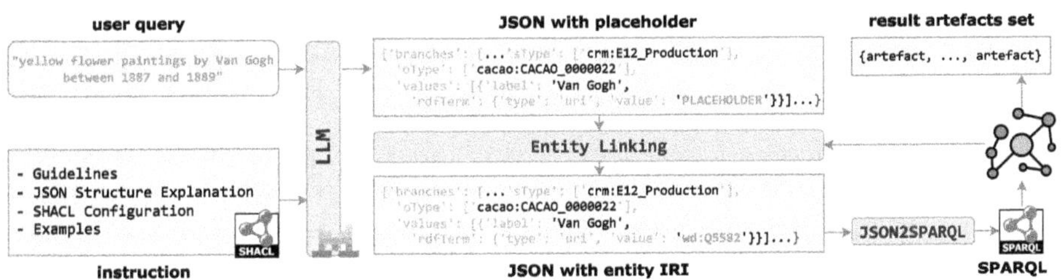

Fig. 4. Multi-stage Text2SPARQL pipeline, inspired by Sparnatural AI [13].

Entity Linking and SPARQL Generation. This second stage resolves placeholder URIs to actual KG entity URIs. For each placeholder, we execute a predefined SPARQL template that searches for entities matching both the entity type and label extracted from the LLM-generated JSON. The lookup uses a case-insensitive string matching against label properties (e.g., `rdfs:label`, `skos:prefLabel`) to maximize recall. When a unique match is found, the placeholder is replaced with the entity's URI; when multiple matches exist, all candidates are retained and expressed via SPARQL UNION in the final query. The final stage converts the placeholder-resolved JSON to executable SPARQL. Given our system is implemented in Python, we adopted and reimplemented Sparnatural's translation logic [11] to ensure technical compatibility, as the original component is built on TypeScript and could not be directly integrated. The generated SPARQL is then executed against the KG's SPARQL endpoint, returning a set of artefact URIs that satisfy the specified constraints.

4.2   Performance Analysis

We evaluate the Text2SPARQL module on the test set of 4,350 synthetic queries, using each query text as input. Unlike CLIP's ranked retrieval, Text2SPARQL returns unranked sets of artefacts, necessitating different evaluation metrics. We employ a three-tier framework (as illustrated in Table 3): *Executability* (technical robustness), *Effectiveness* (retrieval recall), and *Precision* (result quality).

The Text2SPARQL module demonstrates high reliability and stability in deployment, indicating that all generated SPARQL queries are syntactically correct. This execution success suggests every user query can be processed without system failures or error handling overhead. Among successfully executed queries, 80.69% returned non-empty results, indicating that the majority of generated SPARQL queries successfully matched artefacts in the KG. Manual inspection of the 19.31% empty-result cases revealed two primary causes: LLM hallucinations introducing spurious constraints not present in the user query (e.g., fabricating creation time), and encoding visual features from BLIP as metadata constraints (e.g., genre, material) absent in the artefact's triples in the KG.

The module achieved an overall Hit Rate of 79.43%, while 80.69% of queries returned non-empty results, where the small gap (1.3%) represents queries that retrieved artefacts but missed the ground truth. Manual inspection revealed this gap to be primarily due to the same constraint generation issues as above: the LLM encoded visual content as constraints that were absent in the ground truth's KG representation but coincidentally satisfied by other artefacts. For example, given the query "Portrait painting of a lady, made of oil on canvas, the size is 78.5cm in height and 62.5cm in width", the LLM encoded genre, materials, and dimensions in SPARQL; however, while the ground truth lacked the material property, other paintings coincidentally matched all constraints, yielding non-empty but incorrect results. Among queries with non-empty results, 98.43% contained the ground truth artefact, indicating high precision when constraints successfully match. The module excels at precision, with 54.83% of queries achieving perfect precision, returning only the ground truth artefact. Notably, while CLIP achieved higher recall (86.90% R@1) on T2T, its performance benefits from lexical overlap between synthetic queries and ground truth descriptions generated from limited metadata templates. In contrast, Text2SPARQL's hit rate is achieved without access to description texts, demonstrating robust generalization to unseen queries.

Table 3. Text2SPARQL retrieval performance on 4,350 synthetic queries.

Category	Metric	Description	Value
Executability	Execution Success Rate	Queries without syntax errors	100.0%
	Non-Empty Answer Rate	Executed queries returning results	80.69%
Effectiveness	Hit Rate	Queries retrieving ground truth	79.43%
	Hit Rate (non-empty)	Hit rate among queries with results	98.43%
Precision	Exact Match Rate	Only ground truth returned	54.83%

5   System Implementation and Evaluation

This section presents the deployment of our retrieval system. Section 5.1 describes the fusion strategy combining CLIP and Text2SPARQL and reports the corresponding evaluation results, Sect. 5.2 details the system deployment, and Sect. 5.3 presents findings from our user studies.

5.1   Knowledge-Enhanced Fusion Strategy and Evaluation

To enable knowledge-enhanced retrieval, we combine CLIP with Text2SPARQL via weighted linear fusion. Text2SPARQL contributes knowledge reasoning for relational metadata constraints, while CLIP provides robust multimodal ranking through T2I and T2T similarity aggregation with defined T2I and T2T weights. The fusion combines CLIP scores with Text2SPARQL's retrieved artefact set:

$$S_{\text{final}}(q, a) = \alpha_{\text{CLIP}} \cdot S_{\text{CLIP}}(q, a) + \beta_{\text{SPARQL}} \cdot \mathbb{I}(a \in R_{\text{SPARQL}}(q))$$

where q denotes the user query, a denotes an artefact, S_{CLIP} denotes the aggregated T2I and T2T scores, R_{SPARQL} denotes the retrieved artefact set, $\mathbb{I}$ denotes a binary indicator function, and α_{CLIP} and β_{SPARQL} denote the fusion weights.

We optimized the above fusion weights via grid search over $\{0.1, 0.2, \ldots, 0.9\}$ on the validation set. Table 2 presents fusion performance on the test set under two CLIP configurations. Grid search identified optimal weights of $\alpha_{\text{CLIP}} = 0.8$ and $\beta_{\text{SPARQL}} = 0.2$ for both configurations. Knowledge enhancement yields substantial improvements: for balanced T2I/T2T weights ($w_{T2I} = 0.5, w_{T2T} = 0.5$), fusion improves R@1 from 86.90% to 95.03% and MRR from 91.18% to 96.68%; for T2T-prioritized weights ($w_{T2I} = 0.1, w_{T2T} = 0.9$), fusion improves R@1 from 92.07% to 96.37% and MRR from 94.76% to 97.63%. These results demonstrate that symbolic knowledge reasoning provides complementary signals beyond CLIP's learned embeddings. For deployment, we adopt the balanced configuration ($w_{T2I} = 0.5, w_{T2T} = 0.5, \alpha_{\text{CLIP}} = 0.8, \beta_{\text{SPARQL}} = 0.2$), following the same reasoning as in Sect. 3.2.

To provide insight into score distributions, we analyze the similarity scores produced by our CLIP module on queries outside the KG's scope. While the Text2SPARQL module naturally returns empty result sets for irrelevant queries, CLIP as an embedding-based module produces correspondingly lower similarity scores for such queries. We compare top-1 similarity scores on our test set of in-scope queries against an equal-sized sample of out-of-scope queries, observing a median score of 0.72 for in-scope queries versus 0.27 for out-of-scope queries. This score gap suggests that the scores can serve as a practical confidence signal for users.

5.2   System Implementation and Deployment

We implemented the system architecture described in Sect. 2.3 and Fig. 2 as a production deployment serving pilot museums and public users.

The system operates through precomputed embeddings and on-demand query processing. Offline, we precompute CLIP embeddings (images and description texts) for all artefacts (147,497 artefacts so far) in the KG. At runtime, queries are encoded and matched against precomputed embeddings via cosine similarity. For textual queries with knowledge enhancement enabled, CLIP and Text2SPARQL execute in parallel, with results fused as discussed in Sect. 5.1. The system is deployed as a RESTful backend API on CPU-based servers (Intel Xeon Icelake, 4 cores, 4GB RAM). Our API has been integrated into a public CH platform(see footnote 2) for artefact retrieval, where users can input text queries and toggle a knowledge enhancement option to balance precision and response time. Average response times on the test set are 0.12 s without knowledge enhancement retrieval, and 4.8 s for knowledge enhancement retrieval. System scalability is primarily subject to server compute resources and LLM API rate limits, both of which can be addressed through infrastructure scaling and query result caching as demand grows.

5.3   User Study

Beyond the quantitative evaluation presented in Tables 1, 2, and 3, we conducted a two-stage user study to qualitatively assess the system through the integrated CH platform. We first demonstrated the system to CH professionals and collected general feedback for system refinement, then conducted a structured interview comprising questionnaires with two curators from pilot museums, with complementary insights from general users.

During a CH meeting with around 20 professionals (not specifically convened for this system), we demonstrated our initial implementation, which applied knowledge enhancement by default without a toggleable option. Several attendees explored the system and acknowledged its effectiveness in multimodal retrieval and metadata reasoning within the KG scope, while identifying an issue with the score presentation. Users entering out-of-scope queries (e.g., "*shoes*" which was not yet in the KG at that time) received irrelevant results with inflated scores. Inspection revealed that min-max normalization in our fusion implementation amplified weakly related matches. We addressed this by displaying raw cosine similarities and adding example queries to guide users. Discussions with general users suggested that the system could serve as a tool for planning visits to cultural institutions in advance, pointing to directions for broader public accessibility, including more comprehensive KG coverage spanning diverse collections and institutions, and more detailed metadata such as the physical location of artefacts within museums.

Based on this feedback, we refined the system and introduced a toggleable knowledge enhancement option to accommodate varying user needs for precision and response time. We then conducted a one-hour interview with two curators from pilot museums. After demonstrating the refined system and discussing their retrieval needs and current system limitations, curators completed one questionnaire. For each of five self-designed queries, they explored results with and without knowledge enhancement, rating performance across four dimensions: relevance, ranking quality, score-result alignment, and overall satisfaction, on a 5-point Likert scale, and indicating their preferred mode. Since curator responses showed consistency across these dimensions, we report average scores here. Results demonstrated varied impact depending on query characteristics (reported as without knowledge enhancement vs with knowledge enhancement, preferred mode): visual-focused queries (e.g., *"pink dresses"*) showed comparable performance (4.0 vs 4.0, without), while metadata-constrained queries (e.g., *"paintings by Van Gogh on January 1, 1888"*) exhibited significant improvement (2.25 vs 5.0, with). Mixed queries such as *"military costume"* showed moderate gains (3.5 vs 4.5, with). Curators also tested intentional misspellings (e.g., *"Carravagio"* for *"Caravaggio"*), which we addressed by introducing spell correction for query preprocessing. Curators expressed willingness (score 4) to adopt the system in their work and recommend it to colleagues. Open feedback highlighted the need for search result transparency, particularly regarding the criteria and reasoning behind artefact retrieval and ranking, as our future work.

6   Discussion

Impact and Benefits. The deployment of our system provides empirical evidence for integrating vision-language models with semantic web technologies in operational CH environments. Our results demonstrate that domain-adaptive CLIP effectively captures visual and metadata semantics, while the integration of Text2SPARQL significantly enhances retrieval precision through knowledge reasoning over relational metadata constraints, illustrating the benefits of combining neural retrieval with symbolic reasoning. By validating this knowledge-enhanced multimodal architecture under real-world conditions, our work bridges the gap between academic research and operational deployment, offering a practical foundation for applying these methods in CH retrieval systems and potentially other domains requiring multimodal search with knowledge reasoning, such as medical imaging, press archives, and earth observation, where image collections are described by structured metadata in KGs.

Adoption and Uptake. The system has been deployed as a backend API integrated into a public CH platform(see footnote 2), serving museum curators and public users. User studies with museum curators (Sect. 5.3) demonstrated strong adoption intent, with willingness to use the system in daily work and recommend it to colleagues. Curators particularly valued knowledge enhancement for metadata-constrained queries, and expressed satisfaction with the knowledge enhancement toggle design that enables this flexibility. The system's sustainability is ensured through its role as a core retrieval component of an operational CH platform, where the underlying KG continues to grow as pilot museums contribute newly digitized artefacts, while our open-source release enables independent adaptation and broader adoption across domains.

Challenges and Limitations. We encountered four challenges during deployment: training data scarcity, practical limitation of KG, system latency, and typo tolerance. The typo tolerance issue is discussed in Sect. 5.3.

Domain adaptation of neural network models like CLIP in specialized CH domains is hampered by a lack of large-scale training and evaluation datasets. We faced a scarcity of image-text pairs required for CLIP fine-tuning and a lack of diverse, user-like query texts necessary for evaluation. To overcome this, we implemented a synthetic text generation strategy. By leveraging the existing KG metadata, combined with the capabilities of BLIP and template-based sampling, we successfully synthesized a large corpus of diverse artefact descriptions and simulated user queries, demonstrating the feasibility of synthetic dataset creation for training neural network models in data-scarce CH applications.

Text2SPARQL faces a critical challenge when the retrieval target artefact lacks metadata values for KG properties referenced in queries. The problem arises when a property exists in the ontology (e.g., `depicts`) but specific target artefacts lack recorded values of the property in the KG, causing constraint failures even when artefacts are conceptually relevant to other query aspects. This issue is particularly severe for the `depicts` property, which captures visual content but lacks standardized recording practices. When users include any visual description, such as *"paintings of boats on a river next houses in 1889s"*, the LLM may generate multiple `depicts` constraints (*"boat"*, *"river"*, *"house"*), but metadata flexibility means few possibilities that target artefacts satisfy all constraints. Since Text2SPARQL generates SPARQL queries with `AND` logic by default, these constraints frequently result in empty results, wasting its reasoning capacity on other constraints that would otherwise succeed. Our current mitigation disables `depicts` constraints in the Text2SPARQL prompt, though resolving this systematically would further enhance system performance.

The Text2SPARQL pipeline (LLM request, two SPARQL queries, JSON conversion) introduces latency impacting user experience. While our knowledge enhancement toggle allows users to balance precision and response time, further optimization of Text2SPARQL latency would significantly benefit adoption in interactive retrieval systems.

7   Related Work

We review existing CH retrieval approaches along with retrieval systems deployed by CH institutions and their supported queries, and advanced retrieval techniques, including vision-language models and knowledge reasoning.

CH retrieval systems predominantly rely on keyword matching over curated metadata fields. Europeana aggregates metadata from European institutions, offering keyword search, faceted filtering, and entity-based navigation augmented with controlled vocabularies and semantic enrichment [9,27]. Museum collection databases similarly employ keyword-based retrieval with linked open data integration [24,31–33]. However, keyword matching cannot support queries requiring knowledge reasoning, such as temporal constraints like"between 1887 and 1889", nor can it leverage visual information contained in queries, even though collections increasingly contain high-resolution images.

CH retrieval systems supporting queries containing visual content information have been explored in research [8,29,34], but remain scarce in production deployments. Pharos [2,28] and Rijksmuseum [30] enable text-to-image search across digitized collections; however, they lack published technical details [30] on utilized models or evaluation results on retrieval effectiveness [2,28,30]. Furthermore, no deployed system currently integrates visual search with knowledge reasoning.

Vision-language models such as CLIP [29], BLIP [20], and ALIGN [18] have demonstrated strong cross-modal learning capabilities. CLIP has gained particular traction in CH applications. Prior work explored CLIP in the CH domain [7,8], primarily for classification and attribute recognition rather than retrieval. A prototype utilized pretrained CLIP for text-to-image retrieval [21], but without domain-specific fine-tuning or evaluation. To our knowledge, no deployed CH retrieval system employs domain-adaptive CLIP fine-tuning.

Structured knowledge reasoning over KGs requires translating natural language queries into formal query languages such as SPARQL. Text-to-SPARQL approaches include rule-based methods [35], neural network-based methods [3, 22], and LLM-based generation [13,15]. However, LLMs may produce syntactically incorrect queries or hallucinated predicates. Visual query builders such as Sparnatural [11] constrain query construction through SHACL-based configuration [19] and JSON intermediate representation [14], but impose higher usage barriers. Sparnatural AI [13] further employs a Mistral [1] LLM to translate natural language into the JSON representation, lowering the usage threshold.

8 Conclusion and Future Work

We presented and deployed a knowledge-enhanced multimodal retrieval system over a CH KG, integrating a domain-adaptive fine-tuned CLIP model with an LLM-based Text2SPARQL module. We fine-tuned CLIP on synthetic image-text pairs generated from the KG, supporting both text-to-image and text-to-text retrieval to handle queries containing visual and metadata information. To address the limitation of embedding-based retrieval in performing knowledge reasoning, we introduced Text2SPARQL to enhance retrieval results through SPARQL query generation over the KG. Quantitative evaluation demonstrated the effectiveness of domain adaptation and the reliability of generated SPARQL queries. A weighted fusion strategy combines both modules, with results showing that knowledge enhancement significantly further improves retrieval precision. The system has been deployed as a backend API and integrated with a public CH platform, serving both museum curators and public users. User studies with museum curators validated system effectiveness and adoption potential. We further discussed practical deployment challenges, which inform both system refinement and broader technology adoption in CH institutions.

Future work will focus on optimizing Text2SPARQL latency, enhancing retrieval transparency and explainability, and extending support for queries requiring external factual knowledge beyond the KG.

Use of Generative AI. Claude Sonnet 4.5 and Claude Sonnet 4.6 were utilized to polish the text.

Acknowledgments. This work has received funding from the European Union Horizon Research and Innovation programme under grant agreement No 101132389, REEVALUATE project, https://reevaluate.eu/.

Disclosure of Interests. The authors have no competing interests to declare that are relevant to the content of this article.

References

1. AI, M.: Mistral AI Technical Documentation. Mistral AI (2025). https://docs.mistral.ai/. Accessed: Dec 3 2025
2. Alam, M., de Boer, V., Hyvönen, E., Meroño-Peñuela, A., Sack, H., Klic, L.: Linked open images: visual similarity for the semantic web. Semantic Web **13**(3), 347–366 (2022)
3. Banerjee, D., Chaudhuri, D., Dubey, M., Lehmann, J.: Modern baselines for SPARQL semantic parsing. In: Proceedings of the 45th International ACM SIGIR Conference on Research and Development in Information Retrieval, pp. 2260–2265 (2022)
4. Blanco, G., et al.: Cultural Heritage Knowledge Graph Endpoint. Web Service (2025). https://loki.linksfoundation.com/reevaluate-graphdb/. Accessed Dec 3 2025
5. Blanco, G., et al.: ArtKB: A Multimodal Art Knowledge Base for Cultural Heritage. In: The Semantic Web (2026)
6. Borowiecki, K.J., Navarrete, T.: Digitization of heritage collections as indicator of innovation. Econom. Innov. New Technol. **26**(3), 227–246 (2017)
7. Castellano, G., Vessio, G.: Exploiting CLIP-based multi-modal approach for artwork classification and retrieval. In: Proceedings of the International Conference on Image Analysis and Processing (ICIAP) Workshops, pp. 480–488. Springer (2022)
8. Conde, M.V., Turgutlu, K.: CLIP-Art: contrastive pre-training for fine-grained art classification. In: Proceedings of the IEEE/CVF Conference on Computer Vision and Pattern Recognition (CVPR) Workshops, pp. 3956–3960 (2021)
9. Europeana Foundation: Search API documentation. https://europeana.atlassian.net/wiki/spaces/EF/pages/2385739812/ (2024). Accessed 03 Dec 2025
10. Fondazione Aquileia: Official Website. Website (2025). https://www.fondazioneaquileia.it/. Accessed 3 Dec 2025
11. Francart, T.: Sparnatural: a visual knowledge graph exploration tool. In: European Semantic Web Conference, pp. 11–15. Springer (2023)
12. sparna git: LLM instruction file. Sparnatural Platform Repository, GitHub (2025). https://github.com/sparna-git/sparnatural-platform/blob/main/config/dbpedia-en/text2query_postprocess.txt. Accessed 3 Dec 2025

13. sparna git: sparnatural-ai. GitHub Repository (2025). https://github.com/sparna-git/sparnatural-ai. Accessed 3 Dec 2025

14. sparna git: SparnaturalQueryIfc.ts: Intermediate JSON Structure. Sparnatural Repository, GitHub (2025). https://github.com/sparna-git/Sparnatural/blob/master/src/sparnatural/SparnaturalQueryIfc.ts. Accessed 3 Dec 2025

15. Gu, D., Singh, K., Lakshmi Narayanan, A., Rossiello, G., Gliozzo, A., Mihindukulasooriya, N.: Investigating large language models for text-to-SPARQL generation. In: Proceedings of the 1st Workshop on Knowledge Graphs and Large Language Models (KaLLM), pp. 38–53 (2024)

16. Harris, S., Seaborne, A., Prud'hommeaux, E.: SPARQL 1.1 Query Language. W3C Recommendation (Mar 2013), available at: https://www.w3.org/TR/sparql11-query/

17. Haslhofer, B., Isaac, A.: data. europeana. eu: The europeana linked open data pilot. In: Proceedings of the international conference on Dublin Core and metadata applications. Dublin Core Metadata Initiative (2011)

18. Jia, C., et al.: Scaling up visual and vision-language representation learning with noisy text supervision. In: Proceedings of the 38th International Conference on Machine Learning (ICML), pp. 4904–4916 (2021)

19. Knublauch, H., Kontokostas, D.: Shapes Constraint Language (SHACL). W3C Recommendation, World Wide Web Consortium (Jul 2017), [Online]. Available: https://www.w3.org/TR/shacl/

20. Li, J., Li, D., Xiong, C., Hoi, S.: Blip: bootstrapping language-image pre-training for unified vision-language understanding and generation. In: International Conference on Machine Learning, pp. 12888–12900. PMLR (2022)

21. Lockhorst, S., et al.: Semantic Art Search: Exploring the Rijksmuseum collection with CLIP. https://github.com/SjorsLockhorst/sem-art-search (2023), blog post and prototype implementation

22. Luz, F.F., Finger, M.: Semantic parsing natural language into SPARQL: improving target language representation with neural attention. In: Proceedings of the 27th International Conference on Computational Linguistics (COLING), pp. 851–863 (2018)

23. MoMu - Fashion Museum Antwerp: Official Website. Website (2025). https://www.momu.be/. Accessed 3 Dec 2025

24. Musée du Louvre: Collections database. https://collections.louvre.fr/ (2024). Accessed 3 Dec 2025

25. Olympic Museum - Thessaloniki: Official Website. Website (2025). http://www.olympicmuseum-thessaloniki.org/. Accessed 3 Dec 2025

26. van den Oord, A., Li, Y., Vinyals, O.: Representation learning with contrastive predictive coding. In: arXiv preprint arXiv:1807.03748 (2018)

27. Petras, V., Hill, T., Stiller, J., Gäde, M.: Europeana-a search engine for digitised cultural heritage material. Datenbank-Spektrum **17**(1), 41–46 (2017)

28. PHAROS Consortium: https://artresearch.net/resource/search-help. https://artresearch.net/resource/search-help. Accessed 03 Dec 2025

29. Radford, A., et al.: Learning transferable visual models from natural language supervision. In: Proceedings of the 38th International Conference on Machine Learning (ICML), pp. 8748–8763 (2021)

30. Rijksmuseum: Art Explorer. https://www.rijksmuseum.nl/en/art-explorer (2024). Accessed 03 Dec 2025

31. Smithsonian Institution: Smithsonian Open Access. https://www.si.edu/openaccess (2024). Accessed 03 Dec 2025

32. The British Museum: Collection online. https://www.britishmuseum.org/collection (2024). Accessed 03 Dec 2025
33. The Metropolitan Museum of Art: The Metropolitan Museum of Art Collection API. https://metmuseum.github.io/ (2018). Accessed 03 Dec 2025
34. Tsai, C.F.: A review of image retrieval methods for digital cultural heritage resources. Online Inf. Rev. **31**(2), 185–198 (2007)
35. Unger, C., Bühmann, L., Lehmann, J., Ngonga Ngomo, A.C., Gerber, D., Cimiano, P.: Template-based question answering over RDF data. In: Proceedings of the 21st International Conference on World Wide Web (WWW), pp. 639–648 (2012)
36. Vrandečić, D., Krötzsch, M.: Wikidata: a free collaborative knowledgebase. Commun. ACM **57**(10), 78–85 (2014)
37. Winters, J.: AI meets archives: The future of machine learning in cultural heritage. Interview, Council on Library and Information Resources (CLIR) (October 2024). https://www.clir.org/2024/10/ai-meets-archives-the-future-of-machine-learning-in-cultural-heritage/

Towards a Symbolic Representation of the MEMS Development Domain

Florian Diehl[1]([✉])[iD], Ivan Marevic[1][iD], Irlan Grangel-González[2],
Simon Blattner[2][iD], and Maria-Esther Vidal[3][iD]

[1] Robert Bosch GmbH, Mobility Electronics, Reutlingen, Germany
{florian.diehl,ivan.marevic}@de.bosch.com
[2] Robert Bosch GmbH, Corporate Research, Renningen, Germany
[3] TIB Leibniz Information Centre for Science and Technology and Leibniz University
of Hannover and L3S Research Center, Hannover, Germany
vidal@L3S.de

Abstract. Cross-domain data integration remains a critical bottleneck in semiconductor sensor development because siloed data sources, inconsistent semantics, and missing alignment mechanisms restrict unified analysis. We present a general-purpose data integration framework that creates a semantic layer and can be applied across industrial contexts. In this work we applied it to Micro-Electro-Mechanical System (MEMS) development by constructing a domain-specific Knowledge Graph (KG) ecosystem that integrates heterogeneous data from the engineering and manufacturing domains. The KG extends the existing infrastructure and uses a modular and expert-aligned ontology to harmonize formats and units and resolve structural inconsistencies. A lightweight enrichment layer infers additional relations, including temporal process sequences and spatial chip relations. SHACL validation enforces standards and flags missing or inconsistent data. To improve accessibility for non-technical users, we provide a natural-language interface for querying the KG (SPARQL-NLI) that benefits from the KG's unified semantics. The system enables complex cross-domain queries that raw relational data cannot support efficiently. We evaluated the system through competency question coverage, expert feedback, and graph-level analytics. The results show improved data accessibility and more reliable cross-domain querying, although challenges remain in scaling and maintaining the KG. This work demonstrates how a focused semantic integration layer embedded in the existing infrastructure can bridge knowledge gaps and provide reusable and explainable analytics in industrial environments. This supports stakeholders in accessing and exploring cross-domain data without facing common integration hurdles.

Keywords: Knowledge Graph · Industry 4.0 · Smart Manufacturing · Semantic Data Integration · Ontology

1   Introduction

Semiconductor sensors, particularly MEMS devices (accelerometers, gyroscopes, environmental sensors), are fundamental to modern technology from smartphones to vehicles [4,7,27,42]. MEMS enable precise sensing and actuation at the microscale by integrating mechanical and electrical components within a compact form factor. MEMS development, which involves the creation of new sensor generations and improvement of existing products, investigates the full lifecycle from wafer processing to final sensor testing. This requires access to and analysis of data distributed across multiple governance domains, such as manufacturing, assembly, and engineering. Crossing domain borders is often associated with integrating heterogeneous data sources that lack semantic consistency [20,29].

We present a general-purpose data integration system (DIS) that operates within a broader KG ecosystem [8]. It consolidates MEMS development data through ontology-driven semantic integration and produces the MEMS Development KG (*mdevKG*). Our contribution is to show that KGs can extend beyond operational monitoring and support development-centric processes across the MEMS lifecycle. In contrast to manufacturing KGs, which primarily represent production activities, our approach enables holistic traceability and root cause analysis across complex sensor development workflows. We report practical lessons learned and key design choices from the implementation. We evaluated system usability through stakeholder feedback and demonstrated how semantic integration, rule-based reasoning, and SHACL validation further improve robust cross-domain querying, whether performed manually or generated by a SPARQL-NLI.

Key challenges included: scalability issues from explicit materialization (mitigated via graph partitioning), semantic conflicts (resolved through rule-based alignment and SHACL), manual onboarding effort for new sources (requiring future automation), and providing a low-barrier entry point via a SPARQL-NLI.

The subsequent sections are organized as follows: In Sect. 2, we describe a motivating scenario that describes the setting and challenges of data analysis for semiconductor development. Section 3 presents our approach to overcoming them. Section 4 presents an evaluation, Sect. 5 lessons learned. Section 6 discusses related work, and Sect. 7 concludes and presents future directions.

2   Motivating Scenario

Integrated sensor units combine MEMS components with an Application-Specific Integrated Circuit (ASIC). The MEMS component detects mechanical stimuli, such as acceleration, pressure, or rotation, and the ASIC processes the signal before it is sent to the final client at the application layer, such as the motherboard of a smartphone [12]. MEMS production is based on a series of additive and subtractive micro-fabrication processes [3,23,31,41] applied to wafers of single-crystal silicon [25]. Once the structures are finalized, the wafers are diced into individual dies, that is, parts, and packaged into complete sensor units [25]. Testing is performed throughout production, especially during End-of-Line Testing

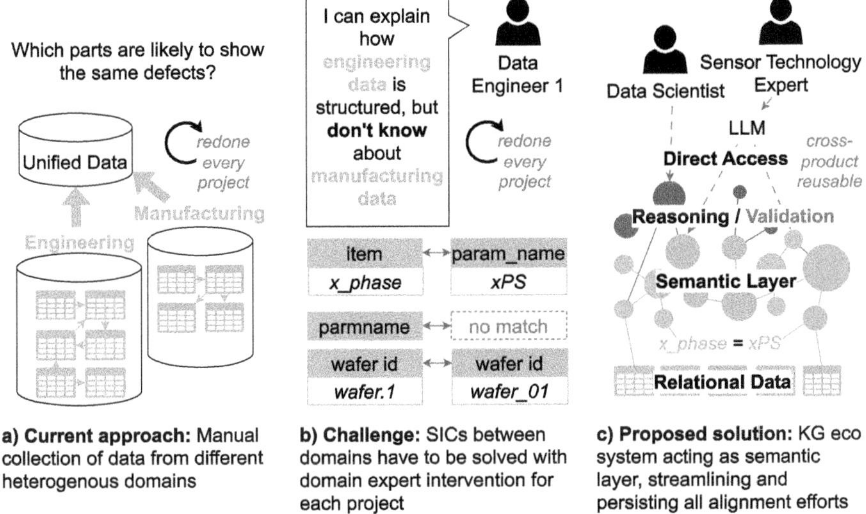

a) Current approach: Manual collection of data from different heterogenous domains

b) Challenge: SICs between domains have to be solved with domain expert intervention for each project

c) Proposed solution: KG eco system acting as semantic layer, streamlining and persisting all alignment efforts

Fig. 1. Motivating Example of KG-augmented Root Cause Analysis. This figure compares a traditional taskforce workflow (a) with a KG-augmented approach (c). In (a), experts manually collect parameters and test values from engineering and manufacturing data from different domains. (b) They must rely on data engineers or other domain experts to overcome SICs. In (c), the KG persists in alignment efforts, shortens data paths, and classifies entities based on logical reasoning. This enables efficient and scalable exploration of complex, heterogeneous data landscapes.

(EOL), where parts are evaluated for functional parameters such as sensitivity. Parts that do not meet the quality thresholds are inked out and excluded [36, 38]. The ratio of inked-out to good parts defines the yield loss. Final Testing (FT) ensures that each assembled sensor meets strict standards before shipment [13].

Manufacturing data provides structured records across EOL-Testing, Inline Control (IC), Process History (PH), Machine Data (MD), assembly and FT. It is locally consistent and supports the monitoring of processes. Engineering data originate from test benches and experimental setups used in the development. It partially overlaps with manufacturing data, but is less consistently annotated and frequently exhibits structural and syntactic inconsistencies due to the highly individual and less standardized nature of the underlying experimental processes.

Consider the following scenario. A MEMS development team must reduce the EOL-Testing yield loss for a customer-critical sensor, *model_123*. Tight delivery schedules require rapid identification of root causes. Sensor technology experts interpret the MEMS architecture and process behavior. MEMS design experts analyze how structural or layout variations influence the performance and reliability. Data engineers manage the ingestion, transformation, and provisioning of raw data, and data scientists develop analytical tools to detect anomalies, identify correlations, and support causal inference. These roles can also be represented in the role model defined by Li *et al.* [24]. Domain experts act as consumers or analysts, data scientists as analysts, and data engineers as builders.

In the development effort for *model_123*, the data scientist identifies impactful EOL failures, prompting sensor technology experts to hypothesize that phase shift effects may cause the observed yield drop. Together with data engineers, the team explored engineering and manufacturing data to assess this hypothesis (cf. Fig. 1). Design and technology experts conduct supporting experiments.

Development teams must continuously extract insights from an expanding pool of data, hindered by a fragmented data landscape. Information is distributed across domain-specific siloed systems, follows different data model conventions, varies by product line, and often lacks semantic consistency [20, 29]. Even focused analyses, such as determining whether the phase shift contributes to EOL test failures in a specific sensor model, require time-consuming manual effort. Relevant data must be located and aligned across domains by resolving Semantic Interoperability Conflicts (SICs), which hinder interpretation and accessibility.

We classify SICs into three categories. (1) Data SICs include data naming conflicts (e.g., *x_phase* vs. *xPS*), data type mismatches, and inconsistent data formats. (2) Structural SICs arise from differences in aggregation (e.g., test-run vs. part), granularity (e.g., *Sensor* vs. *Part* level), synonym/homonym confusion (e.g., *paramname* $\neq$ *param_name*), and synchronization (e.g., resulting from asynchronous data upload). (3) Domain SICs reflect divergent assumptions, including scope differences and unencoded supporting knowledge. These discrepancies hinder consistent interpretation, and reliable cross-source reasoning.

Resolving SICs requires manual effort and collaboration with domain experts to reconstruct context and confirm semantic intent. This slows down the analysis and decision-making. Although SICs may be resolved within a project, limited visibility across projects leads to recurring issues in other development efforts.

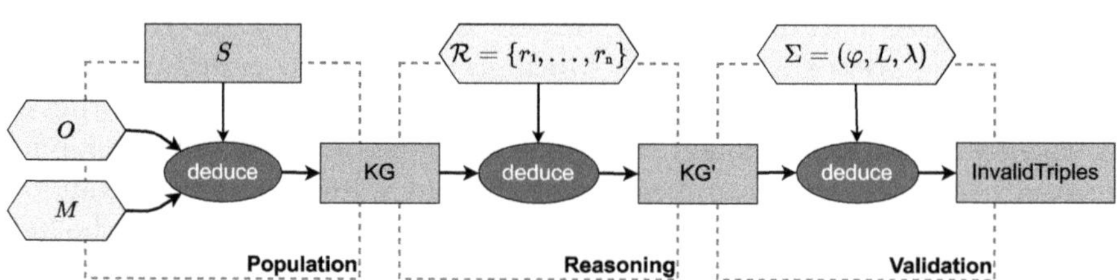

Fig. 2. Structured high-level pipeline for constructing the curated *mdevKG*. Three phases: (1) Population, (2) Reasoning, and (3) Validation. Data sources S are mapped to a unified ontology O using mapping rules M. The result is a KG that is enhanced based on reasoning following a set of rules $\mathcal{R}$ and validated using SHACL constraints to ensure semantic consistency and interoperability. This process enhances robustness, making the resulting KG ecosystem a reliable foundation for industrial reasoning. Color coding: Symbolic inputs (turquoise/square), processes (dark purple/oval), data inputs (orange/square), and formal models (yellow/hexagonal). (Color figure online)

3 Proposed Approach and Implementation

To address the semantic interoperability challenges in MEMS development, we designed a knowledge-driven data integration framework and use it to construct the *mdevKG*. This section presents the formal problem definition, design of our integration system, and implementation pipeline used to construct and validate the KG. We also describe the underlying ontology, mapping and reasoning strategies, data access mechanisms and validation steps for consistent integration.

Problem Statement. MEMS development requires the integration of heterogeneous data across the engineering, manufacturing, and assembly domains. They follow different structures, formats, and semantics, causing SICs to occur. We address the problem of constructing a unified, semantically consistent view of these heterogeneous data sources, as outlined in Sect. 2. This enables cross-domain analysis, ensures traceability, and supports complex reasoning tasks while meeting the practical needs of stakeholders across various roles. This requires a solution for resolving SICs in a scalable, maintainable, and explainable manner.

Proposed Solution. We propose a KG creation process for the *mdevKG* that integrates data collected from a set of data sources and describes them using the MEMS development ontology. Formally, the creation process is defined as [17]: $DIS_{\mathcal{G}} = \langle O, S, M, \Sigma \rangle$ where O is a set of classes and properties of a unified ontology, S is a set of data sources, and M corresponds to mapping rules that define concepts in O as conjunctive queries over sources in S. Σ corresponds to a shape schema and is defined as a tuple $\Sigma = (\Phi, L, \lambda)$, where Φ is a set of shapes, L is a set of shape labels, and $\lambda : L \to \Phi$ is a total function from labels to shapes. The *mdevKG* is created by executing M over the data in S, producing instances of $\mathcal{G}$. Moreover, $DIS_{\mathcal{G}}$ is part of a KG ecosystem [8] with multiple life cycles that allow management, curation, and exploration of the data integrated into the *mdevKG*. This includes a SPARQL-NLI, which enables intuitive access to the KG, consequently lowering the entry barrier for non-expert users.

The structured pipeline depicted in Fig. 2 illustrates the operationalization of the formal system $DIS_{\mathcal{G}}$ Expert teams create a unified ontology O, as well as mapping rules M. The "Population" phase ingests data sources S and applies the mappings M to transform raw data into an initial KG based on a unified ontology. In the "Reasoning" phase, the KG is expanded by applying a rule set $\mathcal{R}$ which is part of the ontology, materializing inferential knowledge. The resulting enriched KG was then validated based on a SHACL shapes schema, ensuring semantic consistency and robustness. This process directly reflects the formal definition, where instances of $\mathcal{G}$ are generated by executing M over S, and Σ governs structural validity and semantic coherence of the *mdevKG*.

Competency Questions. The DIS development process is guided by competency questions (**CQs**) derived from real-world tasks to ensure relevance and validate functionality [11]. For MEMS development they are: **CQ1:** What are the part-level tests most frequently failed for a given part-type? **CQ2:** What are the part-level tests most frequently failed for parts in a given sensor-type? **CQ3:** What wafers did all the parts of a specific sensor come from? **CQ4:** Which specific sensors use a part fabricated on a given wafer? **CQ5:** How do the parameters of a new part design change compared to the established part version? **CQ6:** How long does it take on average to produce a part of specific part-type? **CQ7:** List all parts produced in the neighborhood (square of radius 2) around a given part. **CQ8:** Compare the part-level test results of the parts located at the center of wafers from engineering and manufacturing **CQ9:** Which process immediately followed a given process in the wafer's manufacturing history?

Ontologies. Ontology development followed the TOVE methodology [10]. Entities and relationships were derived directly from the **CQs**, with core concepts modeled as classes and structured hierarchically. For example, **CQ3** introduces `md:Wafer`, `md:Part`, and `md:Sensor`. Object properties such as `md:includesPart` capture relations between entities, data properties such as `md:hasWaferId` are defined from source attributes. Together with sensor technology experts and data scientists, we developed a semiconductor core (SC-Core) ontology. It provides a stable, high-level conceptualization of the semiconductor domain beyond MEMS development and acts as the central semantic reference model to which all other ontologies are aligned. Specialized ontologies correspond to governance domains (Engineering, Manufacturing, Testing, and Assembly). They are not full domain models but capture only aspects relevant to MEMS development. This modular approach enables a focused analysis while preserving consistency from a broader perspective. The ontologies are manually aligned with the Core Information Model for Manufacturing (CIMM) [9] to ensure compatibility with manufacturing frameworks. We also integrated OWL-Time [15] to standardize temporal representations, as well as Quantities, Units, Dimensions, and Types (QUDT) Schema [34] combined with Sensor, Observation, Sample, and Actuator (SOSA) ontology [19] to harmonize measurement units and processes. The network of ontologies and their dependencies is visualized in Fig. 3.

Data Retrieval. All data were provided as relational tables. To support efficient querying, we adopted a two-level partitioning strategy: an aggregated view at the sensor model level, partitioned by *model-month* combinations, and a detailed view at the wafer-level. Data retrieval is a major bottleneck. We applied a dependency-driven approach to reduce the system load without inconsistencies. Sensor-level queries identify relevant wafer instances that trigger wafer-specific retrieval. For this work the scope was limited to a single sensor model and a one-month production window. Traceability data were fully ingested, whereas the EOL test results were sampled during the population. In total, the system processed over 1.2 million rows and 375 columns of raw data across all sources.

KG Population. Population transforms heterogeneous relational data into RDF triples that are linked to ontologies. Each data source was mapped separately using manually defined Resource Mapping Language (RML) mappings. RDFizer [16] was selected based on its performance and scalability. In a benchmark on 500,000 rows and 25 columns of real MEMS development data, RDFizer

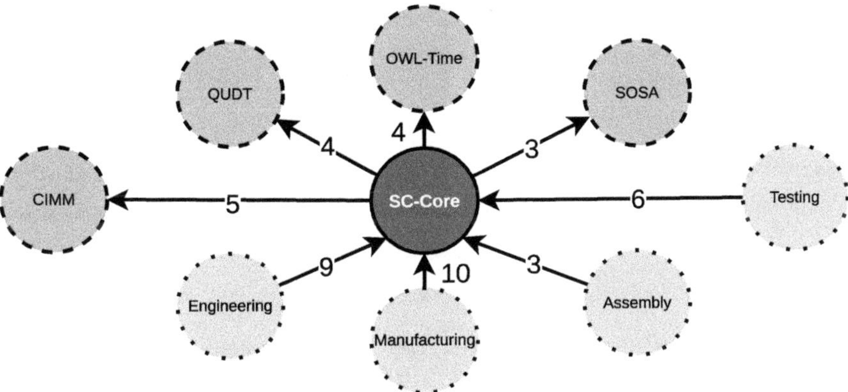

Fig. 3. Ontology dependency network of the *mdevKG*. The visualization depicts the key ontologies incorporated, highlighting their dependency structure and the number of imported concepts. SC-Core serves as the central ontology representing the semiconductor domain, while the specialized ontologies capture a development-oriented perspective on MEMS governance domains. Color coding: Public external ontologies (turquoise/dashed), SC-Core (dark purple/solid), specialized ontologies (light-purple/dotted). (Color figure online)

generated over seven million triples in less than a minute. Subject maps use source-independent IRIs derived from shared domain identifiers to align entities across sources. Virtual graphs were avoided to maintain consistent access and inference, except for high-frequency time-series data, which remain accessible on-demand outside the KG. The materialization process is fragmented by data source, enabling parallel execution. It produces product- and month-partitioned graphs, which improves maintainability. Data-level SICs were resolved using mapping functions embedded in the logic. To resolve data-level SICs, mapping functions are embedded directly into the logic. Parameter names such as x_phase (Engineering) and xPS (Manufacturing) were harmonized using dictionary-based mappings that match patterns via regular expressions. This strategy reliably harmonizes parameters across domains, although it requires manual maintenance. 373 mapping rules and seven mapping functions constitute the *mdevKG*.

KG-Based Reasoning. Reasoning was formalized as a rule set $\mathcal{R}$ in Datalog syntax. Most rules introduce additional paths to enhance data accessibility. For example, a sensor can be directly connected to all wafers on which its parts are fabricated. The following Datalog rule derives a relationship that materializes this implicit connection already present in the underlying production data.

```
includesPartFromWafer(s,w) :- includesPart(s,p),
                              fabricatedOnWafer(p,w).
```

Analogously, test runs were linked to failing parts via `failedForPart(t,p)`.

A more elaborate rule defines neighborhood based on spatial proximity between parts on the same wafer. This is particularly important for identifying potential spatial dependencies and localized defect patterns across adjacent parts.

```
hasNeighbor(p,n) :- fabricatedOnWafer(p,w), fabricatedOnWafer(n,w),
                    hasXCoordinate(p,x), hasYCoordinate(p,y),
                    hasXCoordinate(n,nx), hasYCoordinate(n,ny),
                    |x - nx| ≤ 1, |y - ny| ≤ 1.
```

Of the seven rules used to answer the domain-specific competency questions, two cannot be formalized in SROIQ(D), the logic underlying OWL 2 DL. They require arithmetic operations and datatype comparisons, as in `hasNeighbor(p,n)`. We addressed this by materializing the results using SPARQL CONSTRUCT rules. Infered triples are created in a seperate named graph for better maintainabilty. Another benefit of this approach is improved performance during construction. Inference of `includesPartFromWafer(s,w)` takes about $12\,000$ ms in HermiT and about 7000 ms with SPARQL CONSTRUCT. The computational overhead, is mitigated by our dependency-driven data retrieval strategy, which can be extended into a streaming ETL pipeline. Fianlly in our domain, query-time performance is considered more critical than ingestion latency, which is why we deliberately choose materialization over query-time reasoning.

KG Validation. The KG is validated against a shapes schema $\Sigma = (\Phi, L, \lambda)$ to ensure its structural and semantic integrity. Using SHACL, we perform schema-level checks across the full graph, including data type validation, domain and range enforcement, and cardinality constraints, such as ensuring that each sensor has exactly one sensor ID. Key identifiers, such as wafer and lot IDs, are validated with regular expressions, for example, sensor unique-serial-number USN S00-000. Structural constraints on the sensor composition were verified using qualified value shapes. For example, consumer inertial sensors contain exactly three types of MEMS parts. SHACL combined with OWL-Time enforces consistent time formats and valid intervals where $\text{begin}(I) \leq \text{end}(I)$. Additional checks use `sh:minInclusive` and `sh:maxInclusive` to verify the wafer coordinate ranges and detect misaligned grids. Violations are flagged for review but are not automatically resolved. Manual curation is still required in practice, although SHACL shapes prevent silent failure. The shape schema currently contains 56 node shapes and 261 property shapes, capturing constraints across the *mdevKG*.

SPARQL-NLI. To support intuitive access to the *mdevKG*, we implemented a SPARQL-NLI based on the NL-to-SPARQL framework by Monka *et al.* [28]. We used OpenAI's GPT-4o model via the OpenAI API to generate the responses. The system applies context-aware prompting and injects only the ontology fragments relevant to each question. To prevent bias during the evaluation, we did not provide few-shot domain examples or any additional task-specific guidance.

4 Evaluation

The *mdevKG* contains over 3 million nodes and 15 million edges per sensor model per month. We evaluated its performance using three strategies. First, we assessed effectiveness in addressing real-world needs using formalized competency questions. Second, we report stakeholder feedback from a prototype demonstration to assess usability and value. Third, we analyzed graph statistics before and after reasoning to measure inference's impact on connectivity. Due to confidentiality, the real KG cannot be published. Instead, we provide simulated datasets. Although these datasets do not represent actual MEMS processes, they expose the key SICs described herein. We share a distilled version of the MEMS development ontologies, compressed into a unified model for the simulated data.

Table 1. Evaluation of the competency questions. This table compares how different data access methods answer representative domain CQs: SQL over a relational database, SPARQL (manual/NLI) on a baseline KG, and two enriched KG variants (with reasoning; with reasoning plus SHACL and external ontologies such as OWL-Time, SOSA, and QUDT). For each CQ, we report whether the answer is satisfactory (No, Partly, Yes) and indicate the required implicit knowledge (a–e). This highlights how KG enrichment affects completeness and reduces reliance on domain assumptions.

CQ	SQL	SPARQL (manual)			SPARQL-NLI	
		Base	Reason	SHACL	Base	Reason
CQ1: Most failed EOL (part)	Yes	Yes	Yes	Yes	No	Yes
CQ2: Most failed EOL (sensor)	Yes[a,b]	Yes	Yes	Yes	No	No
CQ3: Wafers used in sensor	Yes	Yes	Yes	Yes	Yes	Yes
CQ4: Sensors with parts from wafer	Partly	Yes	Yes	Yes	Yes	Yes
CQ5: Parameter comparison	Yes[a,b]	Yes	Yes	Yes	Yes	Yes
CQ6: Production time (part)	Yes[c]	Yes[c]	Yes[c]	Yes	No	No
CQ7: Parts in neighborhood	No[e]	Yes[e]	Yes	Yes	Yes	Yes
CQ8: EOL result (center of wafer)	Yes[d]	Yes[d]	Yes[d]	Yes	No	No
CQ9: Process sequence	Yes[c]	Yes[c]	Yes[c]	Yes	No	Yes
Correct result	7.5/9	9/9	9/9	9/9	4/9	6/9
Explicit knowledge only	3/9	5/9	7/9	9/9	9/9	9/9

Implicit Knowledge:

a   Correct join-keys
b   Column name meaning (synonym/homonym mismatches)
c   Awareness of time format & structure
d   Awareness of coordinate-system misalignment
e   Neighborhood semantics

Answering the Competency Questions. This process validates the structural integrity of the KG and demonstrates its applicability to real-world MEMS development scenarios. We compared its performance with a relational SQL baseline and across multiple KG variants (Table 1). The base KG was produced from mappings, one version was enriched through reasoning, and another version incorporated OWL-Time, QUDT, SOSA, and SHACL validation. For each variant, we assessed two interaction modes: manually written SPARQL queries and SPARQL-NLI (cf. Sect. 3). For the SPARQL-NLI, each query was generated at least three times. When executed against the same version of the knowledge graph, repetitions yielded semantically equivalent results. Differences were limited to the syntactic structure of the queries (e.g., ordering of triple patterns or the use of FILTER expressions instead of inline literals). Consequently, all executions produced the same answer set. The first evaluation criterion examined whether answers matched the domain expert's expected results, reported as yes, partly, or no. The second criterion assessed if users needed implicit domain knowledge to formulate correct queries. This metric excludes technical proficiency in SQL or SPARQL and captures only the domain-level assumptions and contextual contextual background knowledge required to derive correct answers. The most prominent form of implicit knowledge is an understanding of core data structures. This knowledge is required to address structural SICs. This includes knowing which foreign keys are relevant for a specific use case and how column names should be interpreted within the domain. An example arises in **CQ2**, where the homonym *prod_name* refers to a part type in the EOL and to a sensor type in assembly. The part type is also recorded in assembly, but it appears in *sub_prod_name*. To aggregate and join data without information loss, a user must understand data granularity. It matters whether a measurement is recorded at the wafer or part level, a caveat in **CQ5**. This effort is unnecessary in the KG system because the ontology and mapping decisions described in Sect. 3 make these relations explicitly available for direct SPARQL-based access.

Following relations in the base KG proved challenging for the SPARQL-NLI. Main failures were hallucinated relations (**CQ1**, **CQ2**, **CQ6**, **CQ8**) and task misunderstandings (**CQ6**, **CQ9**). These issues could be reduced with improved prompts or few-shot examples, though our focus was on improving the results through the KG. Small ontology changes, like introducing fine-grained subclasses with more restrictive domain and range definitions, did not resolve these issues. For **CQ1**, we obtained a correct result via the SPARQL-NLI after adding the reasoning step that infers `md:failedForPart( md:TestRun1, md:Part1)`. Results for **CQ9** improved through inferring `md:subsequentProcess( md:ProcessRun1, md:ProcessRun2)`. Manual SPARQL querying benefits from reasoning as well. The inference of `md:hasNeighborPart( md:Part1, md:Part2)` enabled efficient queries for **CQ7** via property chaining. This advantage over SQL, where equivalent queries need multiple self-joins, becomes especially valuable on the large MEMS datasets. Similarly, **CQ3** cannot be answered with a single query in relational databases. Such a query takes several minutes and results in a memory error. Within the KG system, results can be retrieved without such error.

The full version of the KG system, including SHACL validation and the integration of OWL-Time, SOSA, and QUDT, resolved most of the remaining data-

Table 2. Stakeholder feedback on the *mdevKG* prototype. Responses were collected after a prototype demonstration. The metrics presented are mode, median, agreement (Proportion of Agree/Strongly Agree), disagreement (Proportion of Disagree/Strongly Disagree), and consensus (proportion of responses equal to the mode). The results indicate a generally positive reception of the KG prototype and its applicability in addressing existing data challenges.

Question	Mode	Median	Agreement	Disagreement	Consensus
Q1: *mdevKG* met my expectation	Neutral	Agree	53.3%	0.0%	46.7%
Q2: A broader community in MEMS development should be trained in KGs	Neutral	Agree	53.3%	0.0%	46.7%
Q3: KGs can help integrate data in the MEMS development domain	Agree	Agree	86.7%	0.0%	53.3%
Q4: KGs can augment existing data retrieval (e.g. tables)	Agree	Agree	80.0%	0.0%	46.7%
I perceive a benefit regarding the Demo for ...					
Q5: ... navigating data across multiple sources	Agree	Agree	93.3%	0.0%	60.0%
Q6: ... using KG links instead of table joins	Agree	Agree	93.3%	0.0%	53.3%
Q7: ... resolving semantic conflicts via KG	Neutral	Agree	53.3%	0.0%	46.7%
Q8: ... easy access to sensor/process data	Agree	Agree	93.3%	0.0%	60.0%
Q9: ... expressing knowledge not expressable in relational models	Neutral	Neutral	33.3%	0.0%	66.7%
Q10: *mdevKG* would increase my efficiency	Agree	Agree	53.3%	13.3%	46.7%
Q11: I run a lot of queries per week against existing databases	Disagree	Neutral	20.0%	46.7%	33.3%

level SICs. Users are explicitly informed about mismatches between measurement units, time formats, data formats, coordinate systems, or malformed IDs. Consequently, the full set of **CQs** can be answered without relying on implicit knowledge. We compared query times for **CQ4**. Selecting 100 wafers yields about 500k sensors. SQL fails due to memory limits. SPARQL on the base KG takes about 4000 to 6000 ms, while SPARQL after inferring `includesPartFromWafer(s,w)` takes about 1500 to 1600 ms. Initial executions tend to be on the slower end due to caching, while subsequent runs of the same query are significantly faster.

Feedback of Stakeholders. To validate the relevance of the KG and ensure alignment with real-world needs, we conducted structured feedback with 14 stakeholders from four roles: sensor technology experts (5), MEMS design experts (2), data engineers (3), and data scientists (4). The process included a live demonstration followed by a 12-question survey featuring 11 Likert-scale items [30] (cf. Table 2) and one open-ended question about the drivers and obstacles of KG adoption. The demonstration included a ten-minute introduction to the KG, clarifying that it augments rather than replaces expert work. The stakeholders viewed a prototype GUI that allowed selection from three predefined queries, i.e., **CQ3**, **CQ4**, and a metadata query, executed against the KG. The results were displayed in tabular format with color-coded visualizations showing the provenance of the columns. We compared the SPARQL and SQL query structures for each task, highlighting the KG's ability to express cross-domain relationships. At demonstration time, reasoning capabilities, SHACL validation, and the SPARQL-NLI were not integrated. To interpret responses, we computed metrics from the Likert-scale items aggregated over all responses (cf. Table 2). The mode and median were extracted from the response distributions. Agreement was defined as the share of responses marked as *Agree* or *Strongly Agree*, while disagreement was defined as *Disagree* or *Strongly Disagree*. Consensus was measured as proportion of responses aligned with the most frequent category.

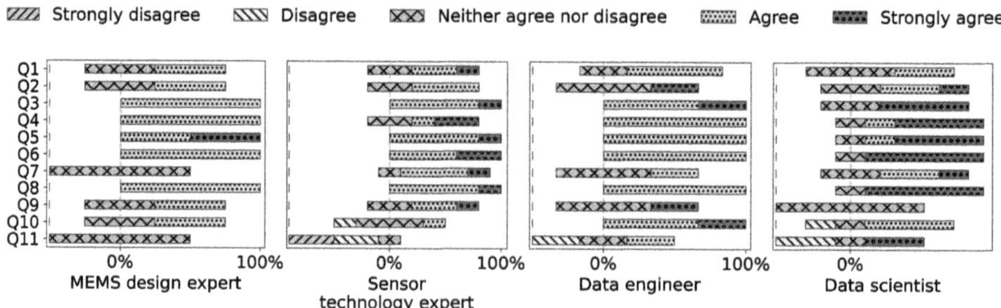

Fig. 4. Results from the questionnaire. The figure summarizes stakeholder feedback from four groups: MEMS design experts, sensor technology experts, data engineers, and data scientists, highlighting overall positive reception of the KG-based approach for improving data linkage, accessibility, and quality in semiconductor development.

As shown in Fig. 4, feedback revealed broad conceptual support for the KG across roles, with strong agreement from data scientists and data engineers. Among the questions **Q5-Q9**, each addressing a specific KG benefit, **Q5** (navigating multi-source data) and **Q8** (access to sensor/process data) showed highest agreement (93.3%) and consensus (60.0%). Indicating stakeholders perceived most value in the KG's ability to unify heterogeneous data and streamline information access. Responses from MEMS design and sensor technology experts were more reserved, likely due to less experience with structured query languages. However, even stakeholders who rarely write queries still recognize the KG's conceptual value. The responses in Free-text revealed key challenges and drivers. Participants emphasized that without accessible interaction mechanisms, the KG's workflow impact would remain limited. The learning curve of structured query languages emerged as a central concern, aligning with the observations of Li et al. [24]. Participants were concerned about the effort required to construct and maintain the KG, particularly at scale and across diverse sources. A factor important for strategic roles. Nonetheless Stakeholders recognized clear adoption benefits. The KG's capability for cross-department linkage and improved traceability was seen as an enabler, even for users less familiar with query languages.

Graph Statistics. We evaluated structural effects of reasoning applied to core EOL-Testing data within the KG, scoped to a single sensor model and one production month[1]. Analytics were generated using *Gephi*[2] (Table 3). Reasoning over EOL-Tests increases edges from 16 089 to 18 252 and average degree from 1.79 to 2.03. This increase reflects a denser graph structure, allowing for more flexible query paths and meaningful connections. Most new edges are `md:failedForPart` between test runs and the parts on which they failed. The impact on the graph topology is measurable. The increase in radius from one to five indicates that formerly peripheral nodes are better integrated, increasing accessibility. This is supported by the decrease in modularity from 0.532 to 0.524, suggesting improved semantic cohesion [6]. The subgraph shows tighter connections between the wafers and the associated tests. These effects reflect a shift from fragmented representations towards a semantically integrated structure that supports localized analytical insights as well as consistent global reasoning across domains.

Table 3. Comparison of graph metrics before and after reasoning. The metrics are computed over the core EOL-Test data for one sensor model. Reasoning increases connectivity (e.g., edge count) and improves cohesion, as indicated by reduced modularity and network diameter [6].

Metrics	pre-reasoning EOL	post-reasoning EOL
Num of nodes	8987	8989
Num of edges	16089	18252
Avg. degree	1.790	2.031
Avg. path length	2.551	2.543
Network diameter	7	7
Network radius	1	5
Eigenvalue Centrality	0.259	0.260
Modularity	0.532	0.524

---

[1] Proportions of failed and passed tests presented in this section are not representative of actual Bosch test outcomes. The dataset was constructed with a predefined ratio of test results and did not reflect real-world distributions.

[2] https://gephi.org/.

5 Lessons Learned

The implementation and evaluation of the *mdevKG* uncovered relevant insights regarding system scalability, integration design, conflict resolution, and user adoption. These lessons reflect the technical and organizational challenges encountered during the deployment of a knowledge-driven system in practice.

Scalability. A key architectural decision was using explicit KG materialization over virtual graphs, enabling reasoning over persistent triples [26] and favoring query speed over creation time. However, this approach introduces scalability challenges. To address this, the KG was partitioned by sensor and time window, with non-core data types offloaded to virtual graphs. This strategy increased integrated data volume, but resulted in a more fragmented access layer. To support scalability, SDM-RDFizer [18] was selected for efficient resource consumption in memory usage and execution time. The scale at which the system was tested reflects typical analysis tasks in the MEMS development domain today.

Data Integration System Design. Modular ontology design is essential for aligning heterogeneous data across governance domains. In practice, we observed that an excessive focus on fully capturing domain complexity from the outset severely hinders productivity during ontology development. A key trap is the tendency towards overanalysis and perfectionism, overlooking one of the most valuable advantages of KGs over traditional structured data: adaptability. Based on these observations, we prioritized rapid initial mapping for each source, followed by iterative refinement aligned with use cases and their requirements. During implementation, we maintained close contact with the providers of the main components used in our system and exchanged technical insights that may inform future improvements for large-scale industrial deployment.

Resolving Semantic Interoperability Conflicts. SICs were handled through several DIS mechanisms. Structural SICs were resolved by mapping each source to harmonized ontologies, ensuring consistent representation across domains. Data-level SICs, such as parameter naming conflicts, for example, x_phase and xPS, were corrected through dictionary-based regex replacements. SHACL validation and the use of OWL-Time and QUDT supported standardized measurement units, time formats, coordinate systems, and IDs. This standardization within the KG improved reliability and consistency and removed common blockers when users navigated cross-domain data. Connectivity within the KG is further strengthened by incorporating reasoning into the KG ecosystem. In the scoped EOL test subgraph, changes observed in graph analytics indicated tighter data integration, which increased the ease of navigation and exploration for human users and improved the results returned via the SPARQL-NLI.

Stakeholder Feedback. The *mdevKG* received broad conceptual support across all stakeholder roles. Data scientists and data engineers expressed particularly strong agreement regarding its benefits. However, stakeholders highlighted the need for more accessible interaction with the KG to make its potential usable even for those without extensive experience in structured query languages. We addressed this by providing a SPARQL-NLI for querying the KG. Another key insight was the concern that manually maintaining and extending the KG may become a barrier. Some stakeholders pointed out that without sufficient automation, scaling and sustaining the ecosystem over time would be difficult. Despite these challenges, they acknowledged the value of the KG. Its ability to support cross-domain data navigation and improve access to development data was considered the most immediate benefit, with 93.3% agreement across all roles.

6   Related Work

The application of semantic technologies in manufacturing and MEMS development has gained increasing attention in recent years. Prior work has addressed aspects such as data structuring, ontology design, explainability, and interoperability. However, challenges remain in integrating heterogeneous sources, aligning knowledge across domains, and enabling scalable and explainable analytics. In this section, we position our contribution in relation to the existing efforts.

Semantic Structuring in MEMS and Manufacturing. In the semiconductor domain, efforts have emphasized the need for machine-interpretable data. Heringhaus *et al.* [13] applied graph neural networks to parameter estimation in MEMS design but did not incorporate semantic representations. Safont-Andreu [35] highlighted limitations of unstructured manufacturing data in AI applications. In contrast, our work provides a formal semantic layer enabling structured and cross-source integration across heterogeneous MEMS development data.

Dynamic and Cross-domain Knowledge Integration. Most existing manufacturing KGs are static, which limits their adaptability to dynamic production contexts. Kejriwal [20] and Moor [29] advocated real-time cross-domain interoperability to improve knowledge utility, accessibility, and reuse. Our approach supports this vision by enabling partitioned, incremental data ingestion, paving the way for scalable, robust, and operationally maintainable integration pipelines.

Ontology Alignment and Semantic Interoperability. The integration of heterogeneous sources remains a core challenge in industrial contexts. Many KGs operate in isolation, constrained by domain-specific ontologies and lacking cross-domain alignment [20,29]. Our work addresses this by co-developing modular sub-ontologies with domain experts and anchoring them to standardized models, ensuring semantic coherence across engineering and manufacturing domains.

Legacy System Integration. Kejriwal [21] noted the difficulty of embedding semantic technologies into legacy IT infrastructures. We mitigate this by deploying a semantic access layer over existing relational systems, preserving backward compatibility, while enabling rule-based reasoning and cross-source navigation.

Explainability and Structural Validation. Explainability is essential for stakeholder trust in automated systems, particularly in safety-critical domains. Kejriwal [20] and Anim [2] advocated SHACL-based validation to ensure structural integrity and facilitate accessible reasoning. Our system integrates SHACL constraints to validate schema compliance and to guide ontology refinement.

Graph-Based Diagnostics and Causality. Koot [22] explores causal inference over traceability data in semiconductor manufacturing, focusing on diagnostics within limited scopes. Their work demonstrated the potential of graph-based representations for root cause analysis. Our approach complements this by enabling broader semantic integration and rule-based abstraction, thereby extending the diagnostic capabilities across domains and production contexts.

7 Conclusion and Future Work

This work presents a KG-driven approach that building a general-purpose data integration framework and applying it to the MEMS semiconductor development domain. As outlined in the introduction, MEMS sensor development relies on heterogeneous, siloed data sources across governance domains. Our contribution addresses this by constructing a scalable, semantically consistent KG enabling unified access and cross-domain analysis in a high-stakes industrial setting. To manage performance and evaluate feasibility, we scoped implementation to one sensor model over a production month. A partitioned and dependency-driven pipeline enabled core data integration, and materialization supported rule-based reasoning. This deployment demonstrated key capabilities, including inference and semantic validation, under realistic conditions. Results confirm that a semantic layer improves data linkage, queryability, and analytical potential for both manual querying and a SPARQL-NLI. Ontologies, mappings, and rules were developed in collaboration with domain experts. This ensured alignment with real-world semantics and requirements, although it introduced scalability limitations. The formalization of domain knowledge and resolution of SICs supported complex cross-domain queries. The framework was adopted by researchers from two departments and fitted to their data needs. After demonstration, four out of seven non-technical users requested access, indicating demand for accessible semantic tooling. Taking into account user feedback, we implemented an SPARQL-NLI that allows querying the *mdevKG* without knowing formal query languages. The SPARQL-NLI benefits from the increased connectivity introduced by reasoning and from standardized data formats ensured through validation. The key impact of our system is the enhanced accessibility of data within and across domain borders. This supports data scientists and data engineers and enables less technical stakeholders to navigate and explore the data. We plan a stable deployment of the system that includes a user-friendly GUI to expand access. The SPARQL-NLI lowers entry barriers, and we expect other AI systems to benefit from the KG system similarly. The semantic layer thus becomes the foundation for more reliable and interpretable agentic systems.

Future Work. We aim to reduce manual effort through KG completion [40], rule mining [5,33], and KG refinement techniques [32,43]. A key step is the systematic integration of non-relational data sources. The KG infrastructure lays the foundation for more advanced reasoning pipelines [1,39] as well as neuro-symbolic [14,37] and agentic AI systems, enabling interpretable and scalable knowledge discovery in semiconductor research and development environments.

Acknowledgments. We used ChatGPT to assist with editing the manuscript.

Supplemental Material Statement. Strict company disclosure regulations prevent the public release of the data, models, and source code used in this work. We provide a simulated mock dataset and a simplified and anonymized version of our ontology with an example mapping file and two SHACL shapes to support the reproduction of key results. They can be accessed via: https://github.com/boschresearch/eswc-mdsim-kg.

References

1. Andresel, M., Tran, T.K., Domokos, C., Minervini, P., Stepanova, D.: Combining inductive and deductive reasoning for query answering over incomplete knowledge graphs. In: Proceedings of the 32nd ACM International Conference on Information and Knowledge Management, pp. 15–24 (2023). https://doi.org/10.1145/3583780.3614816
2. Anim, J., Robaldo, L., Wyner, A.Z.: A shacl-based approach for enhancing automated compliance checking with RDF data. Information **15**(12), 759 (2024). https://doi.org/10.3390/info15120759
3. Bagolini, A., et al.: Development of mems MOS gas sensors with CMOS compatible PECVD inter-metal passivation. Sens. Act. B Chem. **292**, 225–232 (2019). https://doi.org/10.1016/j.snb.2019.04.116
4. Bhatt, G., Manoharan, K., Chauhan, P., Bhattacharya, S.: MEMS sensors for automotive applications: a review, pp. 223–239. Springer Nature Group (2019). https://doi.org/10.1007/978-981-13-3290-6_12
5. Chen, L., et al.: Rule mining over knowledge graphs via reinforcement learning. Knowl.-Based Syst. **242**, 108371 (2022). https://doi.org/10.1016/j.knosys.2022.108371
6. Cherven, K.: Mastering Gephi network visualization. Packt Publishing Ltd. (2015)
7. Fanse, T.: Micro-electro-mechanical system (MEMS) application and prospects in automobile. IOSR J. Mech. Civil Eng. **19**, 17–21 (2022). https://doi.org/10.9790/1684-1901021721
8. Geisler, S., et al.: From genesis to maturity: managing knowledge graph ecosystems through life cycles. Proc. VLDB Endow. **18**(5), 1390–1397 (2025). https://doi.org/10.14778/3718057.3718067
9. Grangel-González, I., Lösch, F., ul Mehdi, A.: Knowledge graphs for efficient integration and access of manufacturing data. In: 25th IEEE International Conference on Emerging Technologies and Factory Automation (ETFA). vol. 1, pp. 93–100 (2020). https://doi.org/10.1109/ETFA46521.2020.9212156
10. Gruber, T.R.: A translation approach to portable ontology specifications. In: Knowledge Acquisition, pp. 199–220 (1993). https://doi.org/10.1007/BFb0024952
11. Grüninger, M., Fox, M.S.: The role of competency questions in enterprise engineering. In: Benchmarking—Theory and practice, pp. 22–31. Springer (1995). https://doi.org/10.1007/978-0-387-34847-6_3
12. el Hak, M.G. (ed.): MEMS: Introduction and fundamentals. Taylor and Francis (2006)
13. Heringhaus, M.E., Buhmann, A., Müller, J., Zimmermann, A.: Graph neural networks for parameter estimation in micro-electro-mechanical system testing. Array **14**, 100162 (2022). https://doi.org/10.1016/J.ARRAY.2022.100162

14. Herron, D., Jiménez-Ruiz, E., Weyde, T.: On the potential of logic and reasoning in neurosymbolic systems using OWL-based knowledge graphs. Neurosymbolic Artifi. Intell. **1**, 29498732251320044 (2025). https://doi.org/10.1177/29498732251320043

15. Hobbs, J.R., Pan, F.: An ontology of time for the semantic web. ACM Trans. Asian Lang. Inf. Process. (TALIP) **3**(1), 66–85 (2004). https://doi.org/10.1145/1017068.1017073

16. Iglesias, E., Jozashoori, S., Chaves-Fraga, D., Collarana, D., Vidal, M.E.: SDM-RDFIZER: an RML interpreter for the efficient creation of RDF knowledge graphs. In: Proceedings of the 29th ACM International Conference on Information & Knowledge Management, pp. 3039–3046. CIKM '20, ACM (2020). https://doi.org/10.1145/3340531.3412881

17. Iglesias, E., Jozashoori, S., Vidal, M.: Scaling up knowledge graph creation to large and heterogeneous data sources. J. Web Semant. **75**, 100755 (2023). https://doi.org/10.1016/J.WEBSEM.2022.100755

18. Iglesias, E., Vidal, M.E., Collarana, D., Chaves-Fraga, D.: Empowering the SDM-RDFIZER tool for scaling up to complex knowledge graph creation pipelines. Semantic Web **16**(2) (2025). https://doi.org/10.3233/SW-243580

19. Janowicz, K., Haller, A., Cox, S.J., Le Phuoc, D., Lefrançois, M.: Sosa: a lightweight ontology for sensors, observations, samples, and actuators. J. Web Semantics **56**, 1–10 (2019). https://doi.org/10.1016/j.websem.2018.06.003

20. Kejriwal, M.: Knowledge graphs: a practical review of the research landscape. Information **13**(4), 161 (2022). https://doi.org/10.3390/info13040161

21. Kejriwal, M., Liu, Q., Jacob, F., Javed, F.: A pipeline for extracting and deduplicating domain-specific knowledge bases. In: 2015 IEEE International Conference on Big Data (Big Data), pp. 1144–1153. IEEE (2015). https://doi.org/10.1109/BigData.2015.7363868

22. Koot, T.R.: Leveraging traceability graph data for causal rule mining in semiconductor manufacturing. Master's thesis, Eindhoven University of Technology (2024). https://research.tue.nl/en/studentTheses/5a4ecf00-b5d2-4ab1-a533-d416c47984cc, supervisors: Dr.ir. H. Eshuis, Dr. A.E. Akcay, Ir. K.J. Braakman, Ir. O.F. Buijs

23. Laermer, F., Franssila, S., Sainiemi, L., Kolari, K.: Chapter 16 - deep reactive ion etching. In: Tilli, M., Paulasto-Krockel, M., Petzold, M., Theuss, H., Motooka, T., Lindroos, V. (eds.) Handbook of Silicon Based MEMS Materials and Technologies (Third Edition), pp. 417–446. Micro and Nano Technologies, Elsevier, third edition edn. (2020). https://doi.org/10.1016/B978-0-12-817786-0.00016-5

24. Li, H., Appleby, G., Brumar, C.D., Chang, R., Suh, A.: Knowledge graphs in practice: characterizing their users, challenges, and visualization opportunities. IEEE Trans. Visual Comput. Graphics **30**(1), 584–594 (2024). https://doi.org/10.1109/TVCG.2023.3326904

25. Liu, C.: Foundations of MEMS. Prentice Hall Press (2022)

26. Makni, B., Ebrahimi, M., Gromann, D., Eberhart, A.: Neuro-symbolic semantic reasoning. In: Neuro-Symbolic Artificial Intelligence: The State of the Art, pp. 253–279. IOS Press (2021). https://doi.org/10.3233/FAIA210358

27. Marek, J.: Mems for automotive and consumer electronics. In: 2010 IEEE International Solid-State Circuits Conference - (ISSCC), pp. 9–17 (2010). https://doi.org/10.1109/ISSCC.2010.5434066

28. Monka, S., et al.: Enhancing manufacturing knowledge access with LLMs and context-aware prompting. In: Endriss, U., et al., (eds.) ECAI - 27th European Conf. on AI, 19-24 October, Santiago de Compostela, Spain - 13th Conf. on Prestigious Applications of Intelligent Systems (PAIS). Frontiers in Artificial Intelligence and Applications, vol. 392, pp. 4665–4672. IOS Press (2024). https://doi.org/10.3233/FAIA241062

29. Moor, M., et al.: Predicting sepsis using deep learning across international sites: a retrospective development and validation study. eClinicalMedicine **62**, 102124 (2023). https://doi.org/10.1016/j.eclinm.2023.102124

30. Moosbrugger, H., Kelava, A. (eds.): Testtheorie und Fragebogenkonstruktion. S, Springer, Heidelberg (2012). https://doi.org/10.1007/978-3-642-20072-4

31. Nguyen, T.T.N., Akagi, D., Okato, T., Ishikawa, K., Hori, M.: Low-temperature atomic layer etching of platinum via sequential wet-like reactions of plasma oxidation and complexation. Appl. Surf. Sci. **687**, 162325 (2025). https://doi.org/10.1016/j.apsusc.2025.162325

32. Paulheim, H.: Knowledge graph refinement: a survey of approaches and evaluation methods. Semantic web **8**(3), 489–508 (2016). https://doi.org/10.3233/SW-160218

33. Purohit, D., Chudasama, Y., Rivas, A., Vidal, M.E.: Sparkle: symbolic capturing of knowledge for knowledge graph enrichment with learning. In: Proceedings of the 12th Knowledge Capture Conference 2023, pp. 44–52. K-CAP '23, Association for Computing Machinery, New York (2023). https://doi.org/10.1145/3587259.3627547

34. QUDT Board: Quantities, units, dimensions and types (2022). https://qudt.org/

35. Safont-Andreu, A., Schekotihin, K., Burmer, C., Hollerith, C., Ming, X.: Artificial intelligence applications in semiconductor failure analysis. EDFA Technical Articles **25**, 16–28 (2023). https://doi.org/10.31399/asm.edfa.2023-2.p016

36. Senturia, S.D.: Microsystem design. Springer Science & Business Media (2005)

37. Shengyuan, C., Cai, Y., Fang, H., Huang, X., Sun, M.: Differentiable neuro-symbolic reasoning on large-scale knowledge graphs. Adv. Neural. Inf. Process. Syst. **36**, 28139–28154 (2023). https://doi.org/10.5555/3666122.3667344

38. Shoaib, M., Hamid, N.H., Malik, A.F., Zain Ali, N.B., Tariq Jan, M.: A review on key issues and challenges in devices level mems testing. J. Sensors **2016**(1), 1639805 (2016). https://doi.org/10.1155/2016/1639805

39. Wan, G., Du, B.: Gaussianpath: a bayesian multi-hop reasoning framework for knowledge graph reasoning. In: Proceedings of the AAAI Conference on Artificial Intelligence. vol. 35, pp. 4393–4401 (2021). https://doi.org/10.1609/aaai.v35i5.16565

40. Yao, L., Mao, C., Luo, Y.: Kg-Bert: Bert for knowledge graph completion (2019). https://doi.org/10.48550/arXiv.1909.03193

41. Ylönen, M., Vähä-Heikkilä, T., Kattelus, H.: Amorphous metal alloy based mems for rf applications. Sens. Actuators, A **132**(1), 283–288 (2006). https://doi.org/10.1016/j.sna.2006.05.033, the 19th European Conference on Solid-State Transducers

42. Yu, H., Cang, S., Wang, Y.: A review of sensor selection, sensor devices and sensor deployment for wearable sensor-based human activity recognition systems. In: 2016 10th International Conference on Software, Knowledge, Information Management & Applications (SKIMA), pp. 250–257 (2016). https://doi.org/10.1109/SKIMA.2016.7916228

43. Zhang, D., et al.: Knowledge graph-based image classification refinement. IEEE Access **7**, 57678–57690 (2019). https://doi.org/10.1109/ACCESS.2019.2912627

Author Index

© The Editor(s) (if applicable) and The Author(s), under exclusive license
to Springer Nature Switzerland AG 2026
M. Acosta et al. (Eds.): ESWC 2026, LNCS 16550, pp. 527–528, 2026.
https://doi.org/10.1007/978-3-032-25159-6